SAPONINS USED IN FOOD AND AGRICULTURE

ADVANCES IN EXPERIMENTAL MEDICINE AND BIOLOGY

Recent Volumes in this Series

Volume 400A
EICOSANOIDS AND OTHER BIOACTIVE LIPIDS IN CANCER, INFLAMMATION, AND RADIATION INJURY, Part A
Edited by Kenneth V. Honn, Santosh Nigam, and Lawrence J. Marnett

Volume 400B
EICOSANOIDS AND OTHER BIOACTIVE LIPIDS IN CANCER, INFLAMMATION, AND RADIATION INJURY, Part B
Edited by Kenneth V. Honn, Santosh Nigam, and Lawrence J. Marnett

Volume 401
DIETARY PHYTOCHEMICALS IN CANCER PREVENTION AND TREATMENT
Edited under the auspices of the American Institute for Cancer Research

Volume 402
AIDS, DRUGS OF ABUSE, AND THE NEUROIMMUNE AXIS
Edited by Herman Friedman, Toby K. Eisenstein, John Madden, and Burt M. Sharp

Volume 403
TAURINE 2: Basic and Clinical Aspects
Edited by Ryan J. Huxtable, Junichi Azuma, Kinya Kuriyama, and Masao Nakagawa

Volume 404
SAPONINS USED IN TRADITIONAL AND MODERN MEDICINE
Edited by George R. Waller and Kazuo Yamasaki

Volume 405
SAPONINS USED IN FOOD AND AGRICULTURE
Edited by George R. Waller and Kazuo Yamasaki

Volume 406
MECHANISMS OF LYMPHOCYTE ACTIVATION AND IMMUNE REGULATION VI: Cell Cycle and Programmed Cell Death in the Immune System
Edited by Sudhir Gupta and J. John Cohen

Volume 407
EICOSANOIDS AND OTHER BIOACTIVE LIPIDS IN CANCER, INFLAMMATION, AND RADIATION INJURY, Volume 2
Edited by Kenneth V. Honn, Lawrence J. Marnett, Santosh Nigam, Robert L. Jones, and Patrick Y-K. Wong

Volume 408
TOWARD ANTI-ADHESION THERAPY FOR MICROBIAL DISEASES
Edited by Itzhak Kahane and Itzhak Ofek

A Continuation Order Plan is available for this series. A continuation order will bring delivery of each new volume immediately upon publication. Volumes are billed only upon actual shipment. For further information please contact the publisher.

SAPONINS USED IN FOOD AND AGRICULTURE

Edited by
George R. Waller
Oklahoma State University
Stillwater, Oklahoma

and

Kazuo Yamasaki
Hiroshima University
Hiroshima, Japan

PLENUM PRESS • NEW YORK AND LONDON

Library of Congress Cataloging-in-Publication Data

Saponins used in food and agriculture / edited by George R. Waller and Kazuo Yamasaki.
p. cm. -- (Advances in experimental medicine and biology ; v. 405)
"Proceedings of an American Chemical Society Symposium on Saponins: Chemistry and Biological Activity, held August 22-25, 1995, in Chicago, Illinois"--T.p. verso.
Includes bibliographical references and index.
ISBN 0-306-45394-0
1. Saponins in agriculture--Congresses. 2. Saponins--Congresses. 3. Food--Congresses. I. Waller, George R. II. Yamasaki, Kazuo. III. American Chemical Society Symposium on Saponins: Chemistry and Biological Activity (1995 : Chicago, Ill.) IV. Series.
[DNLM: 1. Saponins--chemistry--congresses. 2. Food Additives--congresses. 3. Agrochemicals--congresses. 4. Cosmetics--congresses. W1 AD559 v.405 1996 / QU 75 S2409 1996]
S587.5.S35S27 1996
630'.2'47783--dc20
DNLM/DLC
for Library of Congress 96-32063
CIP

Proceedings of the 210th National Meeting of the American Chemical Society Symposium on Saponins: Chemistry and Biological Activity, held August 22 – 25, 1995, in Chicago, Illinois

ISBN 0-306-45394-0

A Division of Plenum Publishing Corporation
233 Spring Street, New York, N. Y. 10013

10 9 8 7 6 5 4 3 2 1

Printed in the United States of America

To Otis C. Dermer
Scientific Editor *par excellence*

PREFACE

The extraordinary technological advances made in saponin research in recent years that have been created and developed by the scientific establishments throughout the world are exciting, and they present intellectually challenging problems which are becoming more solvable using modern techniques. How many of us could have imagined a decade ago what has been accomplished in the (bio)chemical, biological, agricultural, or pharmaceutical laboratories that are being done routinely today. Advanced techniques in NMR spectroscopy, mass spectrometry, x-ray crystallography, use of enzymes, various chromatographic procedures, and new bioassays have been developed; the advances in these areas have been truly phenomenal. On the other hand, we have not progressed nearly so much in natural products, particularly in saponins research, in adopting molecular biology in our scientific endeavors. It is our hope that finding a saponin that can cure a disease or plant traits (e.g. resistance to a virus, microorganisms or insects) can be isolated to a specific gene that will become routine during the 21st century. The advances made in this symposium in our understanding of food and agriculture are outstanding. We must be alert and learn to contribute to the knowledge of agriculture and molecular biology with natural products, and particularly saponins.

Chemists, biochemists, and biologists from more than 25 countries presented their work at the 5-day symposium organized by the American Chemical Society (ACS), Division of Agricultural & Food Chemistry at the 210th Meeting during August in Chicago. Saponins have not been the subject of an ACS symposium before, and scientists in this field have not had an opportunity to consult with each other nor develop plans for future work.

Saponins are a class of natural products which are structurally constructed of aglycones (triterpene or steroidal) and sugars (pentose(s), hexose(s), and/or uronic acid(s)). An appropriate hydrolysis of saponins yields sugars and the aglycone; however, hydrolysis does not necessarily produce the genuine aglycone. Saponins are biological detergents because of glycosylation of the hydrophobic aglycone, and when agitated in water form a soapy lather which gives rise to the name of the group of compounds. This unique ability to cause foaming has been used by mankind throughout the centuries for making cleaning solutions, and indeed it can act as an aid in identification of plant extracts. Triterpenoid, steroid, and steroid alkaloid glycosides are widely spread throughout the plant kingdom, and several have been found in marine animals. Some saponins have cardiac activity, hemolytic activity, activity as fish poisons, cholesterol-reducing ability, bitterness, ability to act as sweeteners, and cosmetics activity, and can also serve as allelochemicals; these are behaviors found in certain saponins rather than in all members of this chemical family. From the biological viewpoint saponins have a diverse group of properties, some deleterious but many beneficial. Their use in plant drugs, folk medicines, etc. has generated great interest in the chemical characterization of these molecules. This has been evident in the Orient (particularly Japan), where the literature on the isolation, purification, separation, structural elucidation, and biological activity of saponins attests to the skill of natural-products biochemists and chemists. It seems that the advantage of saponins to the plant producing them is that they function as protecting agents, growth regulators, and allelochemicals.

Interactions between participants during this meeting helped develop cooperative relationships, and promotes continued assistance of research and development on saponins among scientists from over the world. Such research is now revealing a large array of plants from which to choose, while only a few have been studied at this time. In the future, cell culture techniques will allow the isolation and characterization of individual saponins as well as the enzymes catalyzing their production. The industrial demand for saponins is increasing and with more attention being given to this field, it offers a renewable resource that provides a stable raw material with a practical and dependable supply.

This book is divided into four sections: cosmetics, sweeteners, herbs, and non-alcoholic beverages; regulatory effects on crops, viruses, microorganisms, weeds, and insects; chemical identification using NMR and mass spectrometry; and the nutritional aspects which complete the series of the chapters. Attention is called to the appendixes which contains useful information on the saponin disease network, and NMR and MS data bases. The partial reports of this symposium which appeared in *Chem. & Eng. News*, September 11, **73,** 28-35 (1995) and August 28, **73,** 8-9 (1995) are recommended for reading, although they are not included in this book.

Although this book cannot give a complete account of the meeting, we hope that it will serve as the starting point to help guide the research of others. We especially thank those who contributed chapters but did not attend the symposium. We are pleased that saponin research and development as well as its commercial products will now be the subject of ongoing international meetings.[1]

We acknowledge with sincere appreciation financial support for this symposium from the following organizations/companies: International Science Foundation, Japan Society of Industrial Machinery Manufacturers, Kikkoman Co. Ltd., Maruzen Pharmaceutical Co. Ltd., NPS Pharmaceuticals, Inc., Shimadzu Scientific Instruments, Inc., The Soros Foundations, and the United States Tobacco Company.

George R. Waller
Kazuo Yamasaki

[1]The next meeting will be sponsored by the Phytochemical Society of Europe and will be held in Pulawy, Poland in 1999.

CONTENTS

COSMETICS, SWEETENERS, HERBS, AND NON-ALCOHOLIC BEVERAGES

Application of Saponins in Foods and Cosmetics: Saponins of Mohave Yucca and *Sapindus mukurossi*
Osamu Tanaka, Yukiyoshi Tamura, Hitoshi Masuda, and Kenji Mizutani 1

Novel Sweet-Tasting Saponins of the Cycloartane, Oleanane, Secodammarane, and Steroidal Types
Edward J. Kennelly, Rutt Suttisri, and A. Douglas Kinghorn 13

Intensely Sweet Saponin Osladin: Synthetic and Structural Study
Mugio Nishizawa and Hidetoshi Yamada 25

Steroid Saponins from Fenugreek and Some of Their Biological Properties
Yves Saivaire, Y. Baissac, O. Leconte, P. Petit, and G. Ribes 37

Triterpene Saponins from Maté, *Ilex paraguariensis*
Eloir P. Schenkel, Jarbas A. Montanha, and Grace Gosmann 47

REGULATORY EFFECTS ON CROPS, VIRUSES, MICROORGANISMS, WEEDS, AND INSECTS

Studies of the Phytotoxicity of Saponins on Weed and Crop Plants
Robert E. Hoagland, Robert M. Zablotowicz, and Krishna N. Reddy 57

Regulatory Effects of Saponins in the Pathogenesis of Root Rots in Cereal Crops
Pavel K. Kintia and Galina A. Lupashku 75

Effects of Saponins on the Growth and Activity of Rhizosphere Bacteria
Robert M. Zablotowicz, Robert E. Hoagland, and Stephen C. Wagner 83

Saponins from *Medicago* Spp.: Chemical Characterization and Biological Activity against Insects
Aldo Tava and Miriam Odoardi 97

The Role of Cardenolides in a Crucifer-Insect Relationship
J.A.A. Renwick 111

Can Soyasaponin I and Mono- and Bi- Desmosides Isolated from Mungbeans Serve as Growth Enhancers for Mungbeans and Lettuce?
G.R. Waller, C.F. Yang, L.F. Chen, C.H. Su, R.M. Liou, S.C. Wu, C.C. Young, M.R. Lee, J.S. Lee, C.H. Chou, and D. Kim 123

Oxygen Radical Scavenging Activity of DDMP-Conjugated Saponins and the Physiological Role In Leguminous Plants
Kazuyoshi Okubo and Yumiko Toshiki 141

Alfalfa Saponins: Structure, Biological Activity, and Chemotaxonomy
Wieslaw Oleszek 155

Chemistry and Biological Activity of Triterpenoid Glycosides from *Medicago sativa*
Almira E. Timbekova, M. I. Isaev, and N. K. Abubakirov 171

Saponins and Artifacts
Georges Massiot, Marie-Geneviéve Dijoux, and Catherine Lavaud 183

Triterpenoid Saponins from a Siberian Chemorace of the *Thalictrum* Genus
A.A. Semenov, V.I. Lutsky, A.S. Gromova, T.V. Ganenko, E.A. Khamidullina, and N.N. Trofimova 193

New Preparation of Triterpenoid-3-Sulfates by the Use of SO_3-DMSO Complex
V.I. Grishkovets, A.E. Kondratenko, and V.Ya. Chirva 209

Triterpene and Steroid Saponins from Two *Melilotus* Species
G.V. Khodakov, Yu.A. Akimov, A.S. Shashkov, P.K. Kintia, and V.I. Grishkovets 211

A Novel Triterpenoid Saponin from the Leaves of *Majorana hortensis* (Moench)
R.N. Yadava 223

Chemiluminescence of Oxygen Radical Scavengers Such as DDMP Saponins in the Presence of Radicals and Aldehyde
Y. Yoshiki, K. Igarashi, and K. Okubo 231

Herniaria Saponin B, a Novel Triterpenoid Saponin from *Hernaria fontanesii*
Z. Charrouf, A. Nait-Mbark, D. Guillaume, Y. Leroy, and O. Kol 241

Steroidal Glycosides from *Nicotiana tabacum* L. Seeds and Their Biological Activity
Stephan A. Shvets, Pavel K. Kintia, and Oleg N. Gutsu 247

Novel Microbial Transformations of Steroids
K.M. Madyastha 259

Thermal Behavior of Steroidal Glycosides
Valentin A. Bobeyko and Pavel K. Kintia 271

CHEMICAL IDENTIFICATION USING NMR AND MS

Proton and Carbon-13 NMR Studies of Steroids and Triterpenes
Geoffrey A. Cordell, Long-Ze Lin, and Roberto R. Gil 281

A Systematic NMR Approach for the Determination of the Molecular Structure of Steroidal Saponins
Pawan K. Agrawal 299

Application of Tandem Mass Spectral Approaches to Structural Determinations of Saponins
Catherine E. Costello 317

Structural Determination of Saponins from Mungbean Sprouts by Tandem Mass Spectrometry
M.R. Lee, J.S. Lee, J.C Wang, and G.R. Waller 331

Liquid Secondary Ion Mass Spectrometry and Linked Scanning at Constant B/E/MS for Structure Confirmation of Saponins in *Medicago sativa* (Alfalfa)
P.R. West, G.R. Waller, P.W. Geno, W. Oleszek, and M. Jurzysta 339

Saponins from Alfalfa, Clover, and Mungbeans Analyzed by Electrospray Ionization-Mass Spectrometry as Compared with Positive and Negative FAB-Mass Spectrometry
M.K. Lee, Y.C. Ling, M. Jurysta, and G.R. Waller 353

NUTRITIONAL ASPECTS

Analysis, Heat Stability and Physiological Effects of Saponins from Oats
Gunilla Önning and Nils-Georg Asp 365

Biological Effects of Feed and Forage Saponins and Their Impacts on Animal Production
P.R. Cheeke 377

Effect of Quillaja Saponins on *in Vitro* Rumen Fermentation
Harinder P. S. Makkar and Klaus Becker 387

Do Steroidal Saponins Have a Role in Hepatogenous Photosensitization Diseases of Sheep?
A. Flåyen 395

Steroidal Glycoalkaloids in Andean Potatoes
Isao Kubo and Katsuya Fukuhara 405

APPENDIXES

Appendix I: Plant Saponin Disease Network 419

Appendix II: Saponin Network for Mass Spectrometry and Nuclear Magnetic Resonance Spectrometry 421

Contributors 423

INDEXES

Latin Name Index 429

Subject Index 435

APPLICATION OF SAPONINS IN FOODS AND COSMETICS: SAPONINS OF MOHAVE YUCCA AND *SAPINDUS MUKUROSSI*

Osamu Tanaka,[1] Yukiyoshi Tamura,[2] Hitoshi Masuda,[2] and Kenji Mizutani[2]

[1] Suzugamine Women's College, Inokuchi, Nishi-ku, Hiroshima 733 Japan

[2] Maruzen Pharmaceuticals Co. Ltd., Mukaihigashi-cho, Onomichi 722 Japan

INTRODUCTION

Saponins are known as natural surfactants. In addition to this physical property, saponins exhibit a variety of biological activities, and have been investigated toward the development of new natural medicines and to prove the efficacy of traditional herbal medicines. As to the application of saponins to foods and cosmetics, it is indispensable that sufficient amounts of plant resouces are available, and that the content of saponins must be high. Furthermore, a plant must have a long history of human use as foodstuffs or ingredients of cosmetics, and their safety should be officially guaranteed.

The saponins of Quillaja bark and licorice root are widely utilized in the world. A large amount of Quillaja saponin is utilized in photosensitized film as a surfactant. It is used also in shampoos, liquid detergents, toothpastes and beverages as an emulsifier and a long-lasting foaming agent. Recently, the immunoadjuvant activity of this saponin has attracted much attention as reported by Kensil and Setten in this volume.

Nearly 50,000 tons of licorice roots are consumed on a yearly basis. Licorice extract and its major saponin, glycyrrhizin (yield: more than 3%), are used as a medicine and as a sweetener and flavor enhancer in foods and cigarettes. The sapogenin, glycyrrhetic acid, is utilized in cosmetics. Recently, we have found that the monoglucuronide of glycyrrhetic acid (MGGR) is 941-fold as sweet as sucrose (Mizutani et al., 1994; Kuramoto et al., 1994), and thus promising as a new powerful sweetener and flavor enhancer. Intense in-hibition against two-stage carcinogenesis was also observed for MGGR (Mizutani, 1994).

SCREENING OF ANTIYEAST SAPONINS

It is known that the deterioration of cooked foods is caused mainly by yeasts, and that many skin diseases are due to infection by dermatophytic fungi and yeasts. In an expansion of the utilization of saponins in foods and cosmetics, we have examined antifungal and antiyeast saponins. Table 1 shows the screening results of antiyeast activity tests of crude saponin mixtures from several plants.

Since crude saponin fraction from the pericarps of *Sapindus mukurossi* and the rhizomes of Mohave yucca exhibited significant activity, the active principles of both these materials were further investigated in detail.

MOHAVE YUCCA (*Yucca schidigera*)

Yucca species (Agavaceae), grows widely in North and Central America. Mohave yucca, *Y. schidigera*, has been used as a foodstuff and folk medicine by American Indians as well as by early California settlers, and is recognized as safe for human dietary use by Food and Drug Administration (FDA) (listed in 21 CFR 172.510). The extract of the rhizomes of this plant is now utilized as a long-lasting foaming agent in carbonated beverages and as a flavor enhancer in foods (The Lawrence Review of Natural Products, March 1994, Walters Kluwer Co., St. Louis, Missouri).

Antiyeast and Antifungal Spirostanoid Saponins from the Rhizomes of Mohave Yucca

The presence of steroidal saponins in this plant has been reported previously (Kaneda et al., 1987). However, the isolation and identification of the individual saponins had not been achieved prior to this study.

An extract of the rhizomes was subjected to column chromatography on a highly porous polymer, Diaion HP-20, which is a styrene-divinylbenzene copolymer. After successive elution with water and 60% and 80% MeOH, a saponin fraction which showed significant antiyeast activity against *Saccharomyces cerevisiae* was obtained by elution with 90% MeOH. This fraction was subjected to successive chromatography on silica gel and then octadesylsilylated silica gel (ODS) and was finally separated by HPLC on ODS to give six saponins, yucca saponins 1~6 (YE-1~6). All the saponins showed antiyeast activity.

The structure of the sapogenin of YE-1 was elucidated to be 5β-spirost-25(27)-en-3β-ol by comparison of its carbon-13 NMR signals with those of sarsasapogenin and convallamarogenin. YE-1 afforded 2 mol of glucose and 1 mol of xylose. The structure of the sugar moiety was determined by means of the HMBC-NMR method.

YE-2, the major saponin in the extract, afforded sarsasapogenin (25*S*) and smilagenin (25*R*) in a ratio of about 3:1 on acid hydrolysis. This indicated that YE-2 is a mixture of 25-stereoisomers, YE-2*S* and -2*R*. Inspection of the carbon-13 NMR spectrum revealed the identity of the structure of sugar moiety of YE-2*R* and -2*S* with that of YE-1. YE-2*S* was recently isolated from the rhizome of this plant by Kameyama et al. (H. Kameyama et al.: The 65th Spring National Meeting of the Chemical Society of Japan, 1991).

The structure of the sapogenin of YE-3 was elucidated to be 5β–spirost-25(27)-ene-2β,3β-diol. The structure of the sugar moiety of YE-3 was determined by the HMBC-NMR method.

YE-4 was proved to be a saponin of markogenin, and the structure of its sugar moiety is identical with that of YE-3.

YE-5 has the same sapogenin as YE-1. The structure of the sugar moiety of YE-5 was also elucidated by the HMBC-NMR method.

YE-6 gave sarsasapogenin and smilagenin in a ratio of about 3:1, being a mixture of 25-stereoisomers, YE-6*S* and YE-6*R*. The structure of the sugar moieties of both isomers are identical with that of YE-5. YE-6*R* has previously been isolated from *Yucca gloriosa* (Nakano et al., 1989)

Table 1. Antiyeast Activity of Crude Saponin Fractions.
MIC: minimal inhibitory concentration (μg/ml)
S.c: Saccharomyces cerevisiae; C.u.: Candida utilis

	S.c.	*C.u.*		*S.c..*	*C.u.*
Sapindus pericarp	250	250	*Saponaria* rhizome	>1,000	>1,000
Yucca rhizome	250	NT [a]	Soybean seed	>1,000	NT
Tea seed	500	500	*Quinoa* bran	>1.000	>1,000
Camellia seed	1,000	1,000	*Quillaja* bark	>1,000	>1,000
Ginseng root	1,000	1,000	Licorice root	>1,000	>1,000
Hedera leaf	1,000	1,000	Joazeiro root	NT [a]	>1,000
Marronier seed	1,000	1,000	*Gypsophylla* root	NT [a]	>1,000

[a] NT: not tested

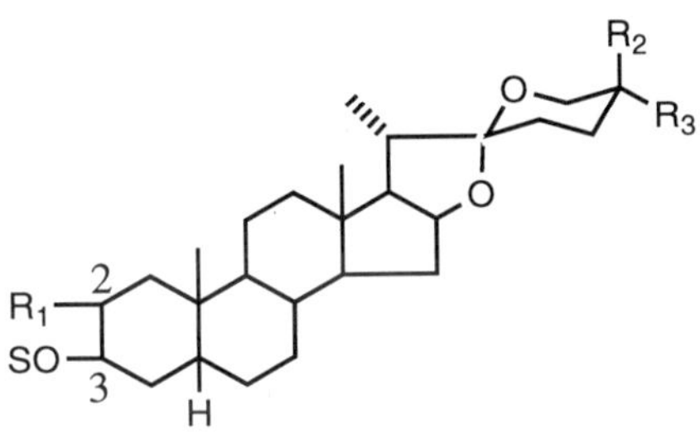

aglycone, S = H

1: sarsasapogenin,
$R_1 = R_3 = H, R_2 = CH_3$
2: smilagenin,
$R_1 = R_2 = H, R_3 = CH_3$
3: markogenin,
$R_1 = OH, R_2 = CH_3, R_3 = H$

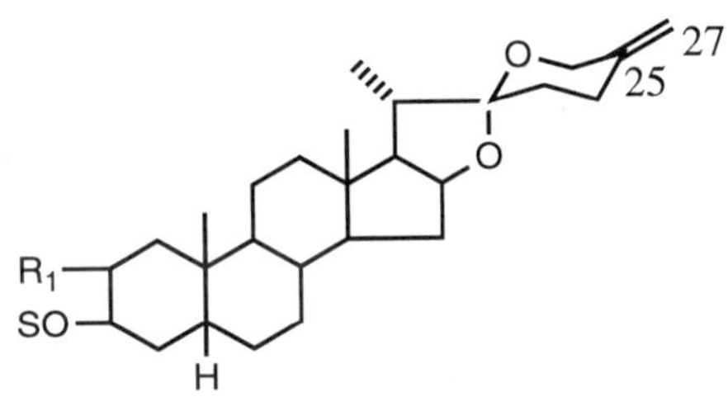

aglycone, S = H

4: 5β-spirost-25(27)-en-3β-ol
$R_1 = H$
5: 5β-spirost-24(27)-ene-2β,3β-diol
$R_1 = OH$

	aglycone	S	yield*	MIC (μg/ml) *S.c.*	*C.a.*	*H.a.*
YE-1	4	A	0.39	12.5	12.5	3.13
YE-2			1.43	12.5	12.5	3.13
YE-2*S*	1	A				
YE-2*R*	2	A				
YE-3	5	B	0.13	100	100	50
YE-4	3	B	0.23	100	100	100
YE-5	4	C	0.17	12.5	25	6.25
YE-6			0.33	25	50	6.25
YE-6*S*	1	C				
YE-6*R*	2	C				

S: sugar moiety

A: –β-Glc-2-β-Glc
3-β-Xyl

B: –β-Gal-2-β-Glc
3-β-Xyl

C: -β-Glc-2-β-Glc
3-β-Glc

* from water extract, % *S.c.: Saccharomyces cerevisiae; C.a. Candida albicans; H.a.: Hansenula anomala*

Figure 1. Saponins from Mohave yucca and their antiyeast activity.

Table 2. Antibacterial Activity of Yucca Saponin Fraction.

Gram-positive bacteria, MIC (μg/ml)			
Staphylococcus		*Bacillus circulans* IFO 3329	>1,000
aureus IID 671	1,000	*Lactobacillus*	
aureus IFO 3060	1,000	*plantarum* IFO 3070	>1,000
epidermidis IID 866	>1,000	*rhamnosus* IFO 12521	>1,000
Bacillus		*Enterococcus faecalis* IFO 3971	>1,000
subtilis IFO 3007	1,000	*Streptococcus mutans* IFO 13955	>1,000
licheniformis IFO 12200	1,000		

Gram-negative bacteria, MIC (μg/ml)			
Escherichia coli HUT 215	>1,000	*Pseudomonas*	
Alcaligenes faecalis IFO 13111	1,000	*aeruginosa* JCM 2776	>1,000
Proteus vulgaris IFO 3851	1,000	*fluorescens* JCM 2779	>1,000
		Klebsiella pneumoniae IFO 14940	1,000

Table 3. Antiyeast and Antifungal Activity of Yucca Saponin Fraction.

Yeast, MIC (μg/ml)			
Saccharomyces		*Pichia*	
cerevisiae IMO 203	62.5	*nakazawae* HUT 1688**	31.3
cerevisiae HUT 2075	31.3	*carsonii** IFO 946	31.3
cerevisiae JCM 2223	62.5	*Debaryomyces*	
Candida famata IFO 664*	31.3	*hansenii* IFO 18**	31.3
Hansenula anomala HUT 7083*	31.3	*hansenii* IFO 27**	62.5
Hansenula sp.*	31.3	*hansenii* IFO 47**	31.3
Cryptococcus sp.*	31.3	*hansenii* IFO 7011**	125
laurentii IFO 609	125	*Zygosaccharomyces*	
Kloeckera apiculata IFO 154	62.5	*rouxii* IFO 845**	31.3
		rouxii IFO 1130**	31.3

* food-deteriorating yeast ** film-forming yeast in soy sauce

Fungi, MIC (μg/ml)			
Aspergillus		*Aspergillus*	
niger IFO 4343	>1,000	*awamori* HUT 2014	>1,000
oryzae HUT 2065	>1,000	*awamori* HUT 2015	>1,000
oryzae HUT 2175	125	*Mucor pusillus* HUT 1185	15.6
oryzae HUT 2188	>1,000	*Rhizopus*	
oryzae HUT 2192	>1,000	*formosaensis* IFO 4756	>1,000
sydowii HUT 4097	>1,000	*nigricans* IFO 4731	>1,000
		Penicillium expansum IFO 5453	>1,000

Dermatophytic yeast and fungi, MIC (μg/ml)			
Candida albicans TIMM 0134	62.5	*Sabouraudites canis* IFO 7863	31.3
Trichophyton		*Epidermophyton floccosum* IFO 9045	31.3
rubrum IFO 5807	15.6		
mentagrophytes IFO 5809	31.3		

Fig. 1 shows the structures of all of these saponins and their antiyeast activity against *Saccharomyces cerevisiae* (brewers yeast), *Candida albicans* (a pathogenic yeast) and *Hansenula anomala* (a food-deteriorating yeast). YE-1, -2, -5 and -6 showed potent growth inhibition against these yeasts, while the activity of YE-3 and -4, saponins of a 2β-hydroxylated sapogenin, is rather weak. The fraction of furostan bidesmosides, which has not been purified, showed no activity.

Antimicrobial Activity of the Saponin Fraction

For the commercial utilization of Mohave yucca, the antimicrobial activity of the saponin fraction which was obtained by column chromatography of the extract on Diaion HP-20 (*vide supra*) was investigated. It showed no or weak growth inhibition against both Gram-positive and Gram-negative bacteria (Table 2).

The antiyeast and antifungal activities are summarized in Table 3. The saponin fraction exhibited potent antiyeast activity. Infection of boiled rice such as "sushi" and "musubi" with *Hansenula anomala* and *Kloeckera apiculata* results in oders smelling like an organic solvent. Infection of cooked beans and processed fish meat with *Candida famata* and *Pichia carsonii* causes oders smelling like kerosene. *Pichia nakazawae, Debaryomyces hansenii* and *Zygosaccharomyces rouxii* are film-forming yeasts, damaging "soy sauce" and "miso", oriental fermented seasonings. The saponin fraction exhibited strong growth inhibition against these food-deteriorating yeasts.

The saponin fraction showed less activity against common fungi, while it significantly inhibited the growth of dermatophytic yeast and fungi.

Potassium sorbate has been utilized in foods as a preservative. Its antiyeast activity depends upon pH. Between pH 5.0-3.0, potassium sorbate completely inhibited the growth of yeasts at the concentration of 0.05%, while at less acidic pH (near neutral), the activity decreased remarkably. In contrast to this, such pH dependence was not observed for the yucca saponin fraction. In the range of pH 6.3-3.0, it entirely inhibited the growth of yeasts at the concentration of 0.03%.

Utilization of the Yucca Extract as an Anti-food-deteriorating Agent

Yucca extract is non-toxic and non-mutagenic. It is recognized as safe for human food use by FDA (listed in 21 CFR 172.510). The extract is tasteless and odorless, exerting no influence on the taste of foods. It is readily soluble in water and stable on heating. Based on the present study, commercial application of the extract for extending the shelflife of cooked foods and fermented seasonings is now under development.

SAPINDUS PERICARPS

Addition of an antifungal and antiyeast ingredient to cosmetics is desirable for the protection of skin against, and prevention of, dandruff generation, dermatomycoses and cutaneous candidiasis.

Significant antiyeast activity was observed for the crude saponin fraction of the pericarps of *Sapindus mukurossi* (Sapindaceae) which grows abundantly in China and Japan. Pericarps of this plant have been used as a natural detergent, and are utilized as foam-stabilizing agents in chemical fire extinguishers in Japan. The pericarps have also been used as an antitussive, anti-inflammatory and anthelmintic agent as well as for treatment of dermatomycoses. In Japan, the pericarp is called "enmei-hi", which means "life prolonging pericarp", and in China, it has been called "wu-huan-zi", which means "non-illness fruit".

Antifungal and Antiyeast Oleanane-saponins of *Sapindus* Pericarps

A number of saponins of hederagenin (monodesmosides and corresponding bidesmosides) have been already isolated from the pericarps of *S. mukurossi* (Kimata et al., 1983). Together with saponins, acyclic sesquiterpene oligoglycosides named mukuroziooside Ia, Ib, IIa and IIb were also isolated from the pericarps in high yields (Kasai et al., 1986). The structures and yields of these constituents are shown in Fig. 2.

The antidermatophytic activities of these constituents are shown in Table 4. All the monodesmosides exhibited strong growth inhibition. It is noteworthy that the activity of sapindoside A is almost as strong as that of griseofulvin, the well-known antidermatophytic antibiotic. Griseofulvin does not show inhibitory activity against a pathogenic yeast, *Candida albicans*, while these monodesmosides exhibited significant inhibition. For the monodesmosides, molluscidal activity was oberved by Hostettman's group (unpublished data). The bidesmoside, mukurozi-saponin-Y1, and the acyclic sesquiterpene oligoglycosides showed no activity.

It was found that while purified monodesmosides of the pericarps are sparingly soluble in water, their solubility was greatly increased in the presence of bidesmosides (Kimata et al., 1983). The acyclic sesquiterpene oligoglycosides also strongly increased the water solubility of the biologically active monodesmosides (Kasai et al., 1986). The antibiotic, sodium ampicillin is hardly absorbed upon oral administration. It was revealed that a solution of monodesmosides, solubilized with the aid of the bidesmosides or the sesquiterpene oligoglycosides, strongly enhanced the absorption of this antibiotic from the small intestine and rectum (Yata et al., 1985, 1986). These phenomena are important for the biological activities of the pericarps.

Structure-Antifungal Activity Relationships

Antidermatophytic activity against *Trichophyton rubrum* was investigated for a variety of oleanane saponins. It was disclosed that for growth inhibition, the presence of free 28-COOH, 23-OH and 3-*O*-glycosyl groups is essential (Fig. 3). All the bidesmosides of hederagenin, such as kalopanaxsaponin B, the 28-COOH of which is glycosylated, showed no activity. Mono- and bidesmosides of oleanolic acid, such as saponin CP4, which lack a 23-OH, also showed no growth inhibition. Saikosaponins, the active principles of *Bupleurum* radix, lack a 28-COOH, exhibiting no activity. Thlandioside H1, a bidesmoside which was isolated from *Thandiatha hookeri* var. *pentaphyla* in a yield of 10% without any chromatography (Nie et al., 1989), showed no activity, while a monodesmoside of hederagenin derived from this bidesmoside, exhibited activity. Activity was also observed for hederagenin 3-*O*-α-L-arabinopyranoside.

Antimicrobial Activity of the Saponin Fraction of *Sapindus* Pericarp

For commercial utilization as an ingredient in cosmetics, the saponin fraction was prepared as follows. The methanolic extract was subjected to chromatography on Diaion HP-20. After removal of sesquiterpene oligoglycosides and other water-soluble constituents by elution with water and then 50% MeOH, the saponin fraction was obtained by elution with 70 % and 80% MeOH.

The saponin fraction showed moderate antibacterial activity against Gram-positive bacteria, while no activity was observed against Gram-negative bacteria (Table 5).

As summarized in Table 6, the saponin fraction exhibited growth inhibition against

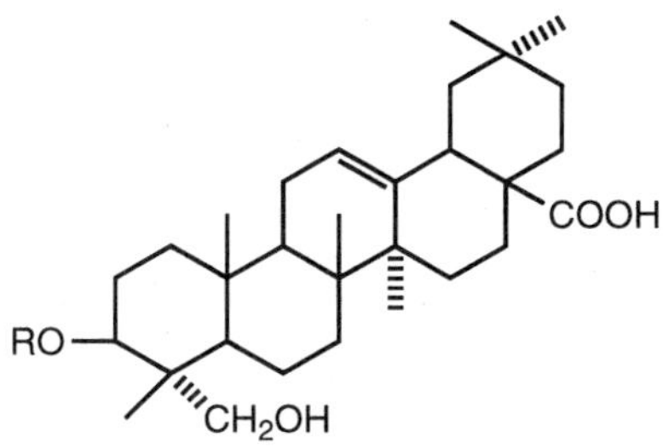

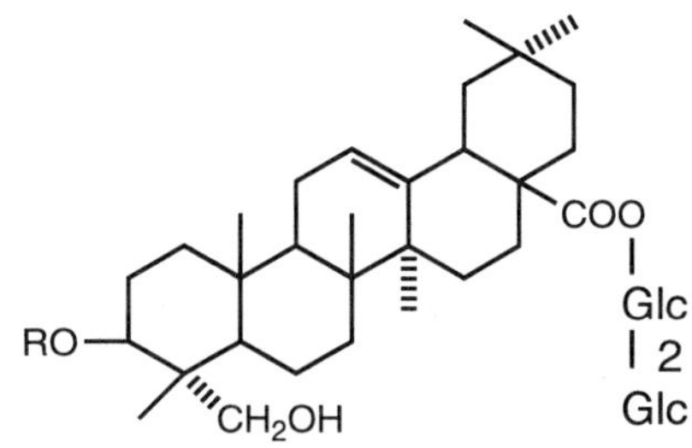

monodesmosides	**R**	**bidesmosides**
Saponin A (0.98)	-Ara(p)-2-Rha-3-Ara(p)	Mukurozi-saponin-Y2 (0.06)
Sapindoside B (0.94)	-Ara(p)-2-Rha-3-Xyl	Mukurozi-saponin-Y1 (0.04)
Saponin C (0.45)	-Ara(p)-2-Rha-3-Ara(f)	
Sapindoside A (0.08)	-Ara(p)-2-Rha	Mukurozi-saponin-X (0.06)
Mukurozi-saponin E1 (0.04)	-Ara-(p)-2-Rha-3-Xyl-4-Ac	
Mukurozi-saponin G (0.04)	-Ara(p)-2-Rha-3-Ara(p)-3,4-Ac_2	

Saponins of Hederagenin

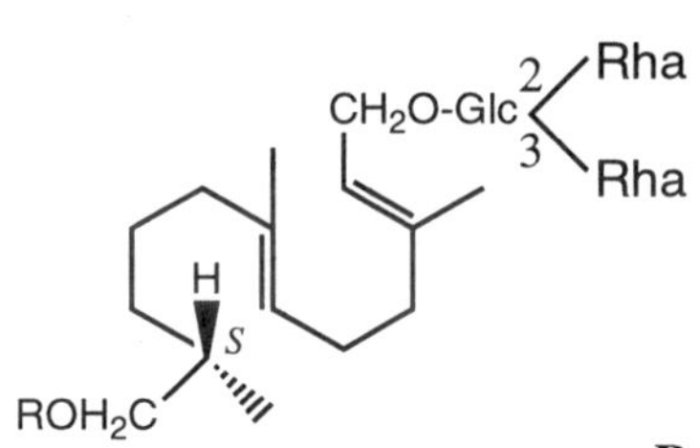

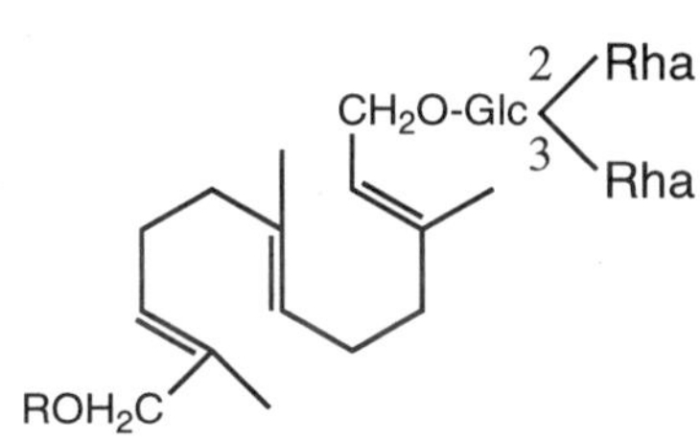

	R	
Mukurozioside Ia (0.6)	-Glc-2-Rha	Mukurozioside Ib (0.4)
Mukurozioside IIa (1.5)	-Glc<2 Rha, 3 Rha	Mukurozioside IIb (2.5)

Sesquiterpene Oligoglycosides

Ara(p): α-L-arabinopyranoside; Ara(f): α-L-arabinofuranoside;
Rha: α-L-rhamnopyranoside; Glc: β-D-glucopyranoside; Xyl: β-D-xylopyranoside
Yield (%) from the dried pericarps is shown in parentheses.

Figure 2. Glycosides from *Sapindus mukurossi*.

Table 2. Antibacterial Activity of Yucca Saponin Fraction.

Gram-positive bacteria, MIC (μg/ml)			
Staphylococcus		*Bacillus circulans* IFO 3329	>1,000
aureus IID 671	1,000	*Lactobacillus*	
aureus IFO 3060	1,000	*plantarum* IFO 3070	>1,000
epidermidis IID 866	>1,000	*rhamnosus* IFO 12521	>1,000
Bacillus		*Enterococcus faecalis* IFO 3971	>1,000
subtilis IFO 3007	1,000	*Streptococcus mutans* IFO 13955	>1,000
licheniformis IFO 12200	1,000		

Gram-negative bacteria, MIC (μg/ml)			
Escherichia coli HUT 215	>1,000	*Pseudomonas*	
Alcaligenes faecalis IFO 13111	1,000	*aeruginosa* JCM 2776	>1,000
Proteus vulgaris IFO 3851	1,000	*fluorescens* JCM 2779	>1,000
		Klebsiella pneumoniae IFO 14940	1,000

Table 3. Antiyeast and Antifungal Activity of Yucca Saponin Fraction.

Yeast, MIC (μg/ml)			
Saccharomyces		*Pichia*	
cerevisiae IMO 203	62.5	*nakazawae* HUT 1688**	31.3
cerevisiae HUT 2075	31.3	*carsonii** IFO 946	31.3
cerevisiae JCM 2223	62.5	*Debaryomyces*	
Candida famata IFO 664*	31.3	*hansenii* IFO 18**	31.3
Hansenula anomala HUT 7083*	31.3	*hansenii* IFO 27**	62.5
Hansenula sp.*	31.3	*hansenii* IFO 47**	31.3
Cryptococcus sp.*	31.3	*hansenii* IFO 7011**	125
laurentii IFO 609	125	*Zygosaccharomyces*	
Kloeckera apiculata IFO 154	62.5	*rouxii* IFO 845**	31.3
		rouxii IFO 1130**	31.3

* food-deteriorating yeast ** film-forming yeast in soy sauce

Fungi, MIC (μg/ml)			
Aspergillus		*Aspergillus*	
niger IFO 4343	>1,000	*awamori* HUT 2014	>1,000
oryzae HUT 2065	>1,000	*awamori* HUT 2015	>1,000
oryzae HUT 2175	125	*Mucor pusillus* HUT 1185	15.6
oryzae HUT 2188	>1,000	*Rhizopus*	
oryzae HUT 2192	>1,000	*formosaensis* IFO 4756	>1,000
sydowii HUT 4097	>1,000	*nigricans* IFO 4731	>1,000
		Penicillium expansum IFO 5453	>1,000

Dermatophytic yeast and fungi, MIC (μg/ml)			
Candida albicans TIMM 0134	62.5	*Sabouraudites canis* IFO 7863	31.3
Trichophyton		*Epidermophyton floccosum* IFO 9045	31.3
rubrum IFO 5807	15.6		
mentagrophytes IFO 5809	31.3		

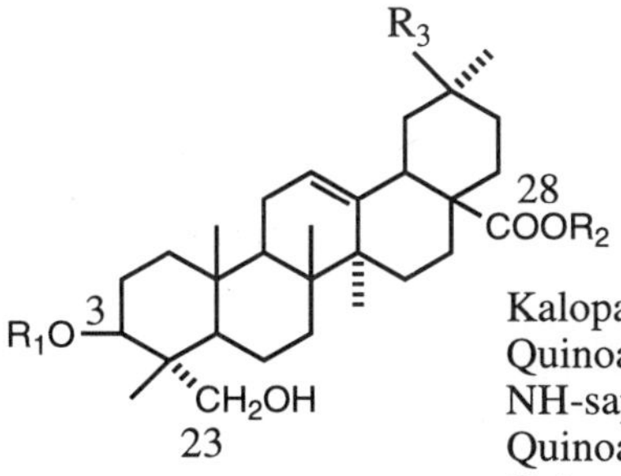

MIC > 400 μg/ml

	R_1	R_2	R_3
Kalopanax-saponin B	-Ara-Rha	-$(Glc)_2$-Rha	Me
Quinoa-saponin 3	-Ara	-Glc	COOMe
NH-saponin F	-Ara	-Glc	Me
Quinoa-saponin 4	-Ara-Glc	-Glc	COOMe
Quinoa-saponin 2	-Ara-Glc	-Glc	Me

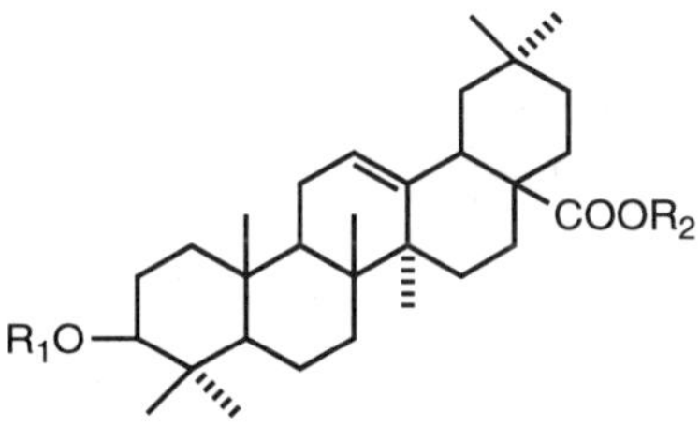

MIC > 400 μg/ml

	R_1	R_2
Saponin CP4	-Ara-Rha-Rib	H
Huzhangoside A	-Xyl-Rha-Rib	H
Huzhangoside B	-Ara-Rha-Rib	-(Glc)2-Rha

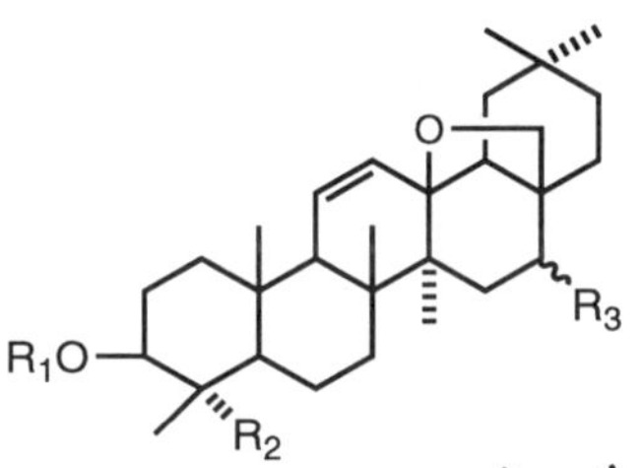

MIC > 400 μg/ml

	R_1	R_2	R_3
Saikosaponin a	-Fuc-Glc	CH_2OH	β-OH
Saikosaponin b	-Fuc-Glc	CH_2OH	α-OH
Saikosaponin c	-Glc-Rha \ Glc	Me	β-OH

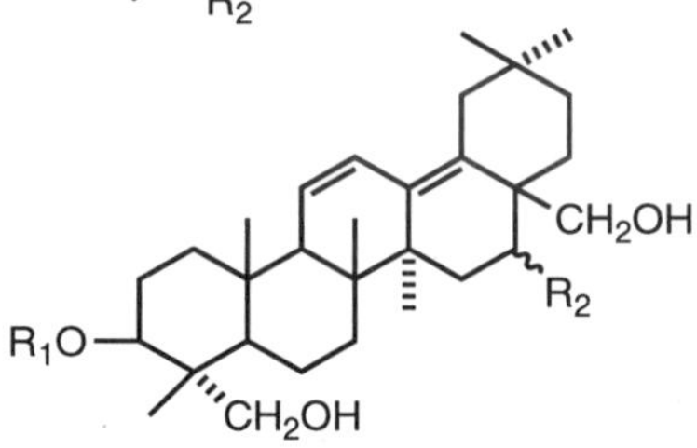

MIC > 400 μg/ml

	R_1	R_2
Saikosaponin b1	-Fuc-Glc	β-OH
Saikosaponin b2	-Fuc-Glc	α-OH

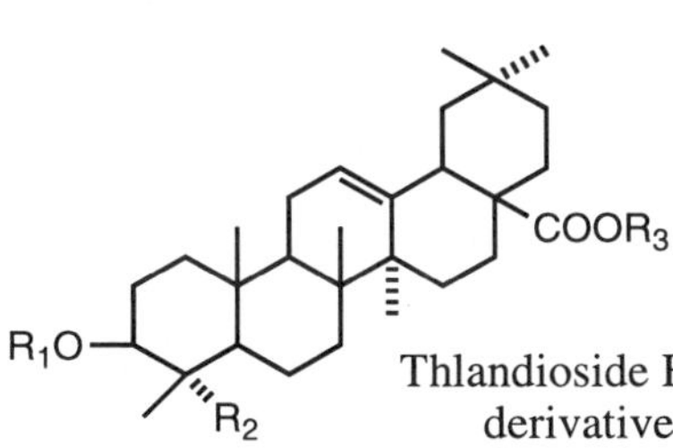

	R_1	R_2	R_3	MIC, (μg/ml)
Thlandioside H1	-GlcA-Gal	CHO	-Xyl-Rha-(Xyl)2	> 400
derivative 1	-GlcA-Gal	CH2OH	-Xyl-Rha-(Xyl)2	> 400
derivative 2	-GlcA-Gal	CH2OH	-H	100
Hederagenin-3-Ara	-Ara(p)	CH_2OH	-H	200

Figure 3. Structure-antifungal activity relationships of saponins against *Tricophyton rubrum*.

the food-deterioratiing yeasts, *Pichia nakazawae, Debaryomyces hansenii* and *Hansenula anomala,* as well as against *Malassezia furfur* which is asssociated with dandruff generation.

The activity of the saponin fraction against common fungi was not so strong, while it exhibited remarkable growth-inhibitory effects against the following dermatophytic fungi and a pathogenic yeast, *Trichophyton rubrum, T. mentagrophytes, Sabouraudites canis,* and *Epidermophyton floccosum* (which are known as dermatophytic fungi) and against *Candida albicans,* a pathogenic yeast which causes cutaneous candidiasis.

Sapindus Saponin Fraction as an Antidermatophytic Ingredient in Cosmetics

It is difficult to use *Sapindus* saponin fraction as a food ingredient without long term toxicity tests, because we have no history of this fraction or *Sapindus* extract as a foodstuff. Furthermore, it tastes very bitter, changing the taste of foods. On the other hand, the extract has been used as a folk detergent, and is listed in the Japanese Cosmetic Ingredient Codex (JCIC), being authorized as an ingredient in cosmetics by the Ministry of Health and Welfare in Japan. We reconfirmed the safety of the saponin fraction by dermal toxicity tests. It did not show primary dermal irritant, sensitization, phototoxicity or photosensitization effects.

Other *Sapindus* Species in Asia

Several *Sapindus* species other than *S. mukurossi* grow in Asia, and their pericarps have been also used as natural detergents. *S. delavayi* (Nakayama et al., 1986; Wong et al., 1991), *S. rarak, S. trifolius* (Kasai et al. 1988) and others. The pericarps of these plants contain similar saponins of hederagenin and sesquiterpene oligoglycosides, being also promising as raw materials for ingredients in cosmetics for the prevention of dermatophytoses.

References Cited

Kaneda, N., Nakanishi, H., and Staba, J.E. 1987. Steroidal constituents of *Yucca schidigera* plants and tissue culture. *Phytochemistry*, 26: 1425, and references cited therein.

Kasai, R., Fujino, H., Kuzuki, T., Wong, W., Goto, C., Yata, N., Tanaka, O., Yasuhara, F., and Yamaguchi, S. 1986. Acyclic sesquiterpene oligoglycosides from pericarps of *Sapindus mukurossi. Phytochemistry,* 25: 871.

Kasai, R., Nishi, M., Mizutani, K., Miyahara, I., Moriya, T., Miyahara, K., and Tanaka, O., 1988. Trifolioside II, an acyclic sesquiterpene oligoglycoside from pericarps of *Sapindus trioliatus. Phytochemistry,* 27: 2209.

Kimata, H., Nakashima, T., Kokubun, S., Nakayama, K., Mitoma, Y., Kitahara, T., Yata, N. and Tanaka, O. 1983. Saponins of pericarps of *Sapindus mukurossi* Gaertn. and solubilization of monodesmosides by bisdesmosides. *Chem. Pharm. Bull.,* 31: 1998, and references cited therein.

Kuramoto, T., Ito, Y., Oda, M., Tamura, Y., and Kitahata, S. 1994. Microbial production of glycyrrhetic acid 3-*O*-β-mono-D-glucuronide from glycyrrhizin by *Cryptococcus magnus* MG-27. *Biosci., Biotech., Biochem.,* 58: 455.

Mizutani, K. 1994. Biological activities, production, and use of chemical constituents of licorice. *Food Phytochemicals for Cancer Prevention II, ACS Symposium Series 547,* C.-T. Ho, T. Osawa, M.-T. Huang, and R.T. Rosen Eds. American Chemical Society, Washington, D.C.

Mizutani, K., Kuramoto, T., Tamura, Y., Ohtake, N., Doi, S., Nakamura, M., and Tanaka, O. 1994. Sweetness of glycyrrhetic acid 3-*O*-β-mono-D-glucuronide and the related glycosides. *Biosci. Biotech. Biochem.*, 58: 554.

Nakano, K., Yamasaki, Y., Imamura, Y., Murakami, K., Takaishi, Y., and Tomimatsu, T. 1989. The steroidal glycosides from the caudex of *Yucca gloriosa*. *Phytochemistry*, 28: 1215.

Nakayama, K., Fujino, H., Kasai, R., Tanaka, O., and Zhou, J. 1986. Saponins of pericarps of Chinese *Sapindus delavayi* (Pyi-shiau-tzu), a source of natural surfactants. *Chem. Pharm. Bull.*, 34: 2209.

Nie, R., Tanaka, T., Miyakoshi, M., Kasai, R., Morita, T., Zhou, J., and Tanaka, O. 1989. Triterpenoid saponin from *Thlandiantha hookeri* var. *pentadactyla, Phytochemistry,* 28: 1711.

Wong, W., Kasai, R., Choshi, W., Nakagawa, Y., Mizutani, K., Ohtani, K., and Tanaka, O. 1991. Acyclic sesquiterpene oligoglycosides from pericarps of *Sapindus delavayi*. *Phytochemistry*, 30: 2699.

Yata, N., Sugihara, N., Yamajo, R., Murakami, T., Higashi, H., Kimata, H., Nakayama, K., Kuzuki, T., and Tanaka, O. 1985. Enhanced rectal absorption of β-lactam antibiotics in rat by monodesmosides isolated from pericarps of *Sapindus mukurossi* (enmei-hi). *J. Pharmacobio-Dyn.*, 8: 1041.

Yata, N., Sugihara, R., Yamajo, R., Murakami, T., Higashi, Y., Kimata, H., Nakayama, K., Kuzuki, T. and Tanaka, O. 1986. Enhanced small intestinal absorption of β–lactam antibiotics in rat in the presence of monodesmosides isolated from pericarps of *Sapinus mukurossi* (enmei-hi). *J. Pharmacobio-Dyn.*, 9: 211.

NOVEL SWEET-TASTING SAPONINS OF THE CYCLOARTANE, OLEANANE, SECODAMMARANE, AND STEROIDAL TYPES

Edward J. Kennelly,[1] Rutt Suttisri,[2] and A. Douglas Kinghorn[1]

[1]Program for Collaborative Research in the Pharmaceutical Sciences
Department of Medicinal Chemistry and Pharmacognosy
College of Pharmacy
University of Illinois at Chicago
Chicago, IL 60612

[2]Department of Pharmaceutical Botany
Faculty of Pharmaceutical Sciences
Chulalongkorn University
Bangkok, 10330 Thailand

INTRODUCTION

For over 15 years we have carried on an active program in the discovery and evaluation of potently sweet compounds of plant origin at the University of Illinois at Chicago (Kinghorn and Soejarto, 1986; Kinghorn and Kennelly, 1995). In the course of this project, we have examined sweet-tasting plants from many parts of the world. While these plants have often been unrelated taxonomically, several of the sweet-tasting compounds discovered have turned out to be triterpenoid and steroidal saponins. This chapter will examine the novel sweet-tasting saponins characterized in our laboratory, which comprise those based on the cycloartane, oleanane, secodammarane, and steroidal nuclei.

Sweet-Tasting Saponins

The knowledge that certain saponins are sweet-tasting may be traced to glycyrrhizin (Figure 1 [**1**]), the well-known oleanane glycoside from the roots of licorice, *Glycyrrhiza glabra* L., and from related species in the same genus. This saponin was first reported over two hundred years ago (see Fenwick et al., 1990 for a general review), but its structure was finalized many years after its initial discovery (Lythgoe and Trippet, 1950; Beaton and Spring, 1955). Ammoniated glycyrrhizin, the fully ammoniated form of glycyrrhizin

(Figure 1 [**1**]), has been accorded "Generally Regarded as Safe" (GRAS) status in the United States, and is widely used as a flavoring and foaming agent in food products (Cook, 1970). In Japan, purified licorice extracts are used for the sweetening of foods (particularly salty and fermented products), beverages, medicines, cosmetics, and tobacco (Kinghorn and Compadre, 1991). Sweet derivatives of glycyrrhizin, including apioglycyrrhizin and araboglycyrrhizin, have been isolated recently from *Glycyrrhiza inflata* Batal (Kitagawa et al., 1993). These two compounds differ from glycyrrhizin (Figure 1 [**1**]) only in the composition of the disaccharide moiety attached to position C-3. Apioglycyrrhizin has been rated as two times the sweetness potency of the parent compound, and araboglycyrrhizin as equal to the sweetness potency of glycyrrhizin; itself about 100 times the sweetness potency of sucrose. Other sweet-tasting licorice constituents are licorice-saponins A3, G2, and H2, and recently a detailed analysis of the structural requirements for sweetness among glycyrrhizin and its analogs has appeared in the literature (Kitagawa et al., 1993).

Glycyrrhizin (Figure 1 [**1**]) exhibits a range of biological activities other than sweetness (Sela and Steinberg, 1989). However, a problem associated with excessive use of this substance is pseudoaldosteronism, resulting in edema, hypertension, mild potassium diuresis, and sodium retention (Conn et al., 1968) Accordingly, a government office in Japan has issued a caution stating that the daily consumption of this oleanane glycoside should be less than 200 mg/day (Kinghorn and Compadre, 1991).

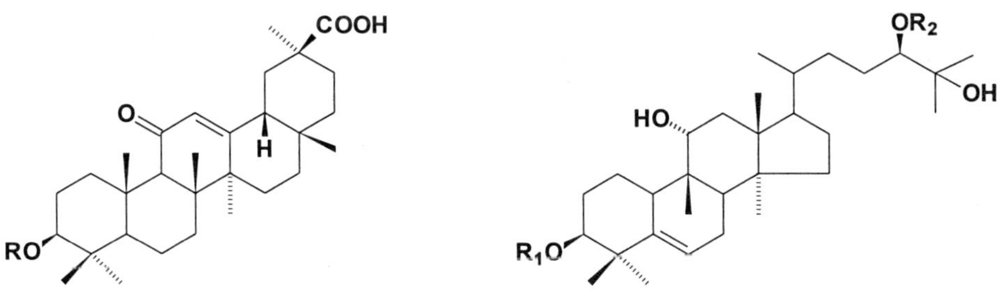

		R			R1	R2
1	Glycyrrhizin	β-glcA^{2}– β-glcA	**2**	Mogroside V	β-glc^{6}– β-glc	β-glc^{2}– β-glc (6: β-glc)

Figure 1. Structures of two sweet-tasting saponins used for sweetening purposes in Japan.

An extract from the dried fruits of *Siraitia grosvenorii* (Swingle) C. Jeffrey (Cucurbitaceae), the Chinese plant "lo han guo", has minor use as a sucrose substitute in Japan. Mogroside V (Figure 1 [**2**]) is the most abundant sweet-tasting principle of *S. grosvenorii* fruits, and possesses about 250 times the sweetness potency of sucrose (Takemoto et al., 1983a-c). Interestingly, cucurbitane triterpenes are more commonly known as bitter-tasting constituents of plants. Although extracts of *S. grosvenorii* fruits have been widely used in the traditional medicine of China for centuries (Kinghorn and Soejarto, 1986; Kinghorn and Compadre, 1991), the plant was not known to modern science until the 1930s, with the structures of the sweet-tasting saponins being elucidated only relatively recently (Takemoto et al., 1983a-c; Matsumoto et al., 1990). Apart from mogroside V (Figure 1 [**2**]), mogroside IV, 11-oxomogroside, and siamenoside I are also sweet constituents of *S. grosvenorii* fruits, with the latter being rated as the most potently sweet cucurbitane glycoside with 563 times the sweetness potency of 5% w/w sucrose (Takemoto et al., 1983a-c; Matsumoto et al., 1990). Sweet cucurbitane glycosides have also been reported to occur in the fruits of *Siraitia sinamensis* (Kasai et al., 1989).

In addition to these commercially important compounds, a number of other sweet-tasting saponins have been reported. Cucurbitane glycosides, including bryoside and bryonoside, isolated from *Bryonia dioica* Jacq. (Cucurbitaceae) (Oobayashi et al., 1992), and carnosiflosides V and VI and scandenoside R6, isolated from *Hemsleya* species (Cucurbitaceae), have been reported to be sweet-tasting (Kasai et al., 1988). A series of oleanane glycosides, purified from the roots of *Periandra dulcis* Mart. (Leguminosae), has been described (Hashimoto et al., 1980, 1982, 1983), and will be mentioned later in this chapter. A sweet-tasting steroidal saponin, osladin, from the fern *Polypodium vulgare* L. (Polypodiaceae), has been isolated and characterized (Jizba et al., 1971; Yamada et al., 1992), and will also be discussed in greater detail later in this paper. A more complete treatment of sweet-tasting saponins has been included in a recent comprehensive review article on natural sweeteners (Kinghorn et al., 1995a).

Sweetness Inhibitors

While the above-mentioned saponins are highly sweet-tasting, another group of saponins is well-known as sweetness inhibitors. These compounds have the ability to suppress the sweet taste of sucrose in humans and animals. Members of a group of about 20 oleanane glycosides, comprised of gymnemic acids I-XVIII and gymnemasaponins III-V, isolated from *Gymnema sylvestre* R. Br. (Asclepiadaceae), have been found to be sweetness inhibitors. For example, gymnemic acid I (Figure 2 [**3**]), obtained by the hot-water extraction of *G. sylvestre*, is an acylated oleanane glucuronopyranoside, and a 0.5 mM solution produces complete suppression of the sweetness induced by 0.4 M sucrose (Yoshikawa et al., 1989).

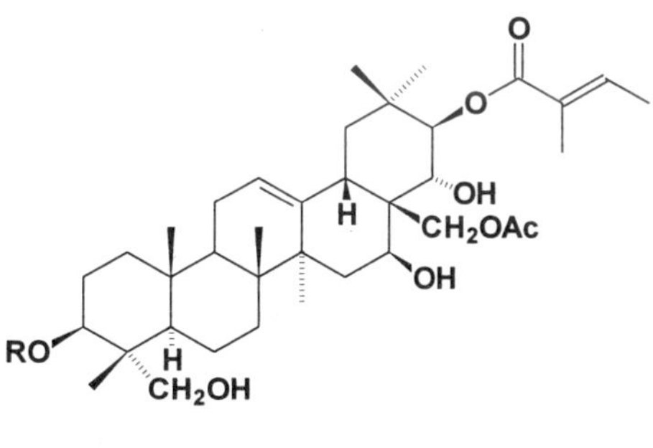

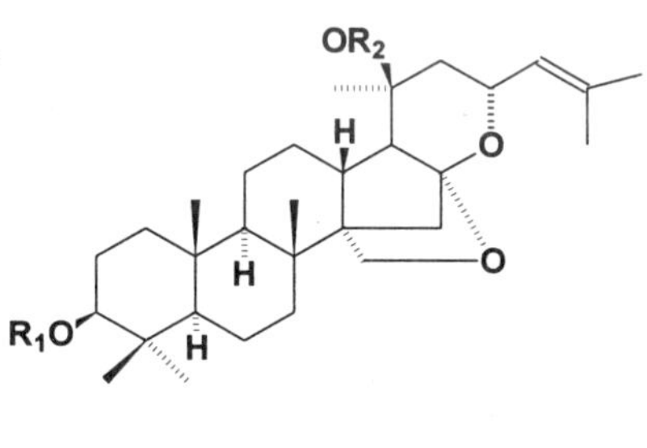

		R		R1	R2
3	Gymnemic acid I	β-glcA	**4** Ziziphin	α-ara^4—α-rha	α-rha-2,3-Ac

Figure 2. Structures of two saponin sweetness inhibitors based on the oleanane and dammarane skeletons.

Ziziphin (Figure 2 [**4**]) is another example of a saponin sweetness inhibitor. This dammarane-type compound was structurally characterized as a constituent of the Chinese jujube tree, *Ziziphus jujuba* P. Miller (Rhamnaceae) (Kurihara et al., 1988; Yoshikawa et al., 1991). Further dammarane saponin sweetness inhibitors have been reported from this same plant (Yoshikawa et al., 1992a). More recently, a series of dammarane glycoside sweetness inhibitors has been reported from the Chinese raisin tree, *Hovenia dulcis* Thunb.

(Rhamnaceae) (Yoshikawa et al., 1992b; 1993). A detailed review of plant-derived triterpenoid sweetness inhibitors has recently been published by our group (Suttisri et al., 1995). While these compounds are not the subject of the present chapter, they are of considerable interest in the sweetness field, especially to help in the understanding of structure-taste relationships among saponins.

APPROACH TO THE DISCOVERY OF SWEET-TASTING NATURAL PRODUCTS

Candidate sweet-tasting plants are chosen for study in our natural sweetener discovery project based primarily on ethnobotanical information and field observations. Ethnobotany is the study of how plants are utilized by society, and, through careful examination of the literature of this discipline, we have found it to be a useful aid in the discovery of novel sweeteners. The ability of humans to taste and appreciate sweetness appears to be notably consistent from culture to culture around the world. Therefore, plants mentioned in the ethnobotanical literature as being sweet tasting in subsequent laboratory investigation have frequently been found to contain high levels of free sugars and polyols, or, much less frequently, potently sweet-tasting natural products. Other plants that we have examined were chosen owing to their intensely sweet taste as a result of field work. We have obtained sweet-tasting plants from around the world, often from botanists who may randomly taste plants. In addition, we have conducted field trips in developing countries, and have found good lead candidate plants in traditional medicine markets. These plants may not always be used as sweeteners *per se*, but their sweet taste is often well known by those who sell the plant (Kinghorn and Soejarto, 1986; Kinghorn and Kennelly, 1995).

Once a plant is selected for study, care must be exercised to insure that the sweet taste is not due simply to high levels of free sugars and polyols. A standard solvent-solvent partition is generally carried out on a methanolic extract of the plant to yield petroleum ether-, ethyl acetate-, 1-butanol-, and water-soluble fractions. If the sweet taste is concentrated into the water-soluble fraction, a TLC examination is initially performed to determine if there are high levels of free sugars or polyols present, backed up by GC/MS examination of the derivatized plant extract. From such GC/MS investigations, we have determined that the free sugar and/or polyol content of a given plant part must be well over 5% w/w in order to exhibit an overtly sweet taste, unless an intense sweetener is also present (Hussain et al., 1990).

Currently, there is no suitable *in vitro* receptor-binding assay to monitor plant extracts, chromatographic fractions, and pure isolates for the presence or absence of sweetness. Therefore, prior to tasting a crude plant extract or to evaluating the sweetness intensity of a sweet plant secondary metabolite by a human taste panel, such samples must first be shown to be innocuous in bacterial mutagenicity and mouse acute toxicity tests (Kim et al., 1988; Choi et al, 1989a; Kennelly et al., 1995b). We have investigated the possiblity of using electrophysiological and behavioral assays with the Mongolian gerbil to substitute for human volunteer subjects in the natural sweetener discovery process. While such methods will probably never have general application for all classes of potently sweet organic compounds, they have been used effectively in our laboratory to evaluate sweet-tasting saponins, both in crude extracts and when pure, from *Abrus precatorius*, *Periandra dulcis*, and *Siraitia grosvenorii* (Jackinovich et al., 1990; Vasquez et al., 1993).

SWEET-TASTING TRITERPENOID SAPONINS

Cycloartane

The most promising sweet-tasting saponins discovered in our program to date have been isolated from *Abrus precatorius* L. (Fabaceae) (the rosary pea or jequirity bean), a plant that is well-known for having toxic seeds owing to the presence of the ribosome-inactivating protein, abrin. Interestingly, the leaves and roots of this weedy pan-tropical vine are extremely sweet-tasting, and do not appear to contain abrin. The roots are well documented as a licorice substitute, and are known as Indian licorice. The leaves, which can be harvested well in advance of seed production, have been used in a sweet Malaysian beverage, and are sold to sweeten betel quid in marketplaces in Java (Choi et al., 1989a; Kinghorn et al., 1995b). During a recent field trip, it was found that *A. precatorius* is cultivated in Indonesia so that the leaves can be used in a medicine used to treat a tropical disease, sprue.

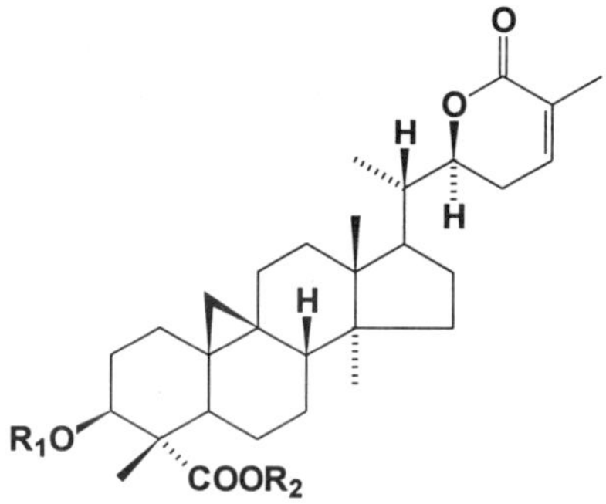

		R_1	R_2
5	Abrusoside A	β-glc	H
6	Abrusoside B	β-glcA-6-CH_3^{2}– β-glc	H
7	Abrusoside C	β-glc^{2}– β-glc	H
8	Abrusoside D	β-glcA^{2}– β-glc	H
9	Abrusoside E	β-glc^{2}– β-glcA	H
10	Abrusoside E 6"-methyl ester	β-glc^{2}– β-glcA-6-CH_3	H
11	Abrusoside E dimethyl ester	β-glc^{2}– β-glcA-6-CH_3	CH_3

Figure 3. Structures of sweet-tasting cycloartane saponins and the derivatives from *Abrus precatorius*.

Early work on this plant indicated that the sweetness was due to the presence of glycyrrhizin (Figure 1 [**1**]). However, in our initial work on the leaves of this plant, we were unable to detect this oleanane glycoside, but did identify four novel potently sweet cycloartane saponins, designated abrusosides A-D (Figure 3 [**5-8**]). These four patented compounds (Kinghorn and Choi, 1993) share a novel aglycone, possessing a cyclopropyl ring, a methylated α,β-unsaturated δ-lactone ring, and a carboxylic acid unit. The structure of the aglycone (abrusogenin) was confirmed by single-crystal X-ray crystallography on its methyl ester (Choi et al., 1989b). Abrusosides A-D (Figure 3 [**5-8**]), which differ from each other only in the saccharide moieties attached to the C-3 position, are not mutagenic or acutely toxic to mice, and have been rated as having about 30, 100, 50, and 75 times the sweetness intensity of 2% sucrose, respectively (Choi et al., 1989a). The compounds may easily be rendered water soluble by making their ammonium salts, and can withstand being heated in the solid form at 180 °C for a week with minimal breakdown.

Recent work in this laboratory has resulted in the isolation of an additional sweet-tasting cycloartane saponin, designated abrusoside E (Figure 3 [**9**]) (Kennelly et al., 1995a). This new glycoside, which has the reversed sugar substitution when compared to abrusoside D (Figure 3 [**8**]), was noted to be only marginally sweet tasting, possibly owing to its poor solubility properties. However, its monomethyl ester, abrusoside E 6”-methyl ester (Figure 3 [**10**]), was judged to be pleasantly sweet tasting, with a preliminary taste-panel rating it as about 150 times the sweetness intensity of 2% sucrose. Interestingly, the dimethyl ester derivative (Figure 3 [**11**]) was not perceived as being sweet tasting, thus lending insight into the structure-activity relationships for sweet-taste perception among this group of compounds.

It is the 11-oxo-12,13-dehydro group of glycyrrhizin (Figure 1 [**1**]) which is responsible for its adrenocorticomimetic effects, as mentioned earlier in this chapter (Kinghorn et al., 1995b). Since the abrusoside sweeteners do not contain an unhindered α,β-unsaturated keto group like that found in glycyrrhizin, it is to be hoped that they will not produce the same undesirable mineralocorticoid effects of this oleanane saponin sweetener (Kinghorn et al., 1995b).

Oleanane

Another sweet-tasting leguminous plant, *Periandra dulcis* Mart., known also as Brazilian licorice, was initially reported to contain glycyrrhizin (Figure 1 [**1**]) (Machado, 1941), but more recent efforts conducted by Hashimoto and co-workers at Kobe Pharmaceutical University in Japan failed to detect any of this oleanane glycoside in the roots of this plant. Research work by Hashimoto’s group in the 1980’s resulted in the isolation of four novel potently sweet oleananes, designated periandrins I-IV (Figure 4 [**12-15**]), and the sweetness potency of each was judged to be about 90 times that of sucrose (Hashimoto et al., 1980, 1982, 1983).

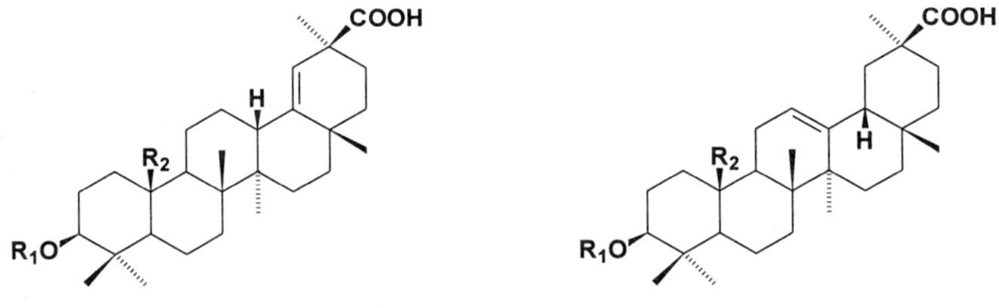

	R_1	R_2		R_1	R_2
12 Periandrin I	β-glcA^{2}—β-glcA	CHO	**13** Periandrin II	β-glcA^{2}—β-glcA	CHO
14 Periandrin III	β-glcA^{2}—β-glcA	CH_2OH	**15** Periandrin IV	β-glcA^{2}—β-glcA	CH_2OH
16 Periandrin V	β-glcA^{2}—β-xyl	CHO			

Figure 4. Structures of five sweet-tasting oleanane saponins from *Periandra dulcis*.

In collaboration with Hashimoto, we further examined the roots of *P. dulcis*, and isolated and characterized periandrin V (Figure 4 [**16**]), an additional sweet-tasting oleanane derivative (Suttisri et al., 1993). This D-xylose-containing glycoside has been rated by a preliminary taste-panel as being about 220 times the sweetness intensity of 2% sucrose, thus making it the sweetest periandrin derivative so far discovered.

Like the abrusosides, the periandrins offer a good opportunity to examine structure-activity relationships for sweetness. A free carboxylic acid group at C-30 appears essential for the exhibition of sweetness in the periandrins, since whenever a methyl or hydroxymethyl group is substituted at this position, the sweetness is lost (Suttisri et al., 1993). The periandrins occur in low yields in *P. dulcis* roots, and are difficult to isolate in pure form, and must be separated from thus far uncharacterized bitter principles which also are biosynthesized in the same plant (Hashimoto et al., 1980, 1982, 1983; Kinghorn and Soejarto, 1986; Suttisri et al., 1993).

Secodammarane

Some of the most promising leads in our sweetener project have come from plants that are actively used by human populations, thus implying some assurance of the relative safety of the plant when ingested. This is the case with *Pterocarya paliurus* Batal. (Juglandaceae), whose leaves are used by local populations in remote areas of Enshi County, Hubei Province, People's Republic of China, to sweeten foods. While our work on this species was in progress, a report appeared describing the isolation of several sweet-tasting compounds, with one being identified as a novel dammarane saponin, cyclocarioside A (Figure 5 [**17**]), judged to be 200 times as sweet as sucrose (Yang et al., 1992).

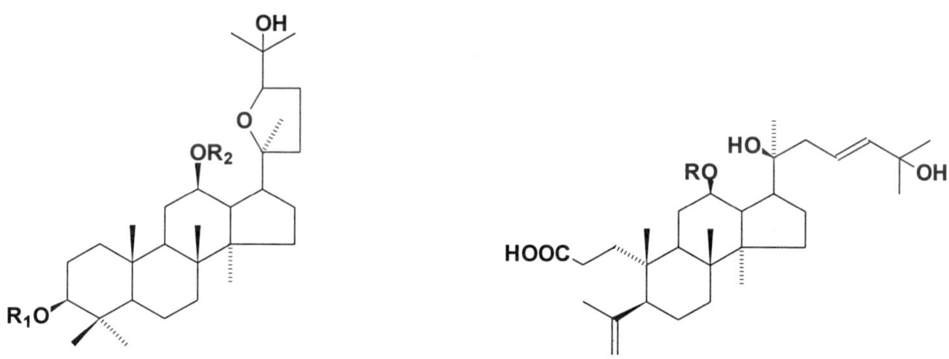

		R1	R2			R
17	Cyclocarioside A	α-ara-5-Ac	α-rha	**18**	Pterocaryoside A	β-qui
				19	Pterocaryoside B	α-ara

Figure 5. Structures of sweet dammarane and secodammarane saponins from *Pterocarya paliurus*.

While no sweet-tasting saponins based on the dammarane skeleton were obtained in our work on *P. paliurus*, two sweet-tasting secodammarane saponins, designated pterocaryosides A (Figure 5 [**18**]) and B (Figure 5 [**19**]) were identified (Kennelly et al.,

1995b). Their structures were characterized with the aid of published data of the model compound (23*E*)-(12*R*,20*S*)-12,20,25-trihydroxy-3,4-secodammara-4(28),23-dien-3-oate, isolated from the male flowers of *Alnus japonica* (Thunb.) Steudel (Betulaceae). Preliminary safety tests established that pterocaryosides A and B were not bacterial mutagens nor acutely toxic to mice, and subsequently, a human taste panel judged these compounds to be about 50 and 100 times as sweet as 2% sucrose, respectively. The onset of sweetness was found to be instantaneous, but both compounds had a persistent, bitter off-taste, which limits their prospects for further development. However, these compounds are the first sweet-tasting secodammarane saponins to have been characterized to date, and the difference in sweetness potency exhibited by the relatively minor saccharide substitution occurring among these two compounds is intriguing.

SWEET-TASTING STEROIDAL SAPONINS

A final group of sweeteners from our project was isolated from the North American licorice fern, *Polypodium glycyrrhiza* DC. Eaton (Polypodiaceae). The bittersweet-tasting rhizomes of this delicate plant were once used by humans in the Pacific northwest region as a foodstuff and medicinal agent. As with *Abrus precatorius* and *Periandra dulcis*, early phytochemical studies (Fischer and Goodrich, 1930) on this fern indicated it contained significant levels of glycyrrhizin (Figure 1 [**1**]). A later report disputed the presence of this oleanane glycoside, and indicated that the sweet taste of the rhizome was due to the presence of high levels of sugars and a small amount of an unidentified substance (Fischer and Lynn, 1933).

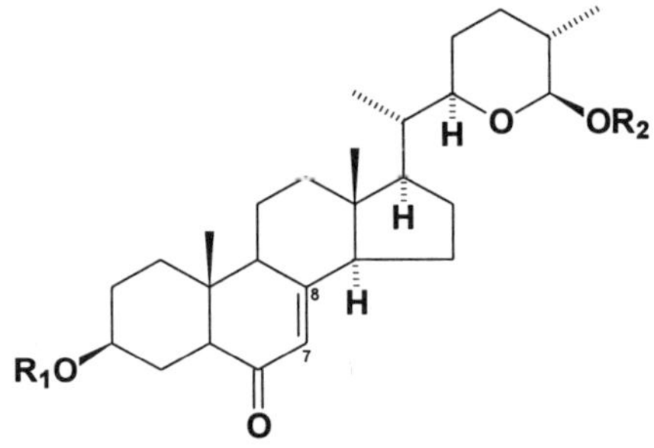

		R_1	R_2	Other
20	Osladin	β-glc$\overset{2}{—}$α-rha	α-rha	7,8-dihydro
21	Polypodoside A	β-glc$\overset{2}{—}$α-rha	α-rha	--
22	Polypodoside B	β-glc	α-rha	--
23	Polypodoside C	β-glc	α-rha-3-CH_3	--

Figure 6. Structures of steroidal saponins from *Polypodium* species.

In an effort to resolve this issue, we examined this plant, and isolated the novel potently sweet steroidal saponin polypodoside A in 0.29% w/w yield (Kim et al., 1988). In our original structure elucidation of polypodoside A, the composition and configuration of the sugar units were definitively established, with the aglycone assigned as polypodogenin, a known compound whose stereochemistry was suggested earlier by Czech workers (Kim et al., 1988). However, recently the stereochemistry of the aglycone of osladin (Figure 6 [**20**]), a structurally related steroidal saponin isolated originally from *Polypodium vulgare* L. (Jizba et al., 1971) was revised from 22*S*,25*R*,26*S* to 22*R*,25*S*,26*R* (Yamada et al., 1992). In collaboration with Nishizawa's group at Tokushima Bunri University in Japan, the stereochemistry of polypodogenin, and hence polypodoside A (Figure 6 [**21**]) was similarly revised (Nishizawa et al., 1994).

Polypodoside A (Figure 5 [**21**]) is the most potently sweet-tasting saponin which has thus far been discovered in our program, being rated on our behalf by an industrial group as about 600 times as sweet as 6% sucrose. However, like the pterocaryosides (Figure 5 [**18,19**]), the hedonic qualities of polypodoside A (Figure 6 [**21**]) were deemed unsatisfactory, having a licorice-like off-taste and a lingering aftertaste (Kim et al., 1988). A second bidesmosidic sweet-tasting saponin, polypodoside B (Figure 6 [**22**]), was isolated from *P. glycyrrhiza*, although it was obtained in insufficient quantities for toxicological and detailed sensory evaluation (Kim and Kinghorn, 1989). Polypodoside C (Figure 6 [**23**]), also isolated from this fern (Kim and Kinghorn, 1989), differs from polypodoside B (Figure 6 [**22**]) only by methylation of the C-26-affixed saccharide moiety, but was devoid of sweetness. This represents another example of how a minor change in the saccharide moiety of a sweet-tasting saponin can dramatically alter its taste profile.

CONCLUSIONS

While saponins are generally regarded as bitter or otherwise unpleasant-tasting substances, two triterpene saponins, glycyrrhizin (Figure 1 [**1**]) and mogroside V (Figure 1 [**2**]), are the principal sweet-tasting components of two plant extracts currently used for sweetening purposes in Japan. In fact, triterpenoid and steroidal saponins together represent the largest single group of the 75 or so known potently sweet-tasting plant secondary metabolites. Novel sweet saponins have emerged from our program of natural sweetener discovery and evaluation, based on the cycloartane (abrusosides A-E, Figure 3 [**5-9**]), oleanane (periandrin V, Figure 4 [**16**]), secodammarane (pterocaryosides A and B, Figure 5 [**18,19**]), and steroidal saponin (polypodosides A and B, Figure 6, [**21,22**]) types. While many of these compounds have unacceptable hedonic properties such as having a lingering aftertaste, the abrusosides seem to have the best taste qualities of the sweet saponins so far obtained in our laboratory. It has been found that subtle variations in the saccharide substituents of the sweet saponins may lead to profound effects on their taste, so these substances have inherent scientific value in helping to provide a better understanding of the role of chemical structure to sweetness.

ACKNOWLEDGMENTS

We would like to thank the following colleagues for the supply of sweet-tasting plant materials: Professor Julia F. Morton (University of Miami, Coral Gables, Florida) for *Abrus precatorius*; Emeritus Professor Yohei Hashimoto (Kobe Women's College of Pharmacy, Kobe, Japan) for *Periandra dulcis*; Professor Bing-Nan Zhou (Shanghai Institute of Materia Medica, Shanghai, China) for *Pterocarya paliurus*; and Professor Frank A. Lang (Southern Oregon State College, Ashland, Oregon) for *Polypodium glycyrrhiza* rhizomes. Early separation work on periandrin V was carried out by A.D.K. in the laboratory of Professor Otto Sticher (Swiss Federal Institute of Technology, Zurich, Switzerland). Professor Mugio Nishizawa (Tokushima Bunri University, Tokushima,

Japan) is acknowledged for collaborative work on the stereochemistry of polypodoside A, and Prof. William Jackinovich, Jr. (Lehman College, City University of New York) for collaborative work on assays using the Mongolian gerbil.

Three colleagues from the University of Illinois at Chicago are thanked particularly for their role in this project: Professor D. Doel Soejarto for the procurement of sweet-tasting lead plants from around the world and for his taxonomic expertise; Professor John M. Pezzuto for the mutagenicity evaluation of sweet-tasting compounds; and Professor Boon K. Teo for the X-ray crystallography performed on abrusogenin methyl ester. We are also especially grateful for the contributions to this project of many excellent graduate students and postdoctorals, whose names are mentioned in the reference section.

This research project has been generously supported by grants R01-DE-08937 and R03-07560 and contract N01-DE-02425 from the National Institute of Dental Research (NIH, Bethesda, Maryland), by a contract with General Foods Corporation (White Plains, New York), and by the Montgomery Foundation (Miami, Florida), who supplied additional quantities of *Abrus precatorius* leaves.

REFERENCES

Beaton, J.M.; Spring, F.S. Triterpenoids. Part XLII. The configuration of the carboxyl group in glycyrrhetic acid. *J. Chem. Soc.* **1955**, 3126-3129.

Choi, Y.-H.; Hussain, R.A.; Pezzuto, J.M.; Kinghorn, A.D.; Morton, J.F. Abrusosides A-D, four novel sweet-tasting triterpene glycosides from the leaves of *Abrus precatorius*. *J. Nat. Prod.* **1989a**, *52*, 1118-1127.

Choi, Y.-H.; Kinghorn, A.D.; Shi, X.; Zhang, H.; Teo, B.K. Abrusoside A: A new type of highly sweet triterpene glycoside. *J. Chem. Soc., Chem. Commun.* **1989b**, 887-888.

Cook, M.R. Ammoniated glycyrrhizin: a useful natural sweetener and flavour potentiator. *Flavour Ind.* **1970**, *1*, 831-832.

Conn, J.W; Rovner, D.R.; Cohen, E.L. Licorice-induced pseudoaldosteronism. Hypertension, hypokalemeia, aldosteronopenia, and suppressed plasma renin activity. *J. Am. Med. Assoc.* **1968**, *205*, 80-84.

Fenwick, G.R.; Lutomski, J.; Nieman, C. Liquorice, *Glycyrrhiza glabra* L. Composition, uses and analysis. *Food Chem.* **1990**, *38*, 119-143.

Fischer, L. and Goodrich, F.J. *Polypodium occidentale*. *J. Am. Pharm. Assoc.* **1930**, *19*, 1063-1071.

Fischer, L. and Lynn, E.V. Licorice fern and wild licorice as substitutes for licorice. *J. Am. Pharm. Assoc.* **1933**, *22*, 1225-1230.

Hashimoto, Y.; Ishizone, H.; Ogura, M. Periandrin II and IV, triterpene glycosides from *Periandra dulcis*. *Phytochemistry* **1980**, *19*, 2411-2415.

Hashimoto, Y.; Ishizone, H.; Suganuma, M.; Ogura, M.; Nakatsu, K.; Yoshioka, H. Periandrin I, a sweet triterpene-glycoside from *Periandra dulcis*. *Phytochemistry* **1983**, *22*, 259-264.

Hashimoto, Y.; Ohta, Y.; Ishizone, H.; Kuriyama, M.; Ogura, M. Periandrin III, a novel sweet triterpene glycoside from *Periandra dulcis*. *Phytochemistry* **1982**, *21*, 2335-2337.

Hussain, R.A.; Lin, Y.-M.; Poveda, L.J.; Bordas, E.; Chung, B.S.; Pezzuto, J.M.; Soejarto, D.D.; Kinghorn, A.D. Plant-derived sweetening agents. Saccharide and polyol constituents of some sweet-tasting plants. *J. Ethnopharmacol.* **1990**, *28*, 103-115.

Jackinovich, Jr., W.; Moon, C.; Choi, Y.H.; Kinghorn, A.D. Evaluation of plant extracts for sweetness using the Mongolian gerbil. *J. Nat. Prod.* **1990**, *53*, 190-195.

Jizba, J.; Dolejs, L.; Herout, V.; Sorm, F. The structure of osladin. The sweet principle of the rhizomes of *Polypodium vulgare* L. *Tetrahedron Lett.* **1971**, 1329-1332.

Kasai, R.; Matsumoto, K.; Nie, R.-L.; Zhou, J.; Tanaka, O. Glycosides from the Chinese medicinal plant, *Hemsleya panicis-scandens*, and structure-taste relationship of cucurbitane glycosides. *Chem. Pharm. Bull.* **1988**, *36*, 234-243.

Kasai, R.; Nie, R.-L.; Nashi, K.; Ohtani, K.; Zhou, J.; Tao, G.-D.; Tanaka, O. Sweet cucurbitane glycosides from fruits of *Siraitia siamensis* (chi-zi luo-han-guo), a Chinese folk medicine. *Agric. Biol. Chem.* **1989**, *53*, 3347-3349.

Kennelly, E.J.; Cai, L.; Kim, N.-C.; Kinghorn, A.D. Abrusoside E, a further sweet-tasting cycloartane glycoside from the leaves of *Abrus precatorius*. *Phytochemistry* **1995a**, in press.

Kennelly, E.J.; Cai, L.; Long, L.; Shamon, L.; Zaw, K.; Zhou, B.-N.; Pezzuto, J.M.; Kinghorn, A.D. Novel highly sweet secodammarane glycosides from *Pterocarya paliurus*. *J. Agric. Food Chem.* **1995b**, *43*, 2602-2607.

Kim, J.; Kinghorn, A.D. Further steroidal and flavonoid constituents of the sweet plant, *Polypodium glycyrrhiza*. *Phytochemistry* **1989**, *28*, 1225-1228.

Kim, J.; Pezzuto, J.M.; Soejarto, D.D.; Lang, F.A.; Kinghorn, A.D. Polypodoside A, an intensely sweet constituent of the rhizomes of *Polypodium glycyrrhiza*. *J. Nat. Prod.* **1988**, *51*, 1166-1172.

Kinghorn, A.D.; Choi, Y.-H. *Natural intense sweeteners*. U.S. Patent 5,198.427, March 30, 1993.

Kinghorn, A.D.; Compadre, C.M. Less common high-potency sweeteners. In *Alternative Sweeteners. Second Edition, Revised and Expanded*; O'Brien Nabors, L.; Gelardi, R.C., Eds.; Marcel Dekker, Inc.: New York, 1991; pp 192-218.

Kinghorn, A.D.; Fullas, F.; Hussain, R.A. Structure-activity relationship of highly sweet natural products, In *Studies in Natural Products: Structure and Chemistry (Part C)*; ur-Rahman, A., Ed.; Elsevier Science BV: Amsterdam, 1995a; Vol. 15, pp 3-41.

Kinghorn, A.D.; Kennelly, E.J. Discovery of highly sweet compounds from natural sources. *J. Chem. Educ.* **1995**, *72*, 676-680.

Kinghorn, A.D.; Soejarto, D.D. Sweetening agents of plant origin. *CRC Crit. Rev. Plant Sci.* **1986**, *4*, 79-120.

Kinghorn, A.D.; Suttisri, R.; Lee, I.-K. Search for noncariogenic sweeteners from medicinal plants, In *Phytochemistry of Plants Used in Traditional Medicine*; Hostettmann, K.; Marston, A.; Maillard, M.; Hamburger, M., Eds., Clarendon Press: Oxford, U.K., 1995b; pp 165-188.

Kitagawa, I.; Hori, K.; Sakagami, M.; Hashiuchi, F.; Yoshikawa, M.; Ren, J. Saponin and sapogenol. XLIX. On the constituents of the roots of *Glycyrrhiza inflata* Batalin from Xinjian, China. Characterization of two sweet oleanane-type triterpene oligoglycosides, apioglycyrrhizin and araboglycyrrhizin. *Chem. Pharm. Bull.* **1993**, *41*, 1350-1357.

Kurihara, Y.; Ookubo, K.; Tasaki, H.; Kodama, H.; Akiyuma, Y.; Yagi, A.; Halpern, B. Studies on the taste modifiers. I. Purification and structure determination of sweetness inhibitory substances in leaves of *Ziziphus jujuba*. *Tetrahedron* **1988**, *44*, 61-66.

Lythgoe, B.; Trippett, S., The constitution of the disaccharide of glycyrrhinic acid. *J. Chem. Soc.* **1950**, 1983-1990.

Machado, A. Chemical study of Brazilian licorice. *Rev. Soc. Brasil. Quim.* **1941**, *10*, 101-103.

Matsumoto, R.; Kasai, R.; Ohtani, K.; Tanaka, O. Minor cucurbitane glycosides from fruits of *Siraitia grosvenorii* (Cucurbitaceae). *Chem. Pharm. Bull.* **1990**, *38*, 2030-2032.

Nishizawa, M.; Yamada, H.; Yamaguchi, Y.; Hatakayama, S.; Lee, I.-S.; Kennelly, E.J.; Kim, J.; Kinghorn, A.D. Structure revision of polypodoside A. Major sweet principle of *Polypodium glycyrrhiza*. *Chem. Lett.* **1994**, 1555-1558, 1979.

Oobayashi, K.; Yoshikawa, K.; Arihara, S. Structural revision of bryonoside and structure elucidation of minor saponins from *Bryonia dioica*. *Phytochemistry* **1992**, *31*, 943-946.

Sela, M.N.; Steinberg, D. Glycyrrhizin: the basic facts plus medical and dental benefits, In *Progress in Sweeteners*; Grenby, T.H., Ed.; Elsevier Applied Science: London, 1989; pp. 71-96.

Suttisri, R.; Chung, M.-S.; Kinghorn, A.D.; Sticher, O.; Hashimoto, Y. Periandrin V, a further sweet triterpene glycoside from *Periandra dulcis*. *Phytochemistry* 1993, *34*, 405-408.

Suttisri, R.; Lee, I.-S.; Kinghorn, A.D. Plant-derived triterpenoid sweetness inhibitors. *J. Ethnopharmacol.* **1995**, *47*, 9-26.

Takemoto, T.; Arihara, S.; Nakajima, T.; Okuhira, M. Studies on the constituents of *Fructus Momordicae*. I. On the sweet principle. *Yakugaku Zasshi* **1983a**, *103*, 1151-1154.

Takemoto, T.; Arihara, S.; Nakajima, T.; Okuhira, M. Studies on the constituents of *Fructus Momordicae*. II. Structure of sapogenin. *Yakugaku Zasshi* **1983b**, *103*, 1155-1166.

Takemoto, T.; Arihara, S.; Nakajima, T.; Okuhira, M. Studies on the constituents of *Fructus Momordicae*. III. Structure of mogrosides. *Yakugaku Zasshi* **1983c**, *103*, 1167-1173

Vasquez, E.; Jackinovich, Jr., W.; Nanayakkara, N.P.D.; Hussain, R.A.; Chung, M.S.; Kinghorn, A.D. Stimulation of the gerbil's gustatory receptors by some potently sweet terpenoids. *J. Agric. Food Chem.* **1993**, *41*, 1305-1310.

Yamada, H.; Nishizawa, M.; Katayama, C. Osladin, a sweet principle of *Polypodium vulgare*. Structure revision. *Tetrahedron Lett.* **1992**, *33*, 4009-4010.

Yang, D.J.; Zhong, Z.C.; Xie, Z.M. Studies on the sweet principles from the leaves of *Cyclocarya paliurus* (Batal.) Iljinskaya. *Acta Pharm. Sin.* **1992**, *27*, 841-844.

Yoshikawa, K.; Amimoto, K.; Arihara, S.; Matsuura, K. Structure studies of new anti-sweet constituents from *Gymnema sylvestre*. *Tetrahedron Lett.* **1989**, *30*, 1103-1106.

Yoshikawa, K.; Nagai, N.; Yoshia, M.; Arihara, S. Anti-sweet natural products. VIII. Structures of hodulosides IV-X from *Hovenia dulcis* Thunb. var. *tomentella makino*. *Chem. Pharm. Bull.* **1993**, *41*, 1722-1725.

Yoshikawa, K.; Shimono, N.; Arihara, S. Antisweet substances, jujubasaponins I-III from *Zizyphus jujuba*: revised structure of ziziphin. *Tetrahedron Lett.* **1991**, *32*, 7059-7062.

Yoshikawa, K.; Shimono, N.; Arihara, S. Antisweet natural products. VI. Jujuba saponins IV,V, and VI from *Zizyphus jujuba* Mill. *Chem. Pharm. Bull.* **1992a**, *40*, 2275-2278.

Yoshikawa, K.; Tumura, S.; Yamada, K.; Arihara, S. Anti-sweet natural products. VII. Hodulosides I, II, II, IV and V from the leaves of *Hovenia dulcis* Thunb. *Chem. Pharm. Bull.* **1992b**, *40*, 2287-2291.

INTENSELY SWEET SAPONIN OSLADIN: SYNTHETIC AND STRUCTURAL STUDY

Mugio Nishizawa and Hidetoshi Yamada

Faculty of Pharmaceutical Sciences
Tokushima Bunri University
Tokushima 770, Japan

INTRODUCTION

Saccharin, dulcin, sodium cyclamate, acesulfame-K, and aspartame are well-known artificial sweeteners that have been discovered accidentally during some synthetic experiments. Intensely sweet tasting substances are also widely distributed in nature particularly in the plant kingdom. These materials have been an exciting research area for natural product chemists and a variety of sweet-tasting natural products have been characterized.[1, 2] Many of them are glycosides of terpenoids or steroids. A sweet principle of the Chinese drug *Glycyrrhiza glabra* (Fabaceae) is the well-known glycyrrhizin (**1**); it has been used as an auxiliary food additive.[3] This glycoside is 150 times sweeter than sucrose. A saponin, mogroside V (**2**), is 250 times sweeter than sucrose and found in the fruit of a Cucurbitaceaeous plant, *Momordica grosvenorii*.[4] The triterpene glycoside abrusoside B (**3**) was recently characterized as the sweet principle of *Abrus precatorius* (Leguminosae),[5] and is 100 times sweeter than sucrose. The sweet diterpene glycoside stevioside (**4**) has been isolated from Paraguayan Composite plant *Stevia rebaudiana*.[6] It is 300 times sweeter than sucrose. Baiyunoside (**5**) is the sweet principle of a Chinese drug *Phlomis betonicoides* (Labiatae),[7, 8] and the labdane-type diterpene glycoside **5** is 250 times sweeter than sucrose. A fern metabolite osladin (**6**) is the intensely sweet steroidal glycoside isolated from the fern *Polypodium vulgare*.[9] Since these sweet-tasting natural glycosides have not been the subject of organic synthesis, we have been interested in developing synthetic approaches to these natural products. We have developed an effective olefin cyclization agent mercury triflate amine complex, $Hg(OSO_2CF_3)_2 \cdot C_6H_5NMe_2$,[10-14] that is useful to construct labdane-type diterpenoids. Therefore, we have investigated the synthesis of baiyunoside (**5**) and a number of analogs. After completing the total synthesis of **5**, we have investigated the tastes of many synthetic analogs. A glycoside of (+)-baiyunol (aglycone of **5**) with α-glucoside-β-glucoside was sweeter than the natural product. Some analogs were bitter.[15-20]

The sweetest glycoside osladin **6** has been our second synthetic target among intensely sweet natural products. When we completed the total synthesis of **6**, we found that it is not sweet at all. Reextraction of the sweet principle of the fern and structural study by X-ray crystallography showed that the real structure of osladin is **7**, in which the stereochemistry at C-22, 25, and 26 is revised.[21] Total synthesis of **7** confirmed that osladin (**7**) is the real sweet principle of *P. vulgare*.[22-24]

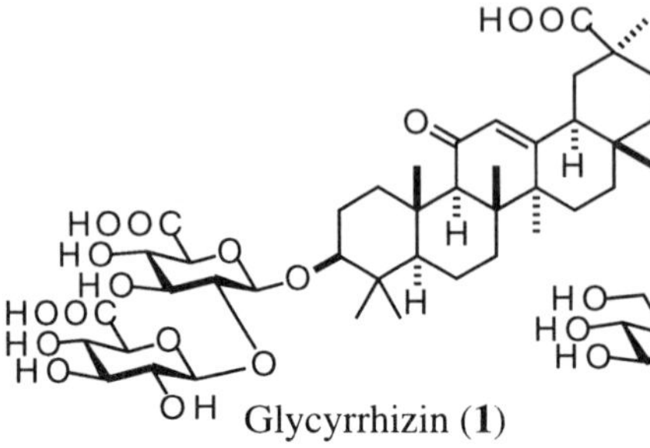

Glycyrrhizin (**1**)
(150 times as sweet as sucrose)

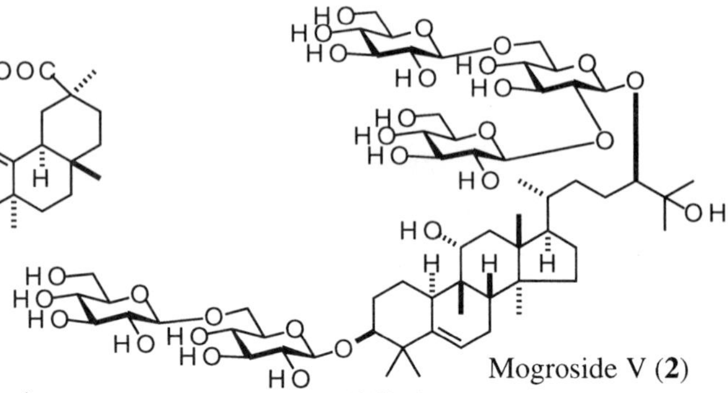

Mogroside V (**2**)
(250 times as sweet as sucrose)

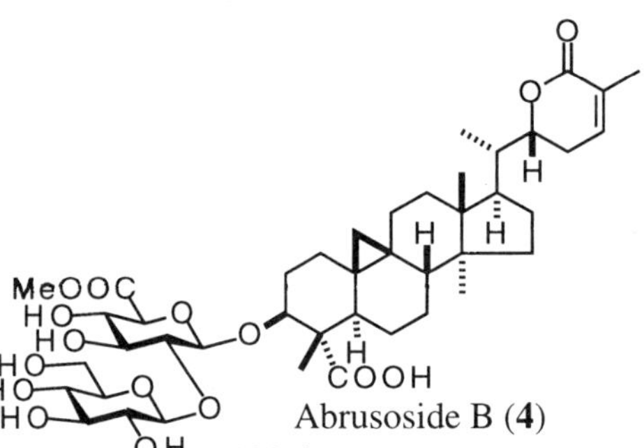

Abrusoside B (**4**)
(100 times as sweet as sucrose)

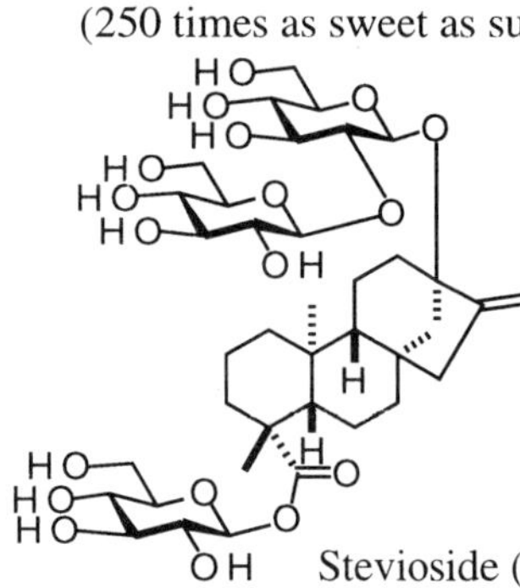

Stevioside (**3**)
(300 times as sweet as sucrose)

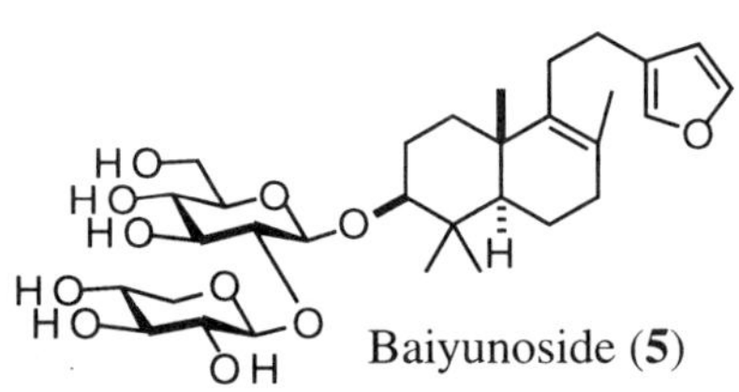

Baiyunoside (**5**)
(250 times as sweet as sucrose)

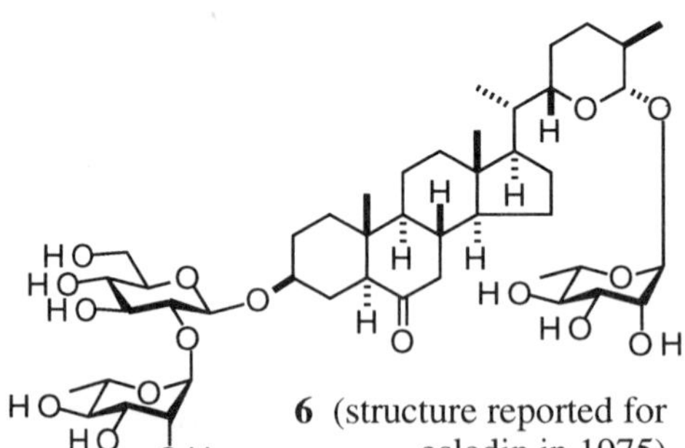

6 (structure reported for osladin in 1975)

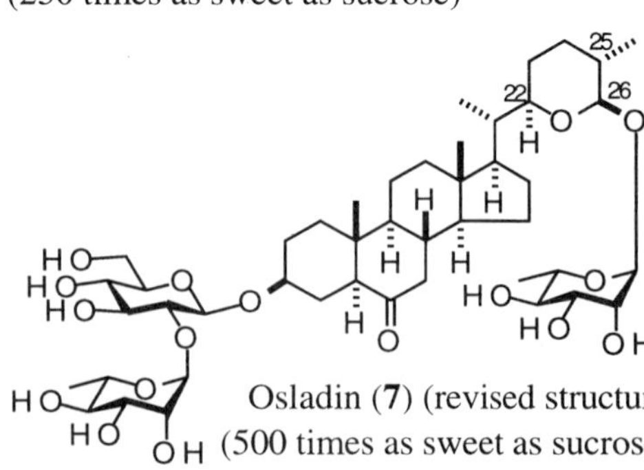

Osladin (**7**) (revised structure)
(500 times as sweet as sucrose)

OSLADIN, A HISTORICAL BACKGROUND

It is well known that the rhizomes of the European fern *Polypodium vulgare* (Polypodiaceae) are intensely sweet. In 1967, Jizba and Herout reported the isolation of a structurally new saponin as the sweet-tasting principle, and named it osladin based on the Czech name of this fern, "osladic".[9] In 1971, they reported its structure.[25] Shortly thereafter, Havel and Cerny achieved a partial synthesis of the aglycone from solasodine, and established the absolute stereochemistry.[26] While the stereochemistry of the glucosidic bond was determined to be β based on an enzymatic hydrolysis using β-glucosidase, the stereochemistry of two rhamnosidic bonds as well as the stereochemistry at C-26 were not determined. Although the structural study of **6** was not completely analized, osladin became quite well-known as the sweetest glycoside after Farnsworth described it as 3,000 times sweeter than sucrose in a review article.[27] We also turned our interest towards the synthesis of osladin since it was reported to be the sweetest glycoside known. If the synthesis were achieved, it would be the first total synthesis of a saponin.[28]

SYNTHESIS OF COMPOUND 6

A retrosynthetic analysis suggested **8** as an advanced intermediate toward the synthesis of **6**. The intermediate **8** will be accessible from disaccharide lactone **9**. Modification of steroidal aldehyde **11** into **10** and subsequent stereoselective glycosylation of **10** were anticipated to produce lactone **9**. The starting material of this synthesis **11** is easily accessible from commercially available (-)-stigmasterol. Sugar residues were used as protecting groups for the C-3 and 26 oxygen functionalities and were introduced in the early stages of the synthesis. Glycosylation of the hemiacetalic hydroxyl group at C-26 must be stereoselective.

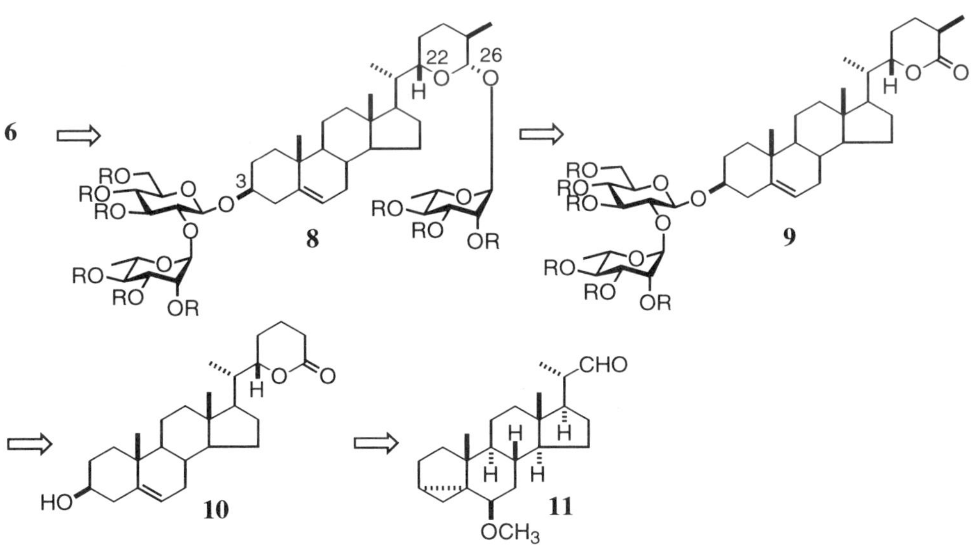

Grignard reaction of 4-pentenylmagnesium bromide with aldehyde **11** afforded alcohols **12** and **13** in 89% yield in a 97:3 ratio. At this point the stereochemistry of the C-22 center of **12** and **13** was not clear. Compound **12** was easily transformed into lactone **14** by ozonolysis followed by an oxidation of the resulting hemiacetal. Triflic acid-catalyzed solvolysis of **14** afforded lactone **10** in 97% yield. The stereochemistry at C-22 of **12** was determined to be *S* by an X-ray analysis of lactone **16** that was prepared by several steps from **12** through **14**. Methylation of **14** afforded two isomers **15a** and **15b**. Triflic acid catalyzed hydrolysis of **15a** afforded **16**. Thus the stereoselectivity of the Grignard reaction of **11** follows Cram's rule to provide C-22 *S* alcohol **12** as the major product.[29]

Since compound **6** has C-25 *R* configuration, we attempted to epimerize the C-25 methyl group of **15ab**. Though neither under kinetic conditions nor thermodynamic conditions the 1:1 ratio of **15ab** did not change, purified DIBALH reduction product **17b** was isomerized into a 10:1 mixture of **18a** and **18b** under thermodynamic conditions.

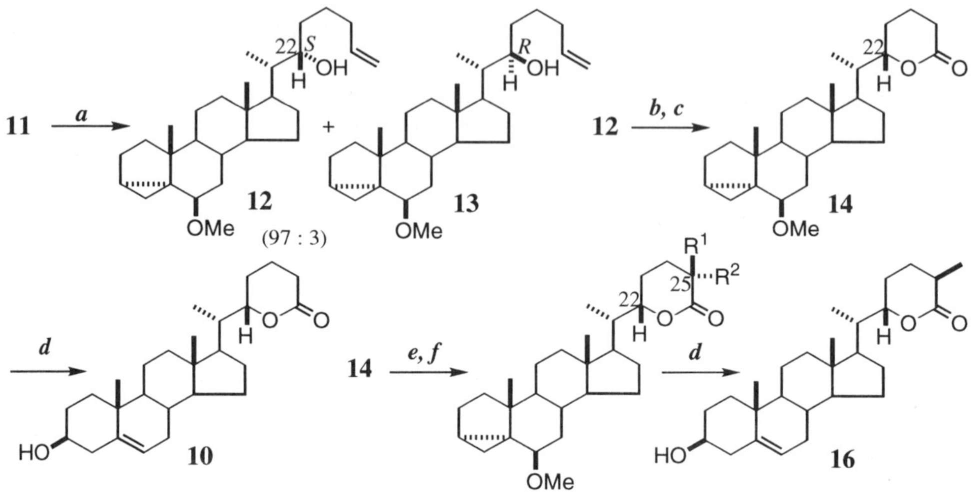

15a: $R^1 = CH_3$, $R^2 = H$ (C-25 *R*)

15b: $R^1 = H$, $R^2 = CH_3$ (C-25 *S*)

a $CH_2{=}CH(CH_2)_3MgBr$, ether *b* O_3, Sudan III, C_2H_5OH-H_2O, then $(CH_3)_2S$ *c* PDC, CH_2Cl_2 *d* TfOH (0.005 eq), dioxane-H_2O *e* LDA, HMPA, THF, then CH_3I *f* HPLC

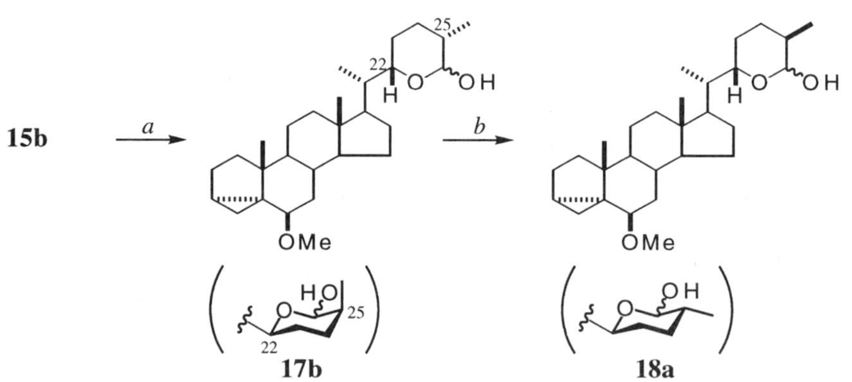

a DIBALH, PhMe *b* NaOMe, MeOH

In this synthetic study we intended to use glycosyl moieties as protecting groups for the C-3 and C-26 hydroxyl functionalities. Glucosyl chloride **19** and rhamnosyl chloride **20** were employed as glycosylation donors. A 2'-discriminating and β-selective glucosylation of **10** was achieved by using glucosyl chloride **19** a catalytic amount of triflic acid (0.05 equiv), and tetramethylurea as an acid scavenger to give β-glucoside **21** in 59% yield. The increased steric hindrance around C-2' of **21** limits the second glucosylation.

The α-rhamnoside linkage was introduced with the highly α-selective thermal glycosylation reaction that we recently developed.[30-34] A mixture of monoglycoside **21**, rhamnopyranosyl chloride **20** and α-methylstyrene (weak acid scavenger) was heated at 80 °C for 60 h without solvent to give disaccharide **22** in 59% yield. The newly formed rhamnoside bond was found to be completely α. The stereochemistry of the original β-glucosidic linkage of **21** was retained during the thermal glycosylation.

A methyl group was introduced at C-25 of **22** by sequential treatment with LDA and methyl iodide to give 1:1 mixture of stereoisomers. After DIBALH reduction of lactone, the hemiacetal was smoothly isomerized into the C-25 *R* compound **23** by treatment with sodium methoxide.

Glycosylation at the C-26 hemiacetal of **23** was achieved by using the classical Koenigs-Knorr glycosylation. Treatment of **23** with rhamnosyl chloride **20** and silver triflate in the presence of tetramethylurea (TMU) provided trisaccharide **24** in 55% yield. Hydroboration followed by PDC oxidation afforded the desired ketone **25**. A palladium(II) hydroxide-catalyzed debenzylation of **25** under hydrogen atmosphere gave **6**. Thus the first total synthesis of a saponin was achieved; however, aqueous solutions of **6** were not sweet at all.[21, 24]

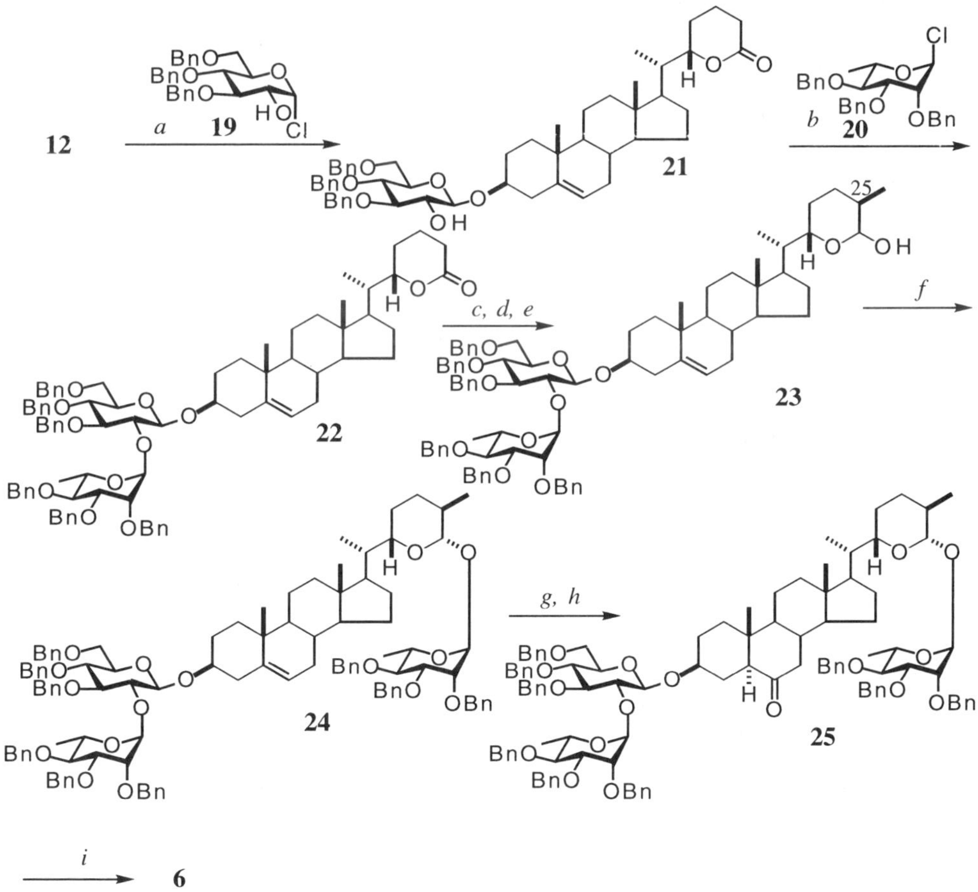

a **19**, TMU, TfOH (0.05 equiv), CH_2Cl_2 *b* **20**, α-methylstyrene, neat, 80°C, 60 h *c* LDA, HMPA, THF, then MeI *d* DIBALH, ether *e* MeONa, MeOH-THF *f* **20**, AgOTf, TMU *g* BH_3•THF,then H_2O_2, NaOH *h* PDC, CH_2Cl_2 *i* H_2, $Pd(OH)_2$

We had to worry about the following things at this stage. 1) We made some mistakes to determine the stereostructure of some intermediates. 2) The reported structure of osladin is not correct. 3) Osladin is not the sweet principle of the fern, but some minor constituent is instead.

ISOLATION AND STRUCTURE REVISION OF OSLADIN

Since the original sample of osladin and spectral data were not available from the Czech original workers, we isolated the rhizomal sweet principle of *P. vulgare* from plants collected in the southern part of Germany by Professors Y. Asakawa and H. Becker. Sweet components of the fern were extracted by ethanol. Successive chromatography and recrystallization afforded the pure sweet compound as colorless crystals (0.02% isolated yield). Since the sweet compound has a mp of 202-204 °C (lit. 201-203 °C)[25] and a molecular formula $C_{45}H_{74}O_{17}$ suggested from HRMS (FAB), this is the same compound, osladin, which was isolated by Jizba and Herout.[9] Both ^{1}H and ^{13}C NMR spectra of the natural osladin were not identical with those of synthetic **6**. Thus, the structure of real osladin differs from that of **6**.

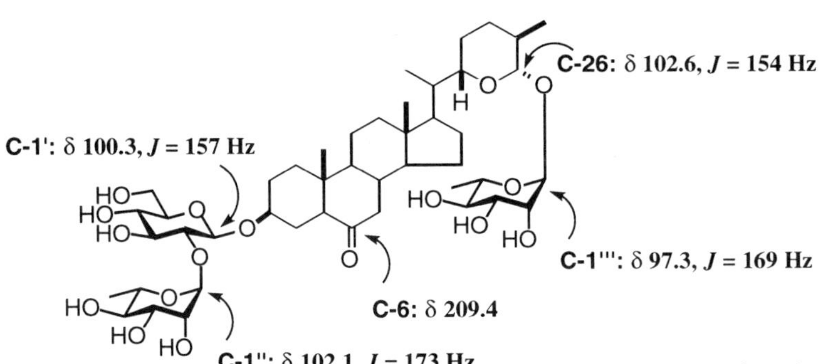

^{13}C NMR Data of Synthetic Glycoside **6** in Py-d_5

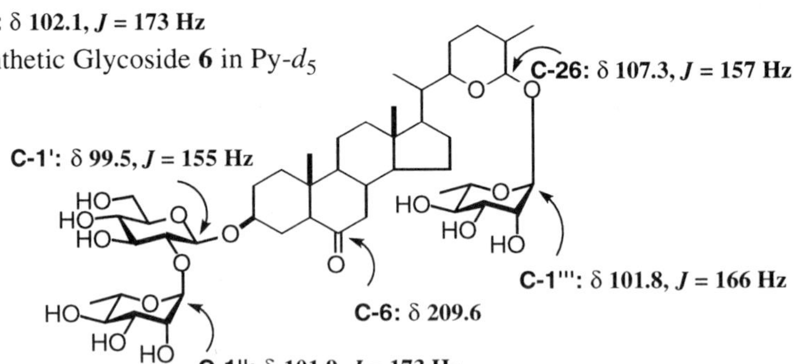

^{13}C NMR Data of Natural Sweet Osladin (**7**) in Py-d_5

It is noteworthy that the ^{13}C NMR chemical shift of C-26 of natural osladin (**7**) occurred at relatively low field (δ 107.3), while synthetic **6** showed δ 102.6. Glycosylation shift is a useful procedure to distinguish the absolute structure of chiral alcohol. Generally, a combination of an *R*-alcohol with an *R*-sugar (β-glucose, β-xylose, or α-rhamnose) or an *S*-alcohol with an *S*-sugar (α-glucose, α-xylose, or β-rhamnose) induces a larger glycosylation shift (Δδ).[36, 37] The *SR* combination results in a smaller Δδ. The C-26 rhamnosidic bond of osladin is a glycoside of a hemiacetalic alcohol. Thus it is interesting to investigate the glycosylation shift for the case of a hemiacetal, namely a 1,1'-disaccharide. Thus we have prepared a variety of (*R,R*)-1,1'-disaccharides such as β-glc-β-glc, β-xyl-β-xyl, α-rha-α-rha, β-glc-β-xyl, β-glc-α-rha and β-xyl-α-rha as well as (*S,S*)-1,1'-disaccharides such as α-glc-α–glc, α-xyl-α-xyl, α-glc-α-xyl, α-glc-β-rha, and α-xyl-β-rha by using AgOTf-promoted glycosylation of glycosyl chloride with a 1-hydroxy sugar. The Δδ values for anomeric carbons (δ value of disaccharide - δ value of original hemiacetal) of all of these disaccharides were less than 3 ppm. The *SR* combinations such as α-glc-β-glc, α-xyl-β-xyl, β-rha-α-rha, α-glc-β-xyl, α-glc-α-rha, β-rha-β-glc, α-xyl-α-rha, and β-rha-β-xyl gave Δδ values larger than 5 ppm. Thus the glycosylation shift may be extended to the glycosides of hemiacetals.[38]

The chemical shift at the anomeric carbon of α-rhamnose itself is δ 95.5. Thus in **6** the glycosylation shift at C-1''' is 1.8 ppm indicating a glycoside originating from *RR* or *SS* combination. Since α-rhamnose itself is *R*, the stereochemistry at C-26 of the aglycone **6** is *R*; C-26 of glycoside **6** itself is *S* owing to change of the priority orders. On the other hand, the glycosylation shift at C-1''' of the sweet osladin (**7**) is 6.3 ppm. The value suggests *R* configuration at C-26 of glycoside **7**.

Further to reveal the structure of osladin (**7**), we have prepared monoclinic single crystals by repeated recrystallization from a mixture of acetone and water. According to the ORTEP drawing as well as the stereostructure, the sweet osladin should be represented by structure **7**. Thus, the C 22*S*, 25*R*, 26*S* stereochemistry assigned by Havel and Cerny[26] needs to be revised to C-22*R*, 25*S*, 26*R*, respectively. In addition, it is important to note that **7** is intensively sweet while **6** is totally free from any taste even though the configurational change is minor.

In the original paper on the structure determination of osladin, Jizba, Dolejs, Herout, and Sorm did not mention the intensity of the sweetness.[9, 25] Although osladin has been claimed to be 3,000 times as sweet as sucrose by Farnsworth,[27] reexamination of sweetness by Ajinomoto Co. suggests it is only 500 times as sweet.[35]

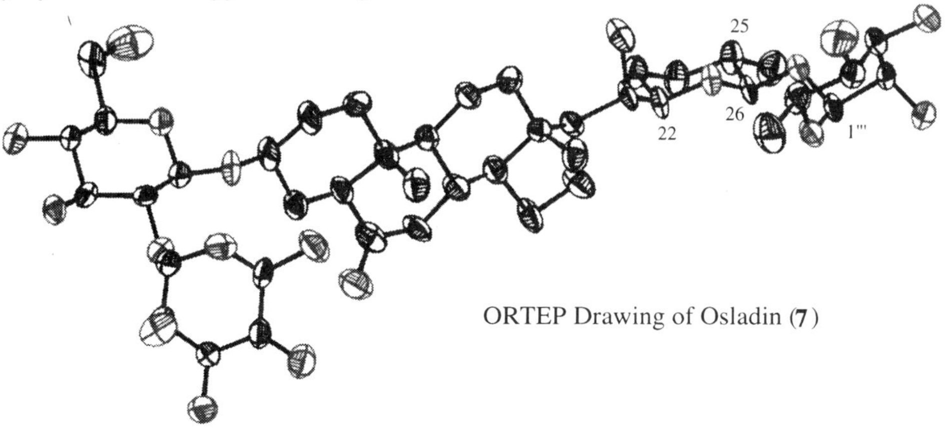

ORTEP Drawing of Osladin (**7**)

TOTAL SYNTHESIS OF OSLADIN

To complete the synthesis of sweet osladin (**7**), we needed to modify the synthesis of **6**. As already discussed, Grignard reaction of the aldehyde **11** generates the C-22 *S* product **12** predominantly. Since the C-22 stereochemistry of real osladin (**7**) is *R*, we needed to invert the stereochemistry at C-22 of **12**. Although a variety of Mitsunobu reaction conditions failed to give any inversion product, Corey's S_N2 reaction of mesylate **26** with KO_2 achieved clean inversion at C-22 to give C-22 *R* carbinol **13** in 88% yield.[39] The alcohol **13** was transformed into lactone **27** by ozonolysis and subsequent PDC oxidation. Methylation at the α-position of the lactone afforded the monomethylated product **28** as a 1:1 mixture of stereoisomers. Since it is possible to isomerize the configuration of the C-25 methyl group at a later stage, this mixture was employed for the solvolysis and glycosylation reactions. Triflic acid-catalyzed solvolysis of **28** afforded homoallylic alcohol **29**. Condensation of **29** and glucosyl chloride **19** catalyzed by triflic acid in the presence of TMU took place in a β-selective manner to give 2' hydroxyl group discriminated glucoside **30** in 57% yield. By means of an α-selective thermal rhamnosylation reaction, the L-rhamnosyl residue was coupled to **30** to give disaccharide **31** in 81% yield.[32] The hemiacetals **32** obtained by DIBALH reduction of **31** were treated with base to give the more stable equatorial methyl product **33**. Glycosylation of the hemiacetal hydroxyl group at C-26 was achieved by using rhamnosyl chloride **20**, AgOTf, and TMU to yield trisaccharide **34** stereoselectively in 61% yield. The characteristic low field shift (δ 106.7) of the C-26 anomeric carbon was observed in the ^{13}C NMR of **34**.[38] The trisaccharide **34** was subjected to hydroboration, and subsequent PDC oxidation to give ketone **35**. The benzyl groups of **35** were cleaved off by $Pd(OH)_2$-catalyzed hydrogenolysis to give osladin **7**. This product was very sweet and showed indistinguishable spectral properties with those of natural osladin. Thus we have shown that the structure **7** represents the real sweet principle of *P. vulgare*.[35]

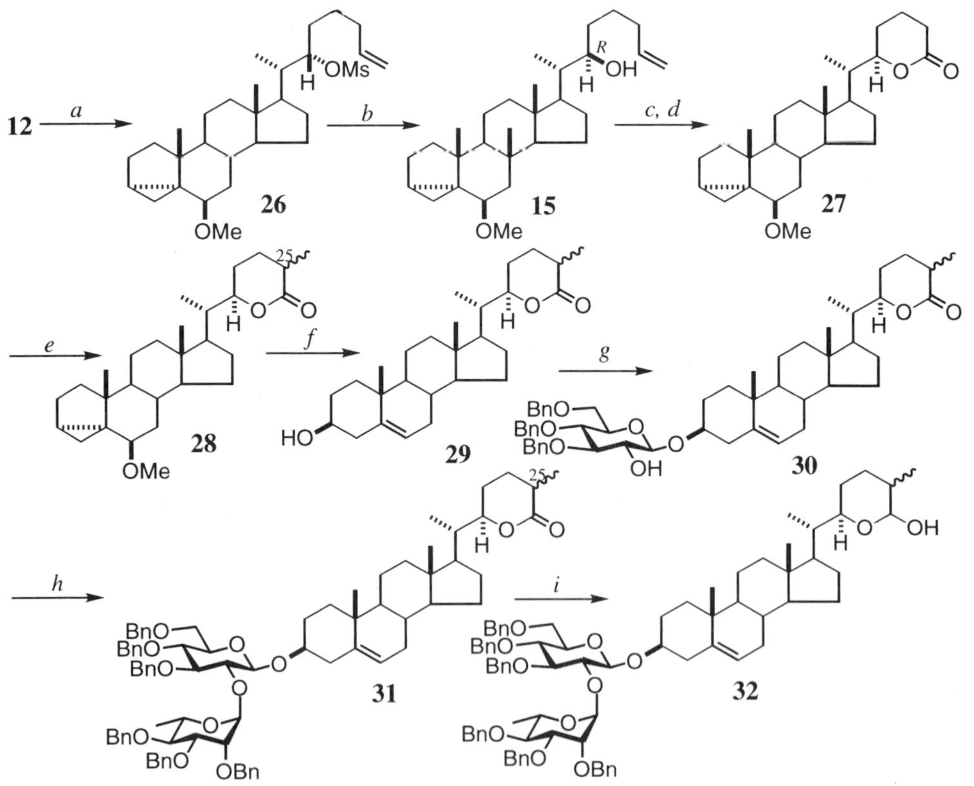

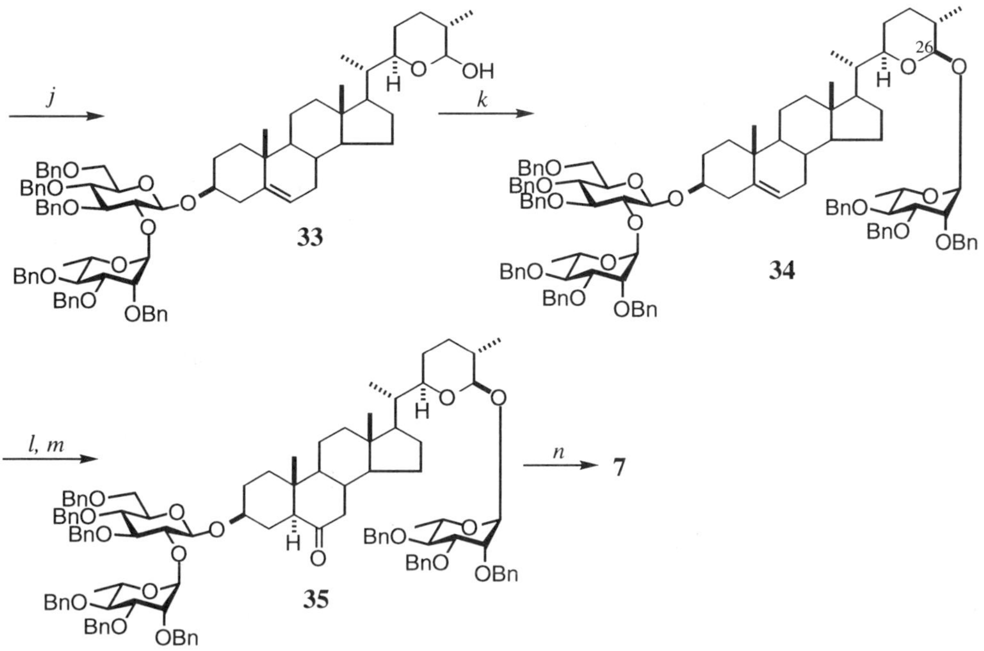

a MsCl/Py *b* KO_2/18-Crown-6/DMSO-DME (1:1) *c* O_3/Sudan III/EtOH-H_2O, then Me_2S
d PDC/CH_2Cl_2 *e* LDA/HMPA/THF then MeI *f* TfOH (0.005 equiv)/dioxane-H_2O (9:1)
g **19**/TfOH/TMU/CH_2Cl_2 *h* **20**/TMU, neat, 80 °C, 56 h *i* DIBALH/ether *j* MeONa/MeOH
k **20**/AgOTf/TMU/CH_2Cl_2 *l* BH_3•THF then H_2O_2/NaOH *m* PDC/CH_2Cl_2
n H_2/Pd$(OH)_2$/MeOH-EtOAc-H_2O

The total synthesis of osladin was therefore completed, and it was finally proved that osladin is the real sweet principle of *Polypodium vulgare*. However it is still mysterious why a mistake happened when Havel and Cerny carried out the chemical derivation of osladin aglycone from solasodine (**36**).[26] They prepared **38a** and **38b** from solasodine through lithium tri-*tert*-butoxyaluminium hydride reduction of ketone **37** followed by nitration to retain the C-25 *R* stereochemistry. They assigned C-22*S* for the minor isomer **38a**, and C-22*R* for the major isomer **38b**. Their criterion for the assignment of stereochemistry at C-22 was optical rotation. According to Burrows' empirical rule,[40] they assigned the C-22*R* configuration to the more dextrorotatory isomer **38b**. Today, it is clear that the reduction of C-22 steroidal ketone like **37** provides the C-22 *S* alcohol predominantly.[41] Thus the structures **38a** and **38b** should be revised to **38a'** and **38b'**, respectively. From these nitrates, they derived **39a** and **39b**. Thus, **39a** and **39b** must be revised into **39a'** and **39b'**. Upon finding that the natural product derivative was identical with **39a**, Havel and Cerny gave the C-22 *S*, C-25 *R* stereostructure for the osladin aglycone. During our synthetic study we found that the C-25 axial methyl group easily epimerizes when C-26 is an hemiacetal. Thus, we suppose that Havel and Cerny isolated **40** (C-22 *R*, C-25 *S*) after isomerization of **39a'**.

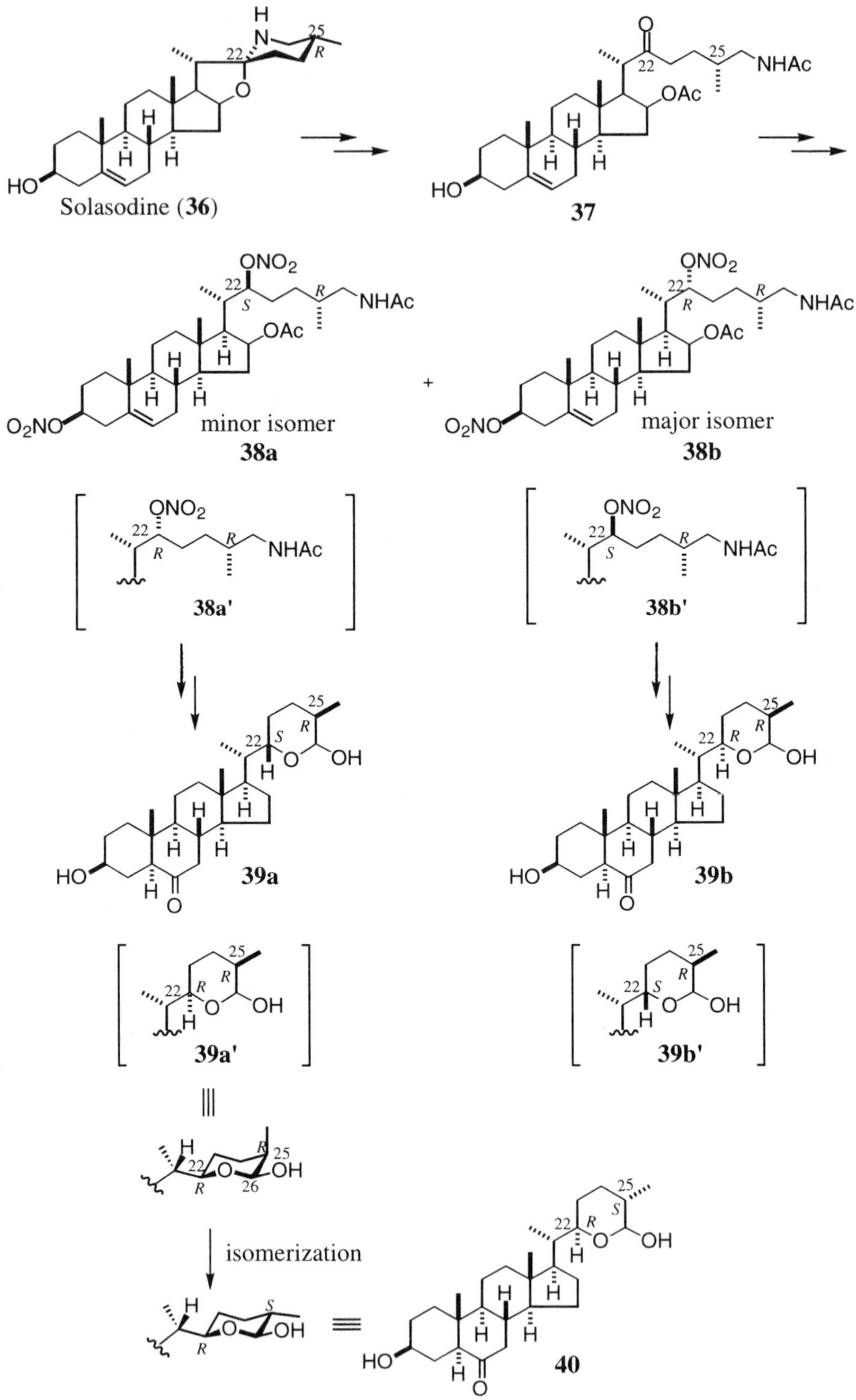
Solasodine (36)
37
minor isomer
38a
major isomer
38b
38a'
38b'
39a
39b
39a'
39b'
isomerization
40

ACKNOWLEDGMENTS

We are deeply indebted to Professor Y. Asakawa of Tokushima Bunri University and Professor H. Becker of Universität Des Saarland for their kind cooperation in the collection of the fern *P. vulgare*. We also thank Dr. C. Katayama for the X-ray diffraction study of osladin. Financial support from Ono Pharmaceutical Co. Ltd, the Ministry of Education Science and Culture, Japanese Government, and Japan Private School Promotion Foundation are also acknowledged.

REFERENCES

1. A. D.Kinghorn and D. D. Soejarto, Intensely sweet compounds of natural origin, *Med. Res. Rev. 9*: 91 (1989).
2. A. D.Kinghorn and J. Kim, Potently sweet compounds from plants: techniques of isolation and identification, *Bioactive Natural Products*, Eds. S. M. Colegate and R. J. Molyneux, CRC Press Inc., Boca Raton, FL. pp 173 (1993).
3. B. Lythgoe and S. Trippett, The constitution of the disaccharide of glycyrrhinic acid, *J. Chem. Soc.* 1983 (1950).
4. T. Takemoto, S. Arihara, T. Nakajima, and M. Okuhira, Studies on the constituents of *Fructus Momordicae*. I. On the sweet principle, *Yakugaku Zasshi* 103:1151 (1983).
5. Y. H. Choi, R. A. Hussain, J. M. Pezzuto, D. D. Soejarto, A. D. Kinghorn, and J. F. Morton, Abrusoside A-D, four novel sweet-tasting triterpene glycoside constituents from the leaves of*Abrus precatorius*, *J. Nat. Prod.* 52:1118 (1989).
6. E. Vis and H. G. Fletcher, Jr., Stevioside IV. Evidence that stevioside is a sophoroside, *J. Am. Chem. Soc.* 78:4709 (1956).
7. T. Tanaka, O. Tanaka, Z. W. Lin, J. Zhou, and H. Ageta, Sweet and bitter glycosides of the Chinese plant drug, Bai-Yun-Shen, *Chem. Pharm. Bull.* 31:780 (1983).
8. T. Tanaka, O. Tanaka, Z. W. Lin, and J. Zhou, Sweet and bitter principles of the Chinese plant drug, Bai-Yun-Shen: Revision of the assignment of the source plant and isolation of two new diterpene glycosides, *Chem. Pharm. Bull.* 33:4275 (1985).
9. J. Jizba and V. Herout, Isolation of constituents of common polypody rhizomes (*Polypodium vulgare* L.), *Coll. Czech. Chem. Commun.* 32:2867 (1967).
10. M. Nishizawa, H. Takenaka, H. Nishide, and Y. Hayashi, A new olefin cyclization agent, mercury(II) triflate/*N,N*-dimethylaniline complex, *Tetrahedron Lett.* 24:2581 (1983).
11. M. Nishizawa, H. Takenaka, and Y. Hayashi, Synthesis and structure determination of isoaplysis-20, *J. Am. Chem. Soc.* 106:4290 (1984).
12. M. Nishizawa, H. Takenaka, and Y. Hayashi, An experimental evidence of the stepwise mechanism in biomimetic olefin cyclization: trapping of cationic intermmediates, *J. Am. Chem. Soc.* 107:522 (1985).
13. M. Nishizawa, H. Takenaka, and Y. Hayashi, Chemical simulation of polycyclic diterpenoid biosynthesis using mercury(II) triflate/*N,N*-dimethylaniline complex: mechanistic aspect of a biomimetic olefin cyclization, *J. Org. Chem.* 51:806 (1986).
14. M. Nishizawa, E. Morikuni, K. Asoh, Y. Kan, K.Uenoyama, and H. Imagawa, Preparetion of mercury (II) triflate in acetonitrile and a biomimetic cyclization of farnesyl sulfone, *Synlett* 169 (1995).
15. M. Nishizawa, H. Yamada, and Y. Hayashi, Biomimetic cyclization of ambliofuran and its analog by using mercury(II) triflate/*N,N*-dimethylaniline complex: synthesis of ambliol-A, *Tetrahedron Lett.* 27:187 (1986).
16. M. Nishizawa, H. Yamada, and Y. Hayashi, Total synthsis of (±)-baiyunol. *Tetrahedron Lett.* 27:3255 (1986).
17. M. Nishizawa, H. Yamada, and Y. Hayashi, Cyclization control of ambliofuran analog: effective total synthesis of baiyunol. *J. Org. Chem.* 52:4878 (1987).

18. H. Yamada and M. Nishizawa, Total synthesis of baiyunoside by a novel 2' discriminated glycosidation, *Tetrahedron Lett.* 28:4315 (1987).
19. H. Yamada and M. Nishizawa, Synthesis of sweet tasting diterpene glycosides, baiyunoside and analogs, *Tetrahedron* 48:3021 (1992).
20. M. Nishizawa and H. Yamada, Chemical synthesis of sweet diterpene glycosides, *Pharmacia* 27:790 (1991).
21. H. Yamada, M. Nishizawa, and C. Katayama, Osladin, a sweet principle of *Polypodium vulgare.* structure revision, *Tetrahedron Lett.* 33:4009 (1992).
22. H. Yamada and M. Nishizawa, Total synthesis of intensely sweet saponin, osladin, *Synlett* 54 (1993).
23. M. Nishizawa and H. Yamada, Synthesis of sweet tasting glycosides, story of osladin, *Chemistry Today* 60 (1993).
24. M. Nishizawa and H. Yamada, Novel synthetic approaches into intensely sweet glycosides: baiyunoside and osladin, *Synlett* 785 (1995).
25. J. Jizba, L. Dolejs, V. Herout, and F. Sorm, The structure of osladin, the sweet principle of the rhizomes of *Polypodium vulgare* L., *Tetrahedron Lett.* 1329 (1971).
26. M. Havel and V. Cerny, Partial synthesis of osladin aglycone from solasodine, *Coll. Czech. Chem. Commun.* 40:1579 (1975).
27. N. R. Farnsworth, Current status of sugar substitutes, *Cosmetics and Perfumery* 88:27 (1973).
28. J. T. Randolph and S. J. Danishefsky, First synthesis of a digitalis saponin. Demonstration of the scope and limitations of a convergent scheme for branched oligosaccharide synthesis by the logic of glycal assembly, *J. Am. Chem. Soc.* 115:8473 (1993).
29. D. J. Cram and D. R. Wilson, Studies in stereochemistry. XXXII. Models for 1,2-asymmetric induction, *J. Am. Chem. Soc.* 85:1245 (1963).
30. M. Nishizawa, Y. Kan, and H. Yamada, A simple metal free 2'-discriminated glucosidation procedure, *Tetrahedron Lett.* 29:4597 (1988).
31. M. Nishizawa, Y. Kan, and H. Yamada, Thermal glycosidation with benzylated glycosyl chlorides: a very simple procedure of glycosidation, *Chem. Pharm. Bull.* 37:565 (1989).
32. M. Nishizawa, Y. Kan, W. Shimomoto, and H. Yamada, α-Selective thermal glycosidation of rhamnosyl and mannosyl chlorides, *Tetrahedron Lett.* 31:2431 (1990).
33. M. Nishizawa, H. Imagawa, Y. Kan, and H. Yamada, Total synthesis of cyclo-L-rhamnohexaose by a stereoselective thermal glycosylation, *Tetrahedron Lett.* 32:5551 (1991).
34. M. Nishizawa, H. Imagawa, K. Kubo, and H. Yamada, Improved synthesis of α-cycloawaodorin, *Synlett* 447 (1992).
35. H. Yamada and M. Nishizawa, Total synthesis and structure revision of intensely sweet saponin osladin, *J. Org. Chem.* 60:386 (1995).
36. S. Seo, Y. Tomita, K. Tori, and Y. Yoshimura, Determination of the absolute configuration of a secondary hydroxy group in a chiral secondary alcohol using glycosidation shifts in carbon-13 nuclear magnetic resonance spectroscopy, *J. Am. Chem. Soc.* 100:3331 (1978).
37. R. Kasai, M. Okihara, J. Asakawa, K. Mizutani, and O. Tanaka, ^{13}C NMR study of α- and β-anomeric pairs of D-mannopyranosides and L-rhamnopyranosides, *Tetrahedron* 35:1427 (1979).
38. M. Nishizawa, S. Kodama, Y. Yamane, K. Kayano, S. Hatakeyama, and H. Yamada, Synthesis and glycosylation shift of 1,1'-disaccharides, *Chem. Pharm. Bull.* 42:982 (1994).
39. E. J. Corey, K. C. Nicolaou, M. Shibasaki, Y. Machida, and C. S. Shiner, Superoxide ion as a synthetically useful oxygen nucleophile, *Tetrahedron Lett.* 3183 (1975).
40. E. P. Burrows, G. M. Horneys, and E. Caspi, Reassignment of configuration to the 22-hydroxycholesterols. Synthesis of (22*S*)- and (22*R*)-^{3}H-cholesterols, *J. Org. Chem.* 34:103 (1969).
41. D. M. Piatak and J. Wicha, Various approaches to the construction of aliphatic side chains of steroids and related compounds, *Chem. Rev.* 78:199 (1978).

STEROID SAPONINS FROM FENUGREEK AND SOME OF THEIR BIOLOGICAL PROPERTIES

Yves Sauvaire[1], Y. Baissac[1], O. Leconte[1], P. Petit[2] and G. Ribes[2]

[1]Laboratoire de Recherche sur les Substances Naturelles Végétales
Université Montpellier II, 34095 MONTPELLIER, FRANCE
[2]Laboratoire de Pharmacologie, Faculté de Médecine et UMR 9921
CNRS, Université Montpellier I, 34095 MONTPELLIER, FRANCE

INTRODUCTION

Fenugreek (*Trigonella foenum-graecum* L., Leguminosae, Trifoliae, Trigonellinae) is an annual plant found primarily in Mediterranean countries, the Middle East, and India. The seeds are most often used as a food spice (curry) and in traditional medicine. Fenugreek seeds are assumed to have restorative and nutritive properties (appetite stimulant) and hypocholesterolemic and antidiabetic effects.
In the last several years, our research has focused on identifying the main elements responsible for these latter properties which were attributed to the high fiber content[1,2] (reducing intestinal absorption of glucose) and the presence of 4-hydroxyisoleucine (increasing insulin secretion)[3].

Moreover, the hypocholesterolemic activity was found to be related to the defatted part of the seed[4] involving saponin-rich subfractions[5,6]. However, the substances responsible for appetite-stimulating activity were unknown before our research[7].

The aim of the present study was to evaluate and describe the role of purified steroid saponins with respect to the following biological properties : antifungal, hypocholesterolemic and feeding behavior.

STEROID SAPONINS FROM FENUGREEK

A review of the literature revealed that the presence of special steroid substances in fenugreek seeds was first reported in 1919 by Wunschendorff[8], a French research scientist working in Algeria. This discovery was confirmed in various follow-up studies notably by Marker et al[9], who detected the presence of diosgenin, gitogenin and traces of tigogenin after hydrolysis of the plant material. Several further studies supported these findings and moreover, using more sophisticated analytical techniques including coupled GC-MS, we detected and identified ten different sapogenins[10] (Table 1).

Sapogenins are found in fenugreek seed in the form of glycosides ; the first research study on these substances was published in the *Comptes rendus à l'Academie des Sciences de Paris* by Heintz[11]. Moreover, fenugreek seed was found to contain at

Saponins Used in Food and Agriculture
Edited by Waller and Yamasaki, Plenum Press, New York, 1996

Table 1. Composition of sapogenins in seeds and in total furostanol saponins of fenugreek[1]

Peak n°	TRR[2]	Sapogenins	Seeds	TFS[3]
1	0.79	Spirosta-3,5-dienes	0.7	2.5
2	0.91	(25*R*)-5ß-spirostan-3ß-ol (smilagenin)	0.6	2.2
3	0.94	(25*S*)-5ß-spirostan-3ß-ol (sarsasapogenin)	1.2	3.3
4	1.00	(25*R*)-spirost-5-en-3ß-ol (diosgenin)	39.0	34.0
5	1.03	(25*R*)-5α-spirostan-3ß-ol (tigogenin) + (25*S*)-spirost-5-en-3ß-ol (yamogenin)	30.3	21.9
6	1.06	(25*S*)-5α-spirostan-3ß-ol (neotigogenin)	7.0	4.9
7	1.50	(25*R*)-spirost-5-en-2α,3ß-diol (yuccagenin) + (25*S*)-spirost-5-en-2α, 3ß-diol (lilagenin)	5.2	4.2
8	1.56	(25*R*)-5α-spirostan-2α,3ß-diol (gitogenin)	10.7	18.8
9	1.62	(25*S*)-5α-spirostan-2α,3ß-diol (neogitogenin)	5.3	8.2

[1] Results expressed in % represent the mean of three determinations.
[2] Retention time of sapogenyl acetate relative to diosgenyl acetate (gas chromatography on DB1 capillary column)
[3] TFS : total furostanol saponins.

Table 2. Structures of steroid saponins of fenugreek seed.

Steroid Saponins	R_F[1]	Structure	References
Trigofoenoside A	0.77	Rha-Glc- [(25*S*)-furostan-3β,22,26-triol]-Glc	Hardman et al.[14] Gupta et al.[17]
Trigofoenoside B	0.69	Rha-Glc-[(25*S*)-furostan-2α,3β,22,26-tetraol]-Glc	Gupta et al.[18]
Trigofoenoside C	0.63	Rha-Glc-[(25*R*)-furostan-2α,3β,22,26-tetraol]-Glc Rha	Gupta et al.[18]
Trigofoenoside D	0.55	Glc-Glc- [(25*S*)-furost-5-en-3β,22,26-triol]-Glc Rha	Gupta et al.[17]
Trigofoenoside E	0.47	Xyl-Glc- [(25*R*)-furostan-3β,22,26-triol]-Glc Rha	Gupta et al.[16]
Trigofoenoside F	0.23	Glc-Glc- [(25*R*)-furost-5-en-3β,22,26-triol]-Glc Rha	Gupta et al.[15]
Trigofoenoside G	0.18	Xyl-Glc-Glc-[(25*R*)-furost-5-en-3β,22,26-triol]-Glc Rha	Gupta et al.[15]
Trigonelloside B (C)		Rha-Glc-[(25*S*)-furost-5-en-3β,22,26-triol]-Glc Rha	Bogacheva et al.[12] Bogacheva et al.[13]

[1] in : $CHCl_3$-MeOH-H_2O : 65, 40, 12 v/v/v

least a dozen different saponins and they were structurally analysed by three main teams : the Russian team of Bogacheva et al.[12,13], the English team of Hardman et al.[14], and the Indian team of Gupta et al.[15,16,17,18].

Eight saponins have been isolated using column chromatography or droplet countercurrent chromatography and structures elucidated by ^{1}H NMR and FAB-MS. These bidesmosidic saponins have two sugar chains, with one bonded at C-3 and one attached through an ether linkage at C-26 with a D-glucose (furostanol saponins, Figure 1). Structures of these are shown in Table 2.

STEROID SAPONIN ISOLATION AND PURIFICATION

Isolation of steroid saponins from fenugreek seed was achieved by extraction and several stages of purification, including a dialysis step (Figure 2).

The product obtained, named total furostanol saponin (TFS) contained only steroid saponins without the other constituents of the whole extract. Saponins occur in furostanic form and fenugreek seeds contain 5-6 % saponins.

ANTIFUNGAL ACTIVITY

Furostanol saponin (FSB) and the corresponding aglycone (SgB) did not show any antifungal activity but its transformation by β-glucosidase into spirostanol-type saponins (SSB) revealed a strong fungicidal activity against fungi and yeast such as *Candida* (Table 3).

The percentage of growth inhibition of several fungi depended on the SSB concentration. For example, 200 $\mu g.ml^{-1}$ of SSB produced strong growth inhibition of *Rosellinia necatrix*, *Trichoderma viride* and *Trichoderma harzianum* , with an MIC_{50} of 25 $\mu g.ml^{-1}$. *Sclerotinia rolfsii* showed lower sensitivity to spirostanol saponins (Figure 3). This activity may have resulted from degradation of the bidesmoside to a fungicidal monodesmoside by fungal glycosidases and acted as a plant protective agent.

The effects of three different SSB concentrations were tested on the viability of *Candida albicans* . With 20 $\mu g.ml^{-1}$, the number of viable cells remained relatively constant during the test (fungistatic activity). At higher concentrations (50 and 100 $\mu g.ml^{-1}$), there was a marked decrease in viability (Figure 4), until around 10 h (fungicidal activity).

The toxic mechanisms of these saponins against fungi appeared to involve the formation of complexes with sterols of plasma membranes and transient disruption of the bilayer shown by measurement of Pi released from *Candida albicans* cells[19].

PHARMACOLOGICAL ACTIVITIES

Hypocholesterolemic effects

Subchronic administration of steroid saponins mixed with food led to a decrease in total plasma cholesterol in normal and streptozotocin diabetic rats (Table 4). As for plasma lipoproteins, the decrease in VLDL-LDL total cholesterol without any change in plasma-free cholesterol, indicated a reduction in cholesterol esters. The absence of any significant modification in triglycerides suggested a decrease in the LDL concentration by saponin and/or sapogenin treatments. Moreover, our previous results showed that approximately 60% of saponins are hydrolyzed into sapogenins in the digestive tract. These aglycones may be implicated together with the parent saponins in the inhibition of cholesterol absorption, the decrease of liver cholesterol concentration and the increased conversion of cholesterol into bile acids by the liver[6].

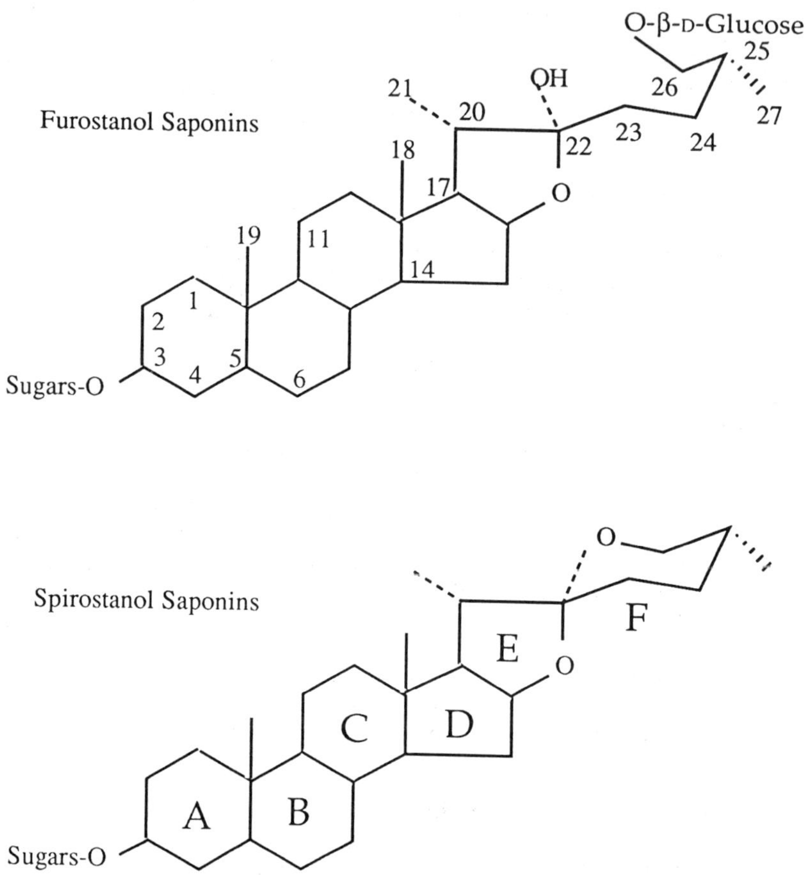

Figure 1. Structure of furostanol and spirostanol saponins.

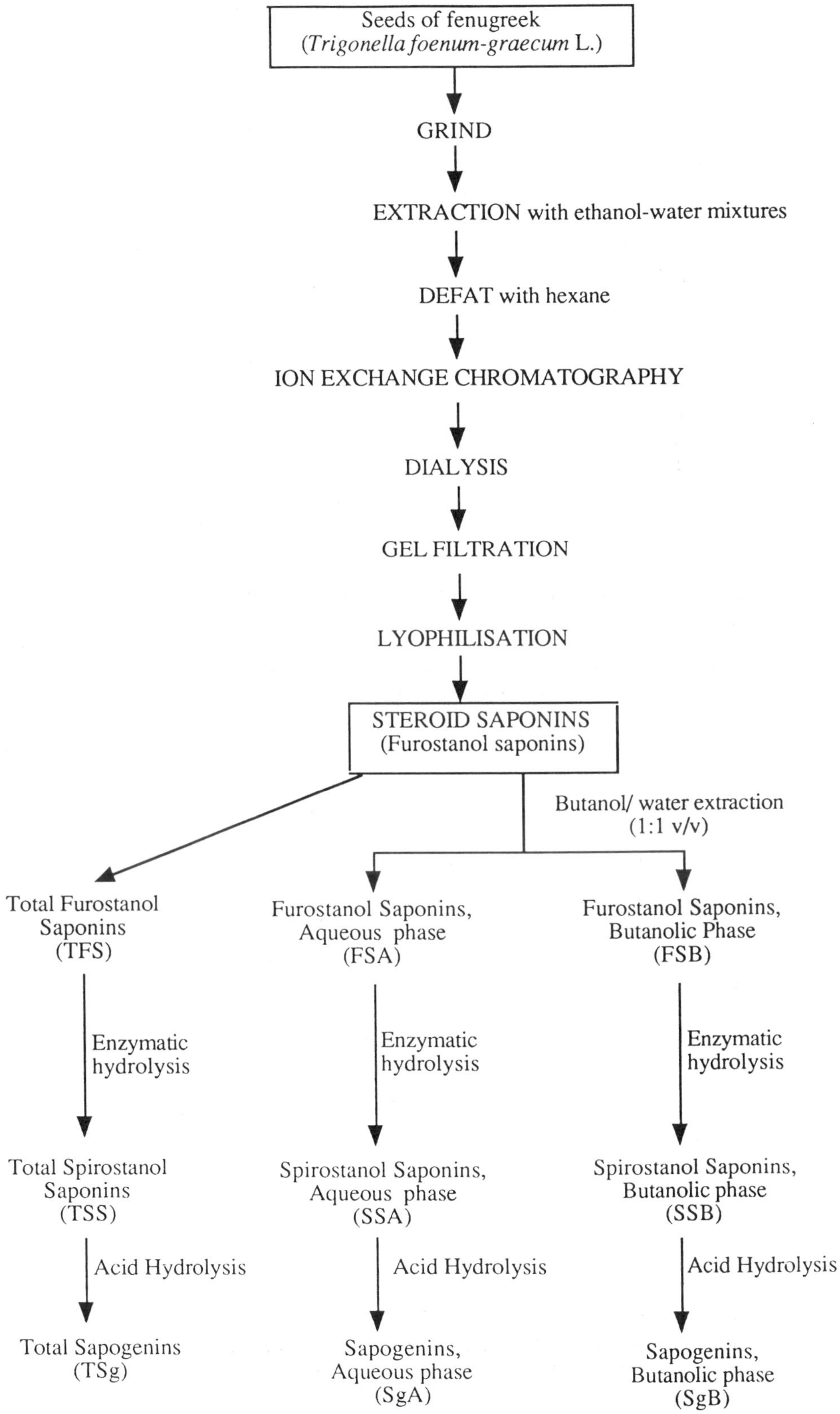

Figure 2. Steroid saponins isolation and purification.

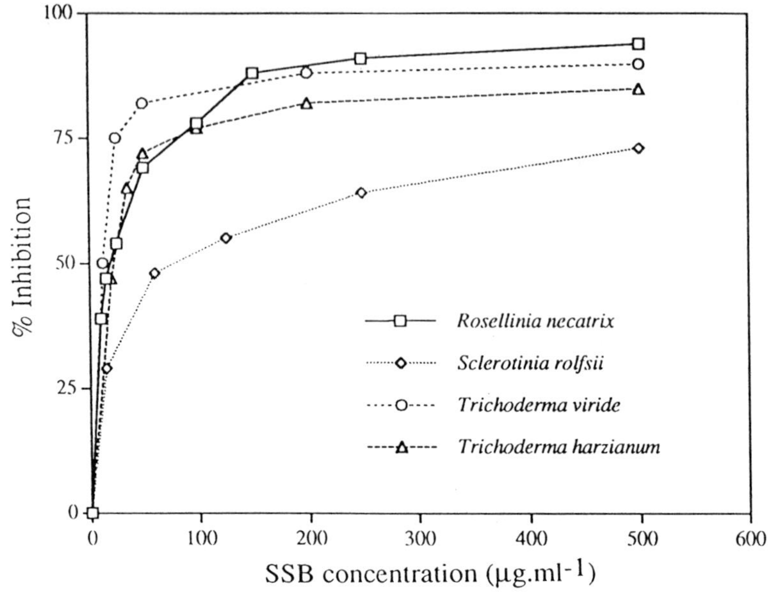

Figure 3. Effects of steroid saponins (SSB) on the growth of *Rosellinia necatrix* (7 days of test), *Sclerotinia rolfsii*, *Trichoderma viride* and *Trichoderma harzianum* (3 days of test).
Growth was assessed by measuring colony diameter. Growth inhibition was expressed as a percentage of the controls. Each point is the mean of five replicates.

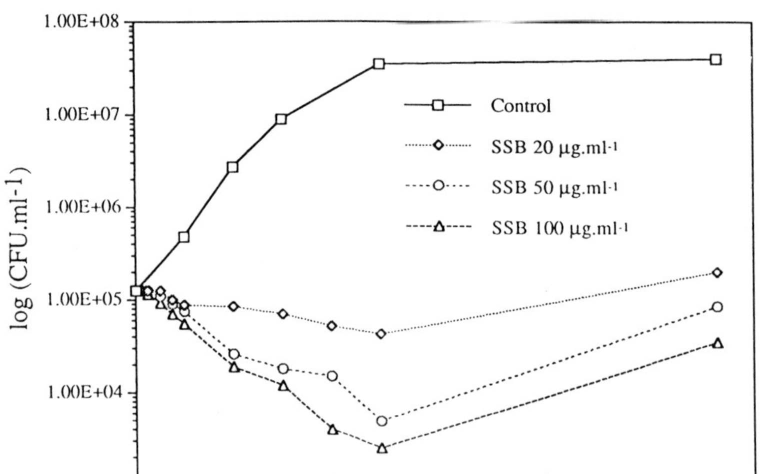

Figure 4. Effects of steroid saponins (SSB) on the growth of *Candida albicans*. Yeast cells of *C. albicans* VW32 were grown in Shadomy medium. CFU.ml^{-1} values were determined by standard dilution and platecount techniques.

Table 3. Antifungal activities of the steroid saponins and sapogenins by agar diffusion test with paper disks.

Organism	FSB[1]	SSB[2]	SgB[3]
Trichoderma viride	-	+	-
Trichoderma harzianum	-	+	-
Sclerotinia rolfsii	-	+	-
Ascochyta imperfecta	-	+	-
Rosellinia necatrix	-	+	-
Candida albicans	-	+	-
Candida guillermondii	-	+	-
Candida pseudotropicalis	-	+	-

[1]Furostanol saponins, [2]Spirostanol saponins, [3]Sapogenins from the butanolic phase.
The codes were assigned as follows : -, no inhibitory zone ; +, inhibitory zone >3mm.

Table 4. Metabolic parameters recorded in normal and streptozotocin diabetic rats after a subchronic administration of steroid saponins (12.5 mg / day per 300 g body weight).

	Glucose (mg/dl)	Plasma TC (mmol/l)	Plasma TG (mmol/l)	HDL TC (mmol/l)	VLDL-LDL TC (mmol/l)
Normal rats					
Controls	76	1.92	1.38	1.68	0.24
(8)	±5	±0.08	±0.20	±0.07	±0.02
Treated	79	1.50***	1.14	1.42	0.16**
(8)	±3	±0.07	±0.09	±0.08	±0.02
Diabetic rats					
Controls	799	1.68	0.70	1.05	0.59
(6)	±102	±0.05	±0.06	±0.10	±0.02
Treated	691	1.39**	0.74	0.91	0.46**
(6)	±33	±0.09	±0.10	±0.06	±0.07

The number of animals is indicated in parentheses ; TC : total cholesterol ; TG : triglycerides ; HDL : high-density lipoprotein ; VLDL : very-low-density lipoprotein ; LDL : low-density lipoprotein.

(**) $P<0.01$; (***) $P<0.001$.

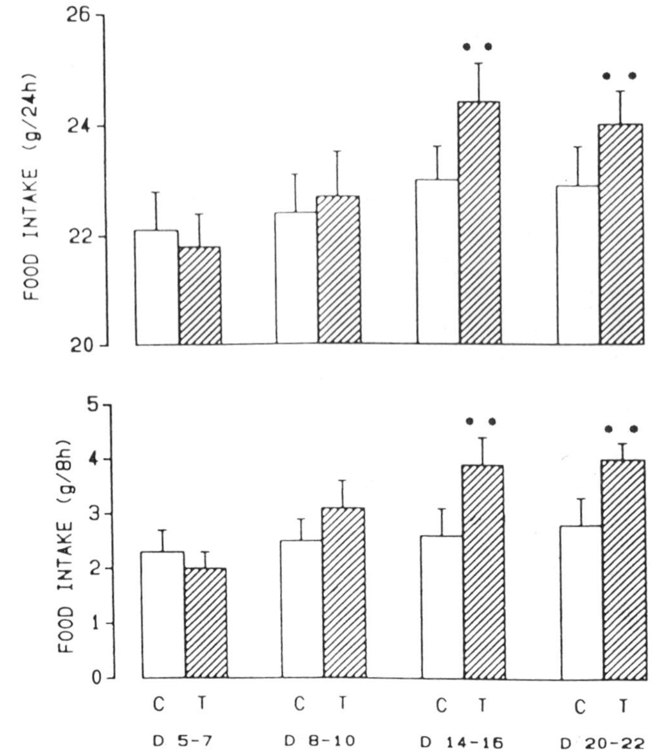

Figure 5. Effect of fenugreek steroid saponin treatment (12.5 mg / day per 300 g body weight) on food intake in normal rats.
Upper panel : food intake/day. Lower panel : food intake during the light phase. D : days ; C : control rats ; T : saponin-treated rats (TPS). Eight animals were used in each set of experiments. Determinations are the mean of three consecutive tests.
**< *P* 0.01. From reference 7.

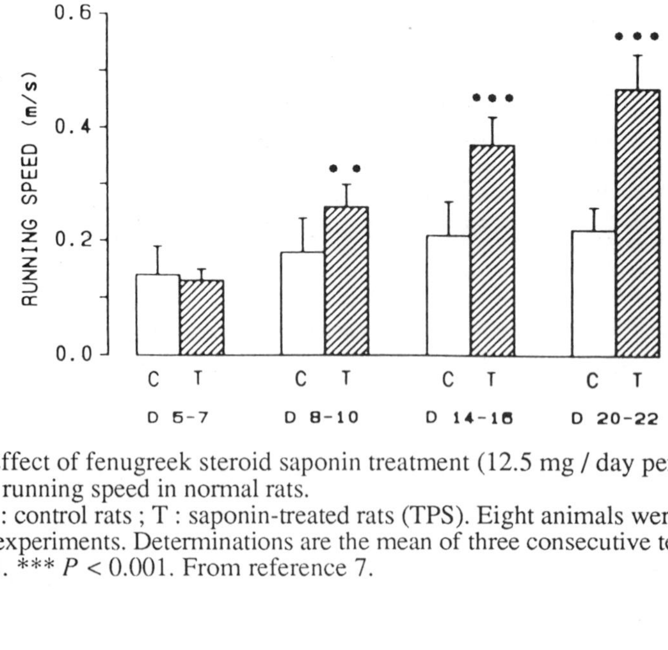

Figure 6. Effect of fenugreek steroid saponin treatment (12.5 mg / day per 300 g body weight) on running speed in normal rats.
D : days. C : control rats ; T : saponin-treated rats (TPS). Eight animals were used in each set of experiments. Determinations are the mean of three consecutive tests.
** $P < 0.01$. *** $P < 0.001$. From reference 7.

Food intake, feeding behavior and weight evaluation

The addition of steroid saponins in the food of normal rats resulted in a progressive increase in daily food intake. This effect was significant after 1 week and persisted after 2 weeks of continuous treatment. The mean total increase was 11% (Figure 5). Normal rats consume 80-90 % of their nutritional intake at night, a consumption pattern which was stable throughout the saponin treatment. In contrast, food intake increased twofold during the daylight phase with the steroid saponin treatment. This suggested that steroid saponin, at the dosage indicated, is capable of altering the circadian rhythm of feeding behavior[7]. In addition, the weight increase pattern during the 2 weeks of treatment was similar in control and treated animals and was not accompanied by an increase in water intake.

Food motivation

The criteria used to evaluate feeding motivation was the rat's running speed to obtain the food[20]. It was shown that the addition of steroid saponin to the rat's food significantly increased the normal rat's speed by 100 % after 2 weeks of treatment. This increase in feeding motivation suggests that steroid saponins are responsible for the appetite stimulation properties of fenugreek seeds.

CONCLUSION

This study has shown that an extract solely containing steroid saponins from fenugreek seeds is endowed with several biological properties : antifungal, hypocholesterolemic and appetite stimulant. Intact saponins are often inactive and transformation into spirostanol forms can partially account for their biological effects[21].

Hence, these compounds extracted from fenugreek seeds are involved in the beneficial effects previously observed with the administration of total extracts of fenugreek seeds.

REFERENCES

1. G. Ribes, Y. Sauvaire, J.C. Baccou, G. Valette, D. Chenon, E.R. Trimble, and M.M. Loubatieres-Mariani, Effects of fenugreek seeds on endocrine pancreatic secretions in dogs, *Ann. Nutr. Metab.* 28:37 (1984).

2. G. Ribes, Y. Sauvaire, C. Da Costa, J.C. Baccou, and M.M. Loubatieres-Mariani, Antidiabetic effects of subfractions from fenugreek seeds in diabetic dogs, *Proc. Soc. Exp. Biol. Med.* 182:159 (1986).

3. Y. Sauvaire and G. Ribes, Composition capable of stimulating insulin secretion intended for the treatment of non-insulin dependent diabetes, *French patent* 92 10 644, *European patent* 93 4021353, *US patent* 5 470 879.

4. G. Valette, Y. Sauvaire, J.C. Baccou and G. Ribes, Hypocholesterolemic effects of fenugreek seeds in dogs, *Atherosclerosis* 50:105 (1984).

5. G. Ribes, Y. Sauvaire, C. Da Costa, J.C. Baccou and M.M. Loubatieres-Mariani, Hypocholesterolaemic and hypotriglyceridaemic effects from fenugreek seeds in alloxan diabetic dogs, *Phytother. Res.* 1:38 (1987).

6. Y. Sauvaire, G. Ribes, J.C. Baccou and M.M. Loubatieres-Mariani, Implication of steroid saponins and sapogenins in the hypocholesterolemic effect of fenugreek, *Lipids* 26:191 (1991).

7. P. Petit, Y. Sauvaire, D. Hillaire-Buys, O. Leconte, Y. Baissac, G. Ponsin, and G. Ribes, Steroid saponins from fenugreek seeds : extraction, purification and pharmacological investigation on feeding behavior and plasma cholesterol, *Steroids* 60:674 (1995).

8. M. H. E. Wunschendorff, La saponine des graines de fenugrec, *J. Pharm. Chim.* 20:183 (1919).

9. R. E. Marker, R.B. Wagner, P.R. Ulshaffer, E.L. Wittbecker, D.D. Goldsmith, and C.H. Ruof, New sources for sapogenins, *J. Am. Chem. Soc.* 69:2242 (1947).

10. P. Brenac and Y. Sauvaire, Accumulation of sterols and steroidal sapogenins in developing fenugreek pods : possible biosynthesis *in situ*, *Phytochemistry* 41:415 (1996).

11. S. Heintz, Les saponosides des graines de fenugrec, *Trigonella foenum-graecum* L., *C. R. Acad. Sci.* 248:283 (1959).

12. N. G. Bogacheva, V.P. Kiselev, and L.M. Kogan, Isolation of 3,26-bisglycoside of yamogenin from *Trigonella foenum-graecum*, *Khim. Prir. Soedin.* 2:268 (1976).

13. N. G. Bogacheva, V.I. Sheichenko, and L.M. Kogan, Structure of yamogenin tetroside from *Trigonella foenum-graecum* seeds, *Khim-Farm. Zh.* 11D:65 (1977). *Chem. Abstr.* 87:180685e (1977).

14. R. Hardman, J. Kosugi, and R.T. Parfitt, Isolation and characterization of a furostanol glycoside from fenugreek, *Phytochemistry* 19:698 (1980).

15. R. K. Gupta, D.C. Jain, and R.S. Thakur, Furostanol glycosides from *Trigonella foenum-graecum* seeds, *Phytochemistry* 23:2605 (1984).

16. R. K. Gupta, D.C. Jain, and R.S. Thakur, Trigofoenoside E-1, a new furostanol saponin from *Trigonella foenum-graecum*, *Indian J. Chem.* 24B:1215 (1985).

17. R. K. Gupta, D.C. Jain, and R.S. Thakur, Furostanol glycosides from *Trigonella foenum-graecum* seeds, *Phytochemistry* 24:2399 (1985).

18. R. K.Gupta, D.C. Jain, and R.S. Thakur, Two furostanol saponins from *Trigonella foenum-graecum*, *Phytochemistry* 25:2205 (1986).

19. O. Leconte, Etude des saponines stéroïdiques du fenugrec (*Trigonella foenum-graecum* L.). Activité antifongique et approches allélopathiques *in vitro*, Thèse, Université Montpellier II, France (1996).

20. T. C. Kirkham and J.E. Blundell, Effect of maloxone and maltrexone on the development of satiation measured in the runway : comparisons with D-amphetamine and D-fenfluramine, *Pharmacol. Biochem. Behav.* 25:123 (1986).

21. K. Hostettmann and A. Marston, *Saponins*, Cambridge University Press, Cambridge (1995).

TRITERPENE SAPONINS FROM MATÉ, *ILEX PARAGUARIENSIS*

Eloir P. Schenkel, Jarbas A. Montanha and Grace Gosmann

Faculdade de Farmácia, Universidade Federal do Rio Grande do Sul
Av. Ipiranga 2752, 90.610.000 Porto Alegre, RS, Brasil

INTRODUCTION

Ilex paraguariensis St. Hil. is a South American native perennial tree belonging to the holly family (Aquifoliaceae). It has been historically used by the Guarani indigenous tribes as a source of a mildly stimulant beverage, called maté, "erva-mate"or "yerba-mate", prepared by infusion of its dried leaves and twigs. The cultivation of *Ilex paraguariensis* started during the XVI century in the Jesuitical settlements, giving origin to the denomination "Jesuits tea". Since then, the agronomic importance of the culture has increased, and it is nowadays an important traditional crop, having a significant economical impact in the southern countries of the South America, with a potential to grow. The yearly production of maté is estimated to be around three hundred tons. In southern Brazil, Paraguay, and Argentina the leaves are also used in popular medicine and included in medicinal herbal products as a tonic, stimulant to the central nervous system, diuretic, and antirheumatic. Knowledge of its chemical composition has remained up to now rather poor. Known secondary metabolites are the xanthines (mainly caffeine), flavonoid glycosides (rutin), and caffeoylquinic acid derivatives (chlorogenic acids).

In previous investigation we demonstrated a high saponin content in the leaves, around 5 to 10 % of the dried plant material [1,2]. In the pharmacological studies, the crude aqueous ethanolic plant extract (1.0 $g.kg^{-1}$, per os) exhibited an antioedematogenic activity on carrageenin-induced oedema in the rat hind paw, after the fourth hour from administration; the same effect was observed for the purified saponin fraction in the dose of 150 $mg.kg^{-1}$ in the second hour after oral administration of 300 $mg.kg^{-1}$ [3]. The crude plant extract (500 $mg.kg^{-1}$) and

Saponins Used in Food and Agriculture
Edited by Waller and Yamasaki, Plenum Press, New York, 1996

the saponin fraction (300 mg.kg^{-1}) did not show antialgesic activity per os, using the hot-plate assay. The crude plant extract (100 mg/ml) and the purified saponin fraction (100 mg/ml) did not show any antimicrobial activity against *Staphylococcus aureus*, *Escherichia coli*, and *Candida albicans* [4]. Chromatographic analysis of the saponin fraction indicated a complex mixture of more than 10 glycosides . The structures of the predominant glycosides, named matesaponin 1, matesaponin 2, matesaponin 3 and matesaponin 4 have already been described [5,6] as ursolic acid 3-*O*-[β-D-glucopyranosyl-(1→3)-α-L-arabinopyranosyl]-(28→1)-β-D-glucopyranosyl ester, ursolic acid 3-*O*-{β-D-glucopyranosyl-(1→3)-[α-L-rhamnopyranosyl-(1→2)]}-α-L-arabinopyranosyl]-(28→1)-β-D-glucopyranosyl ester, ursolic acid 3-*O*-[β-D-glucopyranosyl-(1→3)-α-L-arabinopyranosyl]-(28→1)-β-D-glucopyranosyl-(1→6)-β-D-glucopyranosyl ester, and ursolic acid 3-*O*-{β-D-glucopyranosyl-(1→3)-[α-L-rhamnopyranosyl-(1→2)]}-α-L-arabinopyranosyl]-(28→1)-β-D-glucopyranosyl-(1→6)-β-D-glucopyranosyl ester, respectively. More recently, the structure of matesaponin 5, a minor and highly polar component in the saponin fraction from the leaves, was established as ursolic acid 3-*O*-{β-D-glucopyranosyl-(1→3)-[α-L-rhamnopyranosyl-(1→2)]}-α-L-arabinopyranosyl]-(28→1)-[β-D-glucopyranosyl-(1→4)β-D-glucopyranosyl-(1→6)-β-D-glucopyranosyl] ester [7].

The present paper deals with the structure elucidation of three minor saponins present in the leaves. Spectroscopical analysis revealed each of these saponins as a pair of isomeric glycosides having ursolic acid and oleanolic acid as aglycones. The structures here described for the first time in the maté leaves have not so far been reported for the genus *Ilex*.

RESULTS AND DISCUSSION

TLC analysis of the purified saponin fraction of maté leaves and of its alkaline hydrolysate (Figure 1) indicated that the minor components **J1** and **J2** could be separated in significant amounts from the hydrolysed saponin mixture. After column chromatography, both compounds were peracetylated and chromatographed, in order to enhance their purity. Compound **J3**, in relative higher amount in the mixture, could be isolated by fractionation of the peracetylated saponin mixture. Deacetylation of these compounds, followed by chromatographic comparison, proved that they derive from the genuine saponins present in the maté leaves.

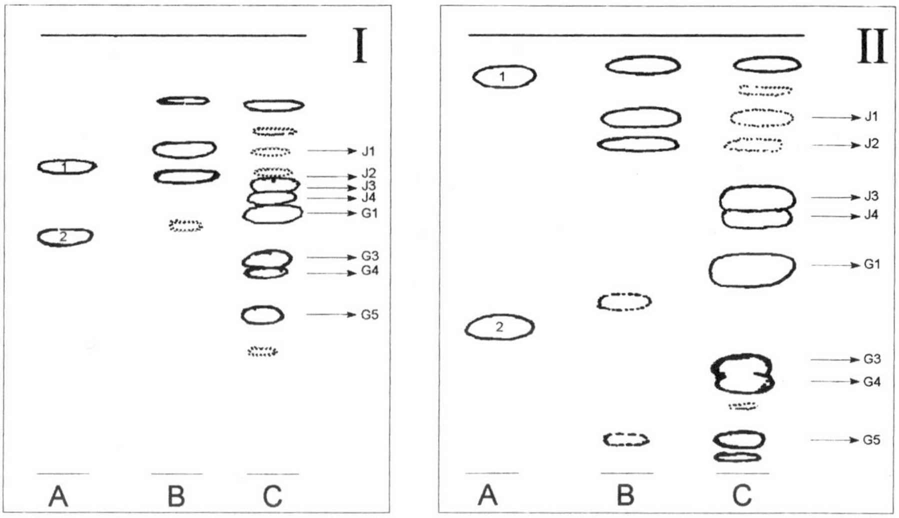

Figure 1.Chromatographic profile of the saponins from *Ilex paraguariensis*. TLC of the purified saponin fraction (**C**), alkaline hydrolysate (**B**) and reference substances (**A**) on silica gel 60 F254 pre-coated plates; **solvent I**: dichloromethane:ethanol:water (80:40:5); **solvent II**: n-butanol:acetic acid:water (12:3:5). **Detection:** anisaldehyde-H_2SO_4. /UV 360 nm. **Reference compounds**: 1. quercetin; 2: rutin; G1: matesaponin 1; G3: matesaponin 2; G4: matesaponin 3; G5: matesaponin 4; for J1-J4 see text.

Compound J1

The mass spectrum (EI) of the peracetylated compound **J1** exhibited peaks at *m/z* 273 (94 %) and 489 (14 %), which were attributed to a terminal methylpentose and to a disaccharide composed of a pentose and a methylpentose, respectively. The peaks at *m/z* 209 (91 %) and 248 (84 %) indicated the presence of an olean-12-ene or urs-12-ene aglycone.

The ^{13}C-NMR spectrum displayed two anomeric carbon signals at δ 104.4 and δ 98.1, attributed to arabinose and rhamnose, respectively. These assignments were supported by the ^{1}H-NMR spectrum, which showed two one-proton signals at δ 4.48 (doublet, J = 6.3 Hz) and δ 5.12 (doublet, J = 1.6 Hz). It was possible to assign the other sugar protons by the ^{1}H-^{1}H COSY experiment.

Compound **J1** was further degraded by oxidation with $NaIO_4$, yielding an aglycone with the same chromatographic characteristics as oleanolic or ursolic acid, indicating thus the positions 2 and 4 of the arabinose moiety as possible linkage site of the terminal sugar. The attachment point of rhamnose was determined to be the 2-position of arabinose by comparison with ^{13}C-NMR spectral data of other acetylated saponins [8].The ^{1}H-NMR data for the sugar chain are in full agreement with literature data for peracetylated 3-*O*-α-L-rhamnopyranosyl-(1→2)-α-L-arabinopyranosyloleanolic acid obtained from *Caltha palustris*[9].

Though the ^{1}H- and ^{13}C-NMR spectra displayed clear signals for only one sugar chain, the spectra demonstrate the presence of two aglycones: oleanolic acid by the signals at δ 123.0 (C-12) and δ 144.0 (C-13) and ursolic acid, by the signals at δ 126.2 (C-12) and δ 138.4 (C-13). The relationship between the integrals of these signals indicates the predominance of the oleanolic acid glycoside.

The above evidences led to the conclusion that compound **J1** is a mixture of two glycosides with the structures 3-*O*-α-L-rhamnopyranosyl-(1→2)-α-L-arabinopyranosyl-oleanolic acid (**J1a**) and 3-*O*-[α-L-rhamnopyranosyl-(1→2)-α-L-arabinopyranosyl]ursolic acid (**J1b**). The small differences between these structures (methyl positions at ring E) explain the difficulty of separation using the current techniques of analysis. On the contrary, it is possible to charaterize the different aglycones by ^{13}C-NMR spectroscopy.

Compound J2

This compound proved to be similar (MS, ^{1}H- and ^{13}C-NMR data, co-tlc) with the glycoside obtained by degradation of matesaponin 1, already reported as 3-*O*-[β-D-glucopyranosyl-(1→3)-α-L-arabinopyranosyl]ursolic acid [5]. The only differences were that the

NMR spectra of **J2** displayed also signals of minor intensity for the oleanolic acid aglycone. Compound **J2** is thus also a mixture of glycosides from ursolic and oleanolic acid, with the same sugar chain, and these compounds are also genuine glycosides of maté leaves.

Compound J3

The FABMS spectrum (positive ion mode) of the peracetylated compound **J3** indicated the molecular mass of 1274 by the pseudomolecular peaks at *m/z* 1281 $(M+Li)^+$ and 1297 $(M+Na)^+$. The peaks at *m/z* 273 (29 %) and 331 (100 %) are indicative of the terminal positions of a peracetylated methylpentose and of a peracetylated hexose; the peak at *m/z* 489 (35 %) can be attributed to an peracetylated fragment of methylpentose-pentose. Alkaline hydrolysis of peracetylated **J3** produced a compound with identical chromatographic behavior to compound **J1**, suggesting a bidesmosidic saponin with the same partial structure.

The ^{13}C-NMR spectrum displayed anomeric carbon signals at δ 104.7 and δ 97.5, which were attributed to arabinose and rhamnose, respectively. The signal at δ 91.6 was attributed to the ester-bonded glucose at C-28. The ^{13}C-NMR spectrum proved also the presence of two aglycones, on the basis of the presence of two signals for the carbonyl ester, at δ 175.7 and δ 175.4 and the signals for C-12 and C-13 at δ 122.9 and 142.9 (oleanolic acid glycoside) and δ 126.2 and 137.3 (ursolic acid glycoside). The assignments were supported by the ^{1}H-NMR spectrum, which displayed anomeric proton signals for arabinose at δ 4.48 (doublet, $J = 6.3$ Hz) and for rhamnose at δ 5.12 (not resolved). For the anomeric proton of the ester-bonded glucose two signals could be seen, at δ 5.58 (doublet, $J = 8$ Hz) and δ 5.55 (doublet, $J = 8$ Hz), both with lower intensities than the other anomeric proton signals. That means that the anomeric proton signals from arabinose and rhamnose are the same for the two glycosides, but the signals for the ester-bonded glucose are different. This fact can be explained by the vicinity of this proton to the site of differentiation of oleanolic and ursolic acids.

All these data indicate that compound **J3** is a mixture of the substances 3-*O*-[α-L-rhamnopyranosyl-(1→2)-α-L-arabinopyranosyl]oleanolic acid (28→1)-β-D-glucopyranosyl ester (**J3a**) and 3-*O*-[α-L-rhamnopyranosyl-(1→2)-α-L-arabinopyranosyl]ursolic acid (28→1)-β-D-glucopyranosyl ester (**J3b**). The integrals for C-12 and C-13 in the ^{13}C-NMR spectrum showed the predominance of the oleanolic acid glycoside.

Separation of such isomeric mixtures of olean-12-ene and urs-12-ene glycosides is a challenging problem [10,11]. Recently, the isolation of the same isomeric pair with the structures here described for **J1a/J1b** and **J2a/J2b** was reported for the seed of *Patrinia scabiosaefolia* (Valerianaceae), performed by repeated separation by preparative HPLC [12].

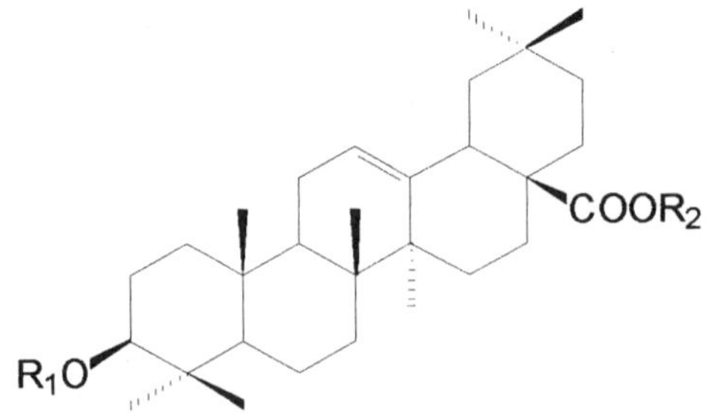

J1a, J2a, J3a

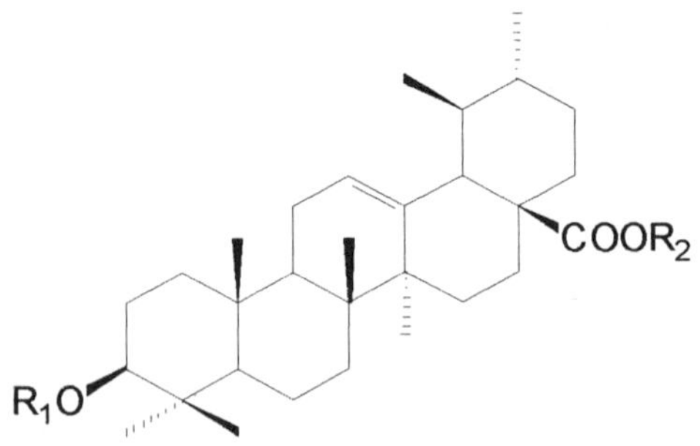

J1b, J2b, J3b

J1a R_1 = α-L-rhamnopyranosyl-(1-2)-α-L-arabinopyranosyl; R_2 = H

J1b R_1 = α-L-rhamnopyranosyl-(1-2)-α-L-arabinopyranosyl; R_2 = H

J2a R_1 = β-D-glucopyranosyl-(1-3)-α-L-arabinopyranosyl; R_2 = H

J2b R_1 = β-D-glucopyranosyl-(1-3)-α-L-arabinopyranosyl; R_2 = H

J3a R_1 = α-L-rhamnopyranosyl-(1-2)-α-L-arabinopyranosyl; R_2 = β-D-glucopyranosyl

J3b R_1 = α-L-rhamnopyranosyl-(1-2)-α-L-arabinopyranosyl; R_2 = β-D-glucopyranosyl

Figure 2. Structure of compounds **J1a, J1b, J2a, J2b ,J3a** and **J3b**

MATERIAL AND METHODS

Plant Material

The leaves of *I. paraguariensis* were collected in Veranópolis, State of Rio Grande do Sul. A herbarium specimen (voucher n° ICN-68648) is deposited in the herbarium of the Botany Department of the Federal University of Rio Grande do Sul.

General Experimental Procedure

Mp's were taken with a Kofler's apparatus and are uncorrected. Optical rotation was recorded with a Perkin-Elmer 241 spectropolarimeter at room temperature (22 ° C). EIMS were performed on a MAT 44 spectrometer and FABMS on a KRATOS MS 80 RFA, using Xe/7-8 kV. ^{1}H and ^{13}C-NMR spectra were obtained with Bruker 400-MHz and Varian 200-MHz spectrometers in C_5D_5N (compound **J2**) or $CDCl_3$ (peracetylated compounds), with TMS as internal standard. TLC was carried out on silica gel 60 F_{254} precoated plates (Merck), employing

BuOH:AcOH:H_2O (12:3:5) and CH_2Cl_2-EtOH-H_2O (80:40:5) for saponins and sugars, light petroleum ether-AcOEt (1:1) for peracetylated saponins, light petroleum ether-AcOEt (2:8) for aglycones and prosapogenins; detection was with anisaldehyde-H_2SO_4/UV 360 nm for saponins and aglycones and aniline phthalate/UV 360 nm for sugars. Acetylation, alkaline hydrolysis, and periodate oxidation were performed as reported by Gosmann [5,6].

Extraction and Purification of the Saponins

Dried and crushed leaves (600 g) were extracted with EtOH-H_2O (4:6) by repeated (3x) decoction. The extract was concentrated, diluted with H_2O, and successively extracted with $CHCl_3$, AcOEt and butanol. The BuOH layer was evaporated to dryness to give a crude saponin fraction (104 g), that was separated from phenolic compounds with an Amberlite XAD-2 column.

Compounds J1 and J2: The saponin fraction (3.0 g) was heated in refluxing 1 N KOH 96% EtOH for 1 h. After neutralization and evaporation of ethanol, the aqueous solution was extracted with AcOEt/BuOH (2:1) and concentrated. This residue (1.5 g) was column chromatographed over silica gel with $CHCl_3$ /EtOH to give compound **J1** (130 mg) and compound **J2** (110 mg).

Peracetylated compound J1: mp 162-167 °C. EIMS *m/z* (rel. int %) 109 (81), 153 (100), 203 (91), 248 (84), 273 (94) [rhamnose$(Ac)_3]^+$, 489 (14) [rhamnose$(Ac)_3$ -arabinose $(Ac)_2]^+$. ^{1}H-NMR ($CDCl_3$): 1.18 (Me, *d*, *J* = 6 Hz, rha H-6), 3.60 (1H, br *d*, *J* = 11.5 Hz, ara H-5), 3.93 (1H, br *d*, ara H-5), 3.97 (1H, *dd*, *J* = 8.7 and 6.4 Hz, ara H-2), 4.25 (1H, *m*, rha H-5), 4.48 (1H, *d*, *J* = 6.3 Hz, ara H-1), 4.99 (1H, *dd*, *J* = 8.8 and 3.5 Hz, ara H-3), 5.17 (2H, *m*, rha H-4 and ara H-4), 5.12 (1H, *d*, *J* = 1.6 Hz, rha H-1), 5.23 -5.33 (1H, *m*, rha H-3), 5.33 (1H,br *d*, rha H-2). ^{13}C-NMR: see Table 1.

Compound J2: mp 275-278 °C, $[\alpha]_D$ + 30.2° (c = 1.46, pyridine). ^{13}C-NMR: see Table1.

Peracetylated compound J2 : mp 148-150 ° C. EIMS *m/z* (rel. int %): 109 (60), 139 (73), 169 (100), 203 (549, 248 (66), 331 (30) [glucose$(Ac)_4]^+$, 547 (44) [glucose$(Ac)_4$-arabinose $(Ac)_2$ $]^+$. ^{1}H-NMR ($CDCl_3$): 4.33 (*d*, *J* = 6.5 Hz, ara H-1), 4.64 (*d*, *J* = 7.8 Hz, glu H-1).

Compound J3: The purified saponin fraction (1.8 g) was acetylated, yielding a pale yellow solid (2.0 g). This mixture was chromatographed over silica gel using light petroleum ether/AcOEt to give compound **J3** peracetylated (85 mg): mp 123-127 °C. FABMS (positive mode) *m/z* (rel. int. %): 109 (53), 169 (88), 273 (30) [rhamnose(Ac)$_3$]$^+$, 331 (100) [glucose(Ac)$_4$]$^+$, 489 (35) [rhamnose(Ac)$_3$ -arabinose (Ac)$_2$]$^+$. ^{1}H-NMR ($CDCl_3$): 4.44 (1H, *d*, *J* = 6.5 Hz, ara H-1), 5.12 (1H, not resolved, rha H-1), 5.58 (1H, *d*, *J* = 8.0 Hz, glu H-1). ^{13}C-NMR: see Table 1.

CONCLUDING REMARKS

The presence of saponins in the leaves of *Ilex paraguariensis* is of considerable interest due to the high content, their influence in the grade of bitterness of the product maté, and the possible physiological significance for the plant. All the saponins identified up to now [**J1a, J1b, J2a, J2b, J3a, J3b**, matesaponins 1-4 [5,6], matesaponin 5 [7] and **J4** (unpublished results)], are glycosides of ursolic or oleanolic acids. In comparison, the saponin pattern for the other South American *Ilex* species is different. Although the production of maté is becoming more industrialized and controlled, adulteration of the product with the introduction of variable amounts of leaves and twigs of others South American *Ilex* species is frequently reported [13]. The adulterations might modify the physiological and pharmacological activity of maté, raising concern about possible health hazards for the consumers. Therefore, the *Ilex* species reported as adulterant and/or substitutes for the genuine maté product are being studied in our laboratory, searching for chemical differences. The saponins from *I. dumosa* are all monodesmosides and bidesmosides of the oleanolic acid with the sugars galactose, glucose and arabinose [14]. The saponins from *I. theezans* [2], *I. pseudobuxus* [15], *I. argentina* [16], *I. integerrima*[17] and *I. taubertiana* [18] all have the aglycones derived from 19 α-hydroxyursolic acid. These are important features, that may have taxonomical significance and may be useful to develop methodologies for the quality control of maté products, besides the possible pharmacological consequences.

ACKNOWLEDGEMENTS

We are grateful to Dr. D. Bergenthal, Dr. W. Werner (Institut für Pharmazeutische Chemie, University of Münster, Germany, and Dr. G. Habermehl (Chemisches Institut der Tieraerzlichem Hochschule, Hannover, Germany) for the measurements of NMR and MS spectra. This work was supported by CNPq fellowships.

Table 1. Selected ^{13}C-NMR data of peracetylated compounds **J1a**, **J1b**, **J3a**, **J3b** ($CDCl_3$) and compounds **J2a** and **J2b** (C_5D_5N)

Carbon	J1a[J1b]	J2b[J2 a]	J3a[J3b]
3-O-sugar			
Ara-1	104.4	107.2	104.7
Ara-2	73.0	71.7	73.0
Ara-3	73.7	84.0	72.9
Ara-4	68.3	69.1	68.1
Ara-5	63.0	66.7	63.0
Rha-1	98.1		97.5
Rha-2	70.1		69.9
Rha-3	68.9		68.0
Rha-4	71.4		71.2
Rha-5	67.0		66.7
Rha-6	17.0		17.0
Glu-1		106.2	
Glu-2		75.5	
Glu-3		78.5	
Glu-4		71.4	
Glu-5		78.2	
Glu-6		62.5	
28-O-sugar			
Glu-1			91.6
Glu-2			70.0
Glu-3			73.0
Glu-4			68.6
Glu-5			72.5
Glu-6			61.6
Aglycone			
C-3	89.9	88.6	89.6
C-12	123.0 [126.2]	125.6 [122.5]	122.9 [126.2]
C-13	144.0 [138.4]	139.3 [144.9]	142.9 [137.3]
C-19	46.0 [39.6]	39.4 [46.4]	46.9 [39.4]
C-20	30.8 [39.5]	39.3 [30.8]	40.6 [38.9]
C-29	33.2 [17.3]	17.2 [33.1]	33.0 [17.3]
C-30	23.7 [21.0]	21.1 [23.6]	23.5 [21.0]

REFERENCES

1. Gosmann, G., 1989, *Saponinas de Ilex paraguariensis St. Hil. (Erva-Mate)*, M.Sc. Thesis, Universidade Federal do Rio Grande do Sul, Porto Alegre, Brasil.
2. Athayde, M.L., 1993, *Saponinas e Triterpenos em Algumas Espécies do Gênero Ilex*, M.Sc. Thesis, Universidade Federal do Rio Grande do Sul, Porto Alegre, Brasil.
3. Montanha, J.A., Heinzmann, B., Schapoval, E.E.S., and Schenkel, E.P., 1990, Atividade antiedematogênica das saponinas de *Ilex paraguariensis, Ciência e Cultura* 7 (suppl.):505-506.
4. Montanha, J. A., 1990, *Estudo Químico e Biológico das Saponinas de Ilex paraguariensis St. Hil. Aquifoliaceae* M.Sc. Thesis, Universidade Federal do Rio Grande do Sul, Porto Alegre, Brasil
5. Gosmann, G., Schenkel, E.P., and Seligmann, O., 1989, A new saponin from mate, *Ilex paraguariensis, J. Nat. Prod.* 52:1367-1370.
6. Gosmann, G., Guillaume, D., Taketa, A.T. C., and Schenkel, E.P., 1995, Triterpenoids saponins from *Ilex paraguariensis, J. Nat. Prod.* 58: 438-441.
7. Kraemer, K.H., Taketa, A.T.C., Schenkel, E.P., Gosmann, G. and Guillaume, D., 1996, Matesaponin 5, a highly polar saponin from *Ilex paraguariensis, Phytochemistry*, in press.
8. Massiot, G., Lavaud, C., Men-Olivier, L.Le., Binst, G.V., Miller, S.P.F., and Fales, H.M., 1988, Structural elucidation of alfafa root saponins by mass spectrometry and nuclear magnetic ressonance Analysis, *J. Chem. Soc., Perkin Trans.* 1: 3071-3079.
9. Bhandari, P., Gray, A.I., and Rastogi, R.P., 1987, Triterpenoid saponins from *Caltha palustris, Planta Med.* 53: 98-100.
10. Sousa, M.P., Matos, M.E.O., Machado, M.I.L., Braz Filho, R., Vencato, I., and Mascarenhas, Y.P., 1984, Triterpenoids from *Guettarda angelica, Phytochemistry* 23: 2589-2592.
11. Srivastava, S.K. and Jain, D.C. ,1989, Triterpenoid saponins from plants of Araliaceae, *Phytochemistry* 28: 644-647.
12. Nakanishi, T., Tanaka, K., Murata, H., Somekawa, M., and Inada, A., 1993, Phytochemical studies of seeds of medicinal plants. Ursolic acid and oleanolic acid glycosides from seeds of *Patrinia scabiosaefolia* Fischer, *Chem. Pharm.Bull.* 41:183-186.
13. Giberti, G.C., 1989, Los parientes silvestres de la yerba mate y el problema de su adulteración, *Dominguezia* 7: 3-21.
14. Heinzmann, B.M and Schenkel, E.P., 1995, Saponins from *Ilex dumosa, J. Nat. Prod.* 58: 1419-1422.
15. Taketa, A.T.C., and Schenkel, E.P., 1994, Saponins from *Ilex pseudobuxus, Acta Farm. Bonaerense* 13: 157-63.
16. Schenkel, E.P., Athayde, M.L., Giberti,G.C., Guillaume, D., 1995, A new saponin from *Ilex argentina. Acta Farm. Bonaerense* 14: 217-221.
17. Costantin, M.B.,1995, *Análise de saponinas em Ilex integerrima e comparação com Ilex theezans*, M.Sc. Thesis, Universidade Federal do Rio Grande do Sul, Porto Alegre, Brasil.
18. Taketa, A.T.C., and Schenkel, E.P., 1995, Saponinas de *Ilex taubertiana, Rev. Bras. Farm.* 76: 9-11.

STUDIES OF THE PHYTOTOXICITY OF SAPONINS ON WEED AND CROP PLANTS

Robert E. Hoagland, Robert M. Zablotowicz, and Krishna N. Reddy

USDA-ARS
Southern Weed Science Laboratory
Stoneville, MS 38776

Several saponins or sapogenins including β-escin, betulin, β-glycyrrhetinic acid, hecogenin, oleandrin, and oleanolic acid were tested in the laboratory, growth chamber, and greenhouse on various weed and crop species. Seed germination, root and shoot growth after root, foliar, or soil application, electrolyte leakage from leaf discs, and greening of etiolated plant tissues were monitored. Esterase activity using fluorescein diacetate (FDA) and *p*-nitrophenyl butyrate (PNPB) was also assayed. The compounds had differential effects on these parameters, depending on the species. The effects of these compounds on electrolyte leakage ranged from no effect to a 10-fold increase above untreated tissue levels after 72 h. Escin increased FDA activity up to 35% above untreated tissue, but other compounds caused no effect or reduced FDA activity. PNPB activity was generally not affected. In greening studies of excised tissue, escin reduced chlorophyll production by 90-100% in several species, with other compounds giving intermediate or no effects. Foliar application (1.0 mM) in the greenhouse had no substantial effect (visible or shoot biomass) on 10 species. However, in short-term (8 to 13 day) tests, β-escin, applied to soil at 88 and 350 μmol/kg soil, drastically reduced emergence in barnyardgrass (*Echinochloa crus-galli* L. Beauv.), hemp sesbania [*Sesbania exaltata* (Raf.) Rydb. ex A.W. Hill], wheat (*Triticum aestivum* L.), and soybean [*Glycine max* (L.) Merr.]. β-Escin also reduced growth in all species but soybean, and barnyardgrass was the most sensitive species tested. Results are discussed in relation to the role of these compounds as plant growth-regulating natural products.

INTRODUCTION

The saponins are comprised of plant glycosides which contain a polar moiety; i.e., water soluble sugar(s) attached to a lipophilic moiety of steroid or triterpenoid nature. The related sapogenins are aglycones that occur in nature or arise from acidic or enzymatic hydrolysis of saponins (32, 54). These compounds are widely distributed in nature and have been found in over 100 plant families (57) encompassing 500 genera (5). The saponins have surfactant or soap-like properties and generally exhibit toxicity to cold-blooded, but not warm-blooded animals (21).

Several saponins have phytotoxic or allelopathic properties. Various triterpenes and sterols have plant-growth-regulating activity or allelopathic properties (14, 16). Of all the various saponins studied as allelopathic agents, root saponins from alfalfa (*Medicago sativa*) have received the most attention. Alfalfa saponins have been linked to yield losses of subsequent crops (29, 35), but some grain crops were insensitive. Purified saponins from *Medicago lupulina* reduced wheat, oat (*Avena sativa* L.), barley (*Hordeum vulgare* L.), and rye (*Secale cereale* L.) growth, whereas similar extracts from *Medicago media* were ineffective as allelochemicals (25). Differences in effects were found to be the result of differences in saponin type in the two species (26). Most allelopathic studies of saponins and sapogenins have been done on crop species, but recently the growth of several weeds was shown to be reduced by such compounds. Alfalfa root saponins exhibited the highest allelopathic effects on barnyardgrass and cheat (*Bromus secalinus* L.) with lesser effects on redroot pigweed (*Amaranthus retroflexus* L.), coffeeweed (also called hemp sesbania), dandelion (*Taraxacum officinale* Weber), and wheat (63, 64). Recently, alfalfa root saponins have been implied to have potential as herbicides (65).

Some compounds related to sapogenins have been shown to have phytotoxic action. The sterol amasterol, isolated from slender amaranth (*Amaranthus viridis* L.), inhibited lettuce (*Lactuca sativa* L.) germination and growth (50). The sterol chondrillasterol, produced by Palmer amaranth (*Amaranthus palmeri* S. Wats.) (17), had differential effects on germination in five *Amaranthus* species (9). Other saponins were found to be seed germination inhibitors (38). Alfalfa saponins were shown to inhibit cotton seed germination (44). A triterpenoid saponin from pea (*Pisum sativum* L.) stimulated lettuce root growth, but did not affect seed germination (58).

Various saponins exhibit antifungal activity, including toxicity to plant pathogenic fungi. Amasterol was toxic to the pathogenic fungus *Helminthosporum oryzi* (50). Other pathogenic fungi may be resistant to saponins because of detoxifying enzymes (1, 43, 46, 52), or because of different membrane composition (52). It has recently been hypothesized that the ability of a phytopathogenic fungus to detoxify a specific plant saponin may determine its host range (7). Release of the saponin avenacin A from oat root vacuoles by fungal invasion is a plant defense mechanism; however, the pathogen *Fusarium avensceum* can detoxify the saponin by producing β-glucosidase and α-arabinosidase, which convert the saponin to the non-toxic aglycone (30, 62). Some saponins exhibit fungicidal activity even at low concentrations (7 to 25 ppm) (33). Saponins can inhibit fungal growth via interactions with sterols contained in the plasma membrane (13, 37). A saponin isolated from *Rapanea melanophloeos* leaves had antifungal activity against the pathogenic fungus *Cladosporium cucumerinum* (40). Saponins can influence the growth of soil microflora, and *Trichoderma* was extremely sensitive to very low saponin concentrations (35). This latter fact was the basis for the development of a sensitive bioassay of saponins (68).

Our interest in saponins arises from efforts to study weed control using conservation tillage/cover crop systems, synthetic herbicides, natural products, plant pathogens (bacterial and fungal), and/or combinations of these methods. Because of the varied interactions of saponins and sapogenins, as outlined above, and the fact that there is only sparse information on the activity of these compounds in soil and plants, we chose to investigate the possible phytotoxic effects of some of these compounds (Fig. 1) on various parameters using several weed and crop plants. Since these experiments were conducted over a long period of time with a variety of plants and chemicals, and because some of the plants lacked sensitivity in some cases, we did not attempt to utilize all compounds on all species in each of the tests.

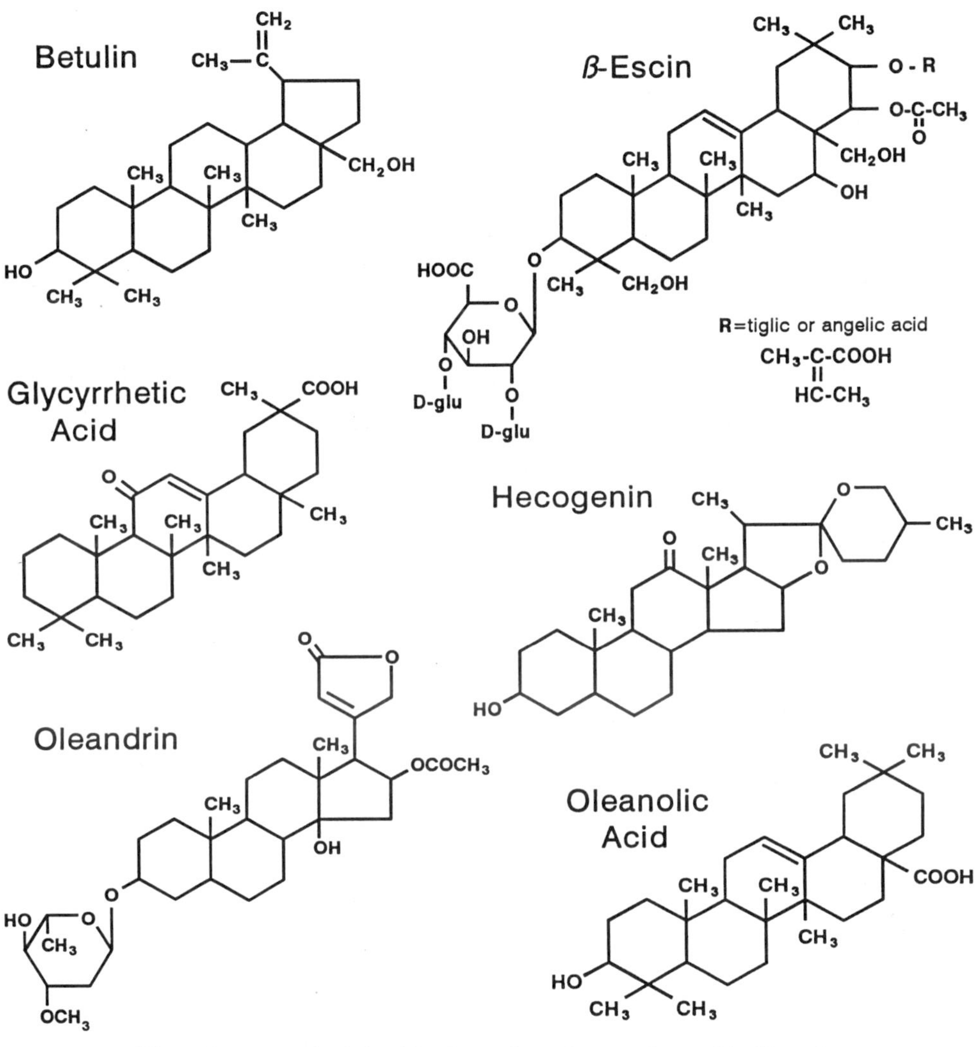

Figure 1. Structural relationship of saponins and sapogenins used in this study.

MATERIALS AND METHODS

Chemicals

β-Escin (92% pure), betulin (lup-20[29]-ene-3β,28-diol), 18β-glycyrrhetinic acid (3β-hydroxy-11-one-18β,20β-olean-12-en-29-oic acid, 95% pure), hecogenin (5α-spirostan-3β-

12-one, 95% pure), oleandrin {5β,20[22]-cardenolide-3β,14,16β-triol-3-[2,6-dideoxy-3-*O*-methyl-α-*L*-arabinohexopyranosyl)oxy]16-acetate}, oleanolic acid (3-hydroxyolean-12-en-28-oic acid), *p*-nitrophenyl butyrate (PNPB), fluorescein diacetate (FDA), and *p*-nitrophenyl-α-D-glucopyranoside (PNPGP) were obtained from Sigma Chemical Co. (St. Louis, MO). The structures of these saponins and sapogenins are presented in Figure 1.

Electrolyte Leakage

Twenty leaf disks (4 mm diam) of each species were cut with a cork borer from greenhouse- or chamber-grown plants (1- to 2-wk-old) and placed in small petri dishes (10 x 35 mm) containing 2-(*N*-morpholino)ethanesulfonic acid buffer (pH 6.5), sucrose (1%), and the test compound. Control solutions contained only buffer and sucrose. Conductivity of test solutions was determined at 0 h and at various intervals thereafter.

Greening Experiments

Seedlings were hydroponically grown etiolated three to five days. Cotyledons and stem tissue were excised and placed immediately into petri dishes containing saponin/sapogenin solutions or suspensions. Dishes were incubated for two h in the dark, then placed in a growth chamber under continuous light [150 $\mu E\ m^{-2}\ s^{-1}$, photosynthetically active radiation (PAR)] at 27 °C. Chlorophyll was extracted 48 h after treatment (HAT) and quantitatively determined spectrophotometrically using dimethyl sulfoxide (4, 23).

Enzyme Assays

Plant tissue was homogenized in potassium phosphate buffer (0.10 M, pH 7.6) with a high-speed Virtis® homogenizer. Homogenates were centrifuged (20,000 x g, 15 min) and supernatants used directly in several enzyme assays. FDA hydrolysis (esterase activity) was measured in plant extracts by determining fluorescein formation spectrophotometrically at 490 nm (20, 34). PNPB was also used to measure esterase activity in plant tissues. Enzyme activity was based on formation of *p*-nitrophenol measured spectrophotometrically at 405 nm. β-Glucosidase activity was measured using PNPGP as substrate. Enzyme activity was quantified as the amount of *p*-nitrophenol produced per unit time. Protein concentration in plant extracts was measured with the Bradford reagent (8).

Germination Tests

Various crop and weed seeds (15 to 35 seeds) were allowed to imbibe aqueous solutions of saponins or sapogenins in small petri dishes (10 x 35 mm) in the dark at 25° C for 24 to 48 h, depending on species. Germination was defined as radicle protusion through the seed coat. Tests were performed in triplicate.

Application of Chemicals to Plant Tissues

Chemicals were applied to seedling roots using hydroponically grown seedlings or via application to the soil. Roots of 3- to 7-day-old hydroponically grown seedlings were placed directly into test tubes containing chemical solutions. Root and shoot elongation and secondary root growth effects were monitored over time. In soil tests, unimbibed seeds or seeds pregerminated in deionized water (24 to 30 h) were placed in Conetainers® (plastic conical pots; 38 mm top diam, 218 mm long) containing soil (Bosket sandy loam) and aqueous solutions of β-escin. Seeds in each Conetainer® were covered with equal volumes of soil and appropriate solution added to give final β-escin concentrations of 0, 88, or 350 μmol/kg soil. Conetainers® were then placed in an enviromental chamber [light:dark cycle of 14:10 h at 30:24°C]. Light (400 μE m^{-2} s^{-1}, PAR) was provided by sodium halide and incandescent lamps. Plant emergence, growth (root and shoot biomass fresh weight), and nodulation (soybean and hemp sesbania) were determined.

In greenhouse studies, the effects of six compounds (betulin, β-escin, β-glycyrrhetic acid, hecogenin, oleandrin, and oleandric acid) applied as foliar sprays were evaluated on six weed species {barnyardgrass, johnsongrass [*Sorghum halepense* (L.) Pers.], hemp sesbania, pigweed, sicklepod (*Cassia obtusifolia* L.), and velvetleaf (*Abutilon theophrasti* Medicus)} and four crop species [corn (*Zea mays* L.), cotton (*Gossypium hirsutum* L.), mungbean (*Vigna radiata* L.), and soybean]. Seedlings were grown for 10 days in the greenhouse prior to treatment. Plants (seven replicates per species/compound combination in a randomized complete block design) were treated in a spray chamber with a 1.0 mM suspension of the above compounds in aqueous ethanol (5%) with 0.25% nonionic surfactant at a rate of 186 L/ha. Aqueous control solutions contained ethanol and nonionic surfactant at the above concentrations. Treatments were evaluated 9 days after treatment (DAT) by determining shoot biomass and visual inspection for phytotoxic effects or injury. Data were analyzed by SAS general linear model and means were separated by Fisher's LSD.

For growth chamber studies, the compounds (aqueous solutions or suspensions without surfactant or ethanol) were applied directly to shoots of 4- to 6-day-old hydroponically grown seedlings using a hand-held compressed air sprayer. Foliage was sprayed to run-off. Elongation and/or fresh weight of roots or shoots were determined several days after treatment.

RESULTS AND DISCUSSION

Plant Growth Responses Caused by Saponins or Sapogenins

Germination Tests. In seed germination tests in petri dishes with betulin, β-escin, glycyrrhetinic acid, and hecogenin at 10^{-3} M, lettuce seed germination was reduced by about 50%, but there was no effect on prickly sida (*Sida spinosa* L.), cucumber (*Cucumis sativus* L.), mungbean, radish (*Raphanus sativus* L.), hemp sesbania, sicklepod, sorghum (*Sorghum vulgare* Pers.), and wheat (data not shown). After three days growth at 26° C in the dark, β-escin treatment reduced growth (shoot elongation by 10 to 20%) in all the above, except cucumber and radish (data not shown). Pure medicagenic acid glycosides, hederagenic glycosides, zahnic acid glycosides, and soyasaponin I at 1000 to 5000 ppm caused no significant effects on germination of wheat seeds (41). A saponin extract from Indian legume (*Albizzia lebbek* Benth.) at low concentrations stimulated germination and growth of wheat, barley, and chickpea (*Cicer arietinum* L.), but was inhibitory at high concentrations (60). Nord and van Atta (38) showed that a saponin extract could lower germination of two plant species, but only at relatively high concentrations (i.e., 1 to 5%). The triterpenoid saponin, chromosaponin I, from pea (59) and soybean (28) was found to stimulate the growth of lettuce roots (58). Concentrations of 1 and 3 mM caused 130 and 190% increases, respectively, in root growth compared to the control seedlings. However, germination of lettuce seeds was not affected at these concentrations.

Tests with Hydroponically Grown Plants. Shoots of mungbean, hemp sesbania, and sicklepod, hydroponically grown in the dark for 4 days, were treated with one of five saponins or sapogenins (10^{-3} M) by spray application, then placed back in the dark for 48 h (Fig. 2). Mungbean shoot elongation was reduced by betulin, glycyrrhetic acid, and oleandrin. Hemp sesbania shoot elongation was reduced by β-escin and oleandrin, and

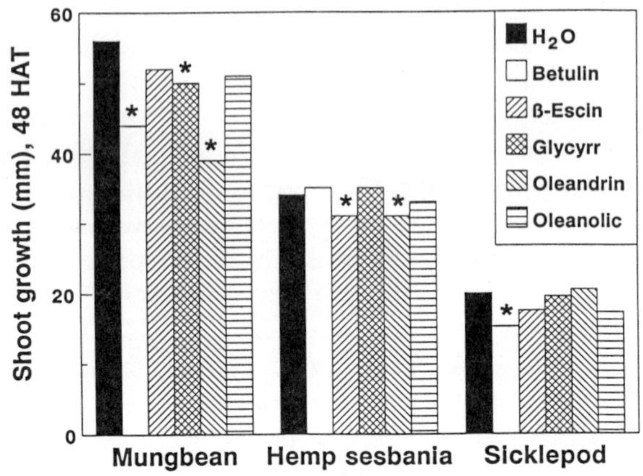

Figure 2. Shoot length of hydroponically grown mungbean, hemp sesbania, and sicklepod seedlings treated foliarly with various saponins, 48 HAT. Asterisks (*) denote that treatment is significantly different from other treatments at the 95% level.

sicklepod shoot elongation was reduced only by betulin. These results suggest some differential specificity of the compounds in these species when applied foliarly. This growth inhibition suggests that some of these compounds are able to penetrate the cuticle of these young seedlings. Two hydroponically grown species (sicklepod and hemp sesbania) were tested for the effects of five saponins/sapogenins (10^{-3} M) via root feeding (data not shown).

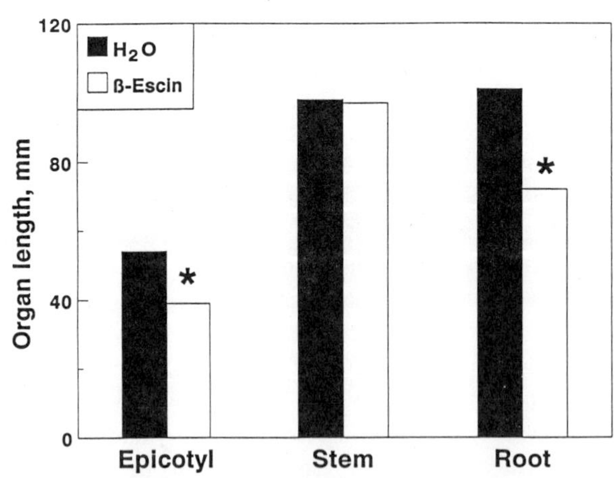

Figure 3. Effects of root-fed β-escin (10^{-3} M) on organ elongation of hydroponically grown mungbean seedings, 72 HAT. Asterisks (*) denote that treatment is significantly different from other treatments at the 95% level.

Root elongation of sicklepod was not affected by any of the compounds 48 HAT. β-Escin was the only treatment that inhibited primary root elongation in hemp sesbania (10%, data not shown). Secondary root formation in hemp sesbania was also inhibited by β-escin. Another species, mungbean, was examined for growth of various plant parts 48 HAT in the light after treatment with β-escin, glycyrrhetic acid, and hecogenin. These three compounds had no effect on primary root elongation of mungbean, but β-escin inhibited the formation of secondary roots (data not shown). β-Escin had no effect on stem elongation of mungbean treated in the light 72 HAT, but epicotyl and root elongation were significantly reduced compared to the water control (Fig. 3).

Soil and Growth Chamber Tests. We examined the effects of β-escin (88 and 350 μmol/kg soil) on emergence and growth of four species in the growth chamber. Both

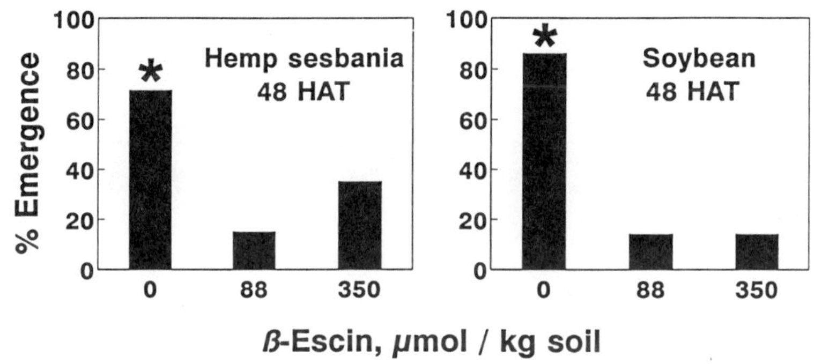

Figure 4. Effects of β-escin on emergence of two legume dicots, 48 HAT. Asterisk (*) denotes treatment is significantly different from other treatments at the 95% level.

concentrations significantly inhibited seedling emergence of the two dicotyledons, hemp sesbania and soybean (Fig. 4). Hemp sesbania shoot fresh weight was significantly reduced by both concentrations of β-escin, but root fresh weight was not significantly altered 13 DAT (Table 1). Nodulation (number/root) was also not altered by either β-escin

Table 1. Effects of β-escin on hemp sesbania shoot and root fresh weight and nodulation, 13 DAT.[1]

β-Escin, μmol/kg soil	Shoot	Root	Nodules
	fresh weight, mg/organ[1]		#/root
0	267 a	79 a	6.8 a
88	240 b	78 a	6.0 a
350	230 b	78 a	5.3 a

[1] Means followed by the same letter within a column do not differ at the 95% level.

concentration. Although there was a trend toward a reduction in soybean shoot fresh weight caused by β-escin, it was not significant at 12 DAT (Table 2). The lower concentration of β-escin significantly reduced soybean root fresh weight accumulation, but there was no significant difference between the high concentration and the untreated plants. High concentrations of some glycosides of medicagenic acid and hederagenin were inhibitory to wheat coleoptile growth, but low concentrations of these compounds accelerated growth (55, 63). Although there was a trend toward nodulation reduction by β-escin at 88 and 350 μmol/kg soil in both hemp sesbania and soybean, the only significant reduction in these relatively short-term soil tests was in soybean at the 350 μmol/kg concentration. Longer time courses to assess effects on nodulation would be useful. In other related work, we have found that 25 μM β-escin could reduce growth of *Rhizobium meliloti* and *Bradyrhizobium japonicum* in culture (66).

Table 2. Effects of β-escin on soybean shoot and root fresh weight and nodulation, 12 DAT.[1]

β-Escin, μmol/kg soil	Shoot	Root	Nodules
	fresh weight, mg/organ		#/root
0	966 a	389 a	19.3 a
88	946 a	229 b	18.6 a
350	828 a	395 a	10.0 b

[1] Means followed by the same letter within a column do not differ at the 95% level.

β-Escin at these two concentrations also significantly reduced seedling emergence of the two monocots, barnyardgrass and wheat, 48 HAT (Fig. 5). Barnyardgrass was

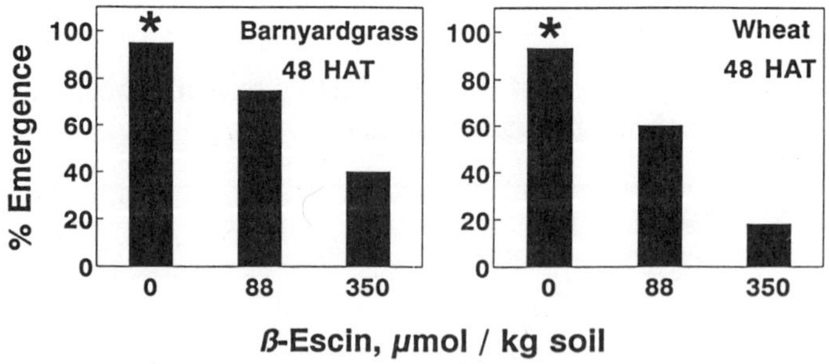

Figure 5. Effects of β-escin on emergence of two monocots 48 HAT. Asterisk (*) denotes that treatment is significantly different from other treatments at the 95% level.

extremely sensitive to β-escin, as evidenced by dramatic reductions in shoot and root fresh weight accumulations 11 DAT (Table 3). Total fresh weight of plants treated with the high concentration of β-escin was only 10% of the untreated plant fresh weight. Both concentrations of β-escin reduced shoot fresh weight accumulation 8 DAT (Table 4). The wheat root plus kernel husk fresh weight was significantly reduced only at the higher β-escin concentration. However, the lack of growth was inversely proportional to the weight of the unmobilized seed storage, and when the kernel husk and unmobilized seed material was removed, it revealed that root fresh weight was also reduced by the lower β-escin concentration. Waller et al. (64) found that the allelopathic activity of root saponins from alfalfa was cheat > barnyardgrass > pigweed > coffeeweed (hemp sesbania) > dandelion ≈ wheat.

Table 3. Effects of β-escin on barnyardgrass root and shoot fresh weight, 11 DAT.[1]

β-Escin, μmol/kg soil	Shoot	Root	Shoot/Root Ratio
0	191.0 a	87.5 a	2.27 a
88	50.4 b	32.5 b	1.55 ab
350	13.6 c	14.6 b	1.35 b

[1] Means followed by the same letter within a column do not differ at the 95% level.

Table 4. Effects of β-escin on wheat shoot and root fresh weight, 8 DAT.[1]

β-Escin, μmol/kg soil	Shoot	Root w/ kernel husk	Root w/o kernel husk
0	117 a	109 a	89 a
88	71 b	96 ab	69 b
350	54 b	85 b	69 b

[1] Means followed by the same letter within a column do not differ at the 95% level.

We observed that β-escin treatment at the higher concentration (350 μmol/kg soil) caused a crusting effect of the soil in the Conetainers®, an effect that was continuous throughout the time course of the emergence and growth tests. Soil texture plays a role in the expression of allelopathic effects of alfalfa saponins (42). The greatest inhibitory effects of saponins on wheat growth were found in lighter soils, and heavier soils reduced activity. Saponin degradation was also more rapid in heavy soils. Other studies on soil texture have shown similar results with other allelochemicals (6, 48). Soil surface watering often resulted in formation of a hard surface layer with reduced oxygen permeability (42). Heavier soils were capable of binding greater amounts of saponins than lighter soils (42). This increased binding in heavier soils presumably lowers the effective saponin concentration available to plant tissues. Thus, such combined interactions may have contributed to the reduced emergence and growth in our studies.

We also observed some browning of roots and increased anthocyanin accumulation in barnyardgrass treated with β-escin. These effects may be due to stress-elevated phenylalanine ammonia lyase and other enzyme activities of the phenylpropanoid pathway, resulting in increased anthocyanin and hydroxyphenolic compound production (11). The presence of alfalfa saponins in the growth medium also caused browning of root tips, followed by decay of the root system (41).

Greenhouse Tests. Foliar application of 1.0 mM saponins or sapogenins produced no visual effects of phytotoxicity on any of ten species evaluated. Likewise, no significant effect on shoot biomass was observed [data for only four species (barnyardgrass, corn, hemp sesbania, and soybean) shown in Table 5; data for cotton, johnsongrass, mungbean, pigweed, sicklepod, and velvetleaf not shown].

Table 5. Effects of foliarly applied saponins and sapogenins on fresh weight accumulation of greenhouse-grown plants.[1]

	Barnyardgrass	Corn	Hemp sesbania	Soybean
	Shoot fresh weight, % change relative to untreated control			
Betulin	6.5	-8.5	-10.2	-3.0
β-Escin	3.5	-1.3	-2.2	0.4
β-Glycyrrhetic acid	9.3	3.4	-6.9	0.1
Hecogenin	33.5	-11.0	-15.8	-0.1
Oleandrin	3.6	3.4	-17.2	-0.9
Oleanolic acid	20.0	-5.5	-14.4	1.8

[1] Aqueous foliar spray (1.0 mM, 5% EtOH, 0.25% nonionic surfactant) at 186 L/ha; 7 replicates/species; treated at 10 days old, evaluated 9 DAT.

Physiological/Biochemical Effects of Saponins and Sapogenins

Electrolyte Leakage. Measurement of cellular leakage caused by membrane permeability effects has been utilized to determine various types of stress in plants, including that induced by phytotoxins (15). Although measurement of various specific compounds that are leaked or excreted can be assessed, conductivity measurement of total electrolytes is generally more advantageous. Leaf discs of six weed and crop species were tested with several saponins or sapogenins (Fig. 6). β-Escin was the most potent compound, and jimsonweed (*Datura stramonium* L.) was the most sensitive species to these compounds. β-Escin enhanced electrolyte leakage in all species except hemp sesbania; the most dramatic increases in conductivity were in wheat and jimsonweed leaf discs. Glycyrrhetic acid increased leakage in jimsonweed and sicklepod, and hecogenin caused increases only in jimsonweed. Other properties of β-escin that demonstrate its alteration of membranes are its hemolytic action on blood cells (10) and its surfactant characteristics (21). Hemolysis by saponins is based on their action that causes swelling and rupture of erythrocytes (2). No significant correlation was found among the allelopathic, hemolytic, and antifungal activities of several alfalfa saponins (41). Thus, the hemolytic index of a particular saponin, or its activity in the sensitive *Trichoderma* bioassay, may not be useful in predicting or identifying compounds that may have allelopathic potential. This further points out that the mechanism(s) of biological activity of these compounds is not fully understood.

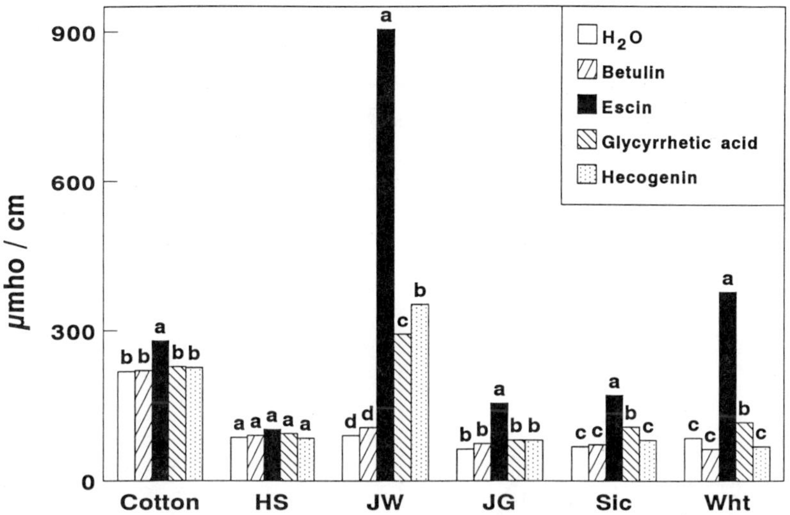

Figure 6. Effects of several saponins/sapogenins on electrolyte leakage from leaf discs of six weed/crop species. Different letters within a species denote significant differences at the 95% level. HS, hemp sesbania; JW, jimsonweed; JG, johnsongrass; Sic, sicklepod; Wht, wheat.

Greening Experiments. Of the four compounds tested, β-escin was the most potent in preventing greening of etiolated tissues (Fig. 7); i.e., it significantly reduced chlorophyll

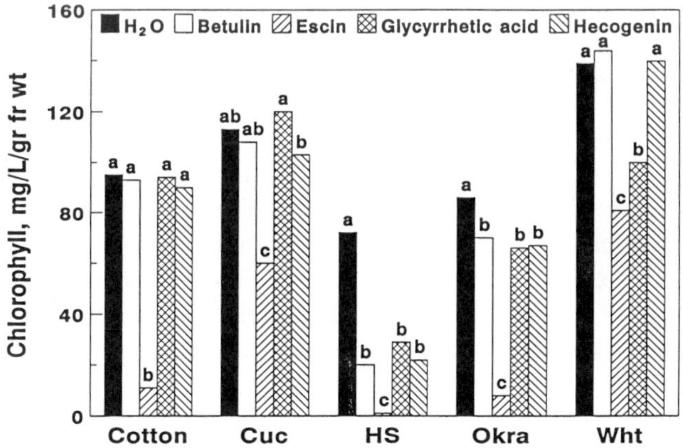

Figure 7. Effects of saponin/sapogenin treatment (10^{-3}M) on greening of etiolated, hydroponically grown seedlings. Different letters within a species denote significant differences at the 95% level. Cuc, cucumber; HS, hemp sesbania; Wht, wheat.

accumulation in all species tested in a range of about 40 to 98% compared to control values.

Hecogenin reduced greening [cucumber, 10%; hemp sesbania, 70%; okra (*Hibiscus esculentus* L.), 22%]. Glycyrrhetic acid inhibited greening in hemp sesbania (22%), okra (23%) and wheat (30%). Betulin-induced greening inhibition was noted in hemp sesbania (72%) and okra (20%). Effects of *β*-escin concentration were determined on seedling cotyledons of two weed species (Fig. 8). At 1 mM, greening was considerably reduced in sicklepod and hemp sesbania cotyledons. Lesser concentrations (0.1 and 0.01 mM) were ineffective in preventing chlorophyll accumulation, except for hemp sesbania treated with 0.1 mM *β*-escin. The relationship between saponin concentration and prolamellar body structure of etioplasts from oat during greening and re-etiolating, and in etioplasts of barley and pea indicates that saponins play a role in prolamellar body structure integrity (27, 31). The mechanism by which externally applied saponins inhibit greening in these plant species is not known.

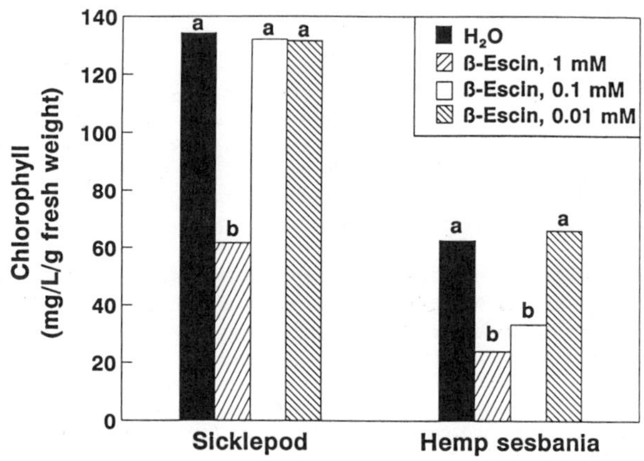

Figure 8. Effects of *β*-escin concentration on greening of etiolated sicklepod and hemp sesbania cotyledons, 48 HAT. Different letters within a species denote significant differences at the 95% level.

Enzyme Tests. Three saponins or sapogenin compounds were examined for possible effects on extractable hydrolytic enzyme activity in three plant species after chemical exposure via root-feeding (Table 6). FDA hydrolysis was increased by hecogenin in all three species and by *β*-escin in mungbean. Other effects of FDA hydrolysis in these species were not significantly different than in untreated plant tissues. Pea pod endocarp tissue treated with saponins exhibited a temporary reduction in cell viability measured as decreased esterase activity by using FDA as a substrate stain (12). *β*-Escin caused significant FDA hydrolysis reductions of 10 to 60% in several diverse bacterial strains, but FDA hydrolytic activity was stimulated in two *Pseudomonad* strains (66).

Esterase activity with PNPB as substrate was unaffected by these three compounds in mungbean and sicklepod, but was lowered by *β*-escin and glycyrrhetic acid in hemp sesbania. PNPGP activity was decreased in mungbean by *β*-escin and hecogenin, but increased by all three compounds in sicklepod. None of the compounds altered enzyme activities of PNPB in mungbean and sicklepod, and PNPGP was affected only in mungbean The differential effects of these compounds might be caused by differences in absorption, translocation, and metabolism in these plants. We are unaware of any definitive reports on these parameters related to external application of saponins to plant tissues. The fact that

Table 6. Effects of saponins or sapogenins on hydrolytic enzyme activity of three plant species on three substrates.[1]

	Treatment			
Species/substrate	None	β-Escin	Glycyrrhetic acid	Hecogenin
Hemp sesbania				
FDA	1813 b	1765 b	1704 b	1988 a
PNPB	17.92 a	15.86 b	15.58 b	18.99 a
PNPGP	1.53 a	1.58 a	1.58 a	1.52 a
Mungbean				
FDA	428 b	577 a	421 b	482 ab
PNPB	18.6 a	17.6 a	18.1 a	18.5 a
PNPGP	3.50 a	2.38 c	3.14 ab	3.10 b
Sicklepod				
FDA	2123 b	2338 ab	2209 ab	2394 a
PNPB	63.7 a	70.1 a	65.5 a	66.4 a
PNPGP	2.58 a	3.26 a	2.88 a	3.05 a

[1] Means followed by the same letter within a column do not differ at the 95% level.

β-escin is a membrane effector as seen from our studies on electrolyte leakage, our bacterial studies (66), and other work (24) suggests that such effects could be responsible for increases or decreases in some of these enzyme activities. Other saponins also alter membranes (3, 18), and several saponins have been reported to affect various enzyme activities. Lipid peroxidation was inhibited by soybean saponins in liver microsomes via possible membrane binding or interaction (36). Soybean saponins also have other antioxidative properties (39). Saponins from ginseng (*Panax quinquefolium* L.) were found to be potent inhibitors of bovine calmodulin-dependent phosphodiesterases, but had no effect on one phosphodiesterase isozyme (53). We are unaware of reports of saponin effects on calmodulin-regulated enzymes in plants. However, since calmodulin is the major Ca^{+2}-receptor protein in plant cells, many calcium-regulated enzymes and/or proteins in plants might be sites of action of certain saponins. Saponins or saponin-like substances inhibited α-galactosidase production in fenugreek (*Trigonella foenum-graecum* L.) seed endosperm (67). One ginseng saponin enhanced acetate incorporation into serum and liver cholesterol of rats, which correlated with increased β-hydroxy-β-methylglutaryl-CoA (HMG-CoA) reductase, a key enzyme in cholesterol biosynthesis (19, 51). This enzyme also occurs in plants, and its inhibition by at least two natural products has been examined as a potential herbicide target site (47). A ginseng saponin could inhibit (*Gypsophila paniculata* cells) or stimulate (*Saponaria officialis* cells) 2,3-oxidosqualene-amyrin cyclase activity, but had no effect on 2,3-oxidosqualene-cycloartenol cyclase activity (22), indicating that 2,3-oxidosqualene-cyclases are regulating steps in the isoprenoid pathway directing biosynthetic flow to either tetracyclic or pentacyclic triterpenes. Overall, there have been few studies of saponins interactions with plant enzymes.

Our studies on possible phytotoxic effects of a variety of saponin/sapogenin compounds have shown that, although several compounds affected plant growth parameters, β-escin was generally the most potent. Thus, we extended the testing to include more tests using β-escin. Our results also suggest that saponins may be more active than their related aglycones, the sapogenins. However, replacement of glycosides of medicagenic acid with sodium to form the sodium salt resulted in a sapogenin with high allelopathic activity toward shoot growth of wheat seedlings and in the browning of wheat roots (42). None of the compounds we tested caused significant effects to a variety of plants when applied as foliar sprays. This suggests that these compounds are either not passing through cuticular barriers, or the compounds are not phytotoxic to foliar tissues at 10^{-3} M. The compounds tested were also not effective seed germination inhibitors, as also reported for alfalfa saponins (42). However, early growth of some species was affected in petri dish tests and in soil treatments by some compounds (especially β-escin). These early growth effects suggest that these and similar compounds may exert effects in soils that can alter growth of weeds and/or crop plants and affect plant competition and ultimately influence crop productivity. Similar effects have been shown to be caused by saponins and related compounds from alfalfa (42, 63, 64). Furthermore, some alfalfa saponins interacted with soil matrices to form complexes that could physically impede growth and emergence or possibly reduce growth by restricting oxygen availability to young developing/emerging seedlings (42). We also observed this soil interaction in our β-escin tests in soils. This phenomenon deserves further attention in order to elucidate the mechanism(s) responsible and to assess the role of such interactions in natural conditions of crop production.

Although most research on allelopathic phenomena of saponins in plants has dealt with alfalfa root and foliar saponins, it is clear that a wide variety of plants produce a multitude of saponins which have diverse chemistries and thus diverse properties and biological activities. The identities of many of these compounds and their distribution, relative concentrations (in plants and soils) and activities have not been ascertained. Even among alfalfa cultivars, there are large differences in the quality and quantity of foliarly produced saponins (45). Under natural conditions, these and related compounds are produced by similar pathways and exist together in the plant and in soils when leached from living plant tissue or when released from dead plant residues. The level of some saponins in soils can be high, and their rate of decomposition can be slow, thus causing reduced growth in subsequent crops (29). Increasing numbers of farmers are using cover crops in combination with reduced- or no-tillage in attempts to reduce soil erosion, increase weed control, reduce herbicide usage, and improve soil quality. Cover crops and their residues can alter microbial populations and enzyme activities responsible for degradation of herbicides in soils (61). These alternations may be caused by the interactions of saponins and other allelochemicals. Residues for typical herbicide-desiccated cover crops can exceed 300 g/m^2 in no-till systems (56). Thus, cover crops can be a source of saponins during their life cycle and when they die and decay. Furthermore, many weeds have been shown to be allelopathic to crops (49), and these sources also need to be examined for the possible roles of saponins/sapogenins in weed-crop competition. Overall, the agronomic potential of saponins as herbicides appears to be quite low, but their allelopathic influence might be important to crop productivity, plant competition, herbicide metabolism, and other factors.

REFERENCES

1. Arneson, P.A. and R.D. Durbin. Sensitivity of fungi to alpha-tomatine. Phytopathology 58:536 (1968).
2. Assa, Y., S. Shang, B. Gestebner, Y. Tencer, Y. Birk, and A. Bondi. Interaction of alfalfa saponin with components of the erythrocyte membrane in hemolysis. *Biochim. Biophys. Acta* 307:83 (1973).

3. Bangham, A.D. and R.W. Horne. Action of saponin on biological cell membranes. *Nature* 196:952 (1962).
4. Barnes, J.D., L. Balaguer, E. Manriques, S. Elvira, and A.W. Davison. A reappraisal of the use of DMSO for the extraction and determination of chlorophylls-A and chlorophylls-B in lichens and higher plants. *Environ. Exptl. Bot.* 32:85 (1992).
5. Basu, N. and R.P. Rastogi. Triterpenoid saponins and sapogenins. *Phytochemistry* 6:1249 (1967).
6. Bhowmik, P.C. and J.D. Doll. Corn and soybean response to allelopathic effects of weed and crop residues. *Agron. J.* 74:601 (1982).
7. Bowyer, P, B.R. Clark, P. Lunness, M.J. Daniels, and A.E. Osbourn. Host range of a plant pathogenic fungus determined by a saponin detoxifying enzyme. *Science* 267:371 (1995).
8. Bradford, M.M. A rapid and sensitive method for the quantitation of microgram quantities of protein utilizing the principle of protein-dye binding. *Anal. Biochem.* 72:248 (1976).
9. Bradow, J.M. "Germination Regulation by *Amaranthus palmeri* and *Ambrosia artemisiifolia,*" pp. 285-300 *in* The Chemistry of Allelopathy. Biochemical Interactions Among Plants. ACS Symp. Ser. 268. A.C. Thompson (Ed.) Am. Chem. Soc. Washington, DC (1985).
10. Budavari, S. (Ed.) "Escin" #3644, p. 58, *in* The Merck Index, 11th Edition. Merck and Co., Inc. Rahway, NJ (1989).
11. Camm, E.L. and G.H.N. Towers. Phenylalanine ammonia-lyase. *Phytochemistry* 12:961 (1973).
12. Christian, D.A. and L.A. Hadwiger. Pea saponins in the pea-*Fusarium solani* interaction. *Exptl. Mycology* 13:419 (1989).
13. Défago, G. and H. Kern. Induction of *Fusarium solani* mutants insensitive to tomatine; their pathogenicity and aggressiveness to tomato fruits and pea plants. *Physiol. Plant Pathol.* 22:29 (1983).
14. Duke, S.O. "Plant Terpenoids as Pesticides" pp. 269-295 *in* Handbook of Natural Toxins, Vol. 6, Toxicology of Plant and Fungal Compounds. R.F. Keeler and A.T.Tu (Eds.) Marcel Dekker, Inc., New York (1991).
15. Duke, S.O. and W.H. Kenyon. "Peroxidizing Activity Determined by Cellular Leakage," pp. 61-66 *in* Target Assays for Modern Herbicides and Related Phytotoxic Compounds. P. Böger and G. Sandmann (Eds.) Lewis Publishers, Boca Raton, FL (1993).
16. Fischer, N.H. and L. Ouijano. "Allelopathic Agents from Common Weeds: *Amaranthus palmeri, Ambrosia artemisiifolia,* and Related Weeds," pp. 133-148 *in* The Chemistry of Allelopathy. Biochemical Interactions Among Plants. ACS Symp. Ser. 268. A.C. Thompson (Ed.) Am. Chem. Soc. Washington, DC (1985).
17. Fischer, N.H. "Plant Terpenoids as Allelopathic Agents," pp. 377-398 *in* Ecological Chemistry and Biochemistry of Plant Terpenoids. J.B. Harborne and F.A. Tomas-Barberan (Eds.) Clarendon Press, New York (1991).
18. Glauret, A.M., J.T. Dingle, and J.A. Lucy. Action of saponin on biological cell membranes. *Nature* 196:953 (1962).
19. Gommori, K., F. Miyamoto, Y. Shibata, T. Higashi, S. Sanada, and J. Shoji. Effects of ginseng saponins on cholesterol metabolism. 2. Effects of ginsenosides on cholesterol synthesis by liver slices. *Chem. Pharm. Bull. Jpn.* 24:2985 (1976).
20. Guilbault, G.G. and D.N. Kramer. Fluorometric determination of lipase, acylase, alpha- and gamma-chymotrypsin and inhibitors of these enzymes. *Anal. Chem.* 36:409 (1964).
21. Harborne, J.B. and H. Baxter (Eds.) "Triterpenoid Saponins," pp. 670-688 *in* Phytochemical Dictionary, A Handbook of Bioactive Compounds from Plants. Taylor and Francis Press, London (1993).
22. Henry, M. , A. Rahier, and M. Taton. Effect of gypsogenin 3,O-glucuronide pretreatment of *Gypsophila paniculata* and *Saponaria officinalis* cell suspension cultures on the activities of microsomal 2,3-oxidosqualene cycloartenol and amyrin cyclases. *Phytochemistry* 31:3855 (1992).
23. Hiscox, J.D. and G.F. Israelstam. A method for the extraction of chlorophyll from leaf tissue without maceration. *Can. J. Bot.* 57:1332 (1979).
24. Hostettmann, K. and A. Marston. "Triterpene Saponins - Pharmacological and Biological Properties," pp. 232-286 *in* Saponins. Cambridge University Press, New York (1995).
25. Jurzysta, M. Effect of saponins isolated from seeds of lucerene on germination and growth of cereal seedlings. *Zesz. Nauk U.M.K. Torum, Ser. Biol.* 13:253 (Polish) (1970).
26. Jurzysta, M. Chemical characteristics of saponins from *Medicago lupulina* seeds. *Proc. 11th Polish Biochem. Soc.* , p. 18. Bialystok, Poland (1973).
27. Kesselmeir, J. and H.G. Ruppel. Relations between saponin concentration and prolamellar body structures in etioplasts of *Avena sativa* during greening and re-etiolating and in etioplasts of *Hordeum vulgare* and *Pisum sativum*. *Z. Pflanzenphysiol.*, Bd. 93:171 (1979).
28. Kudou, S., M. Tonomura, C. Tsukamoto, M. Shimoyamada, M. Uchida, and K. Okubo. Isolation and structural elucidation of the major genuine soybean saponin. *Biosci. Biotechnol. Biochem.* 56:142 (1992).

29. Leshem, Y. and J. Levin. The effect of growing alfalfa on subsequent cotton plant development and on nitrate formation in peat soil. *Plant Soil* 50:323 (1978).
30. Luning, H.U., B.G. Waiyaki, and E. Schlösser. Role of saponins in antifungal resistance. 8. Interactions of *Avena sativa - Fusarium avenaccum*. *Phytopathol. Z.* 92:338 (1978).
31. Lütz, C. and S. Klein. Biochemical and cytological observations on chloroplast development. VI. Chlorophylls and saponins in prolamellar bodies and prothylakoids separated from etioplasts of etiolated *Avena sativum* L. leaves. *Z. Pflanzenphysiol. Bd.* 95:227 (1979).
32. Mahato, S.B., A.M. Ganguly, and N.P. Sahu. Steroid saponins. *Phytochemistry* 21:959 (1982).
33. Marston, A., F. Gafner, S.F. Dossaji, and K. Hostettmann. Fungicidal and molluscicidal saponins from *Dolichos kilimandscharicus*. *Phytochemistry* 27:1325 (1988).
34. Medzon, E.L. and M.L. Brady. Direct measurement of acetylesterase in living protist cells. *J. Bacteriol.* 97:402 (1969).
35. Mishustin, B.N. and A.N. Naumova. Secretion of toxic substances by alfalfa and their effect on cotton and soil microflora. *Akad. Nauk USSR Izvestiya, Ser. Biol.* 6:3 (Russian) (1955).
36. Nishida, K., Y. Ohta, Y. Araki, M. Ito, Y. Nagamura, and I. Ishiguro. Inhibitory effects of "Group A Saponin" and "Group B Saponin" fractions from soybean seed hypocotyls on radical-initiated lipid peroxidation in mouse liver microsomes. *J. Clin. Biochem. Nutr.* 15:175 (1993).
37. Nisius, A. The stromacentre in *Avena* plastids: an aggregation of β-glucosidase responsible for the activation of oat-leaf saponins. *Planta* 173:474 (1988).
38. Nord, E.C. and G.R. Van Atta. Saponin -- a seed germination inhibitor. *Forest Sci.* 6:350 (1960).
39. Ohminami, K., H. Okuda, T. Hami, I. Kitagawa, M. Yoshikawa, S. Arichi, and T. Hayashi. Effect of soyasaponins on liver injury. *J. Jpn. Soc. Nutr. Food Sci.* 34:105 (1981).
40. Okhani, K., S. Mavi, and K. Hostettmann. Mulluscicidal and antifungal triterpenoid saponins from *Rapanea melanophloeos* leaves. *Phytochemistry* 33:83 (1993).
41. Oleszek, W. Allelopathic potentials of alfalfa (*Medicago sativa*) saponins: Their relation to antifungal and hemolytic activities *J. Chem. Ecol.* 19:1063 (1993).
42. Oleszek, W. and M. Jurzysta. The allelopathic potential of alfalfa root medicagenic acid glycosides and their fate in soil environments. *Plant Soil* 98:67 (1987).
43. Osbourn, A.E., B.R. Clarke, J. Dow, and M. Daniels. Partial characterization of avenacinase from *Gaeumannomyces graminis* var. *avenae*. *Physiol. Mol. Plant Pathol.* 38:301 (1991).
44. Pedersen, M.W. Effect of alfalfa saponin on cottonseed germination. *Agron. J.* 57:516 (1965).
45. Pedersen, M.W., B. Berrang, M.E. Wall, and K.H. Davis, Jr. Modification of saponin characteristics of alfalfa by selection. *Crop Sci.* 13:731 (1973).
46. Pegg, G.F. and G.F. Woodward. Synthesis and metabolism of α-tomatine in tomato isolines in relation to resistance to *Verticillium albo-atrum*. *Physiol. Mol. Plant Pathol.* 28:187 (1986).
47. Rademacher, W. "Gibberellins: Metabolic Pathways and Inhibitors of Biosynthesis," pp. 127-145 *in* Target Sites of Herbicide Action. P. Böger and G. Sandmann (Eds.) CRC Press, Boca Raton, FL, (1989).
48. Rice, E.L. Allelopathy -- An Update. *Bot. Rev.* 45:15 (1979).
49. Rice, E.L. "Allelopathic Effects of Weeds on Crop Plants" pp. 38-77 *in* Biological Control of Weeds and Plant Diseases. Univ. of Oklahoma Press, Norman (1995).
50. Roy, S., A.K. Dutta, and D.P. Chakraborty. Amasterol, an ecdysone precursor and a growth inhibitor from *Amaranthus viridis*. *Phytochemistry* 21:2417 (1982).
51. Sakakibara, K., Y. Shibata, T. Higashi, S. Sanada, and J. Shoji. Effect of ginseng saponins on cholesterol metabolism. 1. Level and synthesis of serum and liver cholesterol in rats treated with ginsenosides. *Chem. Pharm. Bull. Jpn.* 23:1009 (1975).
52. Schönbeck, F. and E. Schlösser. "Preformed Substances as Potential Protectants," pp. 653-678 *in* Encyl. Plant Physiol. New Ser. R. Heitefuss and P.H. Williams (Eds.), Springer-Verlag, Berlin (1976).
53. Sharma, S.K. and J. Kalra. Ginsenosides are potent and selective inhibitors of some calmodulin-dependent phosphodiesterase isozymes. *Biochemistry* 32:4975 (1993).
54. Singh, S.B. and R. S. Thakur. Recent advances in the chemistry of steroidal saponins and their genins. *J. Sci. Ind. Res.* 42:319 (1983).
55. Tarikov, G., A.E. Timbekova, N.K. Abubakirov, and R.K. Koblov. Growth regulating activiy of triterpene glycosides isolated from alfalfa (*Medicago sativa* L.). *Uzbeksk. Biol. Zhur.* 6:24 (1988).
56. Teasdale, J.R., C.E. Beste, and W.E. Potts. Response of weeds to tillage and crop cover residues. *Weed Sci.* 39:195 (1991).
57. Tschesche, R. "Advances in the Chemistry of Antibiotic Substances from Higher Plants," pp. 274-289 *in* Pharmacognosy and Phytochemistry, H. Wagner and L. Horhammer (Eds.) Springer-Verlag, Berlin (1971).
58. Tsurumi, S. and Y. Tsujino. Chromosaponin I stimulates the growth of lettuce roots. *Physiol. Plant.* 93:785 (1995).

59. Tsurumi, S., T. Takagi, and T. Hashimoto. A γ-pyronyl-triterpenoid saponin from *Pisum sativum*. *Phytochemistry* 31:2435 (1992).
60. Varshney, I.P. and M.O. Farooq. Etudé l'influence d'une nouvelle saponine d'*Albizzia lebbek*, Benth. sur la germination et la croissance des graines de pois chicke (*Cicer arietinum* L.) et d'orge (*Hordeum vulgare* L.) *Bull. Soc. Chem. Biol.* 35:827 (1953).
61. Wagner, S.C., R.M. Zablotowicz, M.A. Locke, R.J. Smeda, and C.T. Bryson. Influence of herbicide-desiccated cover crops on biological soil quality in the Mississippi Delta. *Proc. 1995 Southern Conservation Tillage Conference for Sustainable Agriculture. MAFES Special Bulletin* 88-7. Mississsissippi State University (1995).
62. Waiyaki, B.G. and E. Schlösser. Role of saponins in antifungal resistance. 9. Species-specific inactivation of avenecin by *Fusarium avenaceum*. *Phytopathol. Z.* 92:346 (1978).
63. Waller, G.R., M. Jurzysta, and R.L. Z. Thorne. Allelopathic activity of root saponins from alfalfa (*Medicago sativa* L.) on weeds and wheat. *Bot. Bull. Acad. Sin.* 34:1 (1993).
64. Waller, G.R., M. Jurzysta, and R.L. Z. Thorne. Root saponins from alfalfa (*Medicago sativa* L.) and their allelopathic activity on weeds and wheat. *Allelopathy J.* 2:21 (1995).
65. Wyman-Simpson, C.L., G.R. Waller, M. Jurzysta, J.K. McPherson, and C.C. Young. Biological activity and chemical isolation of root saponins of six cultivars of alfalfa (*Medicago sativa* L.). *Plant Soil* 135:83 (1991).
66. Zablotowicz, R.M., R.E. Hoagland, and S.C. Wagner. "Effect of Saponins on the Growth and Activity of Rhizosphere Bacteria," this volume (1996).
67. Zambou, K., C.G. Spyropoulos, I. Chinou, and F. Kontos.Saponin-like substances inhibit α-galactosidase production in the endosperm of Fenugreek seeds. *Planta* 189:207 (1993).
68. Zimmer, D.E., M.W. Pedersen, and C.F. McGuire. A bioassay for alfalfa saponins using the fungus, *Trichoderma viride* Pers. ex Fr. *Crop Sci.* 7:223 (1967).

REGULATORY EFFECTS OF SAPONINS IN THE PATHOGENESIS OF ROOT ROTS IN CEREAL CROPS

P.K. Kintia and G.A. Lupashku

Institute of Genetics
Academy of Sciences of Moldova
Chisinau, 2002 Moldova

INTRODUCTION

Fusarium diseases caused by the fungi of the *Fusarium* Link. genus are becoming a serious problem for many agronomic crops, especially cereals, as they induce root rots of plants, shallowness of spikes, moldiness of seed, etc.[1-3]. Though these diseases have been investigated for more than 70 years the problem still remains very acute for a number of reasons: great adaptive potential, high variability and liability of the pathogen genetic apparatus, weak adaptive capacity of crop plants to unfavorable environmental factors, and as a consequence, predisposition to the disease.

Prevalence and increased harmfulness of fusarioses for wheat and triticale, which during the last 10-15 years have repeatedly reached epiphytotic scales in many regions of the world, lead to economic losses and deterioration of alimentary and technological properties of grain; this is hazardous for human health owing to increased content of mycotoxins[4-6].

Mankind has already realized the danger of unilateral application of pesticides to protect plants from diseases, both from the biological viewpoint and involuntary stimulation of new, more aggressive strains of pests.

Summing up the above considerations, new methods to protect plants from diseases with less unfavorable consequence should be employed. It is believed that along with conventional methods, this problem might be solved through involvement of natural protective capacity based on enhancement of plant immunity. Naturally occurring biologically active substances, for example steroidal glycosides, widespread in higher plants, including cereals, could be utilized for this purpose[7,8].

It has been known for a long time that steroidal glycosides can either stimulate or inhibit growth and development of living organisms, including cultivated plants[9,10]. However, information on the nature and mechanisms of influence of steroidal glycoside on plants, including phytopathosystem components, is quite scarce. In this context, data on several steroidal glycosides, tomatin, solanin, chaconin, and avenacin, should be mentioned. As to the mechanisms producing antibiotic activity, saponins are supposed to denature sterols and liquid crystals in cell plasmalema, forming insoluble components and,

Saponins Used in Food and Agriculture
Edited by Waller and Yamasaki, Plenum Press, New York, 1996

as a results, causing cell content release outside[11]. Interestingly, avenacin in oats inhibits the growth of many fungi and bacteria, but is not able to inhibit the growth of *Ophiobolus graminis f. sp. avenae* (one of the agents of root rots) since the pathogen contains avenakinase, an enzyme acting on avenacin and breaking it down, splitting off the final sugar, and thereby inactivates the saponin[11].

Our previous investigations revealed that steroidal glycosides show phytohormonal properties that are concentration dependent[12,13] such as those that are capable of stimulating wheat seedling elongation by 30-60%, photosynthetic pigment accumulation in isolated cucumber cotyledons by 90%, and the formation of secondary roots in stem petioles of beans by 40%, i.e. they possess expressed gibberellin-auxin-, and cytokinin-like properties. It should be noted that phytohormonal effects of steroidal glycosides, as well as their directionality (stimulation/inhibition), correlate directly with the concentration of substance, glycosidic series (furo- or spiro-), and with the reaction of the individual plant.

Our investigations aim at revealing some regulatory mechanisms of steroidal glycosides in the pathogenesis of *Fusarium* root rots in wheat and triticale.

MATERIAL AND METHODS

Pure cultures of the fungi *Fusarium graminearum* Schwabe and *F. oxysporum* (Schlecht) Snyd. et Hans isolated from infected plants of wheat and triticale were used.

Naturally occurring steroidal glycosides isolated following silica gel chromatography of the methanolic extracts and recrystallisation were used in the experiments, among them a sulfated rhamnoglucopyranoside of the furostan series isolated from the above-ground part of *Tribulus terrestris* L. (98% purity)[14], a sulfated bioside of diosgenin of the spirostan series extracted from the above-ground part of *T. terrestris* (100% purity)[14], a spirostanol saponin of diosgenin isolated from *Funkia ovata* Spreng. leaves (95% purity)[15], Moldstim of the furostan series produced from *Capsicum annuum* L. (capsicoside) subsp. seed (90% purity)[15], and Pavstim of the furostan series extracted from the above-ground part of *Digitalis purpurea* L. (purpureaside) (95% purity)[17].

Wheat and triticale seeds were soaked in the cultural liquid (CL) of the fungus (18 h) or sown in preinfected sand (25 g of inoculum per 600 g of sand) to receive infectious material. In order to establish the influence of steroid glycosides on plants, some seeds were kept in the aqueous solutions for 18 h.

Seed emergency was determined by the 7th day (each treatment contained 30 seeds, fourfold replication).

The content of chlorophyll a, chlorophyll b, and total carotenoids in leaves was measured in 80% acetone extract on a "Specord M-40" spectrophotometer.

Cellular membrane (CM) permeability was determined on a "M4100/IT" megometer (100 V). An equal number of segments of the main embryonic root, stem basal part, and the first leaf middle part weighing 1 g was kept in doubly distilled water for 1 h, the contents being mixed 3-4 times. Afterwards, electric resistance of the solutions of electrolytes (R) removed from the plant structures was measured on a megometer in megoohms (moM). The estimates of cell membrane permeability (L) for each treatment were calculated[18] by the formula $L = \frac{1}{R}$. The measurements were replicated 4 times.

The steroidal glycosides influence on growth and development of pure fungal cultures was estimated by measuring the linear increase of colonies and the intensity of conidial sporulation on a mash/agar medium. The latter was measured by using the Goryaev chamber[19] according to the scale we have developed: 201-250 thousand conidia/1 mm^2 - score 4; 151-200 thousand conidia/1 mm^2 - score 3; 101-150 thousand conidia/1

mm^2 - score 2; 51-100 thousand conidia/1 mm^2 - score 1; 25-50 thousand conidia/1 mm^2 - score 0.1; absence of conidia - score 0.

The experiments were replicated 4 times with 3-fold reproduction. The data were analyzed using the one factor variance method. The Student's test at the significance level of 5% was used as a criterion of the reliability between treatments.

RESULTS AND DISCUSSION

Spirostanol steroidal glycosides, sulfated rhamnoglucopyranoside (**1**), and sulfated diosgenin bioside (**2**), earlier established to possess a gibberellin-like activity on flax seedlings[20], were studied in relation to their influence on the sprouting of wheat seeds, presoaked in CL of the pathogenic isolate of the fungus *F. graminearum*. It was shown that the concentrations of 10^{-4} - 10^{-7} M increased seed emergence by 21 - 35% and 40 -48%, respectively for **1** and **2** (Fig. 1).

The increased emergence of seeds treated with the fungal toxins is apparently due to the intensification of cellular division and their elongation under the influence of saponins.

The study of the influence of *F. graminearum* CL on plasmalema permeability (PP) in different structures of wheat and triticale showed that it is more expressed in roots (Table 1). It was revealed that PP in this structure increases more under the influence of the fungus (by 1.7 and 1.3 times in wheat and triticale, respectively), and to lesser extent by the saponin (by 1.2 and 1.1 times). Combined application of the fungus and Moldstim accounted for a decrease in PP as compared with the infected plants, by 1.6 and 1.1 times in wheat and triticale, respectively. The fungus and glycoside when used separately caused, though not equally, an increase in PP. When applied in combination, they caused a decrease, as compared with the efficiency of the fungus alone. This may be explained by inhibition of Moldstim® to a significant extent of the enzymatic activity of *F. graminearum*, which could destroy the plant root tissue, thereby hindering the release of electrolytes outside the cells.

In connection with the expressed cytokinin-like activity of Pavstim, in a preliminary way we established on isolated cucumber cotyledons, the effect of its application on wheat and triticale plants infected with *F. graminearum* (Table 2). The findings show that the fungus contributed to the decrease of "a" and "b" chlorophylls and carotenoids in wheat leaves by 38, 24, and 25%, and in triticale leaves by 11, 26, and 17% , respectively. Pavstim® had a stimulating effect on the above parameters in both crops applied separately and in combination with the fungus. In the latter one, the accumulation of "a" and "b" chlorophylls and total carotenoids was by 1.7, 1.4, and 1.5 and 1.2, 1.5, and 1.2 times higher in wheat and triticale, respectively, as compared with the infected plants.

It was established *in vitro* that the spirostanol saponin diosgenin (SSD) at a concentration of 10^{-2} - 10^{-4}% possessed an inhibiting effect on the growth of *F. graminearum* pure culture and its sporulation. Table 3 shows that at the concentration of 10^{-3}% the colony diameter was by 5.7, 3.5, and 3.2 times lower as compared with the control within 3, 4, and 5 days, and sporulation was at a score level of 0.

® - Registered Trademark is used by former Soviet Union Countries.

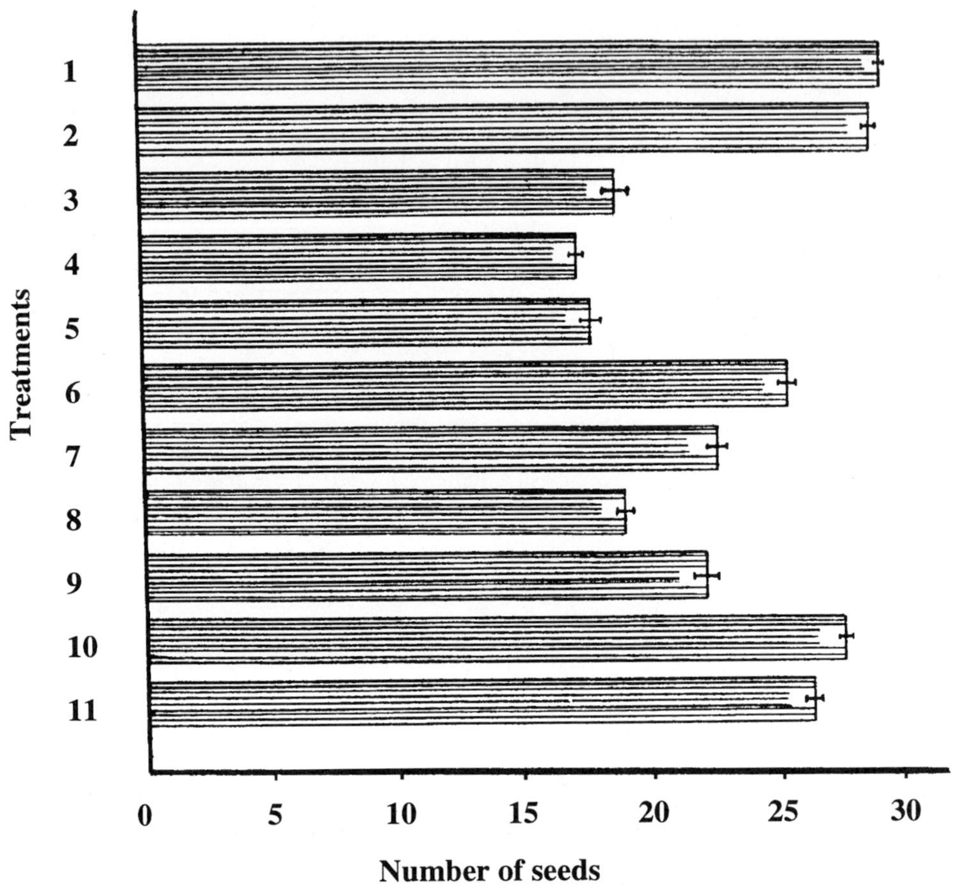

Fig. 1. The influence of steroidal saponins with a gibberellin-like activity on the germinating ability of wheat seeds treated with *F. graminearum* cultural liquid. 1 - Control (H_2O); 2- Chapek's medium; 3 - *F. graminearum* - cultural liquid; 4,5,6,7 - are sulfated rhamnoglucopyranoside at concentrations at 10^{-4}, 10^{-5}, 10^{-6}, and 10^{-7} M, respectively; 8,9,10,11 - are sulfated diosgenin bioside at concentration levels of 10^{-4}, 10^{-5}, 10^{-6}, and 10^{-7} M, respectively.

Table 1. The influence of Moldstim® on cellular membrane permeability in different structures of wheat, triticale, control, and *F. graminearum* infection.

Treatment, Conc., %	Roots, moM $X \pm m_X$	Stems, moM $X \pm m_X$	Leaves, moM $X \pm m_X$
		Wheat	
Control	2.28 ± 0.05°	1.34 ± 0.02°	0.87 ± 0.01°
F. graminearum	3.94 ± 0.16*	2.17 ± 0.07*	1.57 ± 0.06*
Moldstim, 10^{-2}	2.69 ± 0.06°	1.45 ± 0.01°*	0.96 ± 0.03°*
F. graminearum + Moldstim, 10^{-2}	2.44 ± 0.08°	1.41 ± 0.01°*	0.91 ± 0.03°
		Triticale	
Control	1.81 ± 0.04°	1.18 ± 0.01°	0.78 ± 0.01°
F. graminearum	2.30 ± 0.03*	1.42 ± 0.01	0.87 ± 0.01*
Moldstim, 10^{-2}	1.95 ± 0.04°*	1.32 ± 0.03°*	0.81 ± 0.01°*
F. graminearum + Moldstim, 10^{-2}	2.03 ± 0.02°*	1.24 ± 0.01°*	0.81 ± 0.00°*

* - the difference as related to the control is reliable at the significance level of 5%;
° - the difference as related to the *F. graminearum* treatment is reliable at the significance level of 5%.

Table 2. The influence of Pavstim® on the content of photosynthetic pigments in wheat and triticale leaves noninfected and infected with *F. graminearum.*

Treatment, Conc., %	Chlorophyll "a", mg/g $X \pm m_X$	Chlorophyll "b", mg/g $X \pm m_X$	Carotenoids mg/g $X \pm m_X$
		Wheat	
Control	2.49 ± 0.06	1.18 ± 0.03	1.37 ± 0.03
F. graminearum	1.55 ± 0.04*	0.88 ± 0.03*	1.03 ± 0.03*
Pavstim, 10^{-2}	2.74 ± 0.04°*	1.29 ± 0.02°	1.69 ± 0.08°*
F. graminearum + Pavstim, 10^{-2}	2.61 ± 0.03°	1.19 ± 0.02°	1.59 ± 0.05°
		Triticale	
Control	6.05 ± 0.05	2.19 ± 0.05	1.87 ± 0.06
F. graminearum	5.38 ± 0.18*	1.60 ± 0.07*	1.56 ± 0.07*
Pavstim, 10^{-2}	6.66 ± 0.17°*	2.46 ± 0.06°*	2.01 ± 0.07°
F. graminearum + Pavstim, 10^{-2}	6.39 ± 0.18°	2.39 ± 0.06°*	1.86 ± 0.05°

* - the difference as related to the control is reliable at the significance level of 5%;
° - the difference as related to the *F. graminearum* treatment is reliable at the significance level of 5%.

Table 3. The influence of the spirostanol saponin diosgenin (SSD) on the colony growth and sporulation of the fungus *F. graminearum.*

Treatment	Conc., %	Colony diameter, cm ($X \pm m_x$)			Sporulation intensity, score
		3rd day	4th day	5th day	
Control (mash-agar)		5.1 ± 0.05	7.3 ± 0.31	9.0±0.0	4
SSD	10^{-2}	2.1 ± 0.26*	6.9 ± 0.32*	9.0 ± 0.0	0.1
	10^{-3}	0.9 ± 0.28*	2.1 ± 0.14*	2.8 ± 0.29*	0
	10^{-4}	2.2 ± 0.39*	5.0 ± 1.34*	5.7 ± 1.03*	0.1
	10^{-5}	3.4 ± 0.31*	7.4 ± 0.22*	9.0 ± 0.0*	2

* - the difference as related to the control is reliable at the significance level of 5%;

Table 4 shows that concentrations of 10-3 to 10-5% proved to be effective. More expressed effectivity of low concentrations as compared with those of the previous experiment can be explained, perhaps, by the fact that the inherent protective reactions of the plant play an important role in the plant/pathogen system.

Table 4. The influence of the spirostanol saponin diosgenin (SSD) on fusariose infectivity and producing capacity of triticale on an infectious background

Treatment	Conc., %	Infected plants, %	No. grains per main spike
		$X \pm m_x$	$X \pm m_x$
Control (*Fusarium* isolated mixture)		40.1 ± 2.3	45.1 ± 0.7
SSD	10^{-2}	38.8 ± 5.1	48.1 ± 1.1
	10^{-3}	28.1 ± 2.4*	52.8 ± 2.2*
	10^{-4}	18.2 ± 2.8*	64.5 ± 3.0*
	10^{-5}	17.5 ± 3.2*	64.9 ± 2.1*

* - the difference as related to the control is reliable at the significance level of 5%;

This data shows that steroidal saponins are quite effective in inducing the enhancement of cereal crop resistance to *Fusarium* root rots. Owing to their phytohormonal properties, glycosides improve plasmalema permeability of cellular tissues, especially roots, the content of photosynthetic pigments in leaves, seed germinating capacity, and general resistance of plants.

SUMMARY

During the last few years fusarium diseases of cereal crops have led to great economic losses, as well as to deterioration of edible properties of grains owing to increased contents of mycotoxins, which are hazardous for human health. Steroidal saponins can be applied to protect plants from these diseases, caused by the fungi of the *Fusarium* Link. genus owing to their phytohormonal properties. They contribute to enhancement of seed germinating capacity and photosynthetic pigment content, regulate cell membrane permeability in infected plants and inhibit growth and development of fungi, which eventually results in reduction of plant morbidity and increase in their producing capacity.

ACKNOWLEDGEMENTS

Moldstim® and Pavstim® (both furostan series) are commercial products based on the total steroidal glycosides of *Capsicum annum* L. seeds and *Digitalis purpurea* L. from the aerial parts respectively. These products have been registered by the former Soviet Union and put in large scale use in agriculture.

REFERENCES

1. G.A. Lupashku, P.I. Buiucli, and E.C. Veveritsa, The inheritance of the resistance to basal stem fusariose in triticale, In: *Eur. J. Plant Pathol.,* XIII International Plant Protection Congress, The Hague. The Netherlands, Kluwer Academic Publishers: 1029 (1995).

2. V.G. Novokhatka, N.V. Doroshenko, and V.A. Zabolotnaya, Distribution of root rots in winter wheat in the Ukrainian SSR, *Mikol. Fitopat.* 24:352 (1990) (In Russian).

3. A. Bleich, E. Schitzler, and G. Schitzler, Zum Auftreten von Krankheiten bei Triticale, *Pflanzenschutz DDR* 43:204 (1989).

4. K.V. Novozhilov and M.M. Levitin, Investigation directed to solve the problem of fusariose in cereal crop spike, *Vest. S/kh. Nauk.* 10:64 (1990) (In Russian).

5. J. Perkowski and J. Chelkowski, Deoxynivalenol content dependence on kernel size in wheat grain naturally infected by Fusarium, *Microbiol. Alim. Nutr.* 9:217 (1991).

6. B. Rocque, Compagne 1991. Bis repetita. Fusarioses en plus, *Phytoma* 436:38 (1992).

7. A.P. Volynets, S.E. Karoza, P.C. Kintia, and G.A. Lupashku, The influence of steroidal glycosides on the formation and storage of plant pigments, *Dokl. AN Belarus.* 36:265 (1992) (In Russian).

8. A.P. Volynets, S.E. Karoza, P.C. Kintia, and G.A. Lupashku, Auxin-like activity of steroidal glycosides, *Dokl. AN Belarus.* 36:262 (1992) (In Russian).

9. J. Balansard and F. Pellessier, Action des diverses saponines sur le bouturage epiphylle de qualque Begoniacees, *Compt. Rend. Soc. Biol.* 137:454 (1943).

10. G. Helmkamp and J. Bonner, Some relationships of sterols to plant growth, *Plant Physiol.* 28:428 (1953).

11. H. Oku, *Plant Pathogenesis and Disease Control,* Lewis Publishers, Boca Raton, Ann Arbor, London, Tokyo, 193 pp. (1994).

12. V.A. Bobeyko, G.A. Lupashku, I.P. Dikusar, and P.K. Kintia, Phytohormone-like activity of steroidal glycosides in relation to their chemical structures, *Fiziol. i Biokhim. Kulturnykh Rast.* 4:259 (1995) (In Russian).

13. G.A. Lupashku, P.K. Kintia, N.E. Maschenko, S.A. Shvets, and T.I. Prasol, Cluster analysis of cytokinin-like activity of steroidal glycosides, In *3rd Internat. Conf. on the The Regulators of Growth and Development of Plants*, Book of Abstracts, Moscow, 81 (1995).

14. N.E. Maschenko, R. Cyulemelova, P.K. Kintia, and A.S. Shashkov, Sulfated glycoside from the product "Tribeston", *Khim. Prirodn. Soedinen.* 5:649 (1990).

15. N.E. Maschenko, G.V. Lazuryevskiy, and P.K. Kintia, Steroidal glycosides. The structures of funkioside C and D from *Funkia ovata* leaves, *Khim. Prirodn. Soedinen.* 1:123 (1977).

16. R. Tschesche and H. Gutwinski, Capsicosid, ein bisdesmosdisches 22-hydroxyfurostanol glycoside aus dem samen von *Capsicum annuum* L., *Chem. Ber.* 108:265 (1975).

17. R. Tschesche, A.M. Iavellona, and G. Wulff, Steroid Saponine mit mehr als einer Zuckerkette. IX. Purpureagitosiol, ein bisdesmosdisches 22-hydroxyfurostanol-glycoside aus den blättern von *Digitalis purpurea* L., *Chem. Ber.* 107:2828 (1974).

18. S.F. Negrutssky, Yu. D. Dantevich, G.F. Mosunov, and V.A. Popov, Determination of permeability of plant cellular protoplasma using an electric method with the aid of a megometer, *Fiziol. i Biokhim. Kulturnykh Rast.* 3:42 (1971) (In Russian).

19. E.Z. Koval and L.T. Gorbik, Microscopic study of fungi, In *Methods of experimental mycology*, V.I. Bilay, Ed., Naukova Dumka, Kiev, p. 80 (1982).

20. N.E. Maschenko, G.A. Lupashku, and P.K. Kintia, The respiratory function of sulfated glycosides in *Fusarium graminearum* Schwabe infected wheat, In *3rd Internat.Conf. on the The Regulators of Growth and Development of Plants*, Book of Abstracts, Moscow, p. 80 (1995).

EFFECT OF SAPONINS ON THE GROWTH AND ACTIVITY OF RHIZOSPHERE BACTERIA

Robert M. Zablotowicz, Robert E. Hoagland, and Stephen C. Wagner

USDA-ARS
Southern Weed Science Laboratory
Stoneville, MS 38776

INTRODUCTION

Saponins are a group of steroid or triterpenoid glycosides and related chemicals (sapogenins: non-glycosylated) found in roots, shoots, seeds, and flowers of many plant species. Saponins are of agronomic interest because of allelopathic interference with plant growth (Oleszek and Jurzysta, 1987; Waller et al., 1993). Saponins can be released into the soil by secretion from roots and/or leaching from living or decaying plant material (Mishustin and Naumova, 1955; Oleszek and Jurzysta, 1987). Conservation management practices designed to maintain plant residues on the soil surface such as the use of cover crops and reduced tillage are rapidly being adopted by growers; thus the impact of these compounds and other natural products on crop productivity and soil and rhizosphere microbial ecology warrants study.

Early studies by Mishustin and Naumova (1955) indicated that saponins from alfalfa altered the soil microflora. The addition of alfalfa meal to soil decreased the severity of root rot caused by *Phytophthora cinnamoni* (Zentmeyer, 1963) and was associated with the saponin content (Zentmeyer and Thompson, 1967). The effects of saponins from alfalfa and other plants have been widely studied in both phytopathogenic and nonpathogenic fungi and are found to be inhibitory to fungal species such as: *Aspergillus niger, Fusarium oxysporum, Fusarium solani, Phytophthora cinnamoni, Phoma* sp., *Rhizopus mucco, Sclerotium rolfsii, Trichoderma viride*, and *Verticillium albo-atrum* (Zimmer et al., 1967; Levy et al., 1989). Pharmacologically, saponins are also inhibitory to certain medically important yeasts (Polachek et al., 1986). The effects of saponins on fungi has been widely studied; however, little is known about their effects on terrestrial bacteria. Certain alfalfa root saponins are inhibitory to some phytopathogenic bacteria (*Agrobacterium tumefaciens, Corynebacterium michiganense*, and *Pseudomonas lachrimans*), but *Xanthomonas campestris* was inhibited only by saponins derived from aerial parts of alfalfa (Timbekova et al., 1996).

Saponins Used in Food and Agriculture
Edited by Waller and Yamasaki, Plenum Press, New York, 1996

Populations of rhizosphere microflora are influenced by plant species, specific chemicals exuded by roots (Rovira and Davey, 1974), and compounds released from the decomposition of other plants. Our studies were conducted to further explore the effects of saponins on soil and rhizosphere bacteria, using *in vitro* assays. Bacterial isolates selected for study encompass diverse genera and include phytopathogenic species and beneficial bacterial strains (antifungal biological control, plant growth-promoting rhizobacteria, and herbicide-degrading isolates). We selected β-escin as a model saponin because it is economically available in purified form, water soluble, and is phytotoxic to plants in *in vivo* and *in vitro* assays (Hoagland et al., this volume). In addition to bacterial growth, we evaluated microbiological activity using three common enzyme assays. Saponins interact with membrane sterols thereby reducing membrane integrity (Bangham and Horne, 1962; Dourmaskin et al., 1962). However no information on the effects of saponins on membrane integrity of bacteria is available. Thus, we studied the release of extracellular protein as influenced by saponins.

MATERIALS AND METHODS

Chemicals

β-Escin (92% pure), hecogenin (5α-spirostan-3β-12-one, 95% purity), betulin (lup-20[29]-ene-3β,28-diol), 18β-glycyrrhetic acid (3β-hydroxy-11-oxo-18β,20β-olean-12-en-29-oic acid), 95% purity), *p*-nitrophenyl butyrate, fluorescein diacetate, and γ-glutamyl-*p*-nitroanilide were obtained from Sigma Chemical Co. (St. Louis, MO). The structures of these saponins and sapogenins are presented elsewhere (Hoagland et al., this volume).

Bacterial Strains

Many of the bacterial strains used in this study have been previously described (Hoagland et al., 1994; Zablotowicz et al.,1995). All strains with a JM prefix were obtained from J. McInroy, Auburn Univ. Taxonomic identity was provided by the source. Stock cultures were maintained as tryptic soy agar stabs (5 °C) or frozen glycerol suspensions (-5 °C). All strains were grown on tryptic soy broth, except *Bradyrhizobium japonicum* strain 61a101c, which was grown on yeast extract glycerol broth.

Effect of β-Escin on Bacterial Growth

Tryptic soy broth (TSB), Difco (half-strength) was amended with 0, 25, 100, or 500 μM β-escin. The broths were filter sterilized through a 2-μm filter and dispensed into sterile culture tubes (5 ml per tube). Triplicate culture tubes for each concentration of β-escin were inoculated with 100 μl of a 48 h-old TSB culture, and incubated on a rotary shaker for 24 h at 25 °C, 125 rpm. Growth was determined by measuring absorbance at 630 nm in a spectrophotometer. Growth of *B. japonicum* strain 61a101c, *Rhodococcus corallinus* NRRL1554, and *R. meliloti* 102F76 was determined in 50% yeast extract-glycerol broth except that growth was assessed after 48 and 72 h for 102f76 and 61a101c, respectively.

Influence of β-Escin on Bacterial Activity

Bacteria were grown in 100 ml TSB for 48 h, cells harvested by centrifugation (10 min and 6.0 x 10^3g) and washed twice with sterile phosphate buffer (KPi, 0.1 M, pH 7.6). Cells were resuspended in 15 ml sterile KPi (optical density 8.0 to 12.0, 630 nm) and dispensed into sterile glass centrifuge tubes at 2.0 ml per tube. Either 2.0 ml of a 1.0 mM β-escin (KPi, 10% ethanol) or 2.0 ml KPi with 10% ethanol was dispensed into each tube, (three replicates per treatment), and tubes were placed on a rotary shaker (150 rpm, 25 °C). Following a 24-h incubation the cells were harvested by centrifugation (6.0 x 10^3 g, 10 min) and resuspended in 4.0 ml KPi. In cases where the supernatant was studied, the supernatant was filter sterilized (2-μm filter). Protein content of cell suspensions and supernatants were determined according to Bradford (1976).

We studied the hydrolysis of fluorescein diacetate (FDA) as the major indicator of bacterial metabolic activity. FDA is a general substrate for several hydrolytic enzymes i.e. esterases, lipases, and certain proteases (Medzon and Brady, 1969; Guilbault and Kramer, 1964). FDA hydrolytic activity in soil has been correlated with respiratory activity of soil microflora (Schnurer and Rosswall, 1982). The assay consisted of 750 μl of cells or cell supernatant, 750 μl KPi (0.1 M, pH 7.6), and 40 μl FDA (2 mg ml^{-1} in acetone). The assay mixture was vortexed and incubated at 28 °C for 5 to 60 min, depending upon activity. The assay was terminated by extraction with 1.5 ml of acetone followed by centrifugation (5 min, 14 x 10^3 g). Absorbance of the clarified supernatants was determined at 490 nm in 1-cm glass cuvettes.

We studied esterase activity, using *p*-nitrophenyl butyrate (*p*-NPB), to further define esterase activity in select strains. Activity was determined spectrophotometrically in 1-cm methacrylate cuvettes. Briefly, 50 to 200 μl of cell suspensions or 750 μl of filtered supernatants were diluted to a final volume of 1.5 ml in KPi, pH 7.6. The assay was initiated by addition of 40 μl of *p*-NPB (2 mg ml^{-1} in acetone). The cuvette contents were vortexed and changes in absorbance at 405 nm monitored over a 3 to 30 min incubation at 28 °C. In some Gram-negative bacteria, γ-glutamyl transpeptidase activity was determined in cell suspensions by using a technique described elsewhere (Zablotowicz et al., 1995), except that cells instead of cell-free extracts were used.

Effects of β-Escin Compared to other Sapogenins

A comparison of effects of three sapogenins (betulin, hecogenin, and glycyrrhetic acid) as well as β-escin on cellular FDA hydrolytic activity and extracellular protein production was made on four strains: A-136, RA-2, JM1011, and UZ404 (*A. tumefaciens*, *P. fluorescens*, *Curtobacterium flaccumfaciens*, and *B. thuringiensis*, respectively). Bacterial strains (200 ml TSB cultures) were grown for 48 h, and cells washed twice in KPi (pH 7.6) and then resuspended in 60 ml of KPi. Cell suspensions (3.0 ml) were dispensed into 25 ml glass tubes and 150 μl of a 10 mM ethanolic solution of the saponins (or ethanol for untreated controls) added directly to the cell suspensions, with three replicate tubes per strain and treatment. Cells were incubated for 24 h at 25°C at 125 rpm, followed by centrifugation (8 x 10^3g). The supernatant was removed and filtered (2-μM filter) and the cell pellet resuspended in KPi, pH 7.6. FDA hydrolytic activity of cell suspensions and protein content of the supernatant were determined as described above.

Effects of β-Escin on Soil and Soybean Rhizosphere Microflora

The effect of β-escin on soil and rhizosphere microbial populations was assessed in a growth chamber study. The soil was a mixture of Bosket sandy loam and a Dundee silt loam (10:1 w:w), treated with 0, 88, or 350 μmol β-escin kg^{-1} soil, and seeded with two pregerminated (24 h) soybeans (details on plant management are described elsewhere (Hoagland et al., this volume). Nine days after planting the plants were harvested (four replicates per treatment) and soil and rhizosphere populations of total heterotrophic bacteria, Gram-negative bacteria, fluorescent pseudomonads, and total fungi determined by serial dilution plating as described (Hoagland et al., 1994; Wagner et al., 1995). Estimates of colony-forming units are reported based upon oven dry soil or root weight basis. The effects of β-escin on FDA hydrolytic activity of a Dundee silt loam soil was assessed in a laboratory study. Soil (2.0 g, field moisture content 12 %) was added to polypropylene centrifuge tubes, and 0.5 ml of water, 0.5 mM or 1.0 mM β-escin (final concentration added = 0, 125, and 250 μmol kg^{-1}), and the tubes incubated at 28 °C. FDA hydrolysis assays were conducted as described by Wagner et al. (1995) after 1, 24, 72, and 144 h, with six replicates per treatment.

Statistical Analysis

Analysis of variance due to treatment was determined using a general linear model (SAS Institute, Cary, NC) with mean separation at the α = 0.05 level using Fisher's Least Significant Difference test.

RESULTS AND DISCUSSION

Effects of β-Escin on Bacterial Growth

The growth of 25 rhizosphere bacterial strains as affected by β-escin is summarized in Table 1. All bacteria were able to grow at the highest concentration (500 μM), although certain strains were inhibited 11 to 54%, except for certain strains of the family Rhizobiaceae (*Agrobacterium, Bradyrhizobium,* and *Rhizobium*) which were inhibited 62% or greater. In the Gram-positive species tested, 3 of 4 *Bacillus* sp. were inhibited 35 to 39% at 500 μM, while the effects of β-escin on other Gram-positive bacteria was minimal, i.e. only growth of *Arthrobacter sp.* and *Curtobacterium flaccumfaciens* were inhibited 9 and 11% respectively. In the Gram-negative bacteria, the fluorescent pseudomonads were typically sensitive to β-escin, i.e. 3 strains RA-2, M-17, and UA5-40 growth was inhibited at 25 μM , while growth of 5 strains was inhibited 21 to 54% at 500 μM β-escin. Among members of the Enterobacteriaceae family, only *E. cloacae* strain ECMS was inhibited by 25 μM escin, while four of the six strains from this family were inhibited by concentrations of 100 to 500 μM. Increased turbidity was observed in certain strains (JM-443, JM-361, and RB-4) in response to β-escin, which might have been due to increased extracellular polysaccharides rather than a growth stimulation. *Xanthomnas campestris* was inhibited only at 500 μM, while among members of the Rhizobiaceae family, the *A. tumefaciens*, *B. japonicum*, and *R. meliloti* strains were strongly inhibited by 500 μM escin (62, 76, and 92%, respectively), with growth inhibition at 25 μM.

β-Escin had only limited bacteriostatic activity among the strains of rhizosphere bacteria studied. Fungi such as *Trichoderma viride* are inhibited by alfalfa saponins at concentrations as low as 2 μg ml^{-1} (Zimmer, 1967; Oleszek and Jurzysta, 1987). Although the *B. japonicum* strain studied was severely inhibited by 25 μM β-escin, soybean plants grown in

Table 1. Effect of β-escin on growth of rhizosphere bacteria in broth culture, increase in turbidity as determined at 630 mn.

Strain	Growth (% of control)		
	25 μM	100 μM	500 μM
Bacillus			
UW-85 (*B. cereus*)	99[1]	99	65*
BG07 (*B. subtilis*)	94	74*	69*
HD-2 (*B. thuringiensis*)	98	97	98
UZ404 (*B. thuringiensis*)	96	94	61*
Other Gram-positive bacteria			
JM414 (*Anthrobacater sp.*)	101	102	91
JM894 (*Cellulomonas* sp.)	109	115*	96
NRRL1554 (*Rhodococcus corallinus)*	99	97	96
JM1001 (*Curtobacterium* sp.)	98	92	89*
Enterobacteriaceae			
JM-92 (*Citrobacter diversus*)	101	102	91*
SP (*Serratia plymutica*)	92	88*	82*
JM-895 (*Enterobacter asubriae*)	96	85*	86*
ECMS (*Enterobacter cloacae*)	81*	78*	70*
JM-443 (*Klebsiella pneumoniae*)	104	109	115*
JM-361 (*Erwinia caratovara*)	99	100	105
Pseudomonaceae			
RA-2 (*P. fluorescens*)	79*	72*	74*
RB-4 (*P.* sp.)	109	117*	102
PRA-25 (*P. fluorescens*)	88*	57*	46*
JM-1087 (*P. vesicularis*)	101	93	96
UA5-40 (*P. fluorescens*)	86*	79*	79*
BD4-13 (*P. fluorescens*)	95	81*	79*
M-17 (*P. putida*)	83*	73*	69*
Other Gram-negative bacteria			
XC (*Xanthomonas campestris*)	105	103	82*
A-136 (*Agrobaterium tumefaciens*)	80*	50*	38*
102F76 (*Rhizobium meliloti*)	63*	29*	24*
61a101c (*Bradyrhozobium japonicum*)	22*	16*	8*

[1] Mean of three replicates; within a row, asterisk denotes significant difference from the control at the 95% level.

Table 2. Effect of β-escin (500 μM) on FDA hydrolytic activity in rhizobacterial cell suspensions after a 24-h exposure.

	Fluorescein formed (nmol ml^{-1}cells h^{-1})	
Strain	Untreated	β-Escin
Bacillus		
HD-2 (*B.thuringiensis*)	179.4 a[1]	72.2 b
GB-07 (*B. subtillis*)	248.1 a	63.1 b
UW85 (*B. cereus*)	845.5 a	393.4 b
UZ404 (*B. thuringiensis*)	424.6 a	173.6 b
Other Gram-positive bacteria		
JM894 (*Cellulomonas cartae*)	111.2 a	99.1 a
NRRL1554 (*Rhodococcus corallinus*)	504.1 a	323.4 b
JM1011 (*Curtobacterium flaccumfaciens*)	47.2 a	41.8 a
Enterobacteriaceae		
JM92 (*Citrobacter diversus*)	25.1 a	25.8 a
ECMS (*Enterobacter cloacae*)	111.6 a	12.8 b
JM361 (*Erwinia herbicola*)	17.0 a	12.9 b
SP (*Serratia plymutica*)	15.2 a	17.0 a
JM443 (*Klebsiella pneumoniae*)	25.1 a	21.5 a
Pseudomonas		
BD4-13 (*P. fluorescens*)	13.9 a	13.9 a
RA-2 (*P. fluorescens*)	28.9 b	142.8 a
RB-4 (*P.* sp.)	13.4 b	18.0 a
PRA-25 (*P. fluorescens*)	90.5 a	81.0 b
UA5-40 (*P. fluorescens*)	56.5 a	44.0 b
M-17 (*P. putida*)	230.3 a	217.0 a
Other Gram-negative bacteria		
XC (*Xanthomonas campestris*)	92.0 a	61.5 b
A-136 (*Agrobacterium tumefaciens*)	21.5 a	24.2 a

[1]Mean of three replicates; in a row, means followed by the same letter do not differ significantly at the 95% level.

soil treated with 88 and 350 μmol kg^{-1} soil β-escin were nodulated, with a significant reduction in nodulation observed only under the highest concentration (see Hoagland et al., this volume). Growth inhibition of *Agrobacterium tumefaciens* and *Xanthomonas campestris* by alfalfa saponins has been reported (Waller et al., 1993) and our study demonstrates that these strains are also inhibited by β-escin. The significance of growth inhibition by saponins using our model compound (β-escin) indicates the need for further studies with saponins from agronomically important crops and weeds to understand the alleopathic nature of saponins in relation to plant-microbe interactions.

Effects of Saponins on Bacterial Activity

Differential effects upon exposure of resting cells to 500 μM β-escin on FDA hydrolysis were observed among the 20 strains evaluated (Table 2). Significant reductions (10.5 to 59.8%) in FDA hydrolysis were observed in nine strains representing five genera. However, two pseudomonads (RA-2 and RB-4) exhibited increased FDA hydrolytic activity following exposure to β-escin. Among the genera studied, the FDA hydrolytic activity of *Bacillus* as a group were most sensitive.

The effects of β-escin exposure on esterase activity using *p*-NPB was evaluated in 9 strains (Table 3). This esterase activity was not as sensitive as that of FDA, as β-escin inhibited *p*-NPB-esterase activity in 2 strains *Enterobacter cloacae* strain ECMS and *Pseudomonas* sp. RB-4. β-Escin enhanced *p*-NPB esterase activity in 2 strains, *Bacillus cereus* UW-85 and *P. fluorescens* RA-2. The effects of escin of γ-glutamyl transpeptidase was evaluated in several pseudomonads and *Serratia plymutica* SP (Table 4). This peptidase activity was significantly reduced in four of the seven strains tested.

Table 3. Effect of β-escin (500 μM) on *p*-nitrophenyl butyrate esterase activity in rhizobacterial cell suspensions after a 24-h exposure.

	p-Nitrophenol formed (nmol ml^{-1} h^{-1})	
Strain	Untreated	β-Escin
Bacillus sp.		
UW-85 (*B. cereus*)	7.3 b[1]	12.9 a
UZ-404 (*B. thuringiensis*)	10.3 a	9.8 a
Other Gram-positive bacteria		
JM-1011 (*Curtobacterium flaccumfaciens*)	5.8 a	5.3 a
NRRL1544 (*Rhodococcus corallinus*)	34.9 a	38.1 a
Enterobacteriaceae		
JM-92 (*Citrobacter diversus*)	34.7 a	39.0 a
ECMS (*Enterobacter cloacae*)	9.7 a	5.4 b
Pseudomonas sp.		
BD4-13 (*P. fluorescens*)	11.1 a	11.9 a
RA-2 (*P. fluorescens*)	52.9 b	79.3 a
RB-4 (*P.* sp.)	53.4 a	30.6 b

[1]Mean of three replicates; means in row followed by the same letter do not differ significantly at the 95% level.

Table 4. Effect of β-escin (500 μM) on GPNA-γ-glutamyl transpeptidase activity in rhizobacterial cell suspensions after a 24-h exposure.

	p-Nitroaniline formed (nmol ml^{-1} h^{-1})	
Strain	Untreated	β-Escin
BD4-13 (*Pseudomonas fluorescens*)	65.8 a[1]	65.3 a
RA-2 (*P. fluorescens*)	88.9 a	42.2 b
RB-4 (*Pseudomonas* sp.)	131.6 a	130.2 a
PRA-25 (*P. fluorescens*)	172.9 a	154.4 b
UA5-40 (*P. fluorescens*)	107.5 a	83.1 b
M-17 (*P. putida*)	431.9 a	402.4 b
SP (*Serratia plymutica*)	28.8 a	32.1 a

[1] Mean of three replicates; means in a row followed by the same letter to not differ significantly at the 95% level.

In a subset of strains, cells were removed from the incubation buffer via centrifugation and membrane filtration, and extracellular protein content as well as FDA hydrolytic activity of the clarified supernatants determined (Table 5). Extracellular protein content of the supernatants from β-escin-treated cells were significantly greater in six of the ten strains tested. The greatest increase occurred in *E. cloacae* strain ECMS, with a 6-fold greater protein content (ca. 52 % of the total protein) in the supernatant of the β-escin-treated cells

Table 5. Effect of β-escin (500 μM) on extracellular protein levels and FDA hydrolytic activities in rhizobacterial strains after a 24-h exposure.

	Protein (μg ml^{-1})		FDA hydrolysis (nmol ml^{-1} min^{-1})	
Strain	treated	untreated	treated	untreated
JM-361 (*Erwinia caratovara*)	15.7 a[1]	18.9 a	<0.1	<0.1
JM-895 (*Enterobacter asubriae*)	37.2 a	12.9 b	<0.1	<0.1
ECMS (*Enterobacter cloacae*)	505.0 a	74.2 b	7.6 a	1.3 b
JM-443 (*Klebsiella pneumoniae*)	35.9 a	12.6 b	<0.1	<0.1
JM-1087 (*Pseudomonas vesicularis*)	4.1 a	5.8 a	<0.1	<0.1
RA-2 (*P. fluorescens*)	23.1 a	9.8 b	15.6 a	7.1 b
XC (*Xanthomonas campestris*)	211.0 a	142.0 b	4.8 a	0.4 b
NRRL1544 (*Rhodoccus corallinus*)	19.1 a	13.6 b	0.1 a	0.2 a
JM894 (*Cellulomonas cartae*)	107.3 a	53.2 b	<0.1	<0.1
A-136 (*Agrobacterium tumefaciens*)	34.1 a	12.1 b	<0.1	<0.1

[1]Mean of three replicates; in a row, means followed by the same letter do not differ significantly at the 95% level.

compared to the buffer control (ca. 7 % of total protein). The extracellular protein levels in the other β-escin sensitive strains (*Enterobacter asubriae* JM-895, *Erwinia caratovara* JM-443, *P. fluorescens* RA-2, *X. campestris* XC, and *Cellulomonas* sp. JM-894) ranged from 0.5 to 2.9 fold greater in β-escin-treated cells compared to untreated cells. Except for the XC and ECMS increased extracellular protein was less than 5 to 10% of the total protein. In most strains tested, exposure to β-escin had no influence on FDA hydrolysis by supernatants since little or no activity was detected in the supernatant. However, increased levels of FDA hydrolysis was associated with increased extracellular protein in three strains (ECMS, RA-2 , and XC).

Extracellular proteins are commonly observed among many bacteria because certain exoenzymes facilitate hydrolysis of simple and complex molecules. Certain exoenzymes provide a method for uptake of nutrients (amylases, esterases, and peptidases) while others may offer protection from toxic substances (Pollock, 1964). Increased protein excretion may also be associated with factors such as autolysis and membrane leakage. β-Escin is reported to be a potent hemolytic agent with a hemolytic index 1:40,000 (Merck Index). A mechanism of saponin toxicity to eukaryotic species is altered membrane integrity,based on their ability to form complexes with membrane sterols (Bangham and Horne, 1962). Sterols are not present in the membranes of bacteria except for the *Mycoplasma,* which lack cell walls. Increased extracellular protein content upon exposure to β-escin in certain bacterial strains suggests that saponins may be affecting the membrane integrity of these bacteria, but perhaps in a different manner than in eukaryotic organisms.

In several experiments (Table 6), exposure of *P. fluorescens* strain RA-2 to β-escin elicited increases in FDA and *p*-NPB hydrolytic activity. We evaluated the interaction of incubation with the protein synthesis inhibitor chloramphenicol and 500 μM β-escin on hydrolytic enzyme activities using three substrates (FDA, *p*-NPB, and GPNA) and both

Table 6. Effects of β-Escin (500 μM) and/or chloramphenicol (900 μM) on protein levels and FDA hydrolytic, *p*-NPB-esterase, and GPNA-γ-glutamyl transpeptidase activities in cell suspensions and cell-free supernatants of *P. fluorescens* strain RA-2 after a 24-h exposure.

	None	β-Escin	Chloramphenicol	β-Escin + chloramphenicol
Whole cells				
Protein (mg ml^{-1})	1.0 b[1]	1.1 b	1.3 a	1.3 a
FDA (nmol ml^{-1} h^{-1})	20.4 b	71.8 a	12.5 c	17.6 bc
*p*NPB (nmol ml^{-1} h^{-1})	95.1 b	108.8 a	78.9 c	82.1 c
GGTP (nmol ml^{-1} h^{-1})	62.5 a	46.0 c	50.5 b	50.8 b
Supernatant				
Protein (μg $ml^{-1)}$	54.3 b	103.4 a	15.1 c	42.9 b
FDA (nmol ml^{-1} h^{-1})	7.8 b	13.8 a	1.6 d	3.8 c
*p*NPB (nmol ml^{-1} $h^{-1)}$	6.9 a	3.7 b	1.5 c	1.1 c

[1]Mean of four replicates; in a row, means followed by the same letter do not differ significantly at the 95% level.

cellular and extracellular protein. In this experiment β-escin elicited a 3-fold increase in FDA hydrolytic activity, while rates of FDA hydrolysis of cells exposed to β-escin and chloramphenicol were not significantly different than in either, the buffer control or in the chloramphenicol control. Using *p*-NPB as substrate, esterase activity of β-escin-treated RA-2 cells were 15% higher than control cells, but significantly greater than those of cells exposed to either chloramphenicol or chloramphenicol and β-escin. These results suggest that increased hydrolytic enzyme activities in RA-2 after β-escin exposure required *de novo* protein synthesis. As previously described exposure of RA-2 cells to β-escin had significantly lower GNPA-peptidase activity than control cells or cells treated with chloramphenicol or chloramphenicol and β-escin.

Total protein of RA-2 cells exposed to β-escin was similar to that of control cells; however, cells treated with chloramphenicol or chloramphenicol plus β-escin had significantly greater cellular protein than either untreated or β-escin-treated cells. Extracellular protein in the supernatant was significantly greater in cells treated with β-escin compared to all other treatments. Extracellular protein in β-escin and chloramphenicol-treated cells were similar to that of untreated cells although chloramphenicol-treated cells were lower in this respect than all other treatments. These results suggest that extracellular protein exudation/leakage by strain RA-2 caused by β-escin does not require *de novo* protein synthesis, but may be due to effects on membrane integrity allowing leakage of proteins. FDA hydrolytic activity in the supernatant was significantly greater in β-escin-treated cells compared to that untreated control. This was also true in comparing the β-escin plus chloramphenicol-treated cells with cells treated with chloramphenicol alone. However, *p*-NPB esterase activity was greater in supernatants of untreated cells compared to the other treatments. This indicates that the FDA-deacetylating activity in the supernatant may be due to enzymes other than esterase.

Effects of β-Escin Compared to Other Sapogenins

The effects of three sapogenins and β-escin (500 μM) on FDA activity of cell suspensions of four strains were studied. In the two Gram-negative strains, *Agrobacterium tumefaciens* A-136 and *Pseudomonas fluorescens* RA-2, increased FDA activity was associated with exposure to these compounds. In A-136, only betulin, hecogenin, and glycyrrhetic acid increased FDA hydrolytic activity, while in RA-2 all four compounds enhanced FDA hydrolysis. In *B. thuringiensis* strain UZ404 all four compounds reduced cellular FDA activity, while only betulin and hecogenin reduced FDA hydrolysis in JM-1011. In all four strains both glycyrrhetic acid and to a lesser extent β-escin, significantly enhanced extracellular protein accumulation (Table 7). Initial results with β-escin enhancing FDA hydrolysis and *p*-NPB esterase activity in *P. fluorescens* strain RA-2 suggested that ester linkages, i.e. acetate, angelate or tiglate may have been responsible for induction of these activities. The three sapogenins tested are not esterified or glycosylated but still induce FDA hydrolytic activity in both RA-2 or A-136. Thus, this inductive effect may be a generic response to the triterpenoid structure. In all four strains both β-escin and glycyrrhetic acid increased extracellular protein exudation/leakage. In studies with excised leaf discs β-escin caused the greatest electrolyte leakage in five of six plant species tested, while glycyrrhetic acid caused less leakage in three of six plant species (Hoagland et al., this volume), while other compounds were less effective. These results indicate a relationship between chemical structure of the saponin/sapogenin and effects on exudation/leakage of bacterial proteins. As extracellular protein is increased by both β-escin and glycyrrhetic acid, glucose in carbon number 3 may not be required for activity. The three hydroxyl groups and 2 ester linkages in the aglycone ring system of β-escin renders this molecule much more polar than betulin or hecogenin. Polar functional groups of glycyrrhetic acid include one each of hydroxyl, carbonyl, and carboxyl.

Table 7. Effects of saponins or sapogenins (500 μM) on cellular FDA hydrolytic activity and protein leakage in four rhizosphere bacterial strains after a 24-h exposure.

Strain	Untreated	β-Escin	Betulin	Hecogenin	Glycyrrhetic Acid
	FDA hydrolysis (nmol ml^{-1} culture h^{-1})				
A-136 (*Agrobacterium tumefaciens*)	22 c	25 bc	44 a	31 b	33 b
RA-2 (*Pseudomonas fluorescens*)	11 c	23 b	23 b	25 b	46 a
JM1011 (*Curtobacterium flaccumfaciens*)	142 a	130 a	85 c	110 b	134 a
UZ404 (*Bacillus thuringiensis*)	1350 a	1020 c	1020 c	1198 b	253 d
	Protein (μg ml^{-1} supernatant)				
A-136	10 c	31 b	6 c	9 c	48 a
RA-2	117 c	171 b	75 d	117 c	217 a
JM1011	105 c	149 b	105 c	111 c	186 a
UZ404	31 c	52 b	31 c	26 c	366 a

[1]Mean of three replicates; in a row, means followed by the same letter do not differ significantly at the 95% level.

Effects of β-Escin on Soil and Soybean Rhizosphere Microflora

Treating soil with β-escin significantly affected populations of specific soil and soybean rhizosphere microflora in this controlled study (Table 8). Colony-forming units of total bacteria in soil or rhizosphere were unaffected by β-escin. However, increased (3- to 5-fold) colony-forming units of Gram-negative bacteria and fluorescent pseudomonads associated with treatment with β-escin were observed in both the soil and soybean rhizosphere. β-Escin significantly reduced soil fungal propagules. Because colony counts observed on plates from rhizosphere samples were too low for statististical precision, they are not reported. Treating Dundee soil with β-escin significantly reduced FDA hydrolysis 1 h after application; this inhibition continued throughout the seven day incubation (Fig 1; data only shown for 350 μmol kg^{-1}).

Although 500 μM β-escin reduced the growth of 68% of the bacterial strains tested *in vitro*, no reduction in soil or rhizosphere bacterial populations were observed in soil treated with high concentrations (88 to 350 μmol kg^{-1}) β-escin. Increased populations of fluorescent pseudomonads and Gram-negative bacteria observed in both soybean rhizospheres and soil can be due to several factors. The saponin may have increased exudation by soybean seedlings that specificaly favored these bacterial populations. β-Escin restricted soil fungal populations, thereby enhancing the competitiveness of Gram-negative bacteria and

Table 8. Effect of treating soil with two concentrations of β-escin on soybean rhizosphere and soil microbial populations.

Microbial group	Untreated	β-Escin 88 μmol kg^{-1}	β-Escin 350 μmol kg^{-1}
Soybean rhizosphere	log (10) colony-forming units g^{-1} root		
Total bacteria	10.20 a[1]	10.26 a	10.05 a
Gram-negative bacteria	8.86 b	9.31 a	9.41 a
Fluorescent pseudomonads	6.78 b	7.43 a	7.38 a
Root-free soil	log (10) colony-forming units g^{-1} soil		
Total bacteria	7.80 a	7.87 a	7.87 a
Gram-negative bacteria	5.99 b	6.39 a	6.62 a
Fluorescent pseudomonads	4.45 c	4.76 b	4.92 a
Total fungi	3.98 a	3.47 b	3.20 c

[1]Mean of four replicates; in a row, means followed by the same letter do not differ significantly at the 95% level.

pseudomonads. A decrease in soil fungal propagules by β-escin confirms results on other saponins that fungal populations may be more sensitive to saponins than bacteria (Mishustin and Naumova, 1955). These results are preliminary and need to be assessed in other soils, plant species, and saponins of more relevant ecological importance. Alfalfa saponins (medicagenic acid glycosides) can lose allelopathic potential either via metabolism (hydrolysis of glycosides) or sorption to soil (Oleszek and Jurzysta, 1987). Results from pure culture studies may not always be correlated with *in situ* situations due to dynamic biological and chemical processes occurring in soil. The magnitude of inhibition of soil FDA hydrolyitic activity by β-escin supports our *in vitro* findings that FDA hydrolysis was inhibited in 50% of the bacteria studied. Although minimal effects of β-escin on populations of soil and rhizosphere microflora were observed, the inhibition of FDA hydrolysis indicates that saponins may have ecological consequence due to their effects on microbial activity.

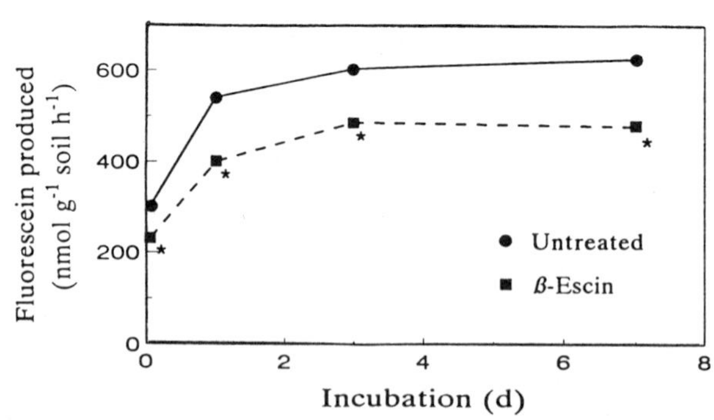

Figure 1. Effect of β-escin (350 μmol kg^{-1} soil) on FDA hydrolysis in a Dundee silt loam soil during a 7-day time course. Asterisk denotes that treatment differs significantly (95% level) from the control at a given sample time.

REFERENCES

Bangham, A.D. and R.W. Horne. Action of saponin on biological cell membranes. *Nature* 196:952 (1962).

Bradford, M.M. A rapid and sensitive method for the quantitation of microgram quantities of protein utilizing the principle of protein-dye binding. *Anal. Biochem.* 72:248 (1976).

Budavari, S. (Ed.) "Escin" #3644, p. 58, *in* The Merck Index, 11th Edition. Merck and Co., Inc. Rahway, NJ (1989).

Dourmashkin, R.R., R.M. Dougherty, and R.J. Harris. Electron microscopic observations on Rous sarcoma virus and cell membranes. *Nature* 194:1116 (1962).

Guilbault, G.G., and D.N. Kramer. Fluorometric determination of lipase, acylase, alpha-and gamma-chymotrypsin and inhibitors of these enzymes. *Anal. Chem.* 36:409 (1964).

Hoagland, R.E., R.M. Zablotowicz, and M.A. Locke. "Propanil Metabolism by Rhizosphere Microflora," pp. 160-183 *in* Bioremediation through Rhizosphere Technology. ACS Symp. Ser. 563, T. A. Anderson and J. R. Coats (Eds.) Am. Chem. Soc., Washington, DC. (1994).

Hoagland, R.E., R.M. Zablotowicz, and K.N. Reddy. "Studies of the Phytotoxicity of Saponins on Weed and Crop Plants," this volume (1996).

Levy, M., U. Zehavi, N. Naim, and I. Polacheck. Isolation, structure determination, synthesis, and antifungal activity of a new alfalfa root saponin. *Carbohydr. Res.* 193:116 (1989).

Medzon, E.L., amd M.L. Brady. Direct measurement of acetylesterase in living protist cells. *J. Bacteriol.* 97:402 (1969).

Mishustin, B.N. and A.N. Naumova. Secretion of toxic substances by alfalfa and their effect on cotton and soil microflora. *Akad. Nauk USSR Ivestiya, Ser. Biol.* 6:3 (Russian) (1955).

Oleszek, W. and M. Jurzysta. The allelopathic potential of alfalfa root medicagenic acid glycosides and their fate in soil environments. *Plant Soil* 98:67 (1987).

Polachek, I., U. Zehari, and N. Naim. Activity of compound G-2 isolated from alfalfa roots against medically important yeasts. *Antimicrob. Agents Chemother.* 30:290 (1986).

Pollock, C.R. "Exoenzymes," *in* The Bacteria, Vol. 1. R. Y. Stainer and I. C. Gunsalus (Eds.) Academic Press, New York (1964).

Rovira, A.D. and C.B. Davey. "Biology of the Rhizosphere," pp. 153-204 *in* The Plant Root Environment. E.W. Carlson, (Ed.) University Press, Charlottesville, Va. (1974).

Schnurer, J, and T. Rosswall. Fluorescein diacetate hydrolysis as a measure of total microbial activity in soil and litter. *Appl. Environ. Microbiol.* 43:1256 (1982).

Timbekova, A.E., M.I. Isaev, and N.K. Abubakirov. "Chemistry and Biological Activity of Triterpenoid Glycosides from *Medicago sativa*." this volume (1996).

Waller, G.R., M. Jurzysta, and R.L. Z. Thorne. Root saponins from alfalfa (*Medicago sativa* L.) and their allelopathic activity on weeds and wheat. *Bot. Bull. Acad. Sin.* 34:1 (1993).

Wagner, S.C., R.M. Zablotowicz, M.A. Locke, R.J. Smeda, and C.T. Bryson. Influence of herbicide-desiccated cover crops on biological soil quality in the Mississippi Delta. *Proc. Conservation Tillage Conference for Sustainable Agriculture, MAFES Special Bulletin* 88:86 (1995)

Zablotowicz, R.M., R.E. Hoagland, M.A. Locke, and W.J. Hickey. Glutathione-*S*-transferase activity and metabolism of glutathione conjugates by rhizosphere bacteria. *Appl. Environ. Microbiol.* 61:1054 (1995).

Zentmyer, G.A. Biological control of *Phytophthora* root rot of avocado with alfalfa meal. *Phytopathology* 53:1383 (1963).

Zentmyer, G.A. and G.R. Thompson. The effect of saponins from alfalfa on *Phytophthora cinnamomi* in relation to control of root rot of avocado. *Phytopathology* 57:1278 (1967).

Zimmer, D.E., M.W. Pederson, and C.F. McGuire. A bioassay for alfalfa saponins using the fungus *Trichoderma viride* pers. ex. Fr. *Crop Sci.* 7:223 (1967).

SAPONINS FROM *MEDICAGO* SPP.: CHEMICAL CHARACTERIZATION AND BIOLOGICAL ACTIVITY AGAINST INSECTS

Aldo Tava and Miriam Odoardi

Istituto Sperimentale per le Colture Foraggere
29 viale Piacenza, 20075 Lodi - ITALY

INTRODUCTION

The occurrence and distribution of secondary metabolites in plants, from an ecological viewpoint, can be associated with a defensive strategy of plants against herbivores. In order to limit damage from microbes, insects, or vertebrate predation, plants have developed complex metabolic pathways to produce specific secondary compounds that are associated with defense against pests and pathogens. Complex antibiotic substances such as alkaloids and terpenes, as well as proteins such as protease inhibitors, provide inducible or constitutive defense both in woody and in herbaceous plants.

Among other compounds, plant saponins have been widely studied for their chemical, physical, and especially physiological properties, as the toxic nature of many saponins has been known for many centuries. Saponins in fact have commonly been considered as antinutritional or toxic constituents of plants consumed by animals and humans.

In the case of alfalfa, one of the best sources of high-quality proteins for animal feeding, the presence of antinutritional compounds such as saponins is detrimental and considered especially harmful to monogastric animals (Cheeke, 1971, 1983; Price et al., 1987; Fenwick et al., 1991). Genetic selection has therefore been carried out by various breeding research groups to obtain new, highly yielding varieties with lower saponin content in the aerial parts of the plants (Pedersen and Wang, 1971; Pedersen et al., 1973; Rotili and Zannone, 1981; Bocsa, 1990). A wide somaclonal variation has been observed for saponin content in a number of clones derived from *in vitro* tissue cultures of *M. sativa* cv. Robot plants: these materials are now being used for selecting new genotypes with low/zero saponin levels and high productivity (Rotili, personal communication).

As in almost all food and forage plants, the saponins found in the genus *Medicago* are triterpene glycosides and a number of them have been isolated in roots, shoots, flowers, and seeds. Alfalfa saponins are reported to be responsible for retardation of growth in chickens and swine (Cheeke et al., 1977; Ameenuddin et al., 1983) and for harmful interactions in polygastric animals with the amount and localization of bacteria in the rumen, where they alter fermentation and affect the site and extent of nutrient digestion (Luc and Jorgensen, 1987; Luc et al., 1987). Saponin toxicity has been demonstrated in a number of biological tests with insects (*Tenebrio molitor*) (Pracros, 1988), fungi (*Trichoderma viride*) (Zimmer et al., 1967), and red blood cells (hemolysis) (Jurzysta, 1979) and allelopathic activity on wheat seedling growth (Oleszek and Jurzysta, 1986). In addition, saponins have a wide range of biological activities; for example, they have been shown to lower plasma cholesterol levels (Malinow, 1984) and to have adverse effects on the lining of the small intestine by increasing its permeability to antigens (Sidhu and Oakenfull, 1986; Gee et al., 1989), or by decreasing the availability of certain essential nutrients (dietary vitamins A and E) in chicken (Jenkins and Atwal, 1994).

EXPERIMENTAL RESULTS

The chemical characterization of the highly biologically active saponins from alfalfa and related species is currently in progress in our laboratory. The scope of our study also includes the identification and isolation of purified saponins in appreciable quantities in order to confirm the most significant findings reported to date with saponin mixtures. The hemolytic activity of saponins is today the simplest and fastest bioassay employed in selecting individual alfalfa plants.

By applying a calibration curve obtained from standard saponin solutions, it is possible to estimate the biologically active fraction in the different plants. In this way several species of the genus *Medicago* have been tested in our laboratory, as well as different cultivars and populations belonging to *M. sativa, M. arborea, M. media, M. varia, M. falcata, M. scutellata, M. truncatula*, and *M. lupulina* (Odoardi et al., 1994;

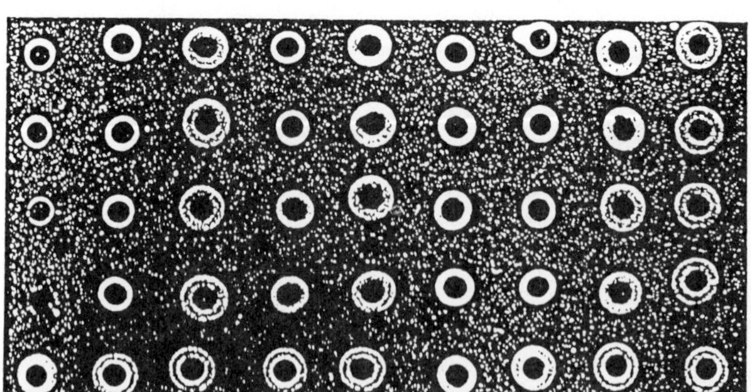

Figure 1. Hemolytic plate showing the hemolytic rings of different *Medicago arborea* genotypes.

Piano et al., 1994). The same procedure is also used when testing for quality characteristics in various forage crops from different peculiar environments, such as bushes from Mediterranean pastures or from tropical sub-arid locations. In Figure 1 an example of a micro-hemolytic plate with *M. arborea* samples is shown.

Even more detailed information on saponin content is obtained by using chemical procedures. Extraction, identification, and quantification of the most abundant saponins and sapogenins, the latter released by acid hydrolysis, have been done mainly on samples from leaves and roots of *M. sativa* cultivars, the most important forage species in Italy, and from leaves of *M. arborea* populations.

Previously reported chemical methods have been utilized (Oleszek, 1988; Tava et al., 1993). The extraction procedure has been optimized by refluxing the defatted samples in 30% aqueous methanol for two hr. The solution obtained is then loaded onto a C_{18} column and saponins are obtained after elution with neat methanol. This procedure allows the removal of interfering extract products such as carbohydrates, simple phenolics, and some flavonoid glycosides (Nowacka and Oleszek, 1994) to yield saponins of sufficient purity to be used for successive analyses and experiments. The repeatability of the method is usually satisfactory, and the saponin content of all plant materials can be determined by using it. The different yields of saponins from *M. sativa* roots and tops, expressed as % of dry matter, are comprised between 2.6-3.8 and 0.3-2.4 respectively, while in *M. arborea* aerial parts saponins ranged from 1.9 to 3.4%. In agreement with the results obtained by the hemolytic test, *M. sativa* shows the highest levels of crude saponins in the roots. This is consistent with the hypothesis of a possible role of saponins as a natural protection for the most vulnerable parts of the plant from insect or microbial attack. Moreover, all the *M. arborea* populations tested show high saponin levels in the aerial parts of the plant.

TLC (Jurzysta and Nowacki, 1979) and HPLC analyses (Oleszek et al., 1990a; Nowacka and Oleszek, 1994) then assisted in the identification of the major components of the saponin extracts by comparison with authentic compounds. The alfalfa saponins have been identified as mono-, bis-, or tridesmosides of medicagenic acid, hederagenic acid, zanhic acid and soyasapogenol B as shown in Figure 2. 3-*O*-ß-(glucopyranosyl) medicagenic acid (3-Glc Med acid, 1) and 3-*O*-ß-(glucopyranosyl)-28-*O*-ß-(glucopyranoside) medicagenic acid (3-Glc, 28-Glc Med acid, 2) are confirmed as the saponins most commonly found in alfalfa. Further, 3-*O*-ß-D-(glucopyranosyl)-28-*O*-[ß-D-xylopyranosyl(1-4)-α-L-rhamnopyranosyl(1-2)-α-L-arabinoside] medicagenic acid (3-Glc, 28-Ara-Rha-Xyl Med acid, 3), 3-*O*-ß-(glucuronopyranosyl)-28-*O*-[ß-D-xylopyranosyl(1-4)-α-L-rhamnopyranosyl(1-2)-α-L-arabinoside] medicagenic acid (3-GlcAc, 28-Ara-Rha-Xyl Med acid, 4), 3-*O*-[ß-D-(glucopyranosyl) (1-2)-ß-D-glucopyranosyl]-28-*O*-[ß-D-xylopyranosyl(1-4)-α-L-rhamnopyranosyl(1-2)-α-L-arabinoside] medicagenic acid (3-Glc-Glc, 28-Ara-Rha-Xyl Med acid, 5), 3-*O*-[α-L-arabinopyranosyl(1-2)-ß-D-glucopyranosyl (1-2)-α-L-arabinopyranoside] hederagenic acid (3-Ara-Glc-Ara Hed acid, 6) and soyasaponin I (7), also reported as common constituents of alfalfa plants (Massiot et al. 1991; Oleszek et al., 1990b) have been identified in our materials. Recently, another major saponin from alfalfa leaves has been identified as zanhic acid tridesmoside, 3-*O*-[ß-D-glucopyranosyl(1-2)-ß-D-glucopyranosyl(1-2)-ß-D-glucopyranosyl]-2ß,3ß,16α-trihydroxyolean-12-ene-23,28-dioic acid-23-*O*-α-L-arabinopyranosyl-28-*O*-[ß-D-apiofuranosyl(1-3)-ß-D-xylopyranosyl(1-4)-α-L-rhamnopyranosyl (1-2)-α-L-arabinoside] (3-Glc-Glc-Glc, 23-Ara, 28-Ara-Rha-Xyl-Api Zan acid, 8) (Oleszek et al., 1992). All these glycosides have been identified in the *Medicago* plants we studied, by comparison with standard compounds. Figure 3 shows HPLC analyses of alfalfa root saponins

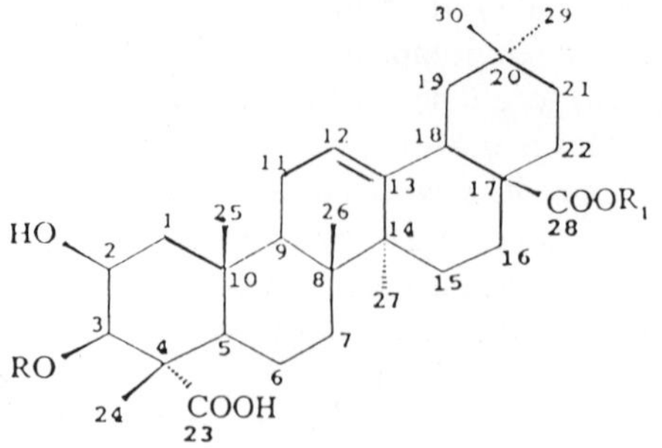

1: R = -Glc
R_1 = H

2: R = R_1 = -Glc

3: R = -Glc
R_1 = -Ara-Rha-Xyl

4: R = -GlcAc
R_1 = -Ara-Rha-Xyl

5: R = -Glc-Glc
R_1 = -Ara-Rha-Xyl

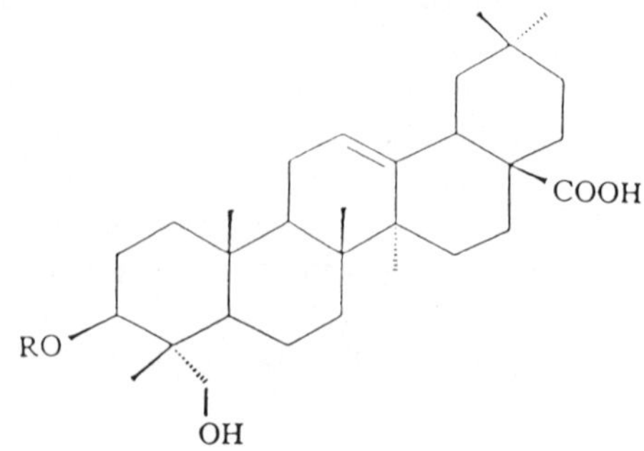

6: R = -Ara-Glc-Ara

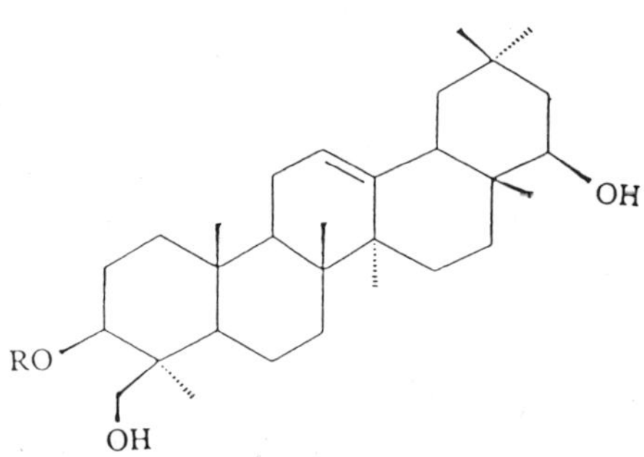

7: R = -GlcAc-Gal-Rha

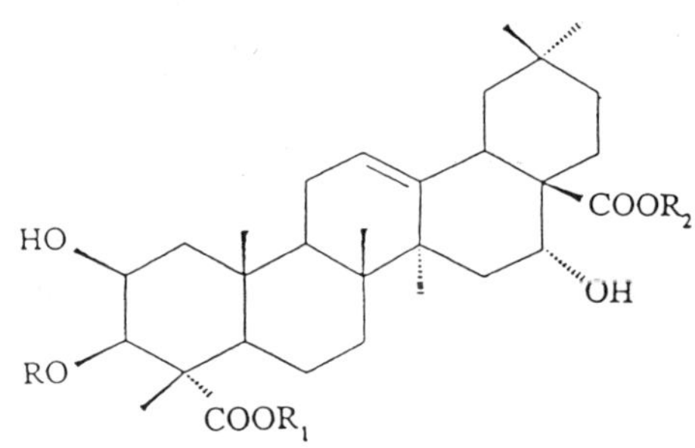

8: R = -Glc-Glc-Glc
R_1 = -Ara
R_1 = -Ara-Rha-Xyl-Api

Figure 2. Structure of identified alfalfa saponins.

derivatized with 4-bromophenacyl bromide. Identified saponins are labelled in the figure. Zanhic acid tridesmoside can only be detected in HPLC by an additional step involving separation from the other saponins and derivatization after alkaline hydrolysis (Nowacka and Oleszek, 1994).

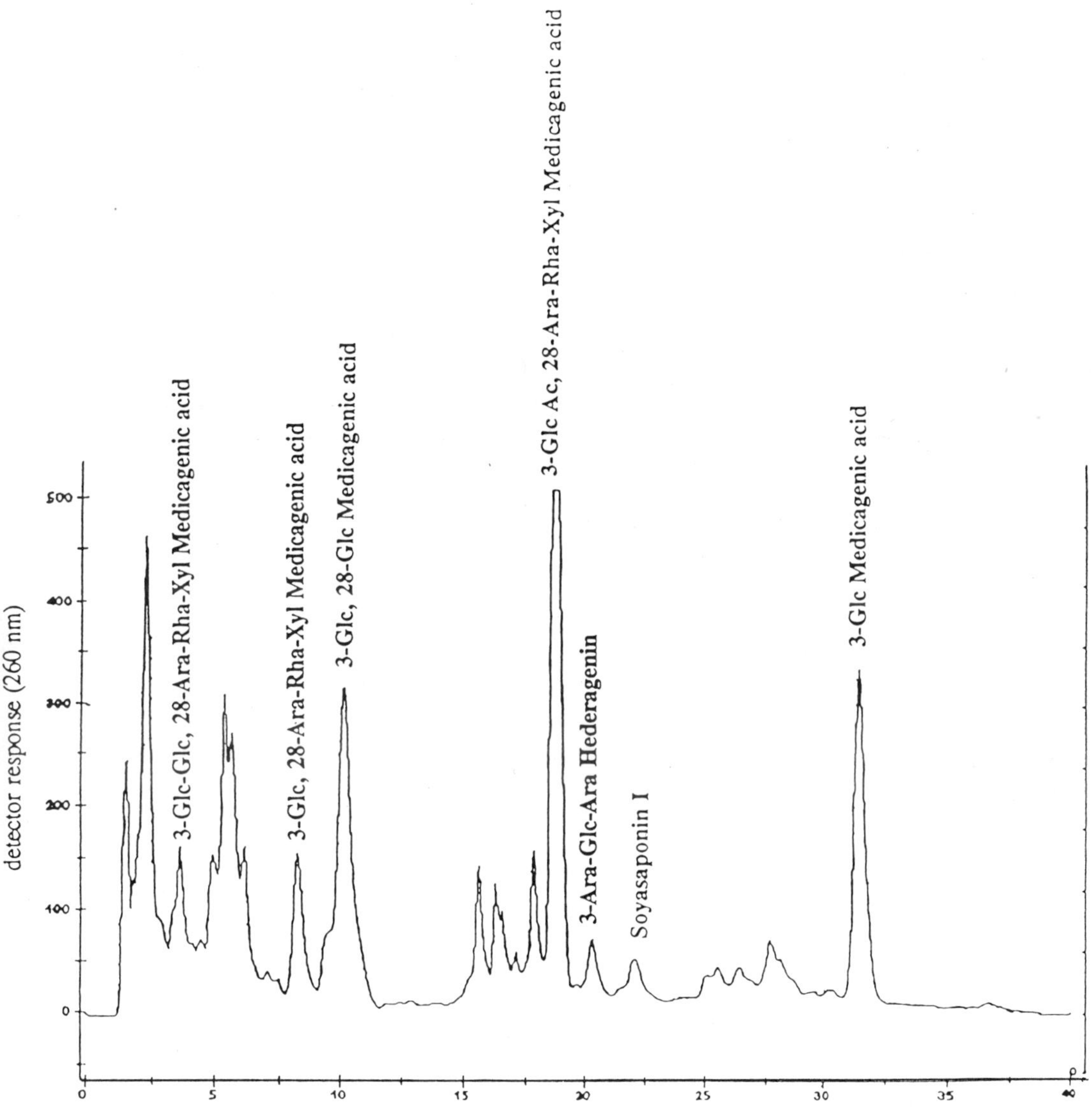

Figure 3. HPLC analyses of derivatized alfalfa root saponins.

On the other hand, information on saponin content can also be obtained by considering the aglycone moieties. For this purpose, identification and quantification of all the sapogenins has been carried out using GC and GC\MS analyses (Jurzysta and Jurzysta, 1978; Brawn et al., 1981; Price et al., 1986; Rao and Boris, 1987). Sapogenins, released after acid hydrolysis of saponins, were methylated and acetylated to give corresponding products that are smoothly eluted under standard GC conditions (Tava et al., 1993). The detection of these aglycone moieties allows the quantitative measurement of differences in sapogenin content among cultivars and varieties. For this purpose, the conditions of the hydrolysis have been tested to determine the optimal reaction conversion of saponins into sapogenins. The best results were obtained upon refluxing for 8 h with 2M HCl in 50% aqueous methanol. More prolonged times gave reduced medicagenic acid yields, owing probably to partial decomposition. Under the conditions of acid hydrolysis, soyasapogenol B was almost completely decomposed into the known artifacts soyasapogenols C, D, E, and F (Jurzysta, 1984; Price et al., 1986). The relative quantities of the soyasapogenols formed varied with hydrolysis time: the yield of soyasapogenol B decreased while other soyasapogenols increased. GC and GC\MS analyses were performed after methylation and acetylation of sapogenins; all the cited aglycones were identified together with oleanolic acid (West, 1979). The identities of individual components were based on retention times and mass spectral data, in agreement with those obtained for the standard compounds and in accordance with literature data (Budzikiewicz et al., 1963). Quantitative determinations of identified aglycones of saponins from *M. sativa* and *M. arborea* are presented in Table 1.

Table 1. Individual sapogenins (mg/g of dry matter) identified in *M. sativa* and *M. arborea* samples.

	M. sativa		*M. arborea*
	roots	tops	tops
Medicagenic acid	4.5 - 5.2	0.2 - 0.6	0.9 - 1.5
Hederagenin	0.8 - 1.3	0.1	<0.1
Zanhic acid	0.1 - 0.4	0.1 - 0.2	0.2 - 1.1
Soyasapogenol B	<0.1	<0.1	<0.1
Soyasapogenol C	0.1 - 0.2	<0.1	0.1
Soyasapogenol D	0.2 - 0.6	0.1	0.1 - 0.3
Soyasapogenol E	0.2 - 0.3	0.1	0.1
Soyasapogenol F	0.2 - 0.3	<0.1	0.1 - 0.2

Analyses were carried out by capillary GC using betulinol acetate as internal standard. This compound has a suitable retention time and gives the correct response factor calculated for different concentrations of standards. Sapogenin content was determined by using a calibration graph. A linear response of derivatized sapogenins (conc./int.std. conc.) *versus* peak area (area/int.std. area) was obtained between 10 and 730 ng per injection. The resolution of derivatives of standard compounds and of the alfalfa sample is shown in Figure 4: all compounds in the complex mixture were baseline separated showing sharp and symmetrical peaks. The GC analyses allowed a practical measura-

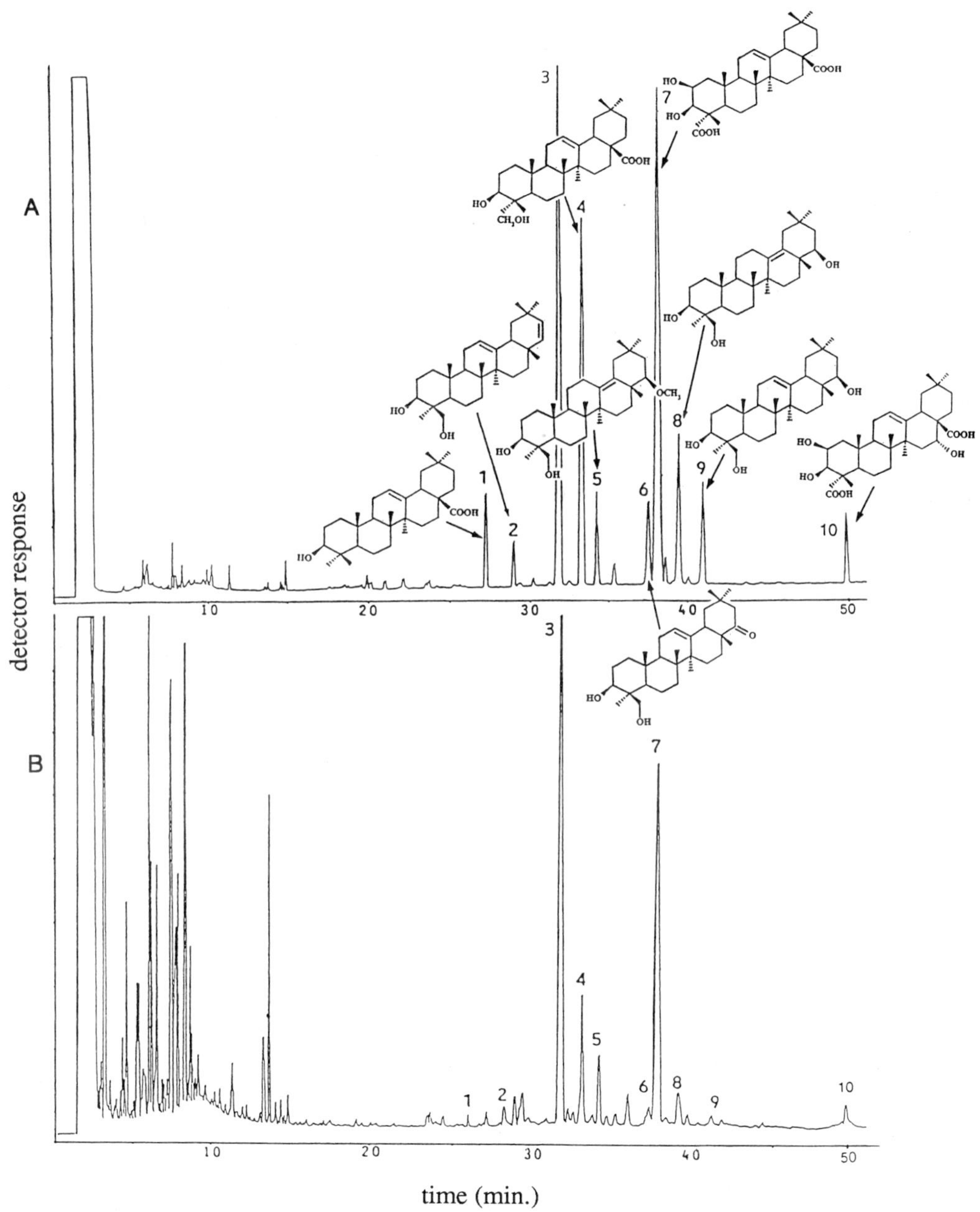

Figure 4. GC analyses of derivatized sapogenins from *M. sativa* roots. A-standard sapogenins; B-from *M. sativa* roots; 1-oleanolic acid; 2-soyasapogenol C; 3-betulinol (internal standard); 4-hederagenin; 5-soyasapogenol D; 6-soyasapogenol E; 7-medicagenic acid; 8-soyasapogenol F; 9-soyasapogenol B; 10-zanhic acid.

ble sensitivity of 10 ng/μl per injection, i.e. a detection limit of 0.003 mg of medicagenic acid/g alfalfa powder. As shown in Table 1, medicagenic acid was the predominant component in the two *Medicago* species studied, followed by hederagenin and zanhic acid.

EFFECT OF SAPONINS ON INSECTS

The toxicity of saponins to insects suggested that they might also provide protection from insect predation (Applebaum et al., 1969). Several investigations have been in fact reported of the effects of saponins from alfalfa and other legumes on numerous classes of insects. Among others, larvae of the grass grub, locust, and flour beetle are strongly affected by alfalfa and other saponin-containing legumes (Applebaum and Birk, 1979). Various alfalfa pests on the contrary are not susceptible to saponins, as they probably have evolved strategies for overcoming the saponin defenses of the plants (Oakenfull and Sidhu, 1989).

The control of insect pests in crops at present largely depends on synthetic chemical compounds, a dependence that will presumably continue for some years. Nevertheless, there are proposals for restriction on the use of synthetic chemical insecticides because of their high cost, impact on the environment, and the increasing frequency of applications due to decreasing effectiveness. All of these factors will make the call for both "soft" alternative chemicals and new technologies to build up plant resistance to insect attack all the more urgent. In addition to the biotechnological approach to producing insect-resistant genetically transformed plants, the isolation, chemical-physical characterization, and biological evaluation of several non-protein compounds that are poisonous to noxious insects may hold promise. The utilization of certain natural, potentially toxic compounds, such as saponins, appears to be attractive and obtainable within a reasonably short time. Our investigations have in fact detected a certain toxic activity of alfalfa saponins on harmful insect pests, e.g. *Lobesia botrana*, not affecting other beneficial insects and mites (Tava et al., 1992).

In more detail, an appreciable quantity of partially purified saponin mixtures from alfalfa leaves was prepared and assayed. Their insecticidal properties were tested on European grape moth (*Lobesia botrana* Den. & Schiff.), Summer fruit tortrix moth (*Adoxophyes orana* F.v.R.), and European corn borer (*Ostrinia nubilalis* Hb.) which are among the most deleterious phytophagous Lepidoptera widespread in Europe.

In Table 2 the insecticidal activity of alfalfa saponins on *L. botrana* is given. The increasing saponin rates added to the artificial diet caused a rise in larval mortality from 11.3% (similar to the control) with 1 ppm, up to 46.1% with 1000 ppm. The contact effect of saponins has been also monitored, with the larvae for one day over a surface treated with saponins: it accounted for a maximum of 22.7% mortality. No appreciable differences in insecticidal activity have been detected between the crude saponins derived from leaves or roots. A further experiment has been done to verify the toxicity mechanism of saponins when ingested by the insects. At least in our experimental conditions, saponin toxicity is thought to be exerted by the well-known ability of certain groups of saponins to form complexes with cholesterol; a decrease in larval mortality has been in fact verified when increasing levels of cholesterol have been added to the artificial diet with saponins: cholesterol overcomes the deleterious saponin effects. The final tests with *L. botrana* have been the determination of the lethal dose (LD_{50}) of saponins, which gave a result of 1,342 ppm per larva, and the control of toxicity against insect eggs, that was negative.

Table 2. Characterization of the insecticidal activity of alfalfa crude saponins on European grape moth (*Lobesia botrana* Den. & Schiff.) larvae.

1 - Ingestion: saponin toxicity when added to the artificial diet

Saponin concentration (ppm)	% mortality (Abbott)
control	12.8
1	11.3
10	33.7
100	38.6
1,000	46.1

2 - Contact: larval mortality after 1 day contact on surfaces treated with a 1,000 ppm saponin solution

	% mortality (Abbott)
control	10.0
contact effect	22.7

3 - Mechanism of action: the insecticidal activity of alfalfa saponins (1,000 ppm saponin in the artificial diet) is inhibited by cholesterol added to the diet

cholesterol (ppm)	% mortality (Abbott)
0	46.1
500	18.6
1,000	16.9
control (0% saponins)	12.8

4 - Lethal dose determined per larva:

$$LD_{50} = 1{,}342 \text{ ppm} \pm 115$$

5 - Toxic effects on eggs: no ovicidal activity was observed

Further experiments carried out with *A. orana* larvae assessed for the combined effects of epidermal contact and ingestion of leaflets treated with alfalfa saponin solutions. Results are summarized in Figure 5: larval viability was drastically affected, leading to mortality as high as 84.9%.

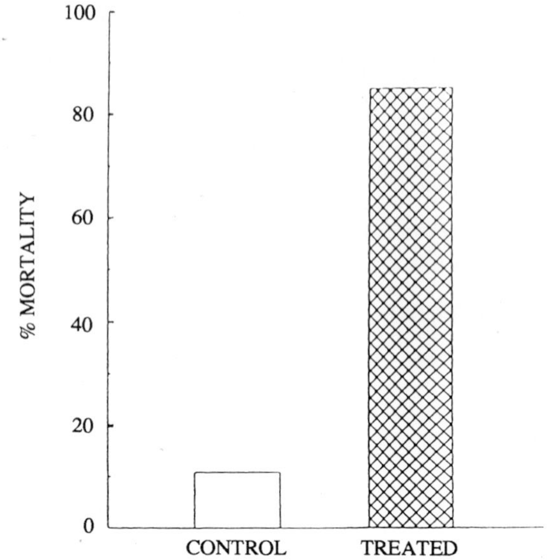

Figure 5. **Insecticidal activity of alfalfa saponins tested on Summer fruit tortrix moth (*Adoxophyes orana* F.v.R.) larvae reared on common bean plantlets (feeding and contact effects)**

Finally, a certain inhibitory effect of alfalfa saponins has been detected on growth and development of European corn borer larvae fed artificial saponin-supplemented diet (Table 3): their larvae developed more slowly, forming fewer chrysalids. Also, a similar inhibitory effect of the alfalfa saponin mixtures has been detected on the larval growth and development of *Tenebrio molitor* and *Spodoptera littoralis* (data not shown). On the other hand, alfalfa saponins turned out to be completely harmless to nematodes: no toxicity at all has been shown by our saponin samples tested on *Heterodera schachtii* Schmidt.

Table 3. Inhibitory activity of alfalfa saponins on European corn borer (*Ostrinia nubilalis* Hb.) larvae when added to the artificial diet.

saponin conc. (ppm)	larval weight (mg)				chrysalids formed	
	4 d	6 d	8 d	% of control	mean days	No.
control	22.9	47.2	85.4	100.0	17.4	23\24
4,000	16.4	33.9	63.7	74.7	18.6	22\24
6,000	11.8	25.2	44.7	52.4	20.5	20\24

CONCLUSIONS

All the toxic effects of saponins are closely related to their chemical structure, and therefore structurally different saponins from different plant sources vary widely in their toxicity. Pharmaceutical application of saponin-rich mixtures are reported, from the old Chinese remedy ginseng to the modern uses of saponins as antiinflammatory and analgesic agents. Our group also tested purified alfalfa saponins as antitumor agents, and the result was an evident inhibition of growth (40-50% of control) of the human leukemic cell line K562 *in vitro* (Figure 6). In view of all the possible applications and uses of alfalfa saponins, with their physiological properties, further and deeper investigation is needed to clarify their activity as individual, pure compounds.

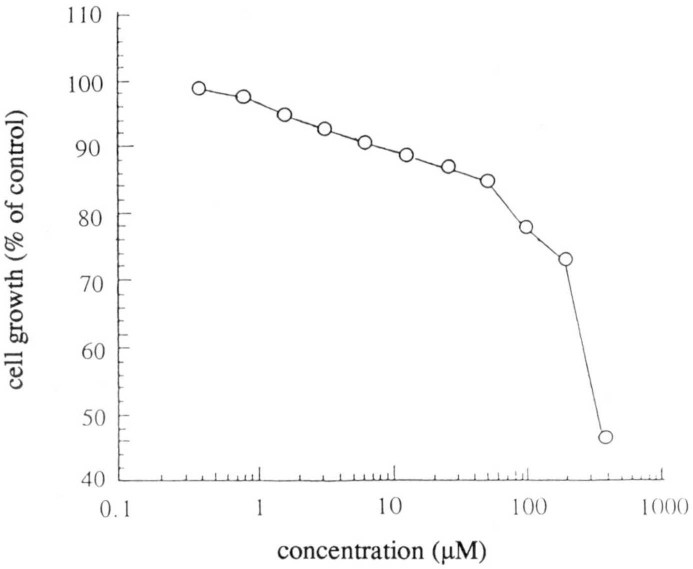

Figure 6. Inhibition of human leukemic cell line K562 growth by alfalfa crude saponins.

ACKNOWLEDGMENTS

We wish to thank Drs. F. Coppolino and D. Forti for providing insect tests used in this study, and Dr. C. Nastruzzi for the antitumoral tests. This research was supported by the Italian Ministry of Agriculture, in the framework of the project "Resistenze Genetiche delle Piante Agrarie agli Stress Biotici ed Abiotici".

REFERENCES

Abbott, W., 1925, A method of computing the effectivenes of an insecticide, *J. Econ. Entomol.* 18:265.

Ameenuddin, S., Bird, H.R., Pringle, D.J., and Sunde, M.L., 1983, Studies on the utilization of leaf protein concentrates as a protein source in poultry nutrition, *Poultry Sci.* 62:505.

Applebaum, S.W., and Birk, Y., 1979 In: *Herbivores - Their Interaction with Secondary Plant Metabolites*, Rosenthal, G.A. and Janzen, D.H., Eds., Academic Press, New York.

Applebaum, S.W., Marco, S., and Birk, Y., 1969, Saponins as a possible factor of resistance of legume seeds to attack of insects, *J. Agric. Food Chem.* 17:618.

Bocsa, I., 1990, Evaluation of the breeding results on low saponin content in lucerne, *Proc. IX Eucarpia Medicago sativa Group Meeting*, Szarvas, Hungary, p. 37.

Brawn, P.R., Lindner, N.M., Miller, J.M. and Telling, G.M., 1981, A gas chromatographic method for the determination of medicagenic acid in lucerne (alfalfa) leaf protein concentrate, *J. Sci. Food Agric.* 32:1157.

Budzikiewicz, H., Wilson, J.M. and Djerassi, C., 1963, Mass spectrometry in structural and stereochemical problems. XXXII. Pentacyclic triterpenes, *J. Am. Chem. Soc.* 85:3688.

Cheeke, P.R., 1971, Nutritional and physiological implications of saponins: a review, *Can. J. Anim. Sci.* 51:621.

Cheeke, P.R., 1983, Biological properties and nutritional significance of legume saponins. In: *Leaf Protein Concentrate*, Telek L. & Graham H., Eds., AVI Publishing, Westport, CT, USA.

Cheeke, P.R., Kinzell, J.H., and Pedersen, M.W., 1977, Influence of saponins on alfalfa utilization by rats, rabbit and swine, *J. Anim. Sci.* 45:476.

Fenwick, G.R., Price, K.R., Tsukamoto, C., and Okubo, K., 1991, Saponins. In: *Toxicant Substances in Crop Plants*, Felix D'Mello, J.P., Duffus, C.M. & Duffus, J.H., Eds., Royal Society of Chemistry, London, UK.

Gee, J.M., Price,K.R., Ridout, C.L., Johnson, I.T., and Fenwick, G.R., 1989, The effect of some purified saponins on transmural p.d. in mammalian small intestine, *Toxicol. in vitro* 3:85.

Jenkins, K.J., and Atwal, A.S., 1994, Effects of dietary saponins on fecal bile acids and neutral sterols, and availability of vitamin A and E in the chick, *J. Nutr. Biochem. Newton* 5:134.

Jurzysta, M., 1979, Haemolytic micromethod for rapid estimation of toxic alfalfa saponin, *Acta Agrobot.* 32:5.

Jurzysta, M., 1984, Transformation of soyasapogenol B into soyasapogenols C,D and F under acidic conditions, *Proc. 14th Int. Symp. Natural Products*, p. 127.

Jurzysta, M. and Jurzysta, A., 1978, Gas-liquid chromatography of trimethylsilyl ethers of soyasapogenols and medicagenic acid, *J. Chromatogr.* 148:517.

Jurzysta, M., and Nowacki, E., 1979, Saponins of the genus *Medicago*, *Acta Agrobot.* 32:13.

Luc, C.D., and Jorgensen, N.A., 1987, Alfalfa saponins affect site and extent of nutrient digestion in ruminants, *J. Nutr.* 117:919.

Luc, C.D., Tsai, L.S., Schaefer, D.M. and Jorgensen, N.A., 1987, Alteration of fermentation in continuous culture of mixed rumen bacteria by isolated alfalfa saponins, *Dairy Sci.* 70:799.

Malinow, M.R., 1984, Saponins and cholesterol metabolism, *Atherosclerosis* 50:117.

Massiot, G., Lavaud, C., Besson, V., Le Men-Olivier, L. and Binst, G., 1991, Saponins from aerial parts of alfalfa (*Medicago sativa*), *J. Agric. Food Chem.* 39:78.

Nowacka, J., and Oleszek, W., 1994, Determination of alfalfa (*Medicago sativa*) saponins by high-performance liquid chromatography, *J. Agric. Food Chem.* 42:727.

Oakenfull, D. and Sidhu, G.S., 1989, Saponins, In: *Toxicants of Plant Origin*, Cheeke, P.R., Ed., CRC Press, Inc., Boca Raton, Florida.

Odoardi, M., Santelli, G., Cremona, R., Berardo. N., Stringi, L. and Giambalvo, D., 1994, Screening *Atriplex halimus* and *Medicago arborea* populations for saponin content, *Proc. Eucarpia Fodder Crop Section Meeting*, Brugge, Belgium, p. 103.

Oleszek, W., 1988, Solid phase extraction-fractionation of alfalfa saponins, *J. Sci. Food Agric.* 44:43.

Oleszek, W., and Jurzysta, M., 1986, Isolation, chemical characterization and biological activity of alfalfa (*Medicago media* Pers.) root saponins, *Acta Soc. Bot. Pol.* 55:23.

Oleszek, W., Jurzysta, M., Ploszynski, M., Colquhoun, I.J., Price, K.R., and Fenwick, G.R., 1992, Zanhic acid tridesmoside and other dominant saponins from alfalfa (*Medicago sativa* L.) aerial parts, *J. Agric. Food Chem.* 40:191.

Oleszek, W., Jurzysta, M., Price, K.R., and Fenwick, G.R., 1990a, High performance liquid chromatography of alfalfa root saponins, *J. Chromatogr.* 519:109.

Oleszek, W., Price, K.R., Colquhoun, I.J., Jurzysta, M., Ploszynski, M., and Fenwick, G.R., 1990b, Isolation and identification of alfalfa (*Medicago sativa*) root saponins: their activity in relation to a fungal bioassay, *J. Agric. Food Chem.* 38:1810.

Pedersen, M.W., Berrang, B., Wall, M.E., and Davis, K.H.,Jr., 1973, Modification of saponin characteristics of alfalfa by selection, *Crop Sci.* 13:731.

Pedersen, M.W., and Wang, L.C., 1971, Modification of saponin content of alfalfa through selection, *Crop Sci.* 11:833.

Piano, E., Berardo, N., Pecetti, L., Valentini, P., and Odoardi, M., 1994, Some observations on the qualitative characteristics of grazing-type lucerne, *Proc. Eucarpia Fodder Crop Section Meeting*, Brugge, Belgium, p. 95.

Pracros, P., 1988, Mesure de l'activité des saponins de la lucerne par les larves du ver de la farine: *Tenebrio molitor* L. (Coléoptère, *Tenebrionidae*). I. - Comparison avec les résultats de divers test biologiques, *Agronomie* 8:257.

Price, K.R., Fenwick, G.R. and Jurzysta, M., 1986, Soyasapogenols - separation, analysis and interconversions, *J. Sci. Food Agric.* 37:1027.

Price, K.R., Johnson, I.T., and Fenwick, G.R., 1987, The chemistry and biological significance of saponins in food and feedingstuffs, *CRC Crit. Rev. Food Sci. Nutr.* 26:27

Rao, D. and Boris, G., 1987, Simple gas chromatographic method for the determination of medicagenic acid in alfalfa (*Medicago sativa*), *J. Chromatogr.* 410:169.

Rotili, P., and Zannone, L., 1981, Alfalfa breeding for forage yield and low saponin content. *Proc. XIV Int. Grassl. Congress*, Lexington, KY, USA, p. 229.

Sidhu, G.S., and Oakenfull, D.G., 1986, A mechanism for the hypocholesterolaemic activity of saponins, *Br. J. Nutr.* 55:643.

Tava, A., Forti, D., and Odoardi, M., 1992, Alfalfa saponins: isolation, chemical characterization and biological activity against insects, *Proc. X Eucarpia Medicago spp Group Meeting*, Lodi, Italy, p. 283.

Tava, A., Oleszek, W., Jurzysta, M., Berardo, N., and Odoardi, M., 1993, Alfalfa saponins: isolation and quantification in two different cultivars, *Phytochem. Anal.* 4:269.

West, L.G., 1979, Identification of oleanolic acid from saponins present in lucerne (alfalfa) roots, *J. Sci. Food Agric.* 30:540.

Zimmer, D.E., Pedersen, M.W., and McGuire, C.F., 1967, A bioassay for alfalfa saponins using the fungus *Trichoderma viride* Pers., *Crop Sci.* 7:223.

THE ROLE OF CARDENOLIDES IN A CRUCIFER-INSECT RELATIONSHIP

J.A.A. Renwick

Boyce Thompson Institute
Tower Road
Ithaca, NY 14853

INTRODUCTION

The cardenolides belong to a group of steroidal compounds, which, along with the bufadienolides, are referred to as cardiac glycosides because of their cardiotonic properties. As a result of this bioactivity, they are widely used to treat congestive heart failure. The bufadienolides are characteristic of toads in the family Bufonidae, which secrete the compounds from specialized glands when they are threatened by predators. Cardenolides, however, are produced almost exclusively by plants and serve to protect these plants from herbivory.[1] The protective qualities of cardiac glycosides stem from their extreme toxicity to most higher animals and insects, due to inhibition of Na^+K^+-ATPase and the resulting effects on the sodium pump. However, at low concentrations, they are valuable medicinal drugs for humans and they have also been shown to have cytotoxic activity against cancer cell lines.[2]

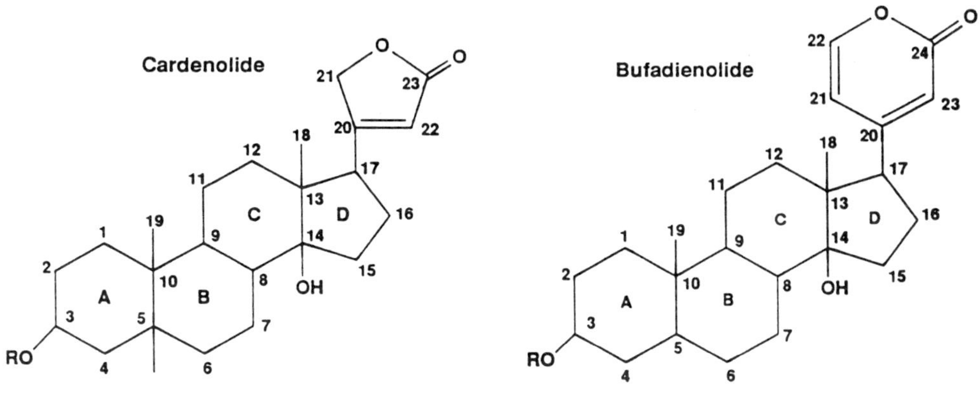

Figure 1. General structures of the cardiac glycosides.

The chemical structures of the two classes of cardiac glycosides differ in the lactone ring attached to C-17 of the basic steroid part of the aglycone (Fig.1). The bufadienolides have a six-membered doubly unsaturated ring, whereas the cardenolides have a five-membered butenolide ring. From 1 to 4 sugars are usually attached through hydroxyl groups at C-2 or C-3 of the aglycone. Solubility and cell penetrability are also affected by the presence of methyl, hydroxyl and aldehyde groups which may occur at different positions on the aglycone.

Cardenolides have been found in at least 15 different plant families (Table 1). The most prominent of these are the dogbane family (Apocynaceae), milkweeds (Asclepiadaceae) and the figwort family (Scrophulariaceae). However, their occasional occurrence in other families such as the Cruciferae may have important biological significance.

Table 1. Plant families and relevant common names of plants producing cardenolides (from Joubert).[1]

Apocynaceae	Dogbane family (*Nerium* spp., e.g. oleander)
Asclepiadaceae	Milkweeds
Crassulaceae	
Celastraceae	
Cruciferae (=Brassicaceae)	Wild mustards
Iridaceae	
Leguminoseae	
Liliaceae	
Melianthaceae	
Moroideae	
Ranunculaceae	
Rosaceae	
Santalaceae	
Scrophulariaceae	Figwort family; e.g. *Digitalis* (foxglove)
Tiliaceae	

The production of cardenolides by plants has undoubtedly evolved as a defense mechanism against potential herbivores. The acute mammalian toxicity of these compounds has long been recognized, and crude extracts of cardenolide-containing plants have been used as poisons for various purposes, and, at medicinal doses, as heart tonics, diuretics and emetics.[1] Many examples of livestock poisoning as a result of grazing on cardenolide-containing plants have been reported.[1,3] However, the bitter taste of these compounds usually deters animals from continued feeding, and they soon learn to avoid the poisonous plants. Most cases of poisoning are a result of plants such as milkweeds being included with the hay that is fed to domesticated livestock. Herbivory by most insects also seems to be prevented as a result of deterrent and toxic effects, although some insects as well as mammals have apparently adapted to this natural defense.

ADAPTATION AND EXPLOITATION

Despite the potent toxicity and bitter taste of cardenolides in plants, many animals are unaffected by their presence. In southern Africa, the spring hare, *Pedetes capensis*, and the Cape porcupine, *Hystrix africae-australis*, readily eat bulbs of the cardenolide-containing *Urginea*, *Ornithoglossum*, *Homeria* and *Moraea* spp., with no ill effect.[1] Many insects have adapted to the presence of cardenolides in their host plants, and at least 8 orders of insects are known to feed on members of the Asclepiadaceae and Apocynaceae.[4] In general, these appear to be specialists on particular plants within these families, and many are brightly colored, presumably to provide a warning to potential predators. Some species, such as the North African grasshopper, *Poekilocerus bufonius*, have defensive glands that are used to secrete cardenolides when attacked. The specific cardenolides that are ejected from the glands are similar to those found in the host plant, and are therefore likely to be directly formed from these.[5]

Sequestration and Tolerance by Insects

The presence of cardenolides in insects that feed on high cardenolide plants has long been recognized. However, differences in feeding habits of insects may give rise to different profiles of the compounds in body tissues. The oleander aphid, *Aphis nerii* contains specific cardenolides that are present in *Nerium oleander* and *Asclepias curassavica* on which they feed. But oleandrin and calactin, which are prominent constituents of the plants, are absent from insect tissues. Rothschild et al.[6] have suggested that this could indicate the absence of these compounds in the phloem on which the aphids feed. Similarly, lygaeid bugs feeding on *Nerium oleander* lacked oleandrin in their tissues, but contained other cardenolides that were presumably obtained from the host plant.[4] The same or similar cardenolides were found in eggs and in unfed larvae, indicating that storage occurs and the chemicals are retained through the different life stages of the insect. Several genera of the lygaeid subfamily Lygaeinae have specialized dorsolateral glands that concentrate the cardenolides for secretion in response to predators.[7] The distaste of insects to predators is thought to vary considerably according to the host species and the plant part where feeding occurs, since a different "palatability spectrum" will be obtained in each situation.[8]

In some insects such as the dogbane tiger moth, *Cycnia tenera*, cardenolide sequestration appears to be unregulated, so that the profile in the insect resembles that of the plant on which it fed.[9] But permeability of the midgut of insects has been suggested as a limiting factor in the uptake and concentration of cardenolides in the diet.[10] This might imply that polarity of the individual compounds is important. However, the lygaeid milkweed bug, *Oncopeltus fasciatus*, exhibits a high degree of selectivity in the compounds that are sequestered. When reared on seeds of two milkweed species that differed in their cardenolide profiles, similar profiles were found in the insects. In this case, it was suggested that polarity of the compounds was not the only factor involved in sequestration, but that metabolism and differential excretion also occurred.[11] The ability to sequester cardenolides appears to be a characteristic of specialist insects, whereas generalists that can tolerate the compounds are likely to excrete them.[12]

The processing and sequestration of cardenolides by larvae of the monarch butterfly, *Danaus plexippus*, have been studied in some detail. When larvae were reared on two different food plants, their cardenolide content reflected their diet, but some cardenolides were selectively sequestered or altered. Differences in storage efficiency and egestion of cardenolides of different polarity were noted.[13] Other studies have shown that larvae can incorporate lower doses of cardenolides more efficiently than higher doses, and that glycosides are favored over the aglycones for storage. Furthermore, the genins of known cardenolides may be converted to polar glycosides for storage.[14] Considerable variation in cardenolide content of monarch butterflies occurs, thus providing a different palatability spectrum, and differences can be related to the geographical location of the larval host plants.[15] Furthermore, the cardenolide pattern in butterflies provides a fingerprint that can be used to determine the plants on which the larvae fed and to map the migration route of the butterflies.[16]

The close association of monarchs with cardenolide-containing plants has resulted in considerable speculation about the possible role of these compounds in guiding the butterflies to their host plants. Most oviposition has been found to occur on plants that contain an intermediate level of cardenolides.[17] However, recent work has shown that cardenolides cannot account for host recognition and that a group of flavonol glycosides seems to be primarily responsible.[18]

Production by Insects

Although most cardenolides in nature are produced by plants, some chrysomelid beetles are known to synthesize their own. Several beetle species were found to secrete complex mixtures of cardenolides from pronotal and elytral glands when feeding on plants that were devoid of cardenolides.[19,20] Although the larvae of these beetles do not have defensive glands, they also produce the cardenolides.[21] The secretions from species of *Chrysolina* beetles are believed to act as a defense against attacking predators, and both deterrent and toxic effects have been reported for *C. herbacea* on ants.[22] This protection presumably exists throughout the whole life cycle of the insects, as cardenolides have also been detected in eggs of *C. fuliginosa*.[23] The biosynthesis of cardenolides in *Chrysolina coerulans* has been shown to occur from cholesterol, by means of a pathway involving a C_{21} intermediate, just as in plants.[24]

Effects on Predators

The benefits of cardenolide sequestration by insects have been demonstrated in several cases. When experimental blue jays were allowed to feed on adult monarchs, they reacted with characteristic vomiting, and quickly learned to avoid these butterflies.[25] Similarly, the effect of sequestered cardenolides in the lygaeid bug, *Caenocoris nerii*, was demonstrated using quails for bioassays. Bugs fed on cardenolide-containing oleander seeds were much less likely to be eaten or attacked than those fed on sunflower seeds.[26]

The massive aggregation of monarch butterflies at their overwintering sites in Mexico would appear to offer a vast food resource for predaceous birds and rodents. However, the stored cardenolides appear to provide effective protection against most of these. Out of thirty-two avian species in the area, only two were capable of selective feeding on the butterflies.[27] In a four year study of rodent feeding and breeding, only one of five species of mice at the location was found to eat significant numbers of the butterflies.[28] The acceptance of butterflies by this species, *Peromyscus melanotis*, was not affected by cardenolide concentration, but the mice have developed a technique for avoiding the cardenolide-laden cuticular material.[29] Feeding deterrent activity was demonstrated for two species that rejected the monarchs, using digitoxin in bioassays with artificial diet.[30] The feeding deterrent activity of cardenolides was further confirmed using five species of *Peromyscus*, and differences in taste sensitivity of these species could explain their differential avoidance of the butterflies.[31]

The presence of cardenolides in an insect can sometimes have unique consequences for the consumer. In addition to feeding deterrent and toxic effects, these compounds may disrupt the normal behavior of a predator. When the orb-web spider, *Zygiella x-notata*, consumes an oleander aphid that has fed on milkweed, it builds a severely disrupted web, while webs of spiders feeding on non-toxic aphids remain normal. This reaction does not appear to depend on the presence of a specific cardenolide, since similar disruptive effects were obtained by feeding digitoxin to the spiders.[32]

CARDENOLIDES IN CRUCIFERS

Certain members of the Cruciferae are known to produce cardenolides, which are believed to constitute a second line of defense against herbivores. The first line of defense is provided by the glucosinolates, a group of sulfur-containing glycosides that are distasteful to most insects.[33] However, many insects have adapted to the glucosinolates and have become specialists on crucifers. Only a few crucifers are refused by these specialist insects, and the presence of cardenolides has been suggested as a possible explanation in some cases. Nielsen[34] was able to show that crucifer-adapted flea beetles were deterred from feeding by selected cardenolides. A general lack of herbivory and the refusal of *Erysimum asperum* by *Pieris* butterflies in a montane region of Colorado provided a basis for some speculation about the chemistry of that plant.[35,36] Several species of *Erysimum* and *Cheiranthus* are known to contain cardenolides.[37] However, it was not until detailed studies of *Pieris* butterfly behavior were conducted that the role of cardenolides in regulating host selection was clearly shown.

The *Erysimum-Pieris* Interaction

The cabbage butterfly, *Pieris rapae*, often referred to as the imported cabbageworm since it was introduced to North America from Europe, will oviposit on most crucifers and a few additional plants that produce glucosinolates. However, a few crucifers are avoided, and the presence of oviposition deterrents in these plants has been demonstrated.[38] *Erysimum cheiranthoides* (wormseed mustard) was avoided by these butterflies, and a potent deterrent was extracted from aqueous extracts into n-butanol. Bioassay-guided isolation of the oviposition deterrent led to a group of cardenolides,[39] and the most active of these were identified as erysimoside and erychroside. An additional cardenolide in the active fraction from HPLC was identified as erycordin, but when tested separately, it had little biological activity.[40] The active glycosides have strophanthidin as the aglycone, attached to a substituted 2,6-dideoxy sugar. However, strophanthidin itself, and helveticoside, which lacks substitution on the dideoxy sugar, were inactive.

Larvae of *P. rapae* also refused to feed on *Erysimum cheiranthoides*, and the presence of a feeding deterrent was demonstrated. Cardenolides were also found to be involved, but the most active compounds were different from the strophanthidin-based glycosides that deterred oviposition.[41] These were subsequently identified as digitoxigenin-based cardenolides. A monoglycoside, digitoxigenin 3-*O*-β-D-glucoside, and two diglycosides, glucodigigulomethyloside and glucodigifucoside, were highly active. An additional compound identified as cheirotoxin was only slightly active.[42] The differential activity of these different groups of cardenolides would suggest that *E. cheiranthoides* has separate as well as overlapping defenses against adults and larvae of the cabbage butterfly and that distinct structure-activity relationships exist.[43]

Comparative studies of *Pieris rapae* and the North American native species, *P. napi oleracea*, have revealed distinct differences in the responses of these insects to cardenolides. *P. napi oleracea* readily accepts *E. cheiranthoides* for oviposition, and actually prefers that plant over cabbage.[44] The native butterfly is much less sensitive to the cardenolide deterrents and is much more sensitive to stimulants in the plant. These results prompted a detailed study of structure-activity relationships for the responses of the two butterfly species to a range of cardenolides[45] (Table 2). Most of the eighteen compounds tested were deterrent to oviposition by both insects, but to significantly different degrees. *P. rapae* butterflies were strongly deterred by K-strophanthin-β, cymarin, erysimoside, convallatoxin and K-strophanthoside. In the case of *P. napi oleracea*, the most deterrent compounds were erychroside, cymarin, erysimoside, convallatoxin and K-strophanthoside. Thus both similarities and differences were found in the responses of the two species. However, it was clear that the strophanthidin-based glycosides were generally more active than the digitoxigenin-based ones, and the number and type of sugars had a significant effect on activity.[45] In general, *P. napi oleracea* was less sensitive than *P. rapae* to most of the cardenolides (Table 2).

Effectiveness of Cardenolides as Deterrents

The different responses of the two *Pieris* species to cardenolides and the role of stimulants in host plant recognition pointed to the importance of identifying the stimulants for oviposition. Analysis of surface extracts of cabbage leaves resulted in the identification of glucobrassicin as the most effective stimulant for *P. rapae* oviposition.[46] However, glucobrassicin is absent from some host plants of this butterfly, and a survey of other crucifers and non-crucifer hosts showed that a major glucosinolate of each plant may serve as the oviposition stimulant. However, aromatic and indolyl members of the group were much more active.[47] Contrary to this, *P. napi oleracea* was more strongly stimulated by the aliphatic glucosinolates, particularly when a sulfinyl or sulfonyl was present in the side chain.[43] As a result of the involvement of stimulants in the acceptance or rejection of host plants, it is important to realize that a balance between stimulants and deterrents may determine the outcome of the insect's taste experience and subsequent behavioral response.[48]

The dynamic nature of the inputs provided by stimulants and deterrents at a leaf surface was emphasized by growing seedlings of *Erysimum cheiranthoides* under different nutrient conditions. At three different levels of nitrogen, considerable variation in the ratio of cardenolides to glucosinolates in the foliage was found.[49] At low levels of nitrogen, elevated C/N ratios were obtained, and the ratio of cardenolides to glucosinolates in the leaf tissue was higher. However, the concentration of cardenolides at the leaf surface was actually lower. This might suggest that a butterfly would be more likely to oviposit on a nutrient stressed plant, but the hatching larvae would be confronted with a higher concentration of deterrent cardenolides after penetrating the surface. Other stresses may also cause induction of allelochemical biosynthesis, but effects on cardenolide production have yet to be reported.

The possible influence of cardenolides on other pests of crucifers have not been studied in any detail. The possible deterrent effect of *Erysimum cheiranthoides* constituents was tested on the diamondack moth, *Plutella xylostella*, and this insect was completely unaffected by the presence of the cardenolides. Analysis of frass from larvae feeding on *E. cheiranthoides* showed that the whole suite of cardenolides passed through the gut essentially unchanged.[50]

Recent experiments with larvae of *Pieris rapae* have demonstrated the effect of dietary experience on the response of larvae to deterrents. When the larvae were reared on nasturtium (*Tropaeolum majus*) or wheat germ diet, they were less sensitive to the deterrents present in nasturtium.[51] When reared on cabbage treated with cardenolides, these insects also remained relatively insensitive to a range of deterrents, including most of the cardenolides.[52] However, digitoxin proved to be an exception. This is the most potent of all the deterrents tested, and the larvae remained just as sensitive to it when they were reared on treated cabbage plants or on wheat germ diet. This might suggest that digitoxin has properties that affect the taste receptors of the insect in some unique way.

CONCLUSIONS

Cardenolides clearly serve to protect plants from a wide array of herbivores, including mammals as well as insects. However, many insects have adapted to this defense and may even sequester the cardenolides for their own use to protect themselves from predators. In crucifers, the cardenolides provide a second line of defense against herbivores, although the degree of effectiveness varies considerably for different insect species. Specific structural

Table 2. Structures and comparative activities of cardenolides as oviposition deterrents for *Pieris rapae* and *P. napi oleracea*. Oviposition deterrent index (ODI) is a measure of activity based on the proportion of eggs laid on control vs treated plants in a choice bioassay. (Data from ref. 45).

Structure	Compound	ODI *P. rapae*	ODI *P. napi*
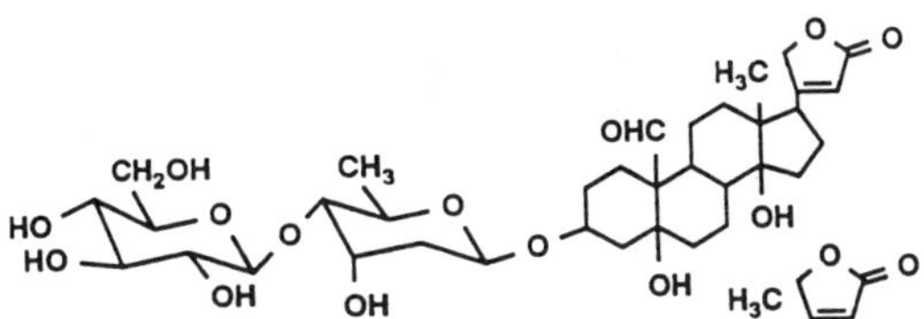	Erysimoside	54	39
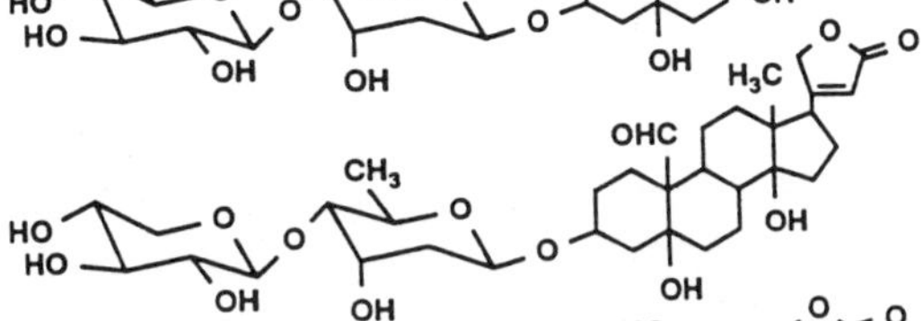	Erychroside	55	43
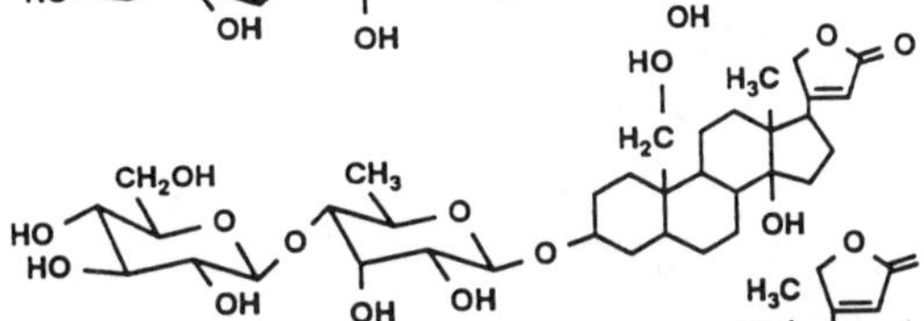	Erycordin	19	1
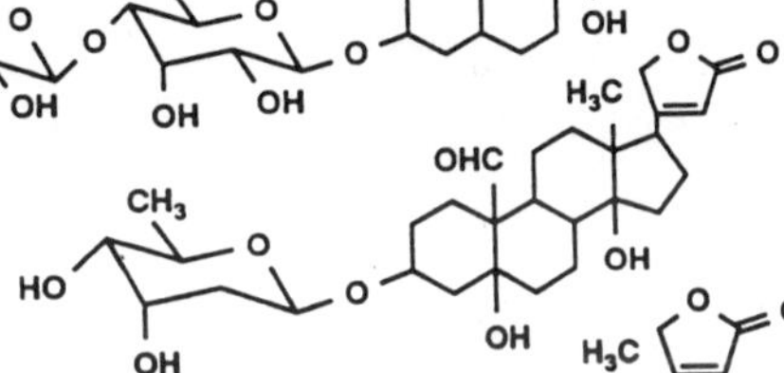	Helveticoside	10	3
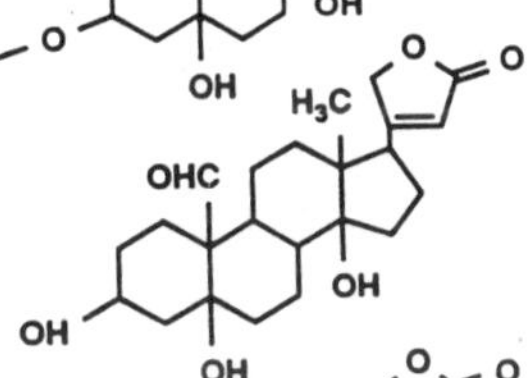	Strophanthidin	0	11
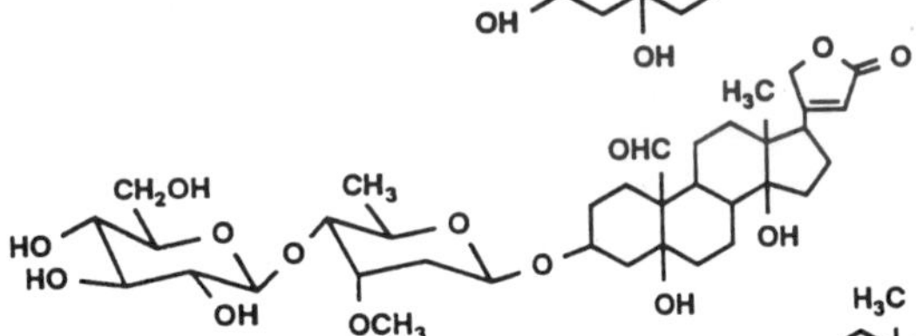	K-Strophanthin-ß	47	10
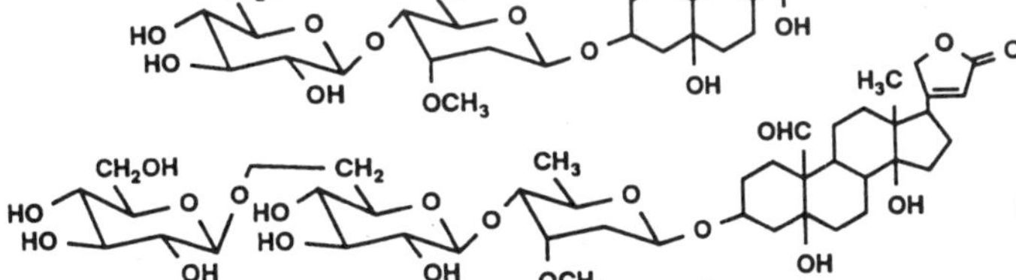	K-Strophanthoside	61	7
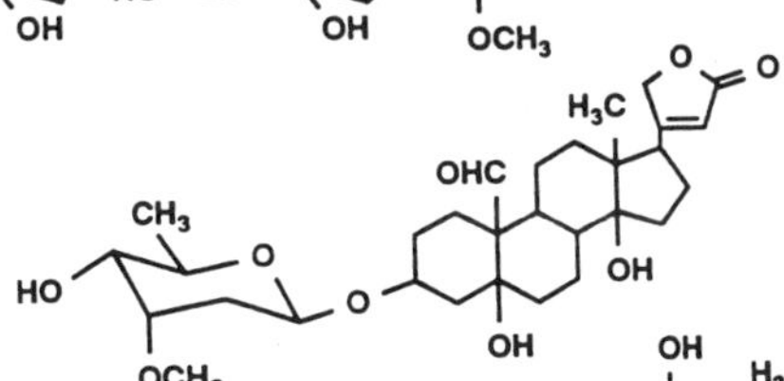	Cymarin	46	5
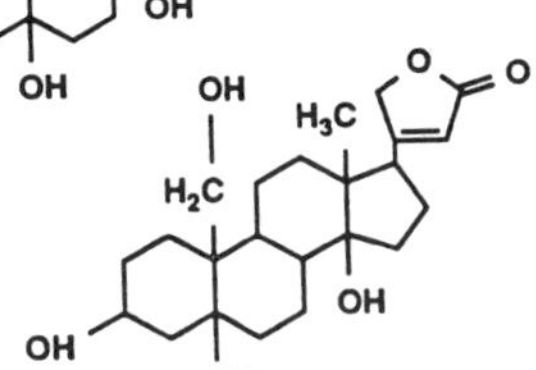	Strophanthidol	15	6

(Cont.)

Table 2. (continued)

		ODI	
		P. rapae	*P. napi*
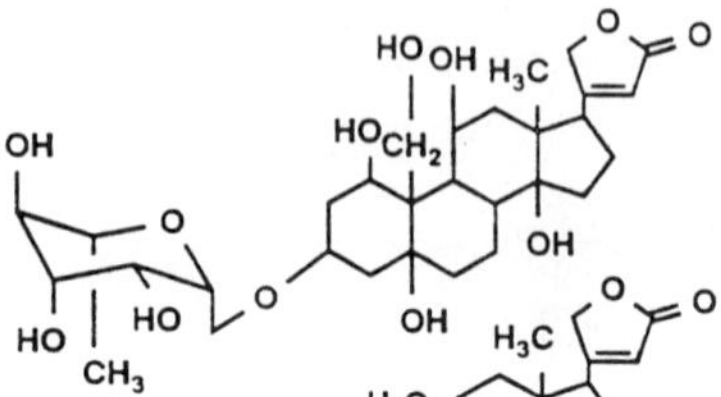	Ouabain	2	16
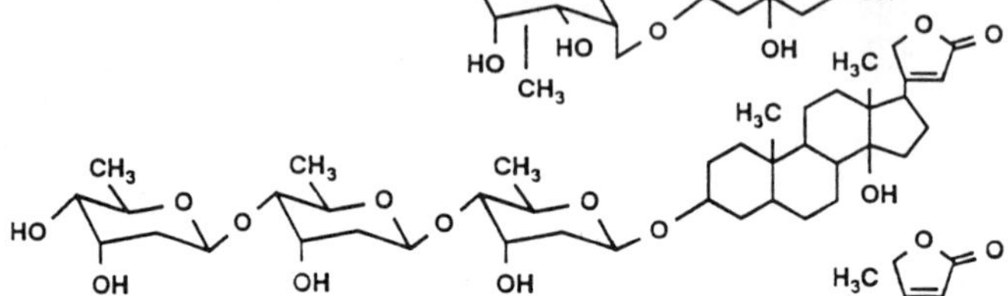	Digitoxin	10	2
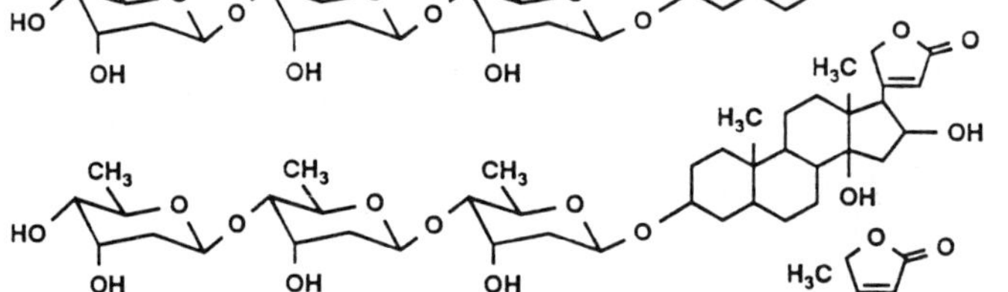	Gitoxin	26	6
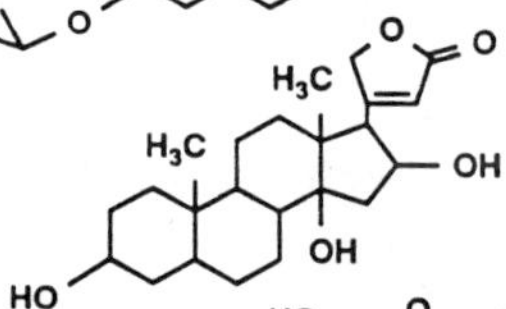	Gitoxigenin	33	8
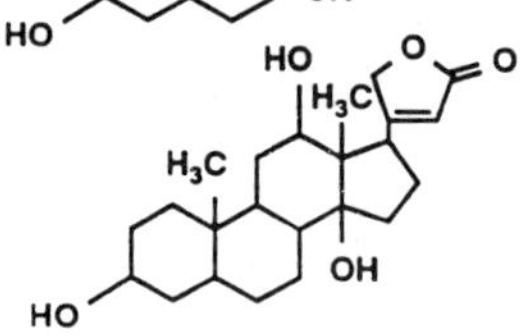	Digoxigenin	9	7
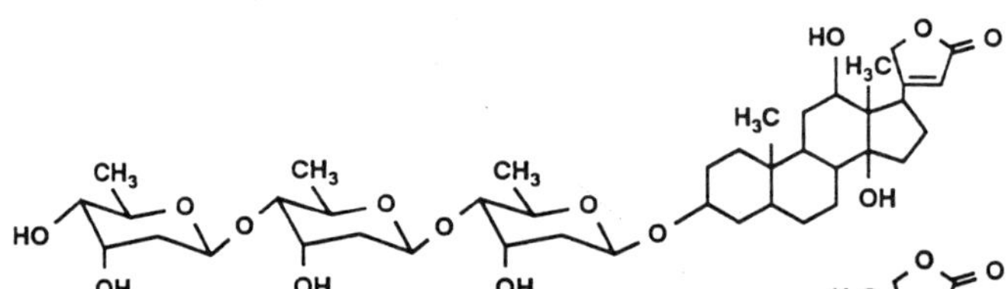	Digoxin	28	6
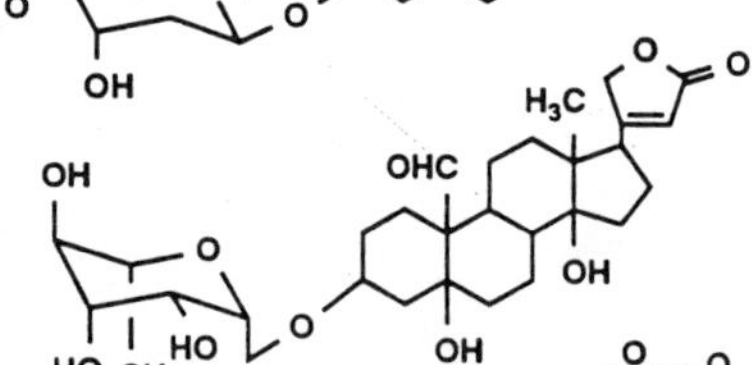	Convallatoxin	41	8
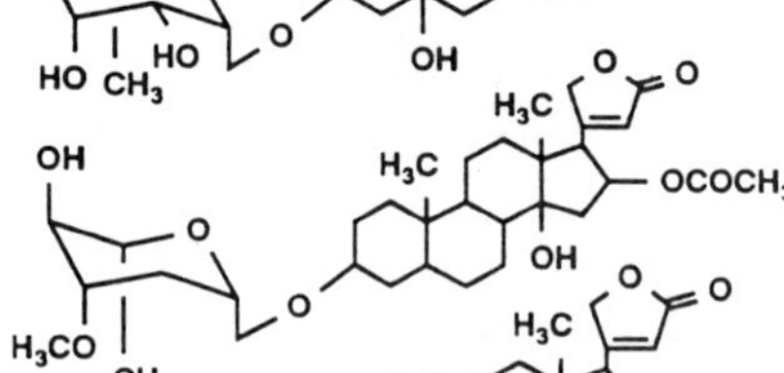	Oleandrin	41	11
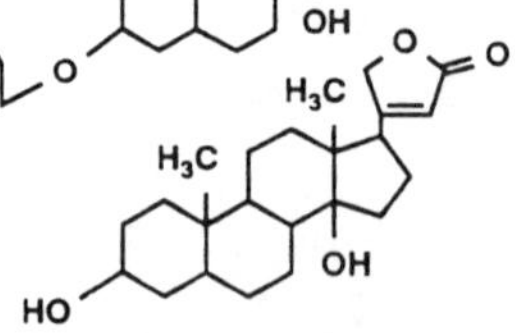	Digitoxigenin	12	6

qualities of the cardenolides are necessary for deterrent activity, and these may be different for closely related species or for the different life stages of an insect.

The production of allelochemicals and the relative quantities of positive and negative cues for insects may be dependent on environmental factors, such as nutrition. Furthermore, the response of insects to cardenolides may depend on their previous dietary experience. The sensitivity of taste receptors can be modified by exposure to deterrents or other compounds in the diet. As a result of these variables in plant production and in insect sensitivity to cardenolides, plant-insect interactions involving this group of compounds must be considered as a dynamic process that is highly unpredictable from one insect or one plant to another.

Acknowledgments

I am indebted to Drs. Lincoln Brower, John Glendinning, Stephen Malcolm, Jacques Pasteels and Larry Thompson for providing information and slides for this presentation. I also thank Nicole Tarnowsky for valuable assistance with the literature. This work was supported in part by NSF Grant DEB-9419797.

REFERENCES

1. J.P.J. Joubert. Cardiac glycosides, in: *Toxicants of Plant Origin*; [Vol. II, Glycosides], P.R. Cheeke, ed., CRC Press, Inc., Boca Raton (1989).
2. N. Kaneda, H. Chai, J.M. Pezzuto, A.D. Kinghorn, N.R. Farnsworth, P. Tuchinda, J. Udchachon, T. Santisuk and V. Reutrakul, Cytotoxic activity of cardenolides from *Beaumontia brevituba* stems, *Planta Medica* 58:429-431 (1992).
3. W. Majak, Mammalian metabolism of toxic glycosides from plants, *Journal of Toxicology - Toxin Reviews* 11:1-40 (1992).
4. J. von Euw, T. Reichstein and M. Rothschild, Heart poisons (cardiac glycosides) in the lygaid bugs *Caeocoris nerii* and *Spilostethus pandurus*, *Insect Biochem.* 1:373-384 (1971).
5. J. von Euw, L. Fishelson, J.A. Parsons, T. Reichstein and M. Rothschild, Cardenolides (heart poisons) in a grasshopper feeding on milkweeds, *Nature* 214:35-39 (1967).
6. M. Rothschild, J. von Euw and T. Reichstein, Cardiac glycosides in the oleander aphid, *Aphis nerii*, *J. Insect Physiol.* 16:1141-1145 (1970).
7. G.G.E. Scudder and S.S. Duffey, Cardiac glycosides in the Lygaeinae (Hemiptera: Lygaeidae), *Can. J. Zool.* 50:35-42 (1972).
8. S.S. Duffey and G.G.E. Scudder, Cardiac glycosides in North American Asclepiadaceae, a basis for unpalatability in brightly coloured Hemiptera and Coleoptera, *J. Insect Physiol.* 18:63-78 (1972).
9. J.A. Cohen and L.P. Brower, Cardenolide sequestration by the dogbane tiger moth (*Cycnia tenera*; Arctiidae), *J. Chem. Ecol.* 9:521-532 (1983).
10. G.G.E. Scudder and J. Meredith, The permeability of the midgut of three insects to cardiac glycosides, *J. Insect Physiol.* 23:689-694 (1982).
11. L.V. Moore and G.G.E. Scudder, Selective sequestration of milkweed (*Asclepias* sp.) cardenolides in *Oncopelus fasciatus* (Dallas) (Hemiptera: Lygaeidae), *J. Chem. Ecol.* 11:667-687 (1985).
12. M. Wink and D. Schneider, Fate of plant-derived secondary metabolites in three moth species (*Syntomis mogadorensis*, *Syntomeida epilais*, and *Creatonotos transiens*), *J. Comp. Physiol. B* 160:389-400 (1990).
13. C.N. Roeske, J.N. Seiber, L.P. Brower and C.M. Moffit, Milkweed cardenolides and their comparative processing by monarch butterflies, *Recent Advances in Phytochemistry* 10:93-167 (1976).
14. J.N. Seiber, P.M. Tuskes, L.P. Brower and C.J. Nelson, Pharmacodynamics of some individual milkweed cardenolides fed to larvae of the monarch butterfly (*Danaus plexippus* L.), *J. Chem. Ecol.* 6:321-339 (1980).
15. L.P. Brower and P. B. McVoy, Variation of cardiac glycoside content of monarch butterflies from natural populations in eastern North America, *Science* 177:426-429 (1972).
16. S.B. Malcolm, B.J. Cockrell and L.P. Brower, Cardenolide fingerprint of monarch butterflies reared on common milkweed, *Asclepias syriaca* L., *J. Chem. Ecol.* 15:819-853 (1989).
17. M.P. Zalucki, L.P. Brower and S.B. Malcolm, Oviposition by *Danaus plexippus* in relation to cardenolide content of three *Asclepias* species in the southeastern U.S.A., *Ecol. Entomol.* 15:231-240 (1990).

18. M. Haribal and J.A A. Renwick, Oviposition stimulants for the momarch butterfly: flavonol glycosides from *Asclepias curassavica, Phytochemistry* 41: 139-144 (1996).
19. J.M. Pasteels and D. Daloze, Cardiac glycosides in the defensive secretion of Chrysomelid beetles: evidence for their production by the insects, *Science* 197:70-72 (1977).
20. S. Dobler and M. Rowell-Rahier, Production of cardenolides versus sequestration of pyrrolizidine alkaloids in larvae of *Oreina* species (Coleoptera, Chrysomelidae), *J. Chem. Ecol.* 20:555-568 (1994).
21. D. Daloze and J.M. Pasteels, Production of cardiac glycosides by chrysomelid beetles and larvae, *J. Chem. Ecol.* 5:63-77 (1979).
22. J.M. Pasteels, J.C. Braekman and D. Daloze, Chemical defense in the Chrysomelidae, in: *Biology of Chrysomelidae*, P. Jolivet, E. Petitpierre and T.H. Hsiao, eds., Kluwer Academic Publishers, The Netherlands (1988).
23. M. Hilker, D. Daloze and J.M. Pasteels, Cardenolide glycosides from the adults and eggs of *Chrysolina fuliginosa* (Coleoptera: Chrysomelidae), *Experientia* 48:1023-1027 (1992).
24. S. van Oyke, J.C. Braekman, D. Daloze and J. M. Pasteels, Cardenolide biosynthesis in chrysomelid beetles, *Experientia* 43:460-462 (1987).
25. L.P. Brower, J.V.Z. Brower and J.M. Corvino, Plant poisons in a terrestrial food chain, *Proc. Nat. Acad. Sci.* 57:893 (1967).
26. D.L. Evans, N. Castoriades and H. Badruddine, Cardenolides in the defense of *Caenocoris nerii* (Hemiptera). *Oikos* 46:325-329 (1986).
27. L.S. Fink, L.P. Brower, R.B. Waide and P.R. Spitzer, Overwintering monarch butterflies as food for insectiverous birds in Mexico, *Biotropica* 15:151-153 (1983).
28. J. Glendinning and L.P. Brower, Feeding and breeding responses of five mice species to overwintering aggregations of the monarch butterfly, *J. Anim. Ecol.* 59:1091-1112 (1990).
29. J.I. Glendinning, Responses of three mouse species to deterrent chemicals in the monarch butterfly. II. taste tests using intact monarchs, *Chemoecology* 1:124-130 (1990).
30. J. Glendinning, L.P. Brower and C.A. Montgomery, Responses of three mouse species to deterrent chemicals in the monarch butterfly. I. Taste and toxicity tests using artificial diets laced with digitoxin or monocrotaline, *Chemoecology* 1:114-123 (1990).
31. J. Glendinning, Effectiveness of cardenolides as feeding deterrents to *Peromyscus* mice, *J. Chem. Ecol.* 18:1559-1575 (1992).
32. S. B. Malcolm, Disruption of web structure and predatory behavior of a spider by plant-derived chemical defenses of an aposematic aphid, *J. Chem. Ecol.* 15:1699-1716 (1989).
33. P. Feeny, Defensive ecology of the Cruciferae, *Annals of the Missouri Botanic Garden* 64:221-234 (1977).
34. J.K. Nielsen, Host plant discrimination within the Cruciferae: feeding responses of four leaf beetles (Coleoptera: Chrysomelidae) to glucosinolates, cucurbitacins and cardenolides, *Entomol. exp. appl.* 24:41-54 (1978).
35. F. Chew, Coevolution of pierid butterflies and their cruciferous foodplants, *Oecologia (Berl.)* 20:117-127 (1975).
36. J.E. Rodman and F S. Chew, Phytochemical correlates of herbivory in a community of native and naturalized Cruciferae, *Biochem. Syst. and Ecol.* 8:43-50 (1980).
37. R. Hegnauer. *Chemotaxonomie der Pflanzen* [Vol. 3: Dicotyledoneae: Acanthaceae-Cyrillaceae] Birkhäuser Verlag, Basel, (1964).
38. J.A.A. Renwick and C.D. Radke, Chemical stimulants and deterrents regulating acceptance or rejection of crucifers by cabbage butterflies, *J. Chem. Ecol.* 13:1771-1776 (1987).
39. J.A.A. Renwick, C.D. Radke and K. Sachdev-Gupta, Chemical constituents of *Erysimum cheiranthoides* deterring oviposition by the cabbage butterfly, *Pieris rapae, J. Chem. Ecol.* 15:2161-2169 (1989).
40. K. Sachdev-Gupta, J.A.A. Renwick and C.D. Radke, Isolation and identification of oviposition deterrents to cabbage butterfly, *Pieris rapae*, from *Erysimum cheiranthoides, J. Chem. Ecol.* 16:1059-1067 (1990).
41. M.B. Dimock, J.A.A. Renwick, C.D. Radke and K. Sachdev-Gupta, Chemical constituents of an unacceptable crucifer, *Erysimum cheiranthoides*, deter feeding by *Pieris rapae, J. Chem. Ecol.* 17:525-533 (1991).
42. K. Sachdev-Gupta, C.D. Radke, J.A.A. Renwick and M.B. Dimock, Cardenolides from *Erysimum cheiranthoides*: feeding deterrents to *Pieris rapae* larvae, *J. Chem. Ecol.* 19:1355-1369 (1993).
43. F.S. Chew and J.A.A. Renwick, Chemical ecology of hostplant choice in *Pieris* butterflies, in: *Chemical Ecology of Insects II*; R.T. Cardé and W.J. Bell, eds., Chapman and Hall, New York (1995).
44. X.P. Huang, J.A.A. Renwick and K. Sachdev-Gupta, A chemical basis for differential acceptance of *Erysimum cheiranthoides* by two *Pieris* species, *J. Chem. Ecol.* 19:195-210 (1993).
45. X.P. Huang and J.A A. Renwick, Cardenolides as oviposition deterrents to two *Pieris* species: structure-activity relationships, *J. Chem. Ecol.* 20:1039-1051 (1994).

46. J.A.A. Renwick, C.D. Radke, K. Sachedev-Gupta and E. Städler, Leaf surface chemicals stimulating oviposition by *Pieris rapae* (Lepidoptera: Pieridae) on cabbage, *Chemoecology* 3:33-38 (1992).
47. K. Sachdev-Gupta, C.D. Radke and J A.A. Renwick, Chemical recognition of diverse hosts by *Pieris rapae* butterflies, in: *Proc. of the 8th Int. Symposium on Insect-Plant Relationships*; S.B J. Menken, J.H. Visser and P. Harrewijn, eds., Kluwer, Dordrecht (1992).
48. J.A.A. Renwick and X. Huang, Interacting chemical stimuli mediating oviposition by Lepidoptera, in: *Functional Dynamics of Phytophagous Insects*, T.N. Ananthakrishnan, ed., Oxford & IBH Publishing Co., New Delhi (1994).
49. U. Hugentobler and J.A.A. Renwick, Effects of plant nutrition on the balance of insect relevant cardenolides and glucosinolates in *Erysimum cheiranthoides*, *Oecologia* 102:95-105 (1995).
50. J.A.A. Renwick, C.D. Radke and K. Sachdev-Gupta, Tolerance of cardenolides in *Erysimum cheiranthoides* by the diamondback moth, *Plutella xylostella*, in: *Symp. Biol. Hung.* [Vol.39], T.Jermy and A. Szentesi, eds., Akademiai Kiado, Budapest (1991).
51. J.A.A. Renwick and X.P. Huang, Rejection of host plant by larvae of cabbage butterfly: diet dependent sensitivity to an antifeedant, *J. Chem. Ecol.* 21:465-475 (1995).
52. X. Huang and J.A.A. Renwick, Cross habituation to feeding deterrents and acceptance of a marginal host plant by *Pieris rapae* larvae, *Entomol. exp. appl.* 76: 295-302 (1995).

CAN SOYASAPONIN I AND MONO- AND BI-DESMOSIDES ISOLATED FROM MUNGBEANS SERVE AS GROWTH ENHANCERS IN MUNGBEANS AND LETTUCE?

G.R. Waller,[1] C.F. Yang,[2] L.F. Chen,[3] C.H. Su, [3] R.M. Liou,[3] S.C. Wu,[3] C.C. Young,[3] M.R. Lee,[4] J.S. Lee,[4] C.H. Chou,[2] and D. Kim[5]

[1]Department of Biochemistry and Molecular Biology
Oklahoma Agricultural Experiment Station
Oklahoma State University
Stillwater, Oklahoma 74078
[2]Institute of Botany
Academia Sinica
Taipei, Taiwan
[3]Department of Soil Science and
[4]Department of Chemistry
National Chung Hsing University
Taichung, Taiwan
[5]Asian Vegetable Research and Development Center
Tainan, Taiwan

INTRODUCTION

Mungbeans (*Vigna radiata* L.) are a crop plant of economic significance in Taiwan and countries around the world. Recently they were found to produce mono- and bi-desmosidic saponins[1-3], with soyasaponin I being dominant. Little is known about the biological activity of saponins on the plant that produces them, although some reports on allelopathic activity (both inhibitory and stimulatory) have been published for mungbeans[4,5]. Research showing the presence of steroid saponins in the plant kingdom and their significance to plant growth was reported[6]. Some of the earliest work was performed on steroid saponins[7], where the growth rate of wheat embryos was approximately doubled by optimum concentration of saponins. The treatment of tomato seed and cereals with dilute solutions of saponins accelerated germination and increased the growth rate. Seeds of pea or corn absorbed water more rapidly in the presence of saponins with a corresponding increase of growth rate. Speculation[7] is to why steroidal saponins increase growth of pea embryos by 40%. These effects may be due to some surface-active activity, which modulate the relation of water to embryos and seedlings in cultivars. It was determined that steroidal saponins exhibit strong auxin activity at very low concentration[8].

Plant sterols influence the cell wall permeability of plants[9]; the review by Price, *et al.*[10] states that the primary action of saponins upon cells is to cause a general increase in the permeability of the plasma membrane. Bidesmosides appear to be primarily a transport form, and when the plant is damaged they could rapidly be converted by enzymes into monodesmosides which tend to be more active[11].

There are many allelopathic reports on controlling microbiological activity[12], especially fungi, and nematodes[13-19]. Allelopathy[2-5,20-28] is a focus of this paper. Allelopathy can be both stimulatory and inhibitory of plant growth; these biochemical reactions occur in all types of plants, soil, microorganisms, and viruses.

Cheng[29] suggested (and we have broadened his suggestion to include plant growth activity of saponins) that the establishment of a specific cause/effect relationship should be sought for allelopathy experiments in which the following steps occur: a) a biochemical is produced by a plant or from plant materials: b) the biochemical is transported from its source to the target plant: c) the target plant is exposed to the biochemical in sufficient quantity and for sufficient time to cause the effect (inhibitory or stimulatory growth). Our objectives were to determine if the saponins present in the mungbean plant can serve as a growth enhancer not only for mungbeans but also lettuce. An explanation of the biological activity using soyasaponin I, as an example, is provided.

METHODS

Plant Material

Mungbeans, cultivar Tainan 5, were grown in the Asian Vegetable Research and Development Center in Tainan, Taiwan throughout the life cycle starting on March 12, 1992 and replicated 3 times. Collection of samples of roots, stems, and leaves at approximately 7-day intervals was completed, and the samples were immediately put in a drying oven at 70° C and ground to pass through a 0.6 mm screen in a Wiley mill. The wet and dry weights were recorded. Mungbean growth stages and other factors are described in Table 1.

Table 1. Mungbean growth stages.

Sample No.	Age of Plants	No. of Plants Collected	Remarks
[Seed Planted March 12, 1992]			
1	7	150	Primary leaves unfolded
2	13	100	First trifoliate leaves appeared
3	28	75	Second trifoliate unfolded and third trifoliate leaves appeared
4	35	50	Third trifoliate unfolded; fourth trifoliate leaves appeared
5	42	25	Fifth trifoliate leaves unfolded; sixth trifoliate leaves appeared
6	49	25	Sixth trifoliate leaves unfolded; pods appeared
7	56	25	Pods were full size
8	64	25	Pods turned black
9	77	25	Maturity with leaves turning yellow
10	91	25	Harvest

Saponin Isolation and Purification

Saponins were isolated from each fraction of leaves, stems, and roots and purified by extraction with 1-butanol according to the procedure shown in Fig. 1.

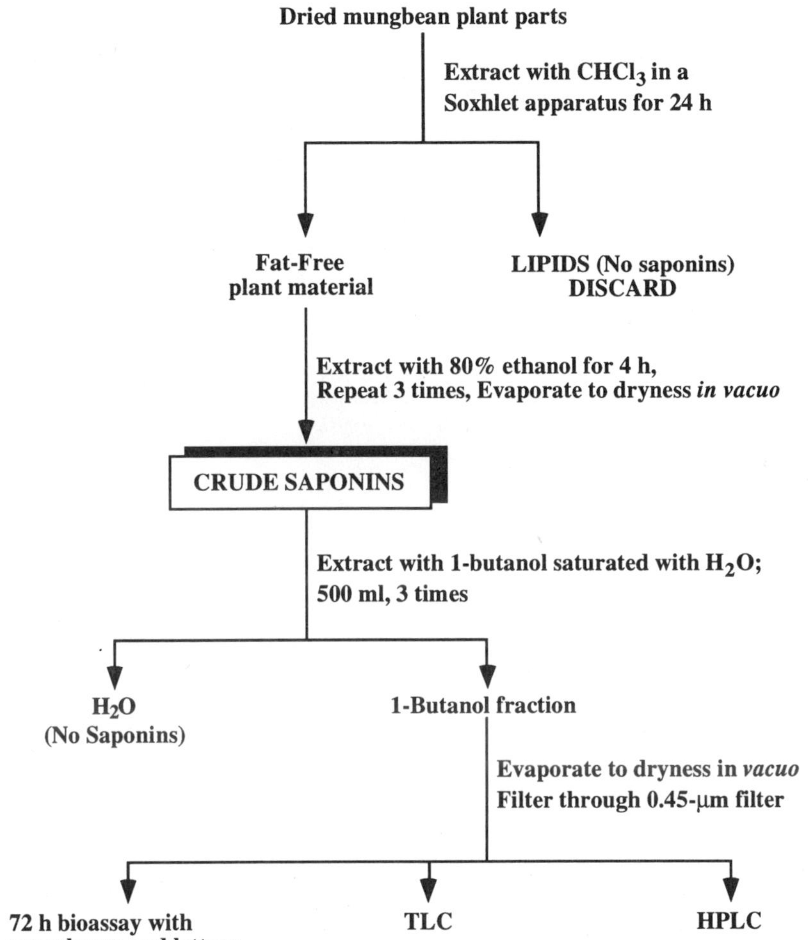

Figure 1. Extraction and partial purification of mungbean saponins from roots, stems, and leaves collected at weekly intervals throughout the life cycle.

Seeds/Plants for Growth Enhancement Tests

Mungbean (*Vigna radiata* L. cultivars Tainan 3 and Tainan 5) and lettuce (*Lactuca sativa* L. cultivar Great Lakes) seeds were obtained from Tainan District Agricultural Improvement Station, Tainan, Taiwan and Taiwan Agriculture Materials Co., Taipei, Taiwan.

Bioassay Procedure

The bioassay experiments were designed to compare the early growth (72 h) of mungbeans and lettuce treated with the test saponin solution and distilled water compared to a control treated with distilled water according to Fig. 2.

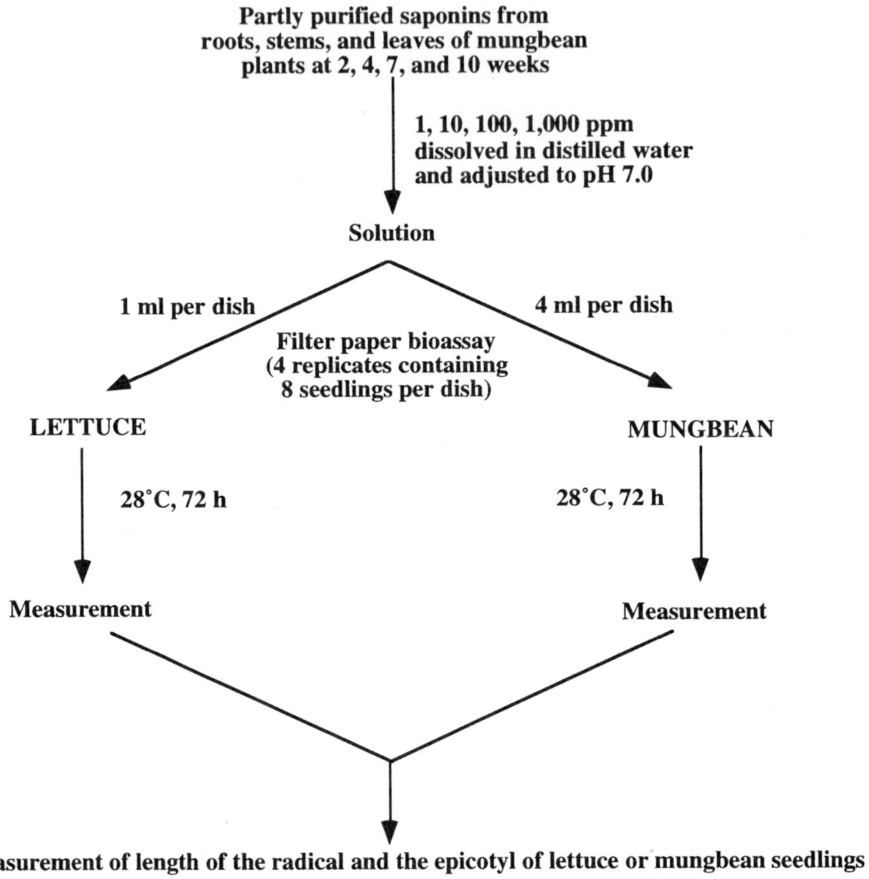

Figure 2. Bioassay procedure used for partially purified saponins from 2, 4, 7, and 10 weeks of the life cycle.

Thin Layer Chromatography (TLC)

C-18 reversed-phase TLC plates (10 x 20 cm) with a coating thickness of 0.25 mm were used for TLC. The prepared samples and standards of soyasaponin I and soyasapogenol B were dissolved in methanol. Two microliters were spotted on plates for identification of the saponins and 1 µl for identification of the aglycones. Standards were prepared at a concentration of 1 µg/µl and various amounts were spotted to assess the concentration of the mungbean saponins. The solvent systems (v:v:v) used separately for identification of the saponins[28] were methanol:water [90:10] and 1-butanol:acetic acid:water [7:2:2]. The plates were sprayed with Liebermann-Burchard reagent[30] and heated to 100° C for 2 min. Fluorescent spots were visualized under UV at 366 nm.

Mass Spectrometry

The partly purified saponins from each sample were analyzed by HPLC[2,4,5] and by mass spectrometry[3]. These techniques have been reported. Soyasaponin I was the reference compound.

RESULTS AND DISCUSSION

Mungbean Analysis Throughout The Life Cycle

The results of one replicate is shown below with respect to increase in dry weight (Fig. 3) and crude saponins isolated (Fig. 4). The dry weight was found to increase mostly for the roots with age of the plant; however, at 60-90 days the stem weight remained relatively constant whereas the leaves continued to gain in dry matter. A slightly different pattern was observed for the crude saponin content, where the roots leveled out at 50-90 days, the stems showed a slight decrease, and the leaves continued to increase dramatically. The timing of saponin biosynthesis, which appeared to have reached a constant value in the roots at 40 days, coincided with the onset of appearance of fungal root disease which has been identified as due to *Rhizoctonis* spp, *Pythium* spp, and *Fusarium* spp[5,32-34]. Leaf biosynthesis of crude saponins appeared to take 3 bursts at approximately 30-40, 55-65, and 78-90 days, and this corresponds to the initiation of flowers followed by leaf development and maturity.

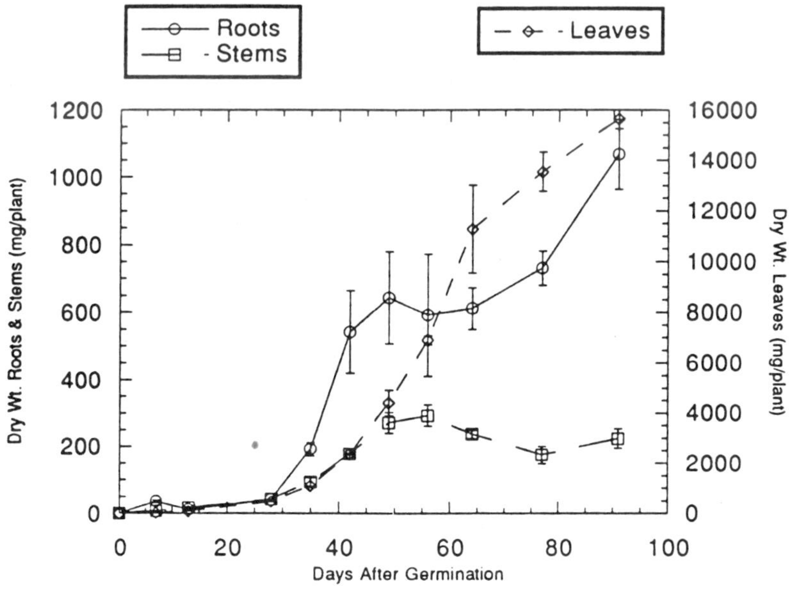

Figure 3. Increase in dry weight of mungbeans throughout germination. Average of 3 replications: Number of plants collected for each age was reported in Table 1.

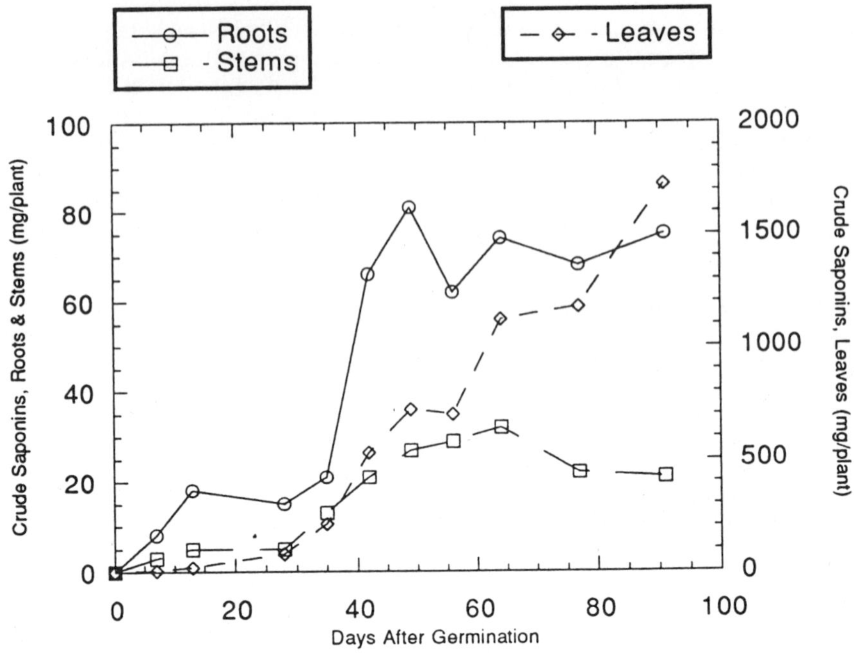

Figure 4. Crude saponins isolated from roots, stems, and leaves of mungbean plants throughout the life cycle.

Occurrence of Soyasaponin I and Mono- and Bidesmosides Studied by Thin Layer Chromatography (TLC)

Preparative TLC was used to fractionate the crude saponins from mungbeans. Fig. 5a shows soyasaponin I which has a molecular weight (MW) of 942 with an R_f value of 0.48 – 0.53. It was found to be the main saponin in all leaf, roots, and stem samples isolated throughout the life cycle of mungbeans. The bidesmosides [R_f = 0.10 – 0.48] and the other monodesmosides (R_f = 0.54 – 0.68) were also found in each fraction. The monodesmosides contained a new saponin (Fig. 5b) possessing MW 780, and Fig. 5c shows soyasaponin III (MW 796). Kitagawa, *et al.*, first identified soyasaponins I and III from soybeans[35]. The upper part of the plate (R_f = 0.70 – 1.00) contained soyasapogenol B as well as other aglycones some not yet identified. Preparative plates were run and the fractions were collected that had R_f = 0 – 0.21, 0.21 – 0.35, 0.35 – 0.48, 0.48 – 0.53, 0.53 – 0.65, 0.65 – 1.00; this information is summarized in Table 2. The data shown in Table 2 shows the distribution of the saponins at 2, 4, 7, and 10 weeks of age with more saponins being formed as the mungbean plants reach maturity. It also indicates that the saponins are being biosynthesized and biodegraded to some extent throughout the life cycle. More bidesmosides and monodesmosides are formed during the last 5–10 week period which corresponds to the pre-flowering, flowering, and pod development conditions.

(a)

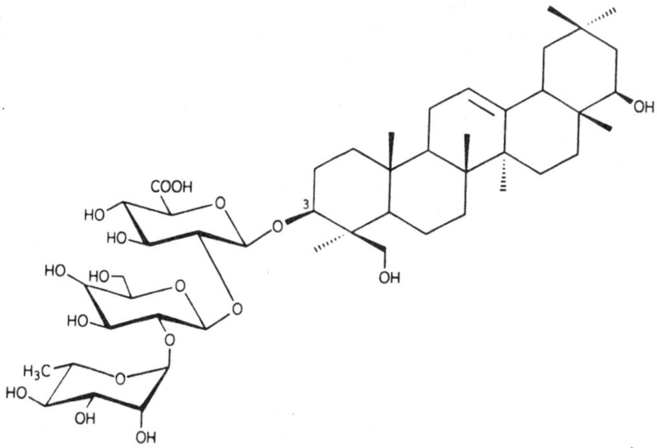

(b)

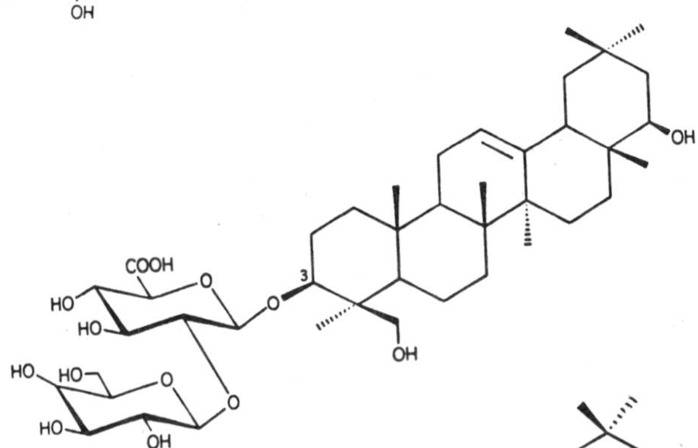

(c)

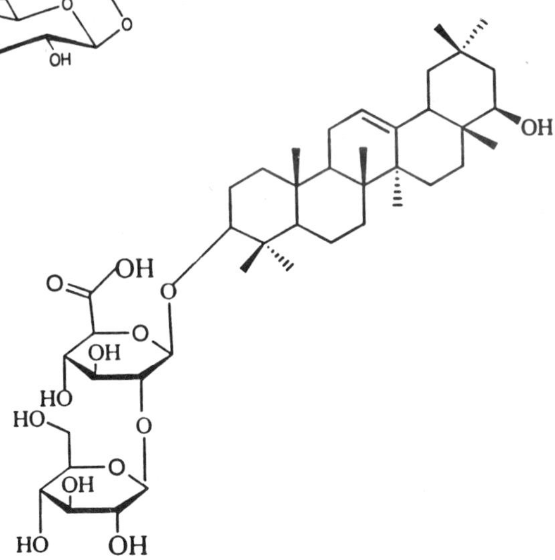

Figure 5. Structures of saponins identified in mungbeans (*Vigna radiata* L.): a) soyasaponin I, 3-O-[α-L-rhamnopyranosyl-(1→2)-β-D-galactopyranosyl-(1→2)-β-D-glucoronopyranosyl] soyasapogenol B; b) newly identified saponin [3-O-[β-D-galactopyranosyl-(1→2)-β-D-glucuronopyranosyl] sophradiol; and c) soyasaponin III, 3-O-[β-D-galactopyranosyl-(1→2)-β-D-glucoronopyranosyl] soyasapogenol B.

Effect on the Growth of Mungbeans and Lettuce

The partially purified saponins at concentrations of 1, 10, 100, and 1,000 ppm were dissolved in distilled water, adjusted to pH 7.0 ±0.1, and used in bioassays for 72 h. After that time, the lengths of mungbean and lettuce seedlings were measured in millimeters and compared to those grown in distilled water. Some of these results are presented for mungbeans in Fig. 6a, b, c and for lettuce in Fig. 7a, b, c; however, most of the results are not shown in this paper. The results with mungbeans were mixed, with the leaves showing the most enhancement of growth, and the stem showing inhibition whereas the root growth was slightly promoted. The results with lettuce show (Fig. 7a, b, c) that mungbean stem saponins were inhibitory at 2 and 4 weeks, but slightly stimulatory at 7 and 10 weeks; the stem saponin produced stimulatory action at all times during the growth season whereas the roots were slightly inhibitory at all times.

Table 2. Quantitative comparison of saponins formed throughout the mungbean's life cycle.[1,2]

	Weeks			
Saponin Spots Corresponding with	2	4	7	10
Bidesmosides				
Leaf	+	+	++	++
Root	–	–	–	–
Stems	++	+	++	++
Soyasaponin I				
Leaf	++	+++	++++[3]	++++[3]
Root	++	++	++	++
Stems	+	+	+++	+++
Other Monodesmosides				
Leaf	+	+	++	++
Root	–	–	–	–
Stems	++	+	++	++

[1]Quantitatively the amount of each series of compounds spotted was similar.
[2]Relative quantities – (absence); + (low); ++ (medium); +++ (high); ++++ (very high); all samples on a dry weight basis.
[3]Many compounds clustered around soyasaponin I.

Fig. 8a and b show the effect of soyasaponin I on the growth of mungbean and lettuce epicotyl and the radicle, and the difference is striking. In the presence of soyasaponin I most of the growth for lettuce radicle occurs at 1 ppm, less so at 10 and 100, and there was inhibition at 1,000 ppm; for the epicotyl an increase in concentration of soyasaponin I shows an increase in growth. The effect of soyasaponin I on the mungbean is slight enhancement of growth of the radical and slightly inhibitory for the epicotyl. A summary of these results is shown in Tables 3 and 4.

The evidence is quite clear for lettuce and mungbean that the presence of partially purified saponins is stimulatory. The leaves showed the greatest enhancement of growth, possibly because they contain more bidesmosides.

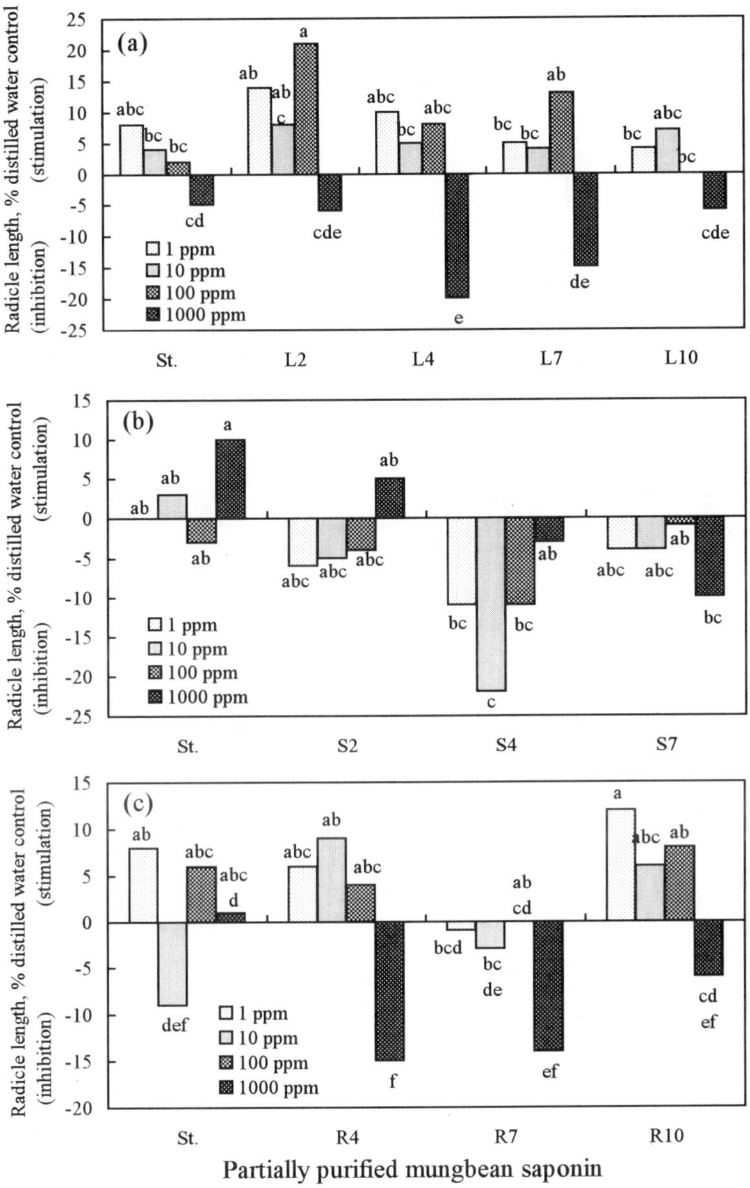

Figure 6a,b,c. The effects of partially purified saponins extracted from the (a) leaves [L], (b) stems [S], and (c) roots [R] of mungbean at 2, 4, 7, and 10 weeks of growth on the growth of mungbean radicle at 72h bioassay. Bars having different letters are significantly different, $P = 0.05$, ANOVA, with Duncan's multiple-range test.

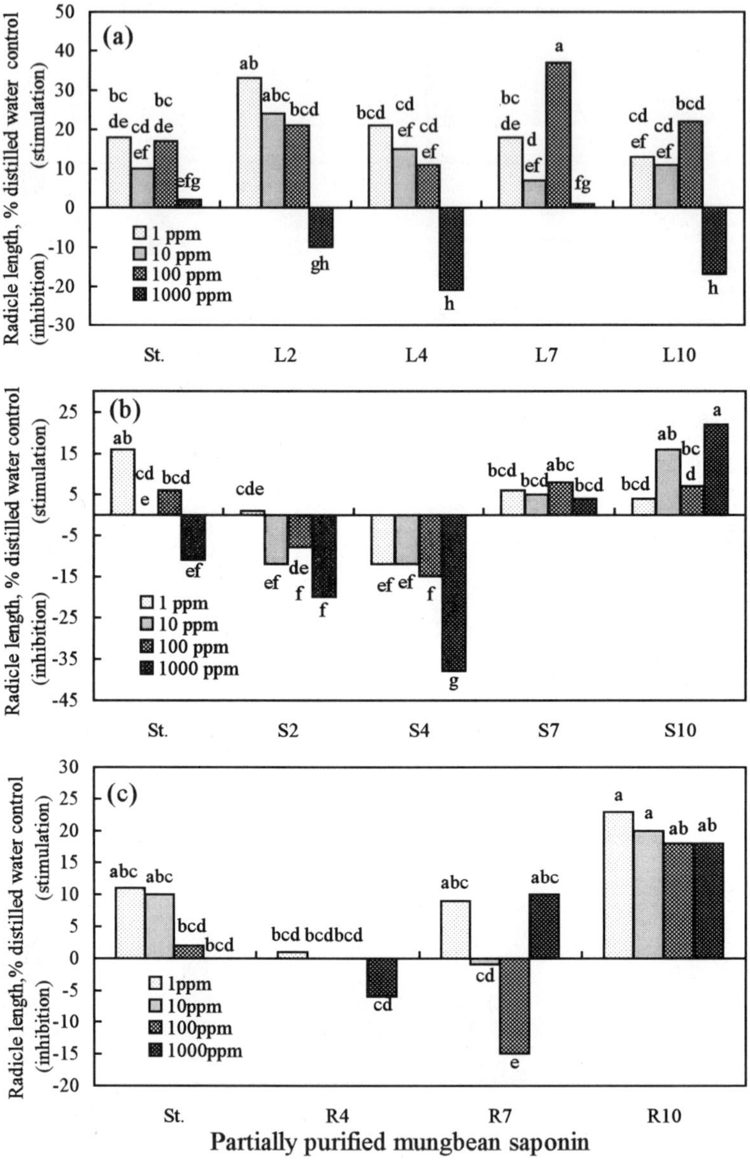

Figure 7a,b,c. The effects of partially purified saponins extracted from the (a) leaves [L], (b) stems [S], and (c) roots [R] of mungbean at 2, 4, 7, and 10 weeks of growth on the growth of lettuce radicle for 72h bioassay. Bars having different letters are significantly different, $P = 0.05$, ANOVA, with Duncan's multiple-range test.

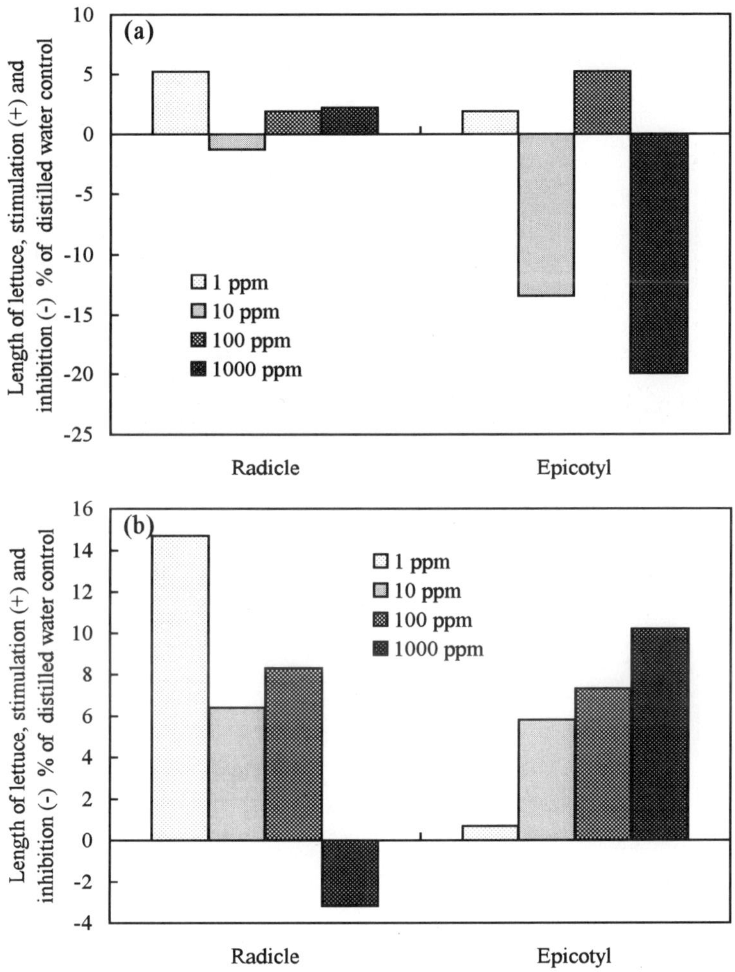

Figure 8a,b. The effects of soyasaponin I on the radicle and epicotyl growth of (a) mungbean and (b) lettuce.

Table 3. Summary of effects of partly purified mungbean saponins on growth of mungbeans.[1]

Field Plant Part	Bioassay Plant Part	Weeks SS1	2	4	7	10	Growth in 72 h
Roots	Radicle	++		NC	NC	+++	Stimulatory
	Epicotyl	NC		NC	NC	——	Inhibitory
Stems	Radicle	+	—	—	+	++	Stimulatory
	Epicotyl	++	++	++	NC	NC	Stimulatory
Leaves	Radicle	+	+++	+++	+++	+++	Stimulatory
	Epicotyl	++	++	+	+	+	Stimulatory

[1]Relative quantities: No Detectable Change - NC; Low (+), Medium (++), High (+++) Stimulatory; Low (–), Medium (—), High (——) Inhibitory

Table 4. Summary of effects of partly purified mungbean saponins on growth of lettuce[1].

Field Plant Part	Bioassay Plant Part	Weeks SS1	2	4	7	10	Growth in 72 h
Roots	Radicle	++		NC	NC	+++	Stimulatory
	Epicotyl	NC		NC	NC	——	Inhibitory
Stems	Radicle	+	—	—	+	++	Stimulatory
	Epicotyl	++	++	++	NC	NC	Stimulatory
Leaves	Radicle	+	+++	+++	+++	+++	Stimulatory
	Epicotyl	++	++	+	+	+	Stimulatory

[1]Relative quantities: No Detectable Change - NC; Low (+), Medium (++), High (+++) Stimulatory; Low (–), Medium (—), High (——) Inhibitory

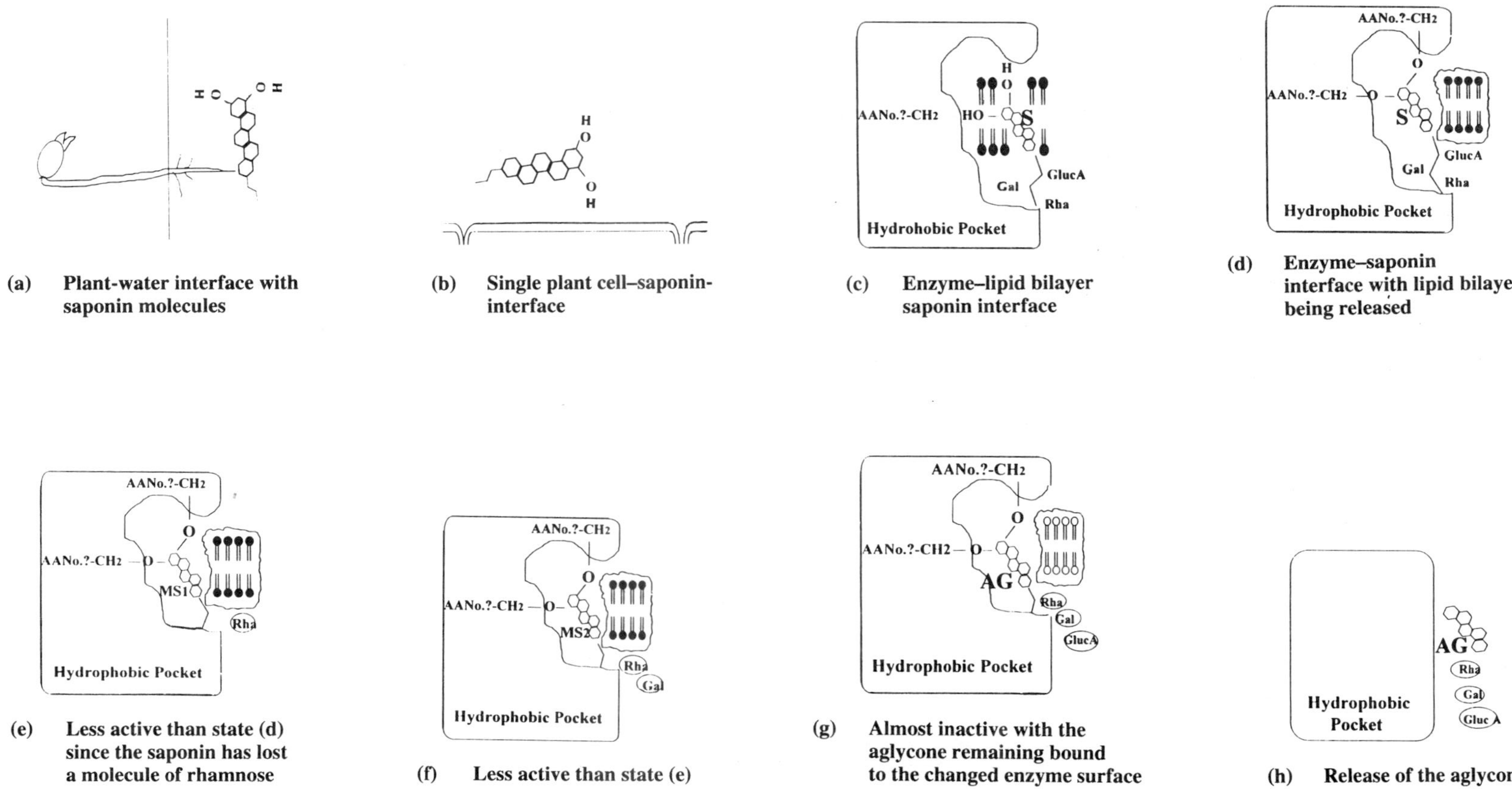

Figure 9. Suggested mechanism of action of mono- and bidesmoside saponins that occur in mungbeans used as an addition to water in the growth of mungbeans and lettuce in a 72–h bioassay. Note: sugars from soyasaponin I were used as an example; which amino acid number is not known for co-valent binding; S = saponin; MS1 = saponin without one molecule of sugar; MS2 = saponin with loss of two molecules of sugar; AG = aglycone with rhamnose, galactose and glucuronic acid free in the medium.

A PROPOSED MECHANISM OF ACTION OF THE MUNGBEAN SAPONINS

Low concentrations of mungbean saponins generally promote the growth of mungbeans and lettuce whereas higher concentrations are inhibitory. A suggested series of events (Fig. 9) which might occur in the seed-plantlet-water interface is proposed for mungbeans and lettuce: i) some saponin molecules may be bound to the outer coating of the seeds but most molecules will be in a matrix[36] or cluster form and uncombined, ii) as the seed begins to imbibe water some of the saponin molecules or clusters become active in transport of water molecules into the plant, i.e., by b and c processes where there is a loose binding to the surface of the seed, iii) as the seed develops into a plant the plant hairs become available to the unreacted molecules of saponin clusters, which will react with the lipid bilayer of the seedling and become attached to an enzyme on the surface, iv) this promotes water going into the plant and section of the Fig. 8d-g shows that the saponin loses sequentially sugar molecules by cleavage by enzymes until what remains on the enzyme surface is the aglycone, which is loosely attached and eventually comes off the root hair. Although the 72–h bioassay was not sterile, the growth of any microorganism will be minimal; however, saponins can arrest the growth of bacteria or fungi. What we do not know is whether the external cell concentration of the saponins in the medium may be high enough to cause the aglycone to remain on the active site; however, it seems that too high a concentration would produce inhibition in plant growth because of the blockage of the enzymatic sites by the aglycone; the concentration of saponin that would be required for this passive diffusion to occur; however, remains an open question. We might ask, is the root hair cell resistant to this series of events? The cell is in constant search of water and nutrients. In this scenario it has no other source of nutrients other than what was present in the seed itself. As the molecules of saponins are synthesized during the 72-h bioassay they may be present as clusters that migrate freely within the plantlet and some are released to the medium. The saponins may be positioned such that the plant cell-saponin may react with concomitant passage of water (and nutrients). As the root grows it may contact more molecules of saponins, and the same cycle will be repeated so that a growth enhancement may be assured as long as saponins are present in the environment.

CONCLUSION

The mungbean saponins showed growth enhancement in 72–h bioassay of around 5-10% throughout the life cycle, with the higher activity being shown as the plants matured. These results confirm and extend the results obtained previously on pot growth experiments on the leaves mixed with soil which indicated a 15-25% increase using mungbeans[4,5].

ACKNOWLEDGMENTS

We thank O.C. Dermer, H.G. Cutler, E. Ollivier, and G. Balansard for critically reviewing this paper. This is published with the approval of the Oklahoma Agricultural Experiment Station, Oklahoma State University, Stillwater, OK. G.R. Waller expresses appreciation to the Department of Soil Science, National Chung Hsing University, Taichung, Taiwan for courtesies during his leave for the 1994 fall and winter months, and expresses appreciation to Academia Sinica, Institute of Botany, Taipei, Taiwan, and the National Science Council, Republic of China in Taiwan for financial support for the sabbatical year, 1991-92 under Grant No. NSC 82-0211-13-001-04 to C.H. Chou.

REFERENCE

1. K.R. Price, J. Eagles, and G.R. Fenwick, Saponin composition of 13 varieties of legume seed using fast atom bombardment mass spectrometry, *J. Sci. Food Agric.* **42**: 183 (1988).

2. G.R. Waller, C.H. Chou, C.S. Cheng, J.C.H. Kuei, and N.N. Lai, Isolation, purification, identification and biological activity of saponins produced by mungbean (*Vigna radiata*) plants. In: *Proc. Int. Symp. on the Adaptation of Food Crops to Temperature and Water Stress,* C.G. Kuo, ed., Asian Vegetable Research and Development Centre, Aug. 13-18, 1992, Taiwan, pp. 419-426 (1992)

3. G.R. Waller, P.R. West, C.S. Cheng, C.H. Chou, and Y.C. Ling, The occurrence of soyasaponin I in *Vigna radiata* (mungbean) sprouts as determined by fast atom bombardment, liquid secondary ion mass spectrometry, and linked scanning at constant B/E MS/MS. *Bot. Bull. Acad. Sinica* (Taiwan) **34**: 323 (1993).

4. G.R. Waller, C.H. Chou, C.S. Cheng, and D. Kim, Autotoxic and allelopathic activity of phytotoxic compounds of mungbeans (*Vigna radiata*) and their surrounding soil, In: *Adaptation of Food Crops to Temperature and Water Stress, Proc. Intern. Symp.*, August 13-18, 1992, C.G. Kuo, ed,. Tainan, Taiwan: Asian Vegetable Research and Development Centre, pp. 427-433 (1992).

5. G.R. Waller, C.H. Chou, C.S. Cheng, and D. Kim, Allelopathic activity of naturally occurring compound(s) from mungbeans (*Vigna radiata*) and their surrounding soil. *Bot. Bull. Acad. Sinica* (Taiwan) **36**: 9 (1995).

6. J. Balansard, F. Pellisser, and S. Conil, Action des saponins de *Quillaya* et de *Sapindus* sur le pouvoir absorbent la germination et le croissance de *Zea mays* L. *Compt. Rend. Soc. Biol.* **140**: 140 (1946).

7. G.K. Helmkamp and J. Bonner, Some relationships of sterols to plant growth, *Plant Physiol.* **28**: 428 (1953).

8. J.C. Vendrig, Growth-regulating activity of some saponins, *Nature* **203**: 1301 (1964).

9. C. Grunwald, Sterol modifications influencing membrane permeability, *Plant Physiol.* **54**: 624 (1974).

10. K.R. Price, I.T. Johnson, and G.R. Fenwick, The chemistry and biological significance of saponins in food and feeding stuffs. *CRC Crit. Rev. Sci. Nutr.* **26**: 27-135 (1987).

11. N.G. Bissett, One man's poison, another man's medicine, *J. Ethnopharmacol.* **32**: 71-81 (1991).

12. M. Takashi, I. Tadao, and S. Takashi, Methods of accelerating action of sludge, Japanese Patent No. 80-118189 (1986).

13. M. Jurzysta and G.R. Waller, Antifungal and hemolytic activity of aerial plants of *Medicago sativa* saponins in relation to saponin composition, In Saponins *Used in Food and Agriculture*, G.R. Waller and K. Yamasaki, eds., Plenum Publishing Co., New York, NY. In press, (1996).

14. R.M. Zablotowiz, R.E. Hoagland, and S.C. Wagner, Effect of saponins on the growth and activity of rhizosphere bacteria, In *Saponins Used in Food and Agriculture*, G.R. Waller and K. Yamasaki, eds., Plenum Publishing Co., New York, NY. In press, (1996).

15. K. Gruiz, Fungitoxic activity of saponins: Practical use and fundamental principles, In *Saponins Used in Food and Agriculture*, G.R. Waller and K. Yamasaki, eds., Plenum Publishing Co., New York, NY. In press, (1996).

16. A.E. Osbourne, Saponin detoxification by plant pathogenic fungi, In *Saponins Used in Traditional and Modern Medicine*, G.R. Waller and K. Yamasaki, eds., Plenum Publishing Co., New York, NY. In press, (1996).

17. C.O. Okunji, M.M. Twu, J.F. Jackson, and J.D. Tally, Biologically active saponins from two *Dracaena* species, In *Saponins Used in Traditional and Modern Medicine*, G.R. Waller and K. Yamasaki, eds., Plenum Publishing Co., New York, NY. In press, (1996).

18. P.K. Kintia, Steroidal saponins and Moldovian plants; Structure and biological activity, In *Saponins Used in Food and Agriculture,* G.R. Waller and K. Yamasaki, eds., Plenum Publishing Co., New York, NY. In press, (1996).

19. W. Oleszek, Alfalfa saponins: structure, biological activity and chemotaxonomy, In *Saponins Used in Traditional and Modern Medicine*, G.R. Waller and K. Yamasaki, eds., Plenum Publishing Co., New York, NY. In press, (1996).

20. H. Molisch, Der Einfluss seiner Pflanzen auf der andere – Allopathie, G. Fischer: Jena (1937).

21. E.L. Rice, *Allelopathy*, 2nd ed. Academic Press, Orlando, FL (1984).

22. E.L. Rice, Allelopathy Growth Stimulation, In *The Science of Allelopathy*, A.R. Putnam and C.S. Tang, eds., John Wiley and Sons, New York, NY (1986).

23. A.M. Grodzinsky, Allelopathy in the Soviet Union, In *Allelochemicals: Role in Agriculture and Forestry*, G.R. Waller, ed., ACS Symposium Series, No. 330, American Chemical Society, Washington, D.C. pp. 39-43 (1987).

24. G.R. Waller, *Allelochemicals: Role in Agriculture and Forestry*, ACS Symposium Series, No. 330, American Chemical Society, Washington, D.C. pp. ix (1987).

25. S.A. Shvets, P.K. Kintia, and O.N. Gutsu, Steroidal glycosides from *Nicotiana tabacum* L. seeds and their biological activity, In *Saponins Used in Food and Agriculture*, G.R. Waller and K. Yamasaki, eds., Plenum Publishing Co., New York, NY. In press, (1996).

26. C.H. Chou and G.R. Waller, *Phytochemical Ecology: Allelochemicals, Mycotoxins, Insect Pheromones, and Allemones*, Institution of Botany, Acad. Sin. Monograph. Series, No. 9, pp. 504 (1989).

27. F.A. Einhellig, Allelopathy: Current status and future goals, In *Allelopathy: Organisms, Processes, and Applications*, I. Inderjit, K.M.M. Dakshin, and F.A. Einhellig, eds., ACS Symposium Series 582, American Chemical Society, Washington, D.C. pp. 1-24 (1995).

28. W. Oleszek, K.R. Price, I.J. Colquhoun, M. Jurzysta, M. Ploszynski, and G.R. Fenwick, Isolation and identification of alfalfa (*Medicago sativa* L.) root saponins: their activity in relation to a fungal bioassay. *J. Agric. Food Chem.* **38**: 1810 (1990).

29. H.H. Cheng, Assessment of the fate and transport of allelochemicals in the soil. In *Phytochemical Ecology: Allelochemicals, Mycotoxins, and Insect Pheromones and Allomones* , C.H. Chou and G.R. Waller, eds., Monograph Series No. 9, Institute of Botany, Academia Sinica, Taipei, ROC, pp. 209-215 (1989).

30. R. Tschesche, E. Lampert, and G. Snatzke, Paper chromatography of triterpenoids. *J. Chromatog.* **5**: 217 (1961).

31. M.R. Lee, J.S. Lee, J.C. Wang, and G.R. Waller, Structural determination of saponins from mungbean sprouts by tandem mass spectrometry, In *Saponins Used in Food and Agriculture*, G.R. Waller and K. Yamasaki, eds., Plenum Publishing Co., New York, NY. In press, (1996).

32. Asian Vegetable Research and Development Center, Tainan, Taiwan, Progress Report. pp 43, 44, and 47 (1980).

33. Asian Vegetable Research and Development Center, Tainan, Taiwan, Progress Report. pp 27-28 (1981).

34. Asian Vegetable Research and Development Center, Tainan, Taiwan, Progress Report. pp 153-154 (1983).

35. I. Kitagawa, H.K. Wang, T. Taniyama, and M. Yoshikawa, Saponin and sapogenol XLI. Reinvestigation of the soyasapogenols, A, B, and E, oleanane-sapogenins from soyabean. Structure of soyasaponins I, II, and III. *Chem. Pharm. Bull.* **36**: 155 (1988).

36. D. Oakenfull, Aggregation of saponins and bile acids in aqueous solutions, *Austral. J. Chem.* **39**: 1671 (1986).

OXYGEN-RADICAL-SCAVENGING ACTIVITY OF DDMP-CONJUGATED SAPONINS AND PHYSIOLOGICAL ROLE IN LEGUMINOUS PLANT

Kazuyoshi Okubo and Yumiko Yoshiki

Faculty of Agriculture, Tohoku University
1-1 Tsutsumitori, Amamiyamavhi, Aobaku
Sendai 981, Japan

INTRODUCTION

Soybean seeds have been consumed by humans for thousands of years because they are rich in nutritional protein and oil. Most Eastern Asians consume soybean seeds from childhood in a variety of soybean products. The incidence of breast and colon cancer in Oriental people is considerably lower than in those living in Western countries,who seldom eat soybean products. Additionally, vegetarians, who are also at decreased risk of breast and colon cancer, frequently consume soybean-based meat substitutes.[1]

Oxygen-radical-scavenging characteristics of DDMP (2,3-dihydro-2,5-dihydroxy-6-methyl -4*H*-pyran -4-one) conjugated saponins and related substances in soybean seeds and the products will be the focus of this chapter. Active oxygen species can cause damage to biomolecules including proteins and DNA.[2-6] Chemiluminescence produced in the presence of aldehydes and superoxide anions, hydroxyl radicals, hydrogen peroxide, or lipid hydroperoxides is a new useful method for detection and establishing characterization of oxygen radical scavengers.

CHEMICAL STRUCTURE OF SAPONINS

Soybean saponins are oleanone-type triterpene saponins. Two nomenclature systems used by Kitagawa and Okubo (Table 1) have served to identify various soyasaponins.[7] For the sake of consistency, only the Okubo system will be used here.

Table 1. Soyasaponin nomenclature assigned by the research groups of Kitagawa and Okubo

	Kitagawa	Okubo
A group - acetylated[a]		
glc-gal-glcUA-A-ara-xyl (2,3,4-triAc)	acetyl A_4	Aa
glc-gal-glcUA-A-ara-glc (2,3,4,6-tetraAc)	acetyl A_1	Ab
rha-gal-glcUA-A-ara-glc (2,3,4,6-tetraAc)	—	Ac
glc-gal-glcUA-A-ara-glc (2,3,4,6-tetraAc)	—	Ad
gal-glcUA-A-ara-xyl (2,3,4-triAc)	acetyl A_5	Ae
gal-glcUA-A-ara-glc (2,3,4,6-tetraAc)	acetyl A_2	Af
ara-glcUA-A-ara-xyl (2,3,4-triAc)	acetyl A_6	Ag
ara-glcUA-A-ara-glc (2,3,4,6-tetraAc)	acetyl A_3	Ah
A group - deacetylated[a]		
glc-gal-glcUA-A-ara-xyl	A_4	deacetyl Aa
glc-gal-glcUA-A-ara-glc	A_1	deacetyl Ab
gal-glcUA-A-ara-xyl	A_5	deacetyl Ae
gal-glcUA-A-ara-glc	A_2	deacetyl Af
ara-glcUA-A -ara-xyl	A_6	deacetyl Ag
ara-glcUA-A-ara-glc	A_3	deacetyl Ah
B group[b]		
glc-gal-glcUA-B	V	Ba
rha-gal-glcUA-B	I	Bb
rha-ara-glcUA-B	II	Bc
gal-glcUA-B	III	Bb′
ara-glcUA-B	IV	Bc′
E group[c]		
glc-gal-glcUA-E		Bd
rha-gal-glcUA-E		Be

[a] Sugar chains on the left of A are linked to 3-O and those on the right to 22-O.

[b] Soyasapogenol B contains a hydroxyl moiety at C-22.

[c] Soyasapogenol E contains a ketone moiety at C-22.

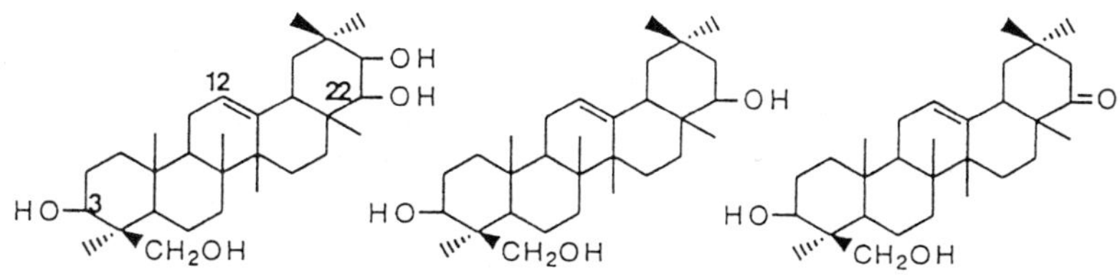

Soyasapogenol A Soyasapogenol B Soyasapogenol E

Figure 1. Structure of soyasapogenol A, B and E

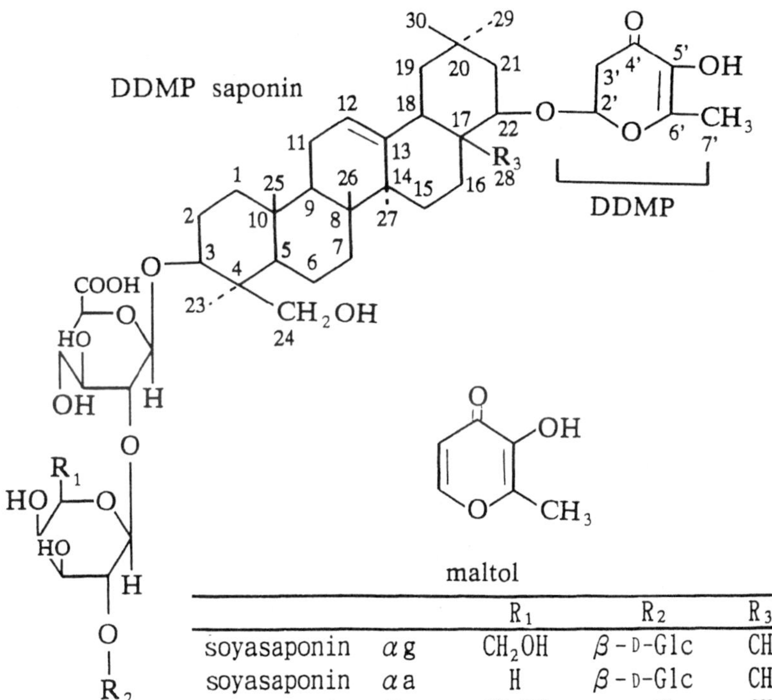

	R_1	R_2	R_3
soyasaponin αg	CH_2OH	β-D-Glc	CH_3
soyasaponin αa	H	β-D-Glc	CH_3
soyasaponin βg	CH_2OH	α-L-Rha	CH_3
soyasaponin βa	H	α-L-Rha	CH_3
soyasaponin γg	CH_2OH	H	CH_3
soyasaponin γa	H	H	CH_3
lablab saponin I	CH_2OH	β-D-Glc	CHO

Figure 2. Structure of DDMP saponins

Many soybean saponins have been isolated and characterized.[8-11] They are divided into three groups A, B, and E according to their respective aglycons, soyasapogenol A, B, or E (Figure 1, Table 1). Group A saponins are bis-desmoside saponins that contain two sugar chains attached at the C-3 and C-22 positions of soyasapogenol A. Group B and E saponins had been thought to be mono-desmoside saponins that contain a sugar chain attached to the C-3 position alone, but recently they were found to contain a 2,3-dihydro-2,5-dihydroxy-6-methyl-4*H*-pyran-4-one (DDMP) moiety at the C-22 position of soyasapogenol B or E (Fig.2).[11] A new type of triterpenoid saponin has recently been isolated from pea [27] and was found to be a conjugate of 3-hydroxy-2-methyl-5, 6-dihydro-4-pyrone and soyasaponin I (Bb). This compound was named chromosaponin I (CSI) which is identical with soyasaponin β g. Saponin Bb belongs to group B saponins, and saponin Be is includes in the group E saponins. Soyasaponin β g (saponin BeE) that contains DDMP at the C-22 position of saponin Bb and Be must be the genuine saponin of "saponins Bb and Be", because the purified soyasaponin β g produces saponin Bb and Be after heating.[12] Soyasaponin β g was stable in acidic pH, but easily changed into Bb at basic pH; the decomposition rate into Bb increased quantitatively with increasing concentration of NaOH (Fig.3).

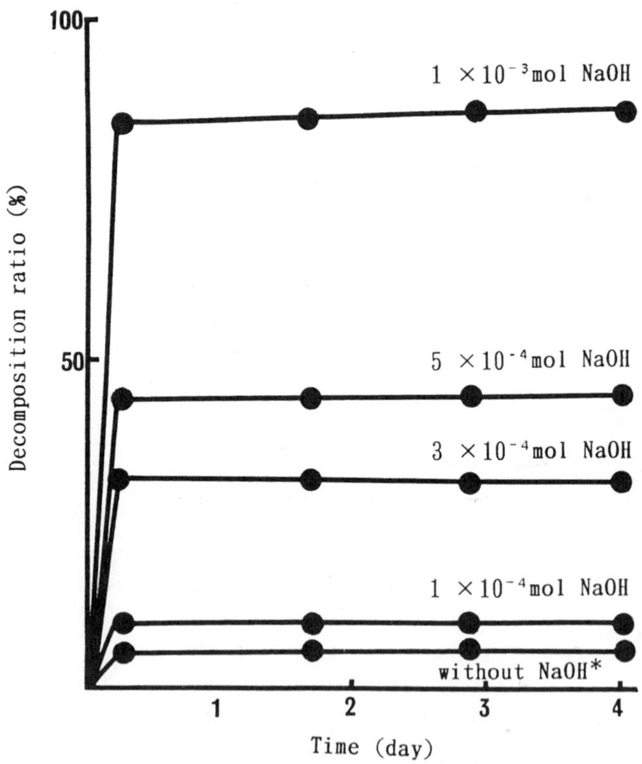

Figure 3. Effect of NaOH concentration on the decomposition of soyasaponin

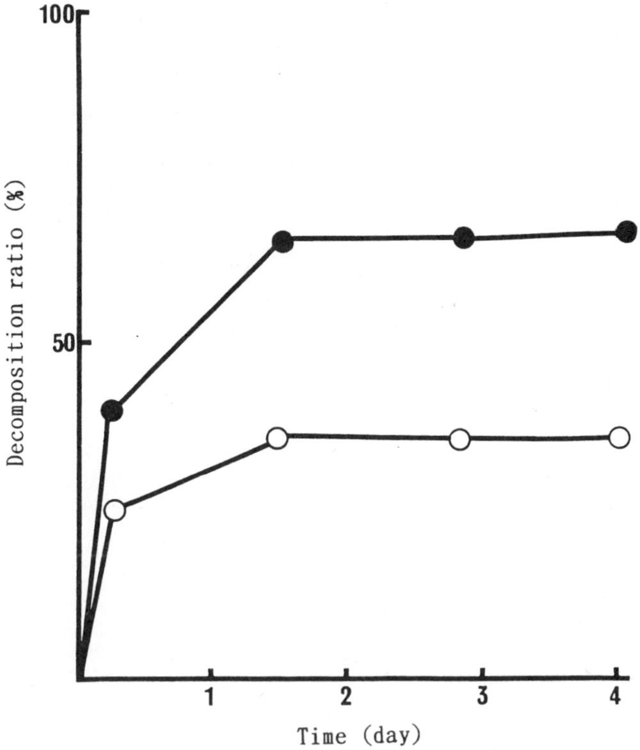

Figure 4. Effect of $FeCl_3$ on the decomposition of soyasaponin β g
soybean saponin Bb(-●-), Be(-○-)

Also soyasaponin β g was comparatively stable at temperature below 90 ℃, but rapidly degraded into Bb when heated at 90 - 100 ℃ for 1 h. In the presence of $FeCl_3$, soyasaponin β g degraded to Bb : Be in approximately 3 : 2 ratio (Fig. 4). Furthermore, other DDMP saponins, α g, β a, γ g, and γ a, which correspond to each of the four group B saponins, Ba, Bc, Bb, and Bc, respectively, have now been been identified.[13] This means that groups B and E saponins must have been derived from the saponins which contain DDMP at the C-22 position of soyasapogenol B.

Recently we have isolated genuine saponins, soyasaponins α g, β g, β a, γ g, and γ a from soybean,[13] and soyasaponin α a from scarlet runner bean (*Phaseolus coccineus*).[14] Also soyasaponin β g was isolated from rootstock of the American groundnut (*Apios americana*).[15] The structure was identified by ^{1}H-NMR and ^{13}C-NMR and by chemical techniques. The distribution of this DDMP saponin in the rootstock was detected by means of the brown color produced by the reaction with $FeCl_3$. A high concentration of DDMP saponin was observed around the cell in

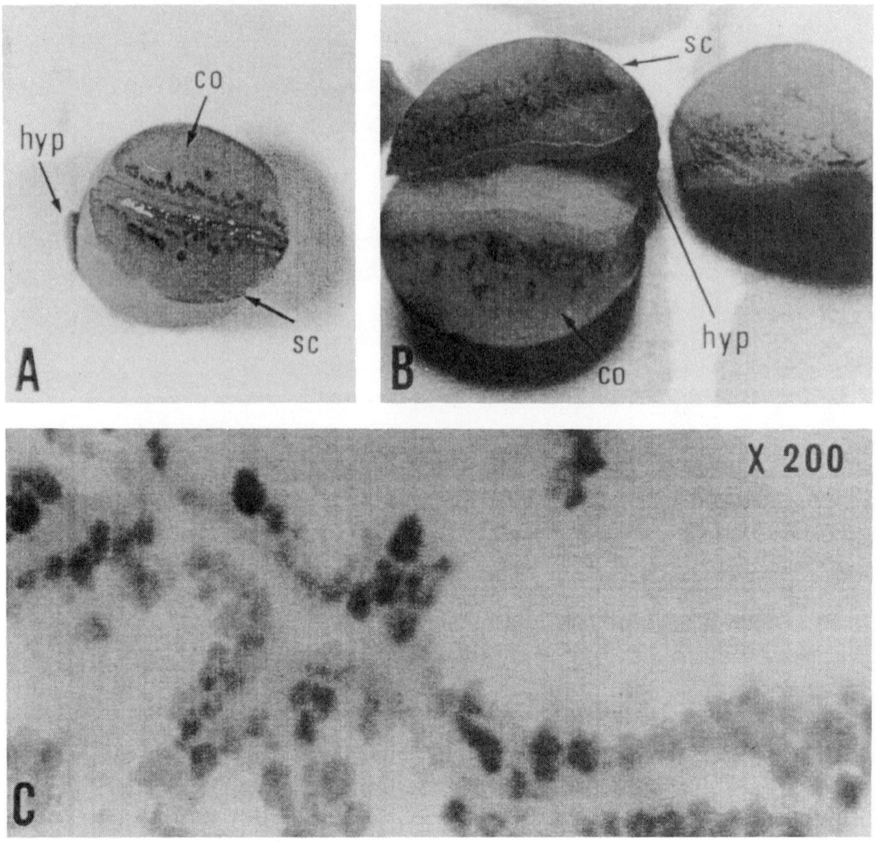

Figure 5. Distribution of DDMP saponins in soybean seeds
A;transection of cotyledon, B;longitudinal section, C;clongated view of vein and minor venation showing DDMP saponin(brown colours on yellow cotyledon), hyp;hypocotyl-radicle axis, co;cotyledon, sc;seed coat

fibrovascular bundles connecting the stem to the plumule. Also a high concentration of DDMP saponin in soybean seed was observed in the fibrovascular bundle of cotyledon and in the hypocotyl (Fig. 5). These data suggest that group B saponins may be widely distributed in legumes as DDMP- conjugated forms. DDMP saponins showed the same UV spectrum with a maximum absorption at 292 nm due to the DDMP moiety. By using this character, DDMP saponin composition in leguminous seeds was investigated. Hypocotyl was extracted with 70% ethanol containing 0.01% EDTA and analyzed by HPLC, directly. As a result, DDMP saponins, especially soyasaponin βg, were found to be widely distributed in legumes, and leguminous seeds have been classified into six types by DDMP saponin composition (Fig. 6). This classification shows

	αg	αa	βg	βa	γg	X
Type I						
Phaseolus vulgaris	●		●	●	●	
Phaseolus lunatus	●		●	●	●	
Pueraria lobata	●		●			
Pycnospora lutescens	●		●			
Glycine max	●		●	●		
Glycine soja	●		●	●		
Type II						
Aeschynomene indica			●			
Alysicarpus vaginalis			●			
Amorpha fruticosa			●	●	●	
Amphicarpaea edgeworthii			●			
Apios americana			●		●	
Arachis hypogaea			●			
Centrosema pubescens			●		●	
Cicer arietinum			●			
Dunbaria villosa			●			
Flemingia prostrata			●			
Galactia tashiroi			●	●		
Glycine tabacina			●		●	
Glycine tomentella			●		●	
Indigofera trifoliata			●			
Lens culinaris			●			
Lotononis bainesii			●			
Lupinus luteus			●			
Medicago lupulina			●			
Medicago sativa			●			
Pisum sativum			●		●	
Rudua aurea			●			
Sesbania grandiflora			●			
Stypholobium fomentosa			●			
Stypholobium japonica			●			
Stylosanthes humilis			●			
Vicia faba			●			
Vicia sepium			●			
Vigna radiata			●			●
Vigna vexillata			●			●
Wisteria floribunda			●		●	
Type III						
Canavalia gladiata					●	
Canavalia maritima					●	

	αg	αa	βg	βa	γg	X
Type IV						
Christia obcordata	●	●	●			
Desmodium heterophyllum	●	●	●			
Desmodium triflorum	●	●	●	●		
Desmodium uncinatum	●	●	●	●		
Phaseolus coccineus	●	●	●	●		
Vigna sinensis	●	●	●	●	●	●
Type V						
Dolichos lablab			●		●	●
Lupinus hirstus						●
Macroptilium atropurpureum	●	●				●
Stylosanthes guyanensis		●				●
Uraria lagopodioides		●				●
Vigna angularis	●		●		●	●
Type VI (no content of DDMP saponins)						
Atylosia scarabaeoides						
Caesalpinia crista						
Caesalpinia pulcherima						
Cajanus cajan						
Cassia alata						
Cassia nomame						
Cassia occidentalis						
Cassia tora						
Ceratonia siliqua						
Cercis siliquastrum						
Cyamopsis tetragonoloba						
Lcucaena lcucocephala						
Maackia amurensis						
Mimosa pudica						
Robinia pseudacacia						
Rhynchosia minima						
Rhynchosia volubilis						
Vicia angustifolia						
Vicia japonica						

Figure 6. Classification of leguminous seeds by DDMP saponins

good agreement with genus or subgenus of Leguminosae, and consequently DDMP saponins may be useful as chemotaxonomical markers.

It has been observed that the saponin composition in soybean seeds is not affected by a difference in cultivation conditions but is peculiar to the variety.[16,26] Soyasaponins Aa and Ab are the major constituents of group A saponins. Most soybean varieties contain either soybean saponin Aa or Ab.[16] Saponin Aa contains a xylose residue at the C-22 position of soyasapogenol A, whereas saponin Ab contains a glucose residue at this position.[17] The existence of soybean saponin Aa and Ab is controlled by codominant alleles at a single locus, and most soybean varieties can be divided into the xylose (X) and glucose (G) types.[18] The variation in saponin composition in soybean seeds can be explained by different combinations of five genes controlling the utilization of soyasapogenol glycosides as substrates (Fig. 7). [27] The function of these genes is variety-specific and organ-specific. Phenotypes of over 1000 soybeans were classified into eight saponin types, and the frequency of phenotypes was different between the cultivated (*Glycine max* L. Merr.) and the wild soybean (*G. soja* Sieb. & Zucc.). The AaBc saponin type predomonated in *G. soja* (58.4% of test collection), but was found in only 0.3% of *G. max*. The mode of inheritance of saponin types is explained by a combination of co-dominant, dominant, and recessive genes. In Fig. 7, the Sg-1a and Sg-1b genes control $\beta(1\rightarrow 3)$xylosylation and $\beta(1\rightarrow 3)$glucosylation of the arabinosyl moiety, respectively, attached at the C-22 position of soyasapogenol A. The sg-10 gene is a null allele at the Sg-1 locus. The Sg-2 gene controls acetylation of the hydroxyl groups of xylosyl and glucosyl moieties attached to Ara at the C-22 position of soyasapogenol A. The Sg-3 gene controls $\beta(1\rightarrow 2)$glucosylation of galactosyl arabinosyl moieties attached to GlcUA and the Sg-4 gene controls $\alpha(1\rightarrow 2)$arabinosylation of GlcUA at the C-3 position of soyasapogenol A and B.

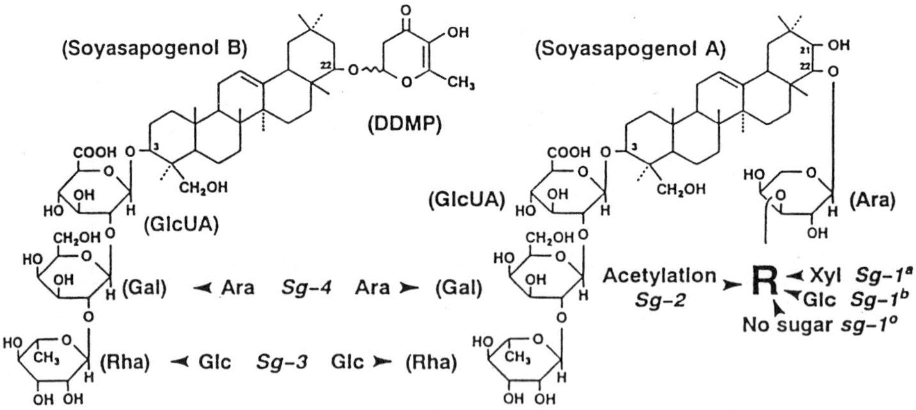

Figure 6. Characterstics of genes controlling elongation and modification of soybean saponin sugar chains attached at the C-3 and C-22 positions of soyasapogenols

CHEMILUMINESCENCE OF PHENOLIC COMPOUNDS IN THE PRESENCE OF ACTIVE OXYGEN SPECIES AND ACETALDEHYDE

The occurrence of chemiluminescence (CL) in the presence of active oxygen species and acetaldehyde (MeCHO) was closely related to the radical-scavenging activity. Phenolic compounds such as flavonoids, anthocyanins and catechins are known to act as superoxide (O_2^-) and hydroxyl radical (HO ·) scavengers.[19,20] The CL of phenolic compounds was measured in the presence of active oxygen species and MeCHO. We investigated the relationship between the chemical structures of some of the phenolics and their CL intensities and/or radical scavenging activities.

The CL of flavonoids was stronger in the presence of MeCHO and HO · which is formed by the Fenton reaction than that in the presence of MeCHO and H_2O_2, but anthocyanins and catechins exhibited stronger CL in the presence of MeCHO and H_2O_2 than that in the presence of MeCHO and HO · .[23] CL decreased following methylation of the hydroxyl groups in the B-ring. The presence of a double bond between C-2 and C-3 contributed only slightly to the stronger CL. The Cl intensity increased with the C-3 sugar chain: hydroxyl (quercetin) < glucose (isoquercetrin) < rhamnosylglucose (rutin).

The CL of anthocyanins and their related compounds in the presence of *tert*-butyl hydroperoxide (*t*-BuOOH) and MeCHO were measured.[24] To investigate the effects of aglycon form, pH 2.6, 7.0, and 9.0 buffers were used in the reaction mixture, respectively, because it was well known that anthocyanins might be converted into pseudo base and chalcone forms depending on the pH of the aqueous medium. The CL intensities of these anthocyanins at basic pH increased about 2 - 7 fold, when compared with those measured at neutral and acidic conditions.These results suggested that the chalcone of anthocyanins was more important for strong CL than formation of the flavylium cation.

The CL of catechins were measured in the presence of active oxygen species and 2% MeCHO. (-)-Epigallocatechin (EGC) exhibited the strongest CL among the eight kinds of catechins.[25] When the reaction mixture composed of EGC, MeCHO, and H_2O_2 was subjected to HPLC and LC-MS, it was confirmed that EGC was almost not lost during the reaction. EGC itself did not show CL. Therefore, this reaction may proceed via a route similar to that described elsewhere.[21] CL intensity of the reaction mixture was dependent on the concentrations of H_2O_2 (X : active oxygen species), phenolic compound (Y : catalytic species) and MeCHO (Z : receptive species) (Fig. 8). We proposed that the CL intensity, *P*, in the presence of H_2O_2 and MeCHO is given by $P = k\ [X]\ \ [Y]\ f\,(Z)$. The calculated photon constants(log *k*) are summarized in Table 2. Log *k* in the presence of HO · and *t*-BuO · was calculated using H_2O_2 and *t*-BuOOH concentrations. In the presence of H_2O_2 as X, *k* of catechins was closely related to the characteristics of their chemical structures, showing good agreement with the flavonoids and anthocyanins. CL of phenolic compounds was increased by the presence of the following partial structures:(a) the pyrogallol structure rather than catechol structure in the B-ring, (b) a free hydroxyl group at C-3 and (c) the stereoscopic structure between C-3 hydroxyl group or glycoside and B-ring.

Radical-scavenging activity of antioxidants is generally expressed from a point of view on how rapidly they react with active oxygen to scavenge them, however, it may be necessary to investigate precisely the mechanism to eliminate high potential energy, which may be produced in the presence of active oxygen and antioxidants in biological systems.

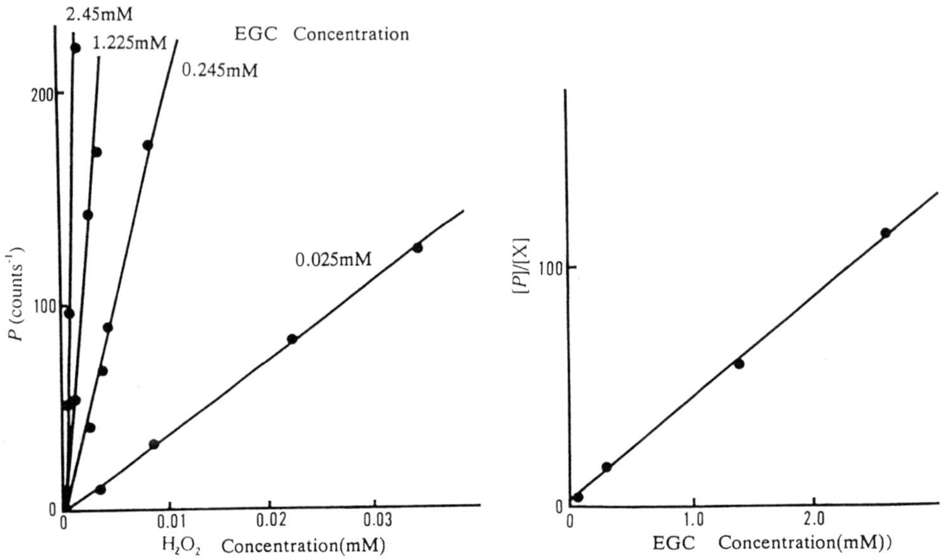

Figure 8. Effect of (-)-epigallocatechin (Y;catalytic species) and hydrogen peroxide (X;active oxygen species) concentration on photon intensity (P) in the presence of acetaldehyde (Z;receptive species)

Table 2. Photon constants of catechins (Y) in the presence of hydrogen peroxide (X) and acetaldehyde (Z)

	k_1	k_2	k_3	k
(-)-Catechin	3.30×10^{6}	3.80×10^{4}	2.13×10^{3}	6.44×10^{4}
(-)-Epicatechin	2.93×10^{7}	1.30×10^{5}	5.14×10^{5}	2.70×10^{5}
(-)-Gallocatechin	9.70×10^{7}	5.75×10^{5}	1.49×10^{4}	4.36×10^{5}
(-)-Epigallocatechin	4.40×10^{10}	5.66×10^{5}	2.24×10^{4}	8.23×10^{6}
(-)-Catechin gallate	2.29×10^{6}	4.50×10^{4}	4.32×10^{2}	3.54×10^{4}
(-)-Epicatechin gallate	7.16×10^{6}	2.80×10^{4}	1.00×10^{3}	5.85×10^{4}
(-)-Gallocatechin gallate	2.48×10^{7}	6.95×10^{5}	5.87×10^{3}	4.66×10^{5}
(-)-Epigallocatechin gallate	2.55×10^{9}	9.09×10^{5}	9.30×10^{3}	2.78×10^{6}
Gallic acid	2.89×10^{6}	2.56×10^{4}	1.48×10^{3}	4.78×10^{4}

k_1; $[P]=k_1[X][Y]f(z)$, k_2; $[P]=k_2[Y][Z]f(x)$, $[P]=k_3[X][Z]f(y)$, k_1, k_2, and k_3; $M^{-2}s^{-1}$, k; $[P]=k[X][Y][Z]$, k; $M^{-3}s^{-1}$

Hydrogen peroxide, catechins or gallic acid, and acetaldehyde were used as factors of X, Y, and Z, respectively.

DDMP saponins showed oxygen-radical-scavenging activity when assayed by using the xanthin oxidase-NH_2OH method, electron spin resonance (ESR) and chemiluminescence.[22] One milligram of DDMP saponin/mg exhibited superoxide-scavenging activity of a degree equivalent to 17.1 units of superoxide dismutase/ml measured by the ESR spin trapping method (Fig. 9). This scavenging activity of DDMP saponin is caused by the DDMP moiety attached to the triterpene aglycon since soybean saponin Bb, which is derived from soyasaponin β g, but which lacks this group, did not show the scavenging activity.

The CL was in good agreement with the radical-scavenging results showed by enzymatic and ESR spin-trapping methods. The CL exhibited in the presence of active oxygen species (X), catalytic species (Y), and receptive species (Z) is most convenient to investigate the scavenging function of active oxygen species. We have found that group A and DDMP saponins are Y equivalent to EGG and Z equivalent to MeCHO, respectively. The photon intensity (*P*) in the presence of 1.12 mM soyasaponin β g increased linearly depending on the concentration of H_2O_2 and soyasaponin Ab (Fig. 10).[26] It is important that soybean saponin emitted CL in the same system; $[P] = k\ [X]\ \ [Y]\ f\ (Z)$. Especially, in the presence of soyasaponin β g exhibited stronger CL (about 100 fold) than MeCHO as Z. This result suggested that the combination of group A and DDMP saponins was important in the scavenging function of active oxygen species, because in the presence EGC as Y, soyasaponin β g exhibited about 10 fold of the CL compared to MeCHO. A scheme of photon emission in the presence of X (active oxygen species), Y (catalytic species), and Z (receptive species) is proposed in Fig. 11.

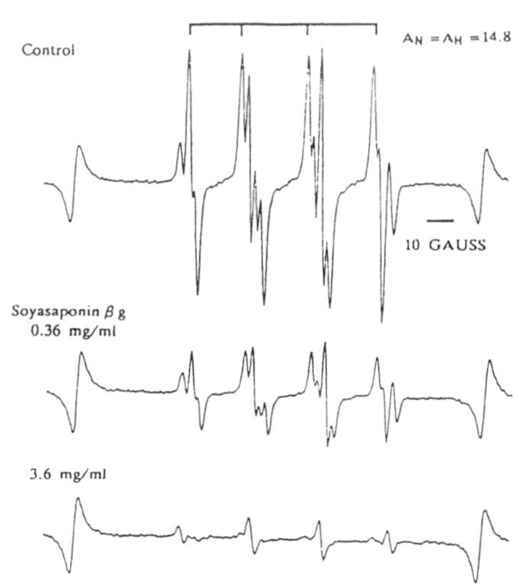

Figure 9. O_2^- scavenging activity of soyasaponin β g measured by ESR spin trapping method The ESR spectra of the DMPO-OOH radical spin adduct in both the absence (control) and in the presence of different concentrations of soyasaponin β g were recorded.

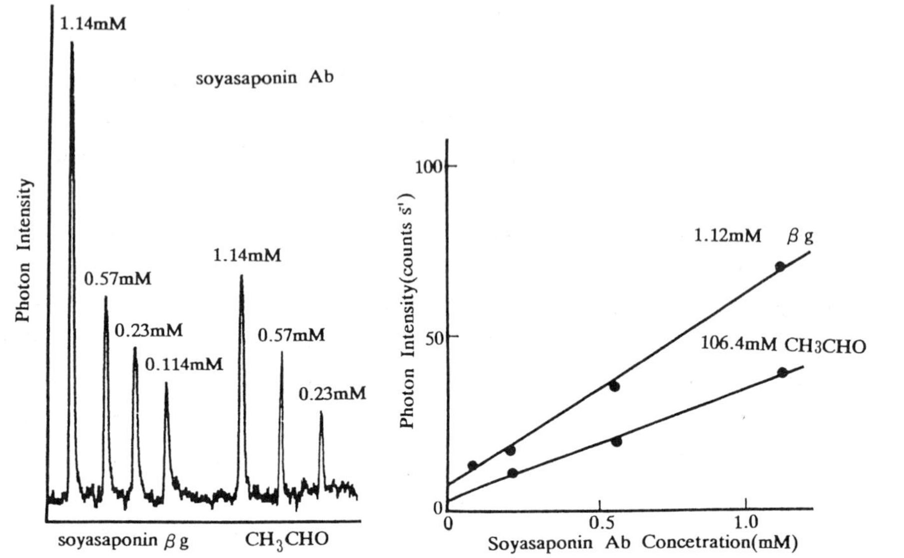

Figure 10. Chemiluminescence of soyasaponin Ab (Y;catalytic species) in the presence of hydrogen peroxide (X;active oxygen species) and soyasaponin β g or acetaldehyde (Z;receptive species)

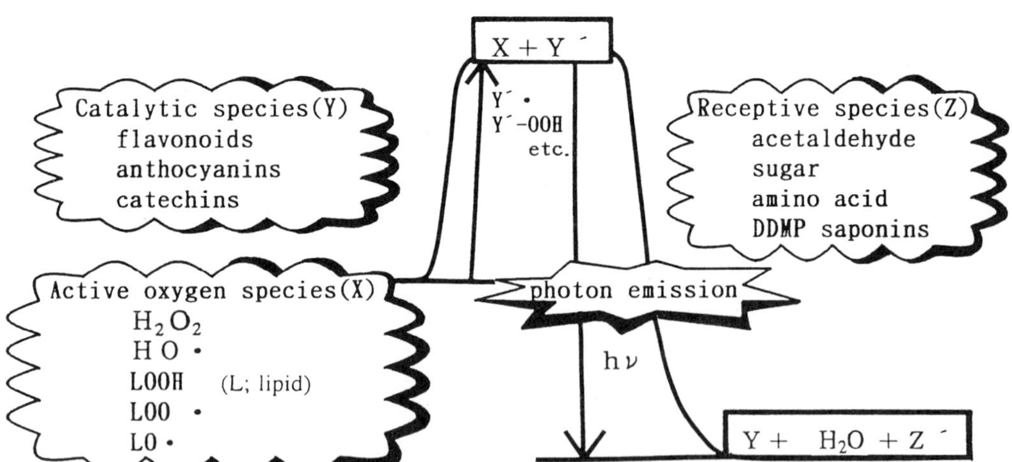

Figure 11. Schematic diagram of XYZ system

REFERENCE

1. P.N.Nair, N.Turjuman, G.Kessie, B.Calkins, G.T.Goodman, H.Davitvitz and G. Nimmagadda, Diet, nutrition intakke, and metabolism in populations at high and loe risk for colon cancer, *Am. J. Clin. Nutr.*, 40:927 (1984).
2. H.S.Basage, Biochemical aspects of free radicals, Biochemical aspects of free radicals, *Biochem. Cell Biol.*, 68:989 (1990).
3. T.Nakayama, M.Kodama and C.Nagata, Free radical formation in DNA by lipid peroxidation, *Agric. Biol. Chem.*, 48:571 (1984).
4. O.T.Aruoma, B.Halliwell, E.Gajewski and M.Dizdaroglu, Damage to the bases in DNA induced by hydrogen peroxide and fwrric ion, *J. Biol. Chem.*, 264:20509 (1989).
5. B.Halliwell and M.Grootveld, Method for the measurement of hydroxy radicals in biochemical system, in: *Methods of BiochemicalAnalysis*, D.Glick, ed., John Wiley & Sons, Inc. New York, p.59 (1988).
6. T.F.Slater, Free-radical mechanisms in tissue injury chelates, *Biochem. J.*, 222:1 (1984).
7. G.R.Fenwick, K.R.Price, C.Tsukamoto and K.Okubo, in *Toxic Substances in Crop Plants*, J.P.F.D ′ Mello, C.M.Duffus and J.H.Duffus, Eds., Royal Society of Chemistry, Cambridge, p. 285 (1993).
8. C.Tsukamoto, A.Kikuchi, S.Kudou, M.Tonomura, K.Harada. T.Iwasaki and K.Okubo, ACS Symposium Series 546, *Food Phytochemicals for Cancer Prevention I*, p. 372 (1993).
9. K.Okubo, K.Kudou, T.Uchida, Y.Yoshiki, M.Yoshikoshi and M.Tonomura, ACS Symposium Series 546, *Food Phytochemicals for Cancer Prevention I*, p. 330 (1993).
10. S.Kudou, M.Tonomura, C.Tsukamoto, M.Shimoyamada, T.Uchida, M.Yoshikoshi and K.Okubo, ACS Symposium Series 546, *Food Phytochemicals for Cancer Prevention I*, p. 340 (1993).
11. S.Kudou, M.Tonomura, C.Tsukamoto, M.Shimoyamada, T.Uchida and K.Okubo, Isolation and structural elucidation of main genuine soybean saponin, BeA, *Biosci. Biotech. Biochem.*, 56:142 (1992).
12. Y.Yoshiki, C.Tsukamoto, K.Harada and K.Okubo, Unstable behavior of DDMP saponins and composition in the subgeneous *soya*, Abstracts of World Soybean Research Conference V, Chiang Mai, Thailand (21-27 Feb.,1994), p.113.
13. S.Kudou, M.Tonomura, C.Tsukamoto, T.Uchida and K.Okubo, Isolation and Structural elucidation of DDMP-conjugated soyasaponins as genuine saponins from soybean seeds, *Biosci. Biotech. Biochem.*, 57:546(1993).
14. Y.Yoshiki, J.H.Kim, K.Okubo, I.Nagoya, T.Sakabe and N.Tamura, Saponin conjugated with 2,3-dihydro-2,5-dihydroxy-6-methyl-4H-pyran-4-one from from Dolichos lablab and the SOD-like activity.*Phytochemistry*, 38:229 (1995).
15. K.Okubo, Y.Yoshiki, K.Okuda, T.Sugihara, C.Tsukamoto and K.Hoshikawa, DDMP-conjugated saponins isolated from groundnut (Apios Americana), *Biosci. Biotech. Biochem*, 58:2248 (1994).

16. M.Shiraiwa, K.Harada and K.Okubo, Composition and content of saponin in soybean seed according to variety, annual cultivation and harvest time, *Agric. Biol. Chem.*, 55:323 (1991).
17. M.Shiraiwa, S.Kudou, M.Shimoyamada, K.Harada and K.Okubo, Composition and structure of "group A" saponin in soybean seed, *Agric. Biol. Chem.*, 55:315 (1991).
18. M.Shiraiwa, F.Yamauchi, K.Harada and K.Okubo, Inheritance of group A saponin in soybean seed, *Agric. Biol. Chem.*, 54:1347 (1990).
19. S.V.Javanovic, S.Steenken, M.Tosic, B.Marjanovic and M.G.Simic, Flavonoids as antioxidants, *J. Am. Chem. Soc.*, 116:4846 (1994).
20. A.Puppo, Effect of flavonoids on hydroxyl radical formation by Fenton-type reaction, *Phytochemistry*, 31:85 (1991).
21. N.Gotoh and E.Niki, Rate of interactions of with vitamin E, vitamin C and related compounds as measured by chemiluminescence, *Biochem. Biophys. Acta*, 1115:201 (1992).
22. Y.Yoshiki and K.Okubo, Oxygen radical scavenging activity of (2,3-dihydro-2,5 -dihydroxy-6-methyl-4*H*-pyran-4-one) conjugated saponin in soybean seed.*Biosci. Biotech. Biochem.*, 59:1556 (1995).
23. Y.Yoshiki, K.Okubo, S.Onuma and K.Igarashi, Chemiluminescence of benzoic acid and cinnamic acid, and flavonoids in the presence of aldehyde and hydrogen peroxide or hydroxyl radical by Fenton reaction, *Phytochemistry*, 39:225(1995).
24. Y.Yoshiki, K.Okubo and K.Igarashi, Chemiluminescence of anthocyanins in the presence of aldehyde and *tert*-butyl hydroperoxide, *J. Biolumi. Chemilumi.*, 10:335 (1995).
25. Y.Yoshiki and K.Okubo, Chemiluminescence of catechins and gallic acid in the presence of aldehyde and active oxygen species at biomedium system*Biochem. Biophys. Acta*, in press (1996).
26. C.Tsukamoto, S.Shimada, K.Igita, S.Kudou, M.Kokubun, K.Okubo and K.Kitamura, Factors affecting isoflavone content in soybean seeds, J. *Agric. Food Chem.*, 43;1184 (1995).
27. S.Tsurumi, T.Takagi and T.Hashimoto, A γ -pyronyl-triterpenoid saponin from *Pisum sativum*, *Phytochemistry*, 31;113 (1992).

ALFALFA SAPONINS: STRUCTURE, BIOLOGICAL ACTIVITY, AND CHEMOTAXONOMY

Wieslaw Oleszek

Department of Biochemistry, Institute of Soil Science and Plant Cultivation, Osada Palacowa, 24-100 Pulawy, Poland

INTRODUCTION

Alfalfa (*Medicago sativa* L) is a valuable source of good quality protein in temperate climates. The aerial parts of the plant may be used as a forage crop in the form of green feed, hay, or condensed pellets. In some countries, such as France[1], considerable amounts of alfalfa are processed to produce leaf protein concentrates (LPC). Human consumption of lucerne is generally very low, but there has been increasing interest in using lucerne as sprouts for green salad, or in the form of tablets or juice, for its effect on serum cholesterol. The hypocholesterolemic activity of lucerne is believed to be related to the presence of chemical principles -- saponins -- occurring in this species[2-5]. On the other hand, the same group of compounds is seen as being detrimental from a nutritional point of view. It has been shown that saponins are especially harmful for monogastric animals[4], affecting their growth by changing the palatability and hence the consumption of feed, or by influencing the digestion and absorption processes. Saponins may also alter fermentation, and affect the site and the extent of nutrient digestion in ruminants[6,7].

Studies on the beneficial or harmful effects of alfalfa saponins conducted before 1990 were limited by the fact that saponin extracts, isolates, and plant materials were poorly defined owing to methodological limitations. Most often the term "biologically active saponins" (those determined with hemolytic or *Trichoderma viride* tests) was the only parameter distinguishing tested samples. In only rare cases were more precise parameters such as the qualitative and quantitative variation of saponins with plant age, variety, the part of plant, environmental and agronomic conditions or structure-dependent biological activities, considered. Moreover, when used in different laboratories, determination of saponins with hemolytic and *T. viride* tests did not provide comparable data. Inter-laboratory trials performed under the EUCARPIA *Medicago* group showed that even quantifications of saponins from the same plant material, using the same procedures and calibration standards, could not produce comparable data. Thus, interpreting most findings on the biological activity of alfalfa saponins is very difficult.

At the end of the 80's and the beginning of the 90's there was renewed interest in alfalfa

Saponins Used in Food and Agriculture
Edited by Waller and Yamasaki, Plenum Press, New York, 1996

saponins. Research groups in Poland, Russia, France, and Israel made significant progress on a number of fronts. Structural determinations were made for most of the dominant alfalfa saponins, and in some cases assessments of their biological activity (hemolytic, antifungal, allelopathic, membrane depolarizing) have been made. These revealed a strong structural dependency. The objective of this paper is to summarize the most important achievements in this field during thge last decade with a critical discussion of some results and prospects for future research.

CHEMICAL STRUCTURES

Early work on the chemical structure of alfalfa saponins has focused mostly on the recognition of the level and occurrence of so called "antifeedant" or "biologically active" principles. The best known of these are saponins with medicagenic acid as an aglycone, which makes up 50-70% of the total aglycone received from the hydrolysis of alfalfa aerial parts or roots[8,9]. Ease of semi-determination of this group of compounds with *T. viride* or hemolytic tests and good correlations between these determinations and nutritional trials left structural work far behind other fields such as breeding and plant selection, nutrition, physiology, and biochemistry. Despite understanding of the chemistry of aglycones such as medicagenic acid, hederagenin, the soyasapogenols, bayogenin, oleanolic acid, and lucernic acid, only a few glycosides had been structurally determined at that time[10,11,12]. In the past decade the structures of several additional compounds of the dominant medicagenic acid glycosides have been determined (Table 1), the majority of which were isolated from the roots. Some of these have been confirmed by several independent research groups, but these also revealed some differences in alfalfas of different origin.

There is an agreement confirmed by all working groups on the presence of two short-chain glycosides, 3-*O*-glucoside and 3,28-diglucoside of medicagenic acid (I, II) in alfalfa roots, and the absence of these compounds in aerial parts of the alfalfa plant. Medicagenic acid glycosides identified in the tops seem to be more complex with bisdesmosidic forms VI, VII, XI, XV dominating. These can be subdivided into two groups, one of which has glucose as the first sugar attached in the 3-*O* position, the other having glucuronic acid in the same position. Trivial names have been given to some of these compounds: Medicoside G (II), H (VIII), J (VI) and L(IX).

Hederagenin glycosides (Table 2) XVI, XVII (medicoside C), XVIII (medicoside I), and XIX were identified only in the roots, in spite of the fact that hederagenin was also identified in the hydrolysates of alfalfa aerial parts[8,9]. But the concentration of hederagenin in hydrolysates from alfalfa tops made up only 0.1-0.4% of the total aglycones[8] or 0.03 mg/g of the dry matter[9] versus 0.82-1.32 mg/g in roots. Thus, hederagenin glycosides do not seem to play a significant role in top saponins, but they may be of fundamental importance in root saponins.

Concentrations of soyasapogenol A, B (together with breakdown products soyasapogenols C, D, and F) and soyasapogenol E in the total amount of aglycones received from the hydrolysis of alfalfa tops were estimated to be 10-15%, 15-30%, and 1% respectively[8]. These results may differ significantly based on plant variety, age, and environmental conditions. Only a few glycosides have been identified, predominantly in the aerial parts of alfalfa plants. These were compounds XX, XXI, XXII, XXIII, and XXIV (Table 3). Compounds XX and XXI, originally isolated and identified as azukisaponin from azuki beans (*Vigna angularis*), were found in American alfalfa together with soyasaponin I (XXII) and dehydrosoyasaponin I (XXIII) (soyasapogenol E) glycoside[26]. Compound XXI has been independently identified in the North American Lahontan variety[20]. The same group has later identified in germinating alfalfa seeds, soyasaponin VI (XXIV), the maltol conjugate of soyasaponin I[27]. It has been suggested that this compound is a natural form present in plant material, which owing to its high lability decomposes easily into soyasaponin I during sample preparation or isolation procedures[27].

Table 1. Medicagenic acid glycosides identified in alfalfa.

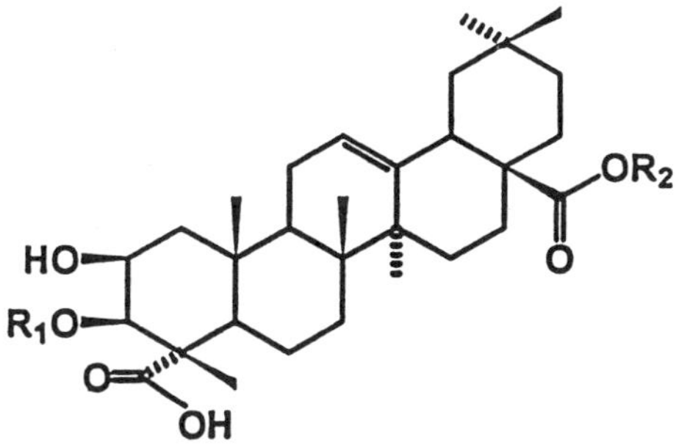

	R_1	R_2	Plant part	Literature cited
I	Glu	H	roots	10,13,14,15,16
II	Glu	Glu	roots	13,14,15,16,17
III	Glu(1→4)Glu	H	roots	18
IV	Glu(1→6)Glu(1→3)Glu	H	roots	12
V	Rha(1→6)GluA(1→2)Glu	H	flowers	11
VI	Glu	Xyl(1→4)Rha(1→2)Ara	roots	16,17,19
			tops	20,21
VII	Glu(1→2)Glu	Xyl(1→4)Rha(1→2)Ara	roots	16,17
			tops	20,21
VIII	Glu	Rha(1→2)Ara	roots	22
IX	Glu(1→2)Glu(1→2)Glu Rha(1→3)┘	Glu	roots	23
X	GluA	H	roots	16
XI	GluA	Xyl(1→4)Rha(1→2)Ara	roots	16
			tops	20,21
XII	Gal(1→2)Glu	Glu	roots	17
XIII	Rha(1→2)Glu(1→2)Glu	Glu	roots	17
XIV	H	Xyl(1→4)Rha(1→2)Ara	tops	20
XV	GluA	Rha(1→2)Ara	tops	21

Table 2. Hederagenin glycosides identified in alfalfa

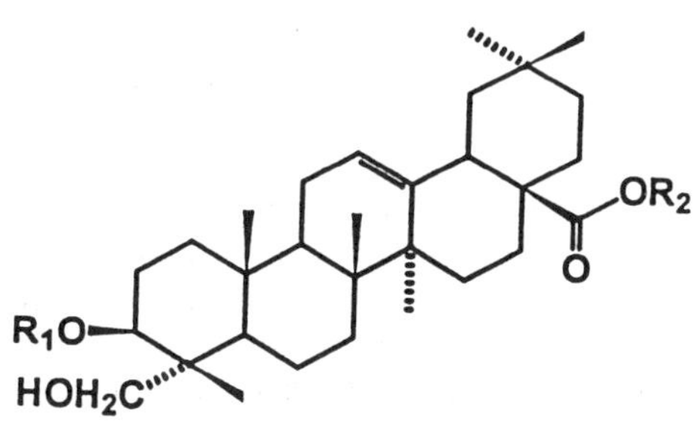

	R_1	R_2	plant part	Literature cited
XVI	Glu(1→2)Ara	H	roots	16,24
XVII	Ara(1→2)Glu(1→2)Ara	H	roots	16,24
XVIII	Ara(1→2)Glu(1→2)Ara	Glu	roots	17,25
XIX	Gal(1→2)Ara	H	roots	17

Table 3. Soyasapogenol B and its dehydro (soyasapogenol E) glycosides

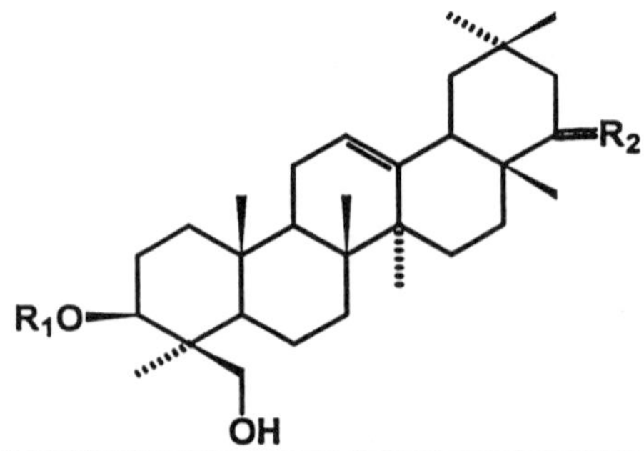

	R_1	R_2	plant part	Literature cited
XX	Glc(1→2)GlcA	β-OH, -H	tops	26
XXI	Rha(1→2)Glc(1→2)GlcA	β-OH, -H	tops	20,26
XXII	Rha(1→2)Gal(1→2)GlcA	β-OH, -H	tops	26
			roots	16
XXIII	Rha(1→2)Gal(1→2)GlcA	=O	tops	26
XXIV	Rha(1→2)Gal(1→2)GlcA	-O-maltol	seeds	27

Table 4. Zanhic acid glycosides identified in alfalfa

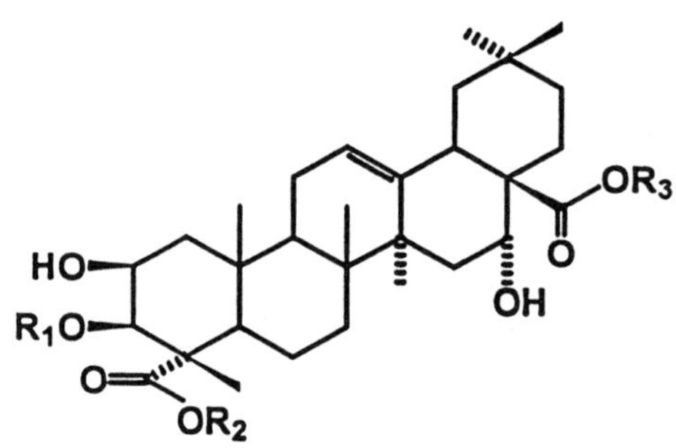

	R_1	R_2	R_3	plant part	Literature cited
XXV	-Glc-Glc-Glc	-Ara	-Ara-Rha-Xyl-Api	tops	21
XXVI	-Glc-Glc-Glc	-Ara	-Ara-Rha-Xyl	tops	21

The last group of compounds, derivatives of zanhic acid, have recently received a lot of attention. As early as 1959, Livingston[28] isolated three compounds. One of these was lucernic acid, but its structure was not fully elucidated. It was later shown that lucernic acid was an artifact, the 13→28 lactone formed by acid-catalyzed cyclization of the γ,δ-unsaturated acid, which in this case proved to be zanhic acid (16α-hydroxymedicagenic acid)[28]. Dimbi[30] coined the trivial name "zanhic acid" for two novel aglycones isolated from *Zanha golugensis,* zanhic acid and zanhic acid γ-lactone. In fact, 16α-hydroxymedicagenic acid had been identified two years earlier from sapogenins produced by the acid hydrolysis of extracts from *Herniaria glabra* and *Herniaria hirsuta*[31]. Zanhic acid and its γ-lactone (lucernic acid) were found together with glucose(1→3) zanhic acid and glucuronic acid (1→3) zanhic acid in acid hydrolysis products of alfalfa leaves, but not in the roots[29]. The first natural saponins containing zanhic acid as aglycone (Table 4) were identified in alfalfa aerial parts only recently[21]. These two, highly polar compounds, being the most bitter principles from alfalfa, are among the most complex saponins ever isolated. One of the two contains apiose, a sugar rarely occurring in nature, as a terminal residue. Its concentration in some cultivars can be as high as 0.5% in dry matter

METHODS OF EXTRACTION, SEPARATION AND STRUCTURAL ELUCIDATION

There is no specific manner for the extraction of saponins from alfalfa plant material. Dried material may be defatted and depigmentated with chloroform[13] or ethyl acetate[8] but this step has not been widely used. Root and aerial parts have been extracted by boiling with MeOH[13,20,21,26], $MeOH:H_2O$ (4:1; 3:7)[20,21], $EtOH:H_2O$ (4:1)[16,18] or H_2O[20]. After the evaporation of the organic solvent, the extract is redissolved in water and crude saponins are removed. The manner of removal depends on the final goal of separation and author requirements for a particular group of saponins. The use of cholesterol, ammonium or sodium salts to fractionate saponins into two classes, precipitable, versus unprecipitable was once widely used[32], but has now been superseded by chemical or chromatographic procedures. The extraction of saponins into 1-BuOH from the aqueous solution has been applied in some cases[13,20,26,27]. This procedure has worked perfectly for monodesmosidic and short-sugar-chain bisdesmosidic saponins. However, bisdesmosidic saponins with several sugar units do not readily go into BuOH and extraction may not be exhaustive. Tridesmosides like zanhic acid tridesmosides (XXV and XXVI) cannot be extracted into BuOH and are totally lost with this procedure. This is probably the reason that until recently tridesmosides were not detected in alfalfa despite high quantities in some varieties. Massiot and co-workers[27] developed a procedure in which saponins were precipitated from a water solution by the use of ether. The precipitate was filtered, and dialyzed against water in seamless cellulose tubing for 4 days. The content of the tubing was freeze-dried to produce a crude saponin mixture.

Oleszek[15] has described a simple method for clean up and separation of alfalfa saponins using C_{18} support. Water solutions of alfalfa root extracts (pH=2.5) or alfalfa sprout extracts (pH=4.8) are loaded on to a C_{18} solid-phase column and saponins are selectively eluted with aqueous methanol. Eluants up to 40% of methanol remove carbohydrates and some phenolics, very abundant interferences in the sample matrix. Eluants containing 50-60% methanol remove only medicagenic acid glycosides, with 70-100% methanol being needed to elute hederagenin and soyasapogenol glycosides. By this technique, saponins can be extracted from plant material with 30% MeOH. The extract needs no evaporation and can be loaded to the C_{18} column and saponins can be extracted-purified in one step, without losing any component of the mixture. The procedure can be further modified by changing the pH of loaded sample to natural (pH=7) with diluted NaOH. Retention of saponins under this condition was changed and some saponins were already eluted with water e.g. 3-*O*-glucuronide of medicagenic acid. By applying a methanol-water linear gradient it is possible to separate the saponin mixture into several classes of compounds, e.g. medicagenic acid glucuronides, glucosides, monodesmosidic forms, and zanhic acid tridesmosides. The procedure can also be used successfully for the preliminary clean-up of plant extracts for saponin quantification[9,33]. A useful technique allowing the preliminary separation of saponins not possessing a free carboxylic function, e.g. hederagenin bisdesmoside (XVIII) or zanhic acid tridesmosides, is to precipitate the remaining saponins with saturated lead acetate solution in water[33].

The separation of individual saponins from the mixture can rarely be achieved in a one-step procedure. Final purification requires a second solvent and column chromatography support, or even in some cases use of the preparative TLC technique. The most frequently used supports are silica gel with eluting systems composed of chloroform-methanol-water (65:35:8; 65:23:4; 60:40:3; 12:8:1; 10:10:1; 5:5:1; 7:3:1)[13,20,26] and ethyl acetate-water-acetic acid (7:2:2)[16,18] or C_{18} column used in reverse phase with methanol-water (1:1; 3:2; 7:3)[16,21]. Column separations are being monitored with TLC using the same solvent systems.

Massiot and co-workers[20] have developed a procedure for the separation of individual compounds based on the acetylation of crude saponins followed by the separation of derivatized compounds on silica gel, using chloroform-methanol (99:1) or combinations of chloroform-hexane -methanol (9:1:0 - 1:0:1). The procedure is very useful for the isolation of compounds for structural work, but it cannot provide pure compounds for biological assays since the reconversion of derivatized compounds into underivatized material requires alkaline hydrolysis. Under these conditions it is possible to cleave some ester-linked sugars.

The structural elucidation of alfalfa saponins has been performed using several different techniques, most often with a combination of them. For many compounds, basic and acidic hydrolysis were performed to yield carbohydrates and prosapogenins or aglycones, which were analyzed independently by physical and spectral techniques[13,16,18,21]. The nature of aglycone-sugar linkages were identified by enzymatic hydrolysis[18], by methylation of the molecule followed by the comparison of products with standards by TLC, GC, and EIMS[13,19,22,23,24], or by inspection of ^{1}H and ^{13}C NMR spectra. Double resonance and other sophisticated techniques have been employed for the analysis of saponin structure without derivatization[16,21] or after peracetylation[20]. The advantages and the philosophy of the methods relying on NMR studies of peracetylated saponins were recently reviewed[1].

Structural confirmation for a group of saponins and sapogenins present in alfalfa roots was obtained by liquid secondary ion mass spectrometry (LSIMS) and linked scanning at constant B/E LSIMS/MS[34]. Positive and negative product ion spectra from protonated, deprotonated, and natriated pseudomolecular precursor ions were compared, including both metastable ions that were generated spontaneously and those generated by high-energy collisionally activated dissociation (CAD), demonstrating the great usefulness of these techniques for structure elucidation.

BIOLOGICAL ACTIVITY

Hemolysis

The capacity of saponins to lyse erythrocytes *in vitro* has been known for a long time, and has been used as means of detecting and quantifying saponins in plant material. As the term "saponins" -- understood widely as a mixture of different glycosides -- progressively developed into a more precise determination of structures, it was found that hemolysis is not an attribute of all saponins. Soybean saponins, for example, which are a mixture of soyasapogenol glycosides, are not hemolytic[35]. For alfalfa saponins, Gestetner and co-workers[31] conducted early, advanced work on hemolytic properties. It was shown that the structure of the aglycone is a determinant feature in hemolytic activity, and that such activity is determined by the aglycone:sugar residues ratio. Saponins with the highest R_f values showed stronger hemolytic activity.

Hemolyzing activity does not seem to be an important feature as regards value of alfalfa for use as feed. It is unlikely that these molecules cross the intestine and enter the blood stream. Using pure medicagenic acid 3-*O*-glucoside, Polacheck et al.[36] did not observe any hemolytic activity in erythrocytes taken even from mice injected intraperitoneally with 0.6 ml of saponin solution (500 μg/ml). But the dependence of hemolytic activity on the chemical structure of a given saponin might be of fundamental meaning for hemolytic methods used for saponin determination. A number of individual compounds were measured for their hemolytic activities via the hemolytic index (HI) and microhemolytic test (Table 5). The hemolytic index in these experiments was understood as the volume in milliliters of 2% (v/v) solution of cows blood in buffered isotonic solution which could be fully hemolyzed by 1 g of saponins. The hemolytic

Table 5. Hemolytic, *T. viride*, and allelopathic activity of some alfalfa saponins

Compounds	Microhemolytic method*	Hemolytic index	*T. viride* ID_{50}**	Allelopathic activity***
$MaNa^+$	17	11896	1.7	29
I	44	18157	1.6	53
II	0	0	33.0	76
VI	17	4294	13.5	34
VII	12	2982	35.0	na
X	33	3581	9.1	110
XI	25	3581	47.5	97
XVII	11	7446	25.5	34
XXII	0	0	0.0	97
XXV	6	2000	0.0	81
3-*O*-GlcGlcGlc				
Zanhic acid	na	na	70.0	42
RS****	27	2982	23.5	68
TS*****	36	3420	85.5	75

na - not analyzed
* Hemolytic ring diameter (mm x 12) for saponin concentration of 2.5 mg/ml
** Saponins concentration (mg/100 ml of medium) resulting in 50% inhibition of fungus growth
*** Growth of wheat seedling roots in the medium containing 100 ppm of saponins
**** Mixture of cholesterol-precipitable alfalfa root saponins
***** Saponin mixture from aerial parts

micromethod test, which is a very useful and efficient screening test for breeders, was performed on glass plates covered with a gelatin-blood mixture[37,38]. It was shown that individual saponins were marked by different hemolytic activities depending on their structures and also on the type of hemolytic test being used. These data clearly proved that hemolytic activity depends on the aglycone structure but also on the sugar composition. The highest hemolytic activity was found in the monodesmosidic 3-*O*-glucoside of medicagenic acid (I), but the activity drastically dropped down when glucose was replaced with glucuronic acid (X), clearly showing the influence of sugar substitution. This influence was more pronounced for the 3,28-diglucoside of medicagenic acid, which surprisingly did not show any hemolytic activity in spite of the fact that other bisdesmosidic medicagenic acid glycosides showed activity. It was generally observed that medicagenic acid glycosides substituted with glucose in the 3-*O* position showed higher activities than their analogues substituted with glucuronic acid.

Relatively high activity was showed by hederagenin monodesmoside (XVII), low activity was expressed by zanhic acid tridesmoside (XXV), and no hemolysis was caused by soyasaponin I (XXII). These differences clearly show that the results of the determination of alfalfa saponins by hemolytic methods can be strongly influenced by qualitative differences in their structures, especially in case of roots or seedlings, which are rich with monodesmosidic forms. They also show that such saponins as zanhic acid tridesmoside or soyasaponin I can not be determined with these procedures.

Effects on fungi

Off all the fungi tested, *Trichoderma viride* showed the highest sensitivity to the presence of saponins in the growth medium[39,40,41]. Researchers have taken advantage of this sensitivity to quantify saponins in alfalfa[40,42]. Scardavi and Elliot[43] claimed that the sugar chain of the saponin moiety is an important factor in determining its water solubility, an essential condition for activity against *T. viride*. Support for this view was later found by Nonaka[44] who showed that medicagenic acid itself has low activity towards *T. viride* as compared to its glycosides. As a result of comparing the activity of a number of saponin fractions isolated from alfalfa, the author concluded that structural features other than the aglycone were responsible for the variable growth inhibition. As shown above for hemolysis, also activity towards *T. viride* was positively correlated to the ratio of aglycone:sugars[32]. Aglycones were more toxic towards *Sclerotium rolfsii* than were parent saponins[45]. Later it was suggested that the presence of two carboxylic acid residues and one hydroxyl group in medicagenic acid were a major, but not the only, factor determining the strong growth-impairing effects of alfalfa saponins toward fungi[32,46]. In series of experiments performed with medicagenic acid, its 3-*O*-glucoside and 3,28-diglucoside, and their methyl and acetyl derivatives, it was shown that blockage of some functional groups decreased activity[47]. In all cases blockage of carboxylic acid groups led to impaired growth, so that 55-60% of the growth was maintained even at concentrations of 10 mg/100 ml. Blocking the -OH groups in the aglycone or sugar moieties produced very significant lowering of antifungal activity. For example, 2,3-diacetylmedicagenic acid at a concentration of 6 mg/100 ml inhibited fungal growth by 40%, a result similar to that found for medicagenic acid with both -COOH groups esterified. The blocking of both aglycone and sugar -OH groups reduced the inhibition of *T. viride* to 5% at the same concentration. This indicates that the activity of medicagenic acid and its derivatives is markedly dependent upon free -COOH and -OH groups. Similar results have also been obtained in studies of the relative activity of medicagenic acid and its synthetic derivatives against plant-pathogenic fungi[48]. For compounds where the hydroxyl functions of the aglycone remained unchanged, activity against *Sclerotium rolfsii* was higher than for compounds where these functions were modified by acetylation or methylation. It was concluded that a potential free hydroxyl at position C-3 is essential for antimycotic activity, e.g. *Sclerotium rolfsii*, *Fusarium oxysporum* f. sp. *lycopersici*, *Rhizoctonia solani*, and *Aspergillus niger*.

The effect of glycosylation on antifungal activity is pronounced (Table 5). Blockage of the 3-OH function with glucose did not change overall activity[16]. Thus, ID_{50} had similar values for medicagenic acid and its 3-*O*-glucoside; the 3-*O*-glucuronide had slightly lower activity. Further glycosylation at position 3-*O* had no influence on overall activity and medicagenic acid 3-*O*-maltoside showed activity like that of medicagenic acid and its 3-*O*-glucoside against some pathogenic fungi[18]. However, for the bisdesmosidic forms, activity against *T. viride* was reduced 10-27 times as compared to medicagenic acid aglycone or its glucoside. The composition of the ester-linked sugar chain did not seem to be significant. Clearly, bisdesmosidic medicagenic acid glycosides substituted with glucuronic acid at the 3-*O* position again showed lower activities than their structural analogues substituted with glucose.

Hederagenin monodesmoside (XVII) showed activity comparable to that of medicagenic acid bisdesmosides[16]. Soyasaponin I and zanhic acid tridesmoside showed no activity against *T. viride*, indicating that determination of these saponins with the fungal test is not possible and that other analytical procedures need to be used[21]. Basic hydrolysis of zanhic acid tridesmoside produced 3-*O*-glu-glu-glu-zanhic acid, which displayed antifungal activity, but the level of this activity was several times lower than the activity of medicagenic acid bisdesmosides. This may indicate that the 16-OH function is not essential for antifungal activity, and it may have significant meaning with regard to the bitterness of the compound; zanhic acid glycoside was the most bitter and throat-irritating compound of all alfalfa saponins tested[21].

Of all the tested glycosides, medicagenic acid 3-*O*-glucoside was certainly superior in activity to the others. This suggested the idea of using this compound as the basis for developing a new group of antimycotic or pesticidal agents. It was documented that this compound displayed considerable activity against ten medically important yeasts, e.g. *Candida*, *Torulopsis* and *Geotrichum* spp. MICs obtained by both agar and broth dilution methods ranged from 3 to 15 μg/ml[36]. The results were not fully confirmed by other research groups (Oleszek, unpublished) in which pure compounds were assayed for *in vitro* activity against *Candida albicans* B311 by using the agar-well diffusion assay at a concentration of 1 mg/ml with the standard antifungal antibiotic amphotericin B and rifamycin as positive controls. No essential activity was found for medicagenic acid, its sodium salt, and 3-*O*- and 3,28-diglucosides of medicagenic acid and for a mixture of alfalfa root saponins. Similarly, no essential activity against *T. viride*, *Candida albicans*, *Aspergillus flavus*, *A. fumigatus*, *Saccharomyces cerevisiae*, *E. coli*, *Staphylococcus aureus*, *Bacillus subtilis*, *Pseudomonas aeruginosa*, *Trichophyton mentagrophytes*, and *Mycobacterium intracellulare* was found for zanhic acid tridesmoside and its basic hydrolysis product at a concentration of 1 mg/ml[49].

Inhibitory activity of alfalfa saponins against soil-borne fungal pathogens of cereals *Gaeumannomyces graminis* v. *tritici* and *Cephalosporium gramineum* were tested[50]. The most potent inhibitors of Ggt growth were medicagenic acid sodium salt (ID_{50}=3.5 μg/ml) and its 3-*O*-glucoside (ID_{50}=7.0 μg/ml). These two compounds were also most inhibitory to *C. gramineum* growth but they also strongly influenced sporulation, which was reduced almost completely at a concentration of 50 μg/ml .

Effects on animals

The overall influence of legume saponins on animals has been widely reviewed[4]. One mechanism pointed out in this review, which might account for the growth-depressing effects of saponins, was the lowering of feed intake because of unpalatability (bitterness of saponins). Recent sensory test trials performed with laboratory staff volunteers using saponins isolated from alfalfa aerial parts showed that zanhic acid tridesmoside is the most bitter astringent and throat-irritating compound of a number of compounds tested[21]. This proves that zanhic acid tridesmoside may be a major principle responsible for bitterness of alfalfa and may influence a palatability of the feed. Once swallowed, saponins may irritate the membranes of the mouth and the digestive tract. This may primarily influence the absorption of nutrients, but perhaps also allergens, xenobiotics, and other toxic dietary components. In this matter zanhic acid seems to be a very important compound too. In a series of experiments, the effect of a range of structurally divergent alfalfa saponins on the potential difference (PD) across the rat small intestine was examined *in vitro*[51]. Typically, there was an immediate reduction in PD, although there was also considerable variation in response to particular compounds. Among the glycosides of medicagenic acid, bisdesmosides containing four sugar moieties had activities equal to a monodesmoside and to medicagenic acid itself. The exception was 3,28-diglucoside medicagenic acid, which again -- as with hemolytic activity -- did not affect membrane potential. Zanhic acid tridesmoside again showed the highest activity of all saponins tested. The high depolarizing activity of zanhic acid glycosides suggests that a lot of work done on the nutritional value of alfalfa and the role of alfalfa saponins as antinutritional factors needs to be revised.

Intraruminal administration of alfalfa saponins to sheep was performed[6,7]. It was shown that saponins reduced nutrient degradation and microbial fermentation in the rumen. In the presence of saponins, fractional digestive coefficients of organic matter, hemicellulose, cellulose, and nitrogen were reduced in the stomach, but increased in the small intestine. The weak point of these studies was poor characterization of saponin preparation and results cannot be related to any structural features. Similarly, in the work of Malinow et al.[52] on the safety of alfalfa saponins for human consumption for hypercholesterolemia treatment, no information was provided on the purity and composition of the saponin samples used. These studies showed that rats fed alfalfa saponins at levels of 1% in the diet for up to 26 weeks showed no ill effects, although a potentially beneficial reduction in serum cholesterol and triglycerides was observed. No adverse effects were observed in the prime species *Macaca fascicularis*[53] following consumption of an undefined mixture of alfalfa-top saponins for up to 78 weeks.

Quite different results were obtained in a series of experiments on the effect of zanhic acid tridesmoside on small animals. The compound was administered to hamsters via a stomach tube and the animals were observed for several days (Oleszek, unpublished). It was shown that during the first few hours after administration, animals had breathing problems. After 10-15 h animals suffered nervous system perturbations followed by death after 24. The LD_{50} value for the tridesmoside was calculated at 562 mg/kg of body weight and its toxicity rating can be classified as toxic/moderately toxic[54]. At necropsy bloat syndromes were observed. The intestines were heavily filled with gas. Since prior to administration of saponins animals were not fed for several hours, any other reason for this phenomenon apart from the influence of saponins can be reasonably excluded. Some authors concluded that saponins are not significantly involved in the bloat syndrome[4,55], but present observation again raises the question of the contribution of saponins in bloating. This might be particularly important if we realize that early work was performed without knowledge of the appearance and possible role of zanhic acid glycosides.

Activity against insects

Some preliminary work has been performed and aimed at the possible practical application of alfalfa saponins as a modern insecticide. Alfalfa saponin mixtures have been tested against summer fruit tortrix moth (*Adoxophyes orana* F.V.R) and European grape moth (*Lobesia botrana* Schiff)[56]. The increasing amounts of saponins added to the diet caused larval mortality to increase from 11.3% at 1 ppm to 46.1% at 1000 ppm. The contact effect accounted for a maximum of 22.7% mortality. No appreciable differences have been detected in the insecticidal activity exerted by crude saponins derived from two cultivars nor by the part of the plant, e.g. leaves and roots.

Crude alfalfa root saponins, their prosapogenins produced by alkaline hydrolysis, and medicagenic acid sodium salt were tested in field trials against spider mite (*Tetranychus urticae* Koch.) and hop aphid (*Phoron humuli* Schrank). Plants were sprayed with a 0.1 and 0.2% solution of saponin products. It was shown that prosapogenins were the most active against both phytophages. Crude saponins and medicagenic acid sodium salt were less active[57].

Saponins isolated from the aerial parts of alfalfa were tested against the Colorado potato beetle (*Leptinotarsa decemlineata*) together with glucosinolates from rape seeds and lupin alkaloids[58]. The larvae of the Colorado potato beetle were fed on potato leaves sprayed with 0.5 and 1% solutions. There was no repellant effect observed for any of the tested compounds, but feeding proved to be least intense on saponin-treated leaves. The larvae which fed on treated leaves had lowest body weight gains with the saponin treatment, suggesting the antifeedant activity of these compounds. The mortality on saponin-treated leaves was 100% at both concentrations tested.

Allelopathic activity

The allelopathic functions of alfalfa saponins are not completely understood. There are divergent opinions on whether they are released into the environment by leaching or secretion from intact plants[59,60]. Despite these divergences, there is evidence that saponins may harm plants when introduced into the soil with alfalfa plant material[61,62]. Early work done on these effects has been previously summarized in some detail[63]. Recent developments in this problem are covered by several papers[62,64,65,66,67].

In experiments performed by Wyman-Simpson et al.[62] it was shown that alfalfa root saponins inhibited wheat and cheat (*Bromus secalinus* L.). The authors imply that these compounds thus have potential to be used as herbicides. They compared the inhibitory activities of saponins isolated from alfalfa varieties grown under divergent seasonal conditions and concluded that the amount, structure, and type of saponins present in alfalfa roots vary with time. Later, it was shown that purified alfalfa root saponins affect germination and seedling growth of some weeds and wheat in Petri dish tests[65,66]. The effect was concentration dependent, showing for wheat (*Triticum aestivum* L.), dandelion (*Taraxacum vulgare* L.), coffeweed (*Sesbania exaltata* L.) and pigweed (*Amaranthus retroflexus* L.) some growth stimulation in the range of 10-100 ppm, and inhibition at higher doses. The most sensitive were barnyard grass (*Echinochloa Crus-galli* L.) and cheat (*Bromus secalinus* L.), the growth of which was inhibited even at concentrations of 10 ppm. It was also suggested that cultivars of *Medicago sativa* grown in different parts of the world possess slight differences in the composition of root saponins. However, these differences are not enough to prevent them from having the same relative allelopathic activities[65].

In the work of Tarikov et al.[64] it was shown that individual glycosides 3-*O*-glucoside (I), 3,28-diglucoside (II) of medicagenic acid, monoglucoside of hederagenin (XVII), and a mixture of root saponins possessed structure- and concentration-dependent activities against the germination and growth of wheat and cotton (*Gossypium arboreum* L.). Wheat was more sensitive than cotton, and mono- and diglucosides of medicagenic acid showed similar activity and were the most potent compounds.

Ten structurally divergent alfalfa saponins together with a mixtures of root and top saponins were tested against the germination and seedling growth of wheat[67]. Comparing the allelopathic activities of individual compounds (Table 5) permutted some general conclusions about structure-activity relationships. First, glycosides of medicagenic acid, hederagenin, and zanhic acid were inhibitory, whereas soyasapogenol B glycoside was slightly stimulatory for wheat seedlings growth. Among medicagenic acid glycosides, those having glucose in the C-3 position showed higher activity than similar compounds substituted with glucuronic acid. Zanhic acid glycosides were more active than glycosides of medicagenic acid substituted at C-3 with glucuronic acid. In general, monodesmosides were more active than bis- or tridesmosides, but evident exceptions could be registered.

The above results proved, however, that the allelopathic potentials of most active alfalfa saponins are not much different from the activities of other groups of plant secondary metabolites[68]. Thus, these saponins seem to have little potential as herbicides. However , their possible protective function in plants cannot be ignored.

DETERMINATION OF ALFALFA SAPONINS

The majority of work on the determination of saponins in alfalfa was done using biological tests such as *Trichoderma viride* and hemolysis. But as shown above, in spite of their simplicity, these methods are at best only approximate and do not distinguish between different saponins. In addition, they do not show changes in individual components of the saponin mixture whose biological activity may vary according to their chemical structure. Some of the components, e.g. soyasaponin I and zanhic acid tridesmosides, cannot be determined with these procedures.

Early analytical methods developed to replace biological tests were in general based on the hydrolysis of the plant extract, followed by titrimetric analysis[69], or gas (GC)[70,71,72] or high performance liquid chromatography (HPLC)[29] of the released aglycones. All these procedures are not very useful owing to the potential loss of material during hydrolysis and derivatization, and to the impossibility of recognition of structural differences in the parent glycosides. Recent availability of pure standards of intact saponins permitted work on the development of analytical HPLC procedures allowing determination of individual compounds. Due to the lack of a UV-chromophore, alfalfa saponins have to be derivatized prior to analysis. For this purpose derivatization with 4-bromophenacyl bromide has been successfuly employed[73]. Derivatization with this reagent can be completed with any saponin containing a free carboxylic group. Most of alfalfa saponins fulfill this requirement. All medicagenic acid glycosides contain at least one free carboxyl group, soyasaponin I possesses a COOH group on the glucuronic acid moiety and hederagenin monodesmosides have this group free at the C-28 position. These compounds can be derivatized and HPLC determined without any additional treatment. A problem arises with hederagenin bisdesmoside (XVIII) and zanhic acid tridesmosides. These compounds must undergo alkaline hydrolysis, deblocking carboxylic groups, prior to derivatization[74]. Care has to be taken when applying this procedure since the alkaline hydrolysis of zanhic acid tridesmoside generates different prosapogenins depending on the solvent used[74].

A HPLC procedure allowed determination of the dominant glycosides in alfalfa roots and tops[32] (Table 6) and alfalfa sprouts (Oleszek, unpublished). It was shown that the dominant saponins in roots were medicagenic acid glycosides substituted at the C-3 position with glucuronic acid or glucose and hederagenin glycosides. There was, however disagreement found between the *T. viride* test and HPLC determination which showed concentrations of 5 and 2.4% in dry matter, respectively. Thus, application of the fungal test in the analysis of alfalfa roots seemed to cause overestimation of saponin concentration, probably due to the exceptional activity of 3-*O*-glucoside medicagenic acid. The antifungal activity of this compound is about 40 times that of any other compound. Small variations in the amount of this glucoside may dramatically change the final result.

Table 6. Saponin concentration in alfalfa roots and aerial parts as determined by HPLC and *T. viride* test.

Compound	saponin content (% of dry matter)	
	roots	tops
I	0.25	-
II	0.39	-
VI	0.40	-
XI	0.76	0.37
XV	-	0.11
XVII	0.11	-
XXII	0.05	0.26
XXV	-	0.30
unidentified	0.45	0.29
Total	2.41	1.53
T. viride test	5.0	0.95

A similar situation occurred when saponins in alfalfa sprouts were determined. According to literature sources the concentration of saponins in alfalfa sprouts may reach the level of 8.7-9.5% in dry matter[60,74]. The HPLC determination showed that saponin concentration in sprouts indeed grows from 2 mg/g at the beginning of germination (soyasaponin I) up to 6-8 mg/g (0.6-0.8% in dry matter) at the 8-16th day, but this is a level of saponin in mature alfalfa plants. HPLC determination gives a result which is ten times lower than that cited in literature for biological determination. The reason for these differences is again the biosynthesis of the highly active 3-*O*-glucoside of medicagenic acid in seedling roots.

This analytical procedure also allows a more detailed registration of seasonal changes in saponin concentration in alfalfa. Several reports using biological methods showed the highest concentration of saponins in the middle of the summer or in the second cut[76]. Application of HPLC confirmed to some degree these findings and proved a high correlation between the concentration of dominant compounds 3GlcA,28AraRhaXyl medicagenic acid and *T. viride* evaluation (Oleszek, unpublished). But at the same time there was no correlation found between the concentration of total saponins evaluated with *T. viride* and the zanhic acid tridesmoside level. The concentration of zanhic acid tridesmoside was the highest late in the season when the level of medicagenic acid glycosides dropped drastically. These two groups of compounds are closely related biosynthetically and it seems that at the end of the vegetation season, conditions are better for the synthesis of zanhic acid, which is a higher-oxidized form than medicagenic acid. It still remains unknown if this oxidation plays any physiological role in the plant and if medicagenic acid is a biosynthetic precursor of zanhic acid.

CHEMOTAXONOMIC SIGNIFICANCE OF ALFALFA SAPONINS

Glycosides of medicagenic acid seldom occur in the plant kingdom. Medicagenic acid has been found only in two species of *Herniaria*[31], in *Castanospermumu australe*[77], in *Trigonella monspeliaca*[78] -- now called *Medicago monspeliaca* -- and in *Dolichos kilimandscharicus*[79]. It was found, however in all the tested species of the genus *Medicago*[80]. Advantage has been taken from this finding to use medicagenic acid as a taxonomic marker. As shown by Small[81], a satisfactory circumscription of *Medicago* has not been available, no single character having gained broad acceptance as diagnostic of the genus because of a lack of strongly confirmatory characters. The delimitation of *Medicago* from its relative *Trigonella* has been controversial.

Medicagenic acid from leaf tissue of 44 species of tribe Trigonellae Schulz and hemolytic effects of leaf tissues, together with myricetin occurrence in flowers, were studied by Jurzysta et al.[82]. These compounds were found in all species of the genus *Medicago*, including five species of debated taxonomic position, *M. radiata*, *M. hybrida*, *M. cretacea*, *M. arborea*, and *M. lupulina*, giving further support for the placement of these questionable species into *Medicago*. The same chemical support was found for placing three species of the *Buceras* section of the genus *Trigonella*, *T. monantha*, *T. monspeliaca*, and *T. polyceratia*, in the genus *Medicago*. In other studies[83] it was found that leaves of the *Medicago* species are hemolytic, although there are considerable differences in the strength of the reaction among species and population of species. The leaves of species of six related genera, *Trigonella*, *Melilotus*, *Trifolium*, *Parochetus*, *Ononis*, and *Lotus*, consistently proved not to be hemolytic. Thus, hemolysis by leaves was recognized as a synapomorphic discriminator, supporting controversial expanded circumscription of *Medicago*.

The level of hemolytic saponin content in leaves of 1213 plants of different variants of *Medicago sativa* proved to be useful for tracing the origin of some cultivated and wild alfalfas[84]. *Medicago sativa* ssp. *caerulea*, which is the most important ancestor of alfalfa, had a very low saponin content. The most primitive forms of cultivated alfalfa from Turkey and wild *Medicago sativa* spp. *sativa* of Turkey both also had a very low hemolytic saponin content. This is consistent with and most likely explained by a direct origin of these two Turkish groups from sympatric *M. sativa* spp. *caerulea*. The second most important ancestor of alfalfa, *Medicago sativa* spp. *falcata*, had the highest saponin content. It has been suggested that the modern European and American cultivars have intermediate levels of hemolytic saponins owing to hybridization and introgression of *Medicago sativa* spp. *falcata* and low-saponin forms discussed above.

Hemolytic capacities and the composition of saponin aglycone has been found to be more or less characteristic of certain infrageneric groups[85]. Only about a third of *Medicago* species had hemolytic seeds. Three interrelated species of *Trigonella* proved to be the only species of genera other than those of *Medicago* with hemolytic seeds, but in these cases the hemolysis does not originate from medicagenic acid and the reason remains unknown. Inside *Medicago*, the four major subsections of the section *Spirocarpos* can be distinguished by presence or absence of medicagenic acid and its 16-OH derivative, zanhic acid. The latter sapogenin was found only in three of the subsections of *Spirocarpos* and thus seems to reflect the phylogenic recency of these groups within *Medicago*. The distribution of zanhic acid in combination with other considerations proved useful in establishing that *Medicago granadensis* seems to be the oldest species in subsection *Intertextae*. It is not easy to judge whether the presence or absence of medicagenic acid and zanhic acid represents the evolutionarily derived conditions. More work is needed to determine if medicagenic acid is the biosynthetic precursor of zahnic acid.

The seed saponin profile of *Medicago hybrida*, most recently assigned to section *Medicago*, strongly supports the placement of this species and its relative *Medicago sufruticosa* in a separate section. These findings clearly show that more extensive qualitative analysis of saponins in different parts of all species of *Medicago* may clarify more of the unsolved systematic puzzles in this economically important genus.

Acknowledgements. The author is grateful for a grant (5 S302 054 04) from the State Committee for Scientific Research (KBN), Poland, which supported part of these studies.

REFERENCES

1. G. Massiot and C. Lavaud, Digestibility-inhibiting substances from alfalfa. in: *Phytochemistry and Agriculture*. T.A. Van Beek and A.K. Stobart, eds., Clarendon Press, Oxford (1994).
2. M.R. Malinow, P. McLaughlin, L. Papworth, C. Stafford, G.O. Kohler, A.L. Livingston, and P.R.Cheeke, Effect of alfalfa saponins on intestinal cholesterol absorption in rats, *Am. J. Clin. Nutr.* 30:2061 (1977).
3. M.R. Malinow, P. McLaughlin, G.O. Kohler, and A.L. Livingston, Prevention of elevated cholesterolemia in monkeys by alfalfa saponins, *Steroids*. 29:105 (1977).
4. P.R. Cheeke, Biological properties and nutritional significance of legume saponins, in: *Leaf Protein Concentrate*, L. Telek and H. Graham, eds., AVI Publishing, Westport, CT., USA, (1983).
5. M.R. Malinow, Saponins and cholesterol metabolism, *Atherosclerosis* 50:117 (1984).
6. C.D. Lu and N.A. Jorgensen, Alfalfa saponins affect site and extent of nutrient digestion in ruminants, *J. Nutr.* 117:919 (1987).
7. C.D. Lu, L.S. Tsai, D.M. Schafer, and N.A. Jorgensen, Alteration of fermentation in continuous culture of mixed rumen bacteria by isolated alfalfa saponins, *Dairy Sci.* 70:799 (1987).
8. M. Jurzysta, *Investigations of Saponins in Native Lucerne Populations (Medicago media* Pers.), (in Polish), R(170), IUNG, Pulawy (1982).
9. A. Tava, W. Oleszek, M. Jurzysta, N. Berardo, and M. Odoardi, Alfalfa saponins and sapogenins: Isolation and quantification in two different cultivars, *Phytochem. Anal.* 4:269 (1993).
10. R.J. Morris, W.B. Dye and D.S. Gisler, Isolation, purification and structural activity of an alfalfa root saponins, *J. Org. Chem.* 26:1241 (1961).
11. R.J. Morris and E.W. Hussay, A natural glycoside of medicagenic acid. An alfalfa blossom saponin, *J. Org. Chem.* 30:166 (1965).
12. B. Gestetner, Structure of saponin from lucerne, *Phytochemistry* 10:2221 (1971).
13. A.E. Timbekova and N.K. Abubakirov, Triterpene glycosides of alfalfa. I. Medicoside G -- a novel bisdesmoside from *Medicago sativa*, *Khim. Prir. Soed.* 4:451 (1984).
14. M. Levy, U. Zehavi, M. Naim, and J. Polacheck, An improved procedure for the isolation of medicagenic acid 3-*O*-glucopyranoside from alfalfa roots and its antifungal activity on plant pathogen, *J. Agric. Food Chem.* 34:960 (1986).
15. W. Oleszek, Solid-phase extraction-fractionation of alfalfa saponins, *J. Sci. Food Agric.* 44:43 (1988).
16. W. Oleszek, K.R. Price, I.J. Colquhoun, M. Jurzysta, M. Ploszynski, and R.G. Fenwick, Isolation and identification of alfalfa (*Medicago sativa*) root ssaponins: their activity in relation to a fungal bioassay, *J. Agric. Food Chem.* 38:1810 (1990).
17. G. Massiot, C. Lavaud, L. Le Men-Olivier, G. Van Binst, S.F. Miller, and H.M. Fales, Structural elucidation of alfalfa root saponins by mass spectrometry and nuclear magnetic resonance analysis, *J. Chem. Soc., Perkin Trans.* 1:3071 (1988).
18. M. Levy, U. Zehavi, M. Naim, and I. Polacheck, Isolation, structure determination and antifungal activity of a new native alfalfa root saponin, *Carbohydr, Res.* 193:115 (1989).
19. A.E. Timbekova and N.K. Abubakirov, Triterpene glycosides of alfalfa. III. Medicoside I, *Khim. Prir. Soed.* 5:607 (1986).
20. G. Massiot, C. Lavaud, V. Besson, L. Le Men-Olivier, and G. Van Binst, Saponins from aerial parts of alfalfa (*Medicago sativa*), *J. Agric. Food Chem.* 39:78 (1991).
21. W. Oleszek, M. Jurzysta, M. Ploszynski, I.J. Colquhoun, K.R. Price, and G.R. Fenwick, Zanhic acid tridesmoside

and other dominant saponins from alfalfa (*Medicago sativa* L.) aerial parts, *J. Agric. Food Chem.* 40:191 (1992).
22. A.E. Timbekova, M.F. Larin, M.P. Jagubaev, and N.K. Abubakirov, Triterpene glycosides of alfalfa. V. Medicoside H, *Khim. Prir. Soed.* 5:673 (1989).
23. A.E. Timbekova, A.L. Bereschagin, A.A. Semenov, and N.K. Abubakirov, Triterpene glycosides of alfalfa. Medicoside L, *Khim. Prir. Soed.* 2:221 (1990).
24. A.E. Timbekova and N.K. Abubakirov, Triterpene glycosides of alfalfa. II. Medicoside C, *Khim. Prir. Soed.* 6:805 (1985).
25. A.E. Timbekova and N.K. Abubakirov, Triterpene glycosides of alfalfa. IV. Medicoside J, *Khim. Prir. Soed.* 5:610 (1986).
26. I. Kitagawa, T. Taniyama, T. Murakami, M. Yoshihara, and M. Yoshikawa, Saponin and sapogenol. XLVI. On the constituents in aerial part of American alfalfa, *Medicago sativa* L. The structure of dehydrosoyasaponin I, *Yakugaku Zasshi*, 108:547 (1988).
27. G. Massiot, C. Lavaud, M. Benkhaled, and L. Le Men-Olivier, Soyasaponin VI, a new maltol conjugate from alfalfa and soybean, *J. Nat. Prod.* 55:1339 (1992).
28. A.L. Livingstone, Lucernic acid, a new triterpene from alfalfa, *J. Org. Chem.* 24:1567 (1959).
29. G. Massiot, C. Lavaud, D. Guillaume, and L. Le Men-Olivier, Reinvestigation of the sapogenins and prosapogenins from alfalfa (*Medicago sativa*) *J. Agric. Food Chem.* 36:902 (1988).
30. M.Z. Dimbi, R. Warin, C. Delaude, and R. Huls, Structure de l'acide zanhique et de l'acide zanhique-γ-lactone, triterpenoides nouveaux isoles de *Zanha golugensis* et de *Ganophyllum giganteum* *Bull. Soc. Chim. Belg.* 93:323 (1984).
31. G. Klein, J. Jurenitsch, and W. Kubelka, Structur der Sapogenine von Herba Herniariae (*Herniaria glabra* L. und *Herniaria hirsuta* L.), *Sci. Pharm.* 50:216 (1982).
32. B. Gestetner, Y. Assa, Y. Henis, Y. Birk, and A. Bondi, Lucerne saponins. IV. Relationship between their chemical constitution, and haemolytic and antifungal activities. *J. Sci. Food Agric.* 22:168 (1971).
33. J. Nowacka and W. Oleszek, Determination of alfalfa (*Medicago sativa*) saponins by high-performance liquid chromatography, *J. Agric. Food Chem.* 42:727 (1994).
34. P.R. West, G.R. Waller, P.M. Geno, W. Oleszek, and M. Jurzysta, Liquid secondary ion mass spectrometry and linked scanning at constant B/E LSIMS/MS for structure confirmation of saponins in *Medicago sativa* (alfalfa), *Saponins Used in Food and Agriculture*, G.R. Waller and K. Yamasaki, eds., Plenum Publishing Co., New York, NY. In press, pp____1996.
35. B. Gestetner, Y. Birk, and Y Tencer, Soybean saponins. Fate of ingested saponins and the physiological aspect of their hemolytic activity, *J. Agric. Food Chem.* 16:1031 (1968).
36. I. Polacheck, U. Zehavi, M. Naim, M. Levy, and R. Evron, Activity of compound G2 isolated from alfalfa roots against medically important yeasts, *Antimicrob. Agent Chemother.* 30:290 (1986).
37. M. Jurzysta, Hemolytic micromethod for rapid estimation of toxic alfalfa saponins, *Acta Agrobot.* 32:5 (1979).
38. W. Oleszek, Structural specificity of alfalfa (*Medicago sativa*) saponin haemolysis and its impact on two haemolysis-based quantification methods, *J. Sci. Food Agric.* 53:477 (1990).
39. M.W. Pedersen, D.E. Zimmer, J.O. Anderson, and C.F. McGuire, A comparison of saponins from DuPuits, Ranger and Quinta alfalfas, *Proc. 10th Int. Grasslands Congress*, pp. 693-698, Helsinki (1966).
40. D.E. Zimmer, M.W. Pedersen, and C.F. McGuire, A bioassay for alfalfa saponins using the fungus *Trichoderma viride*, *Crop Sci.* 7:223 (1967).
41. A.L. Livingston, L.C. Whitehand, and G.O. Kohler, Microbial assay for saponin in alfalfa products, *J. Assoc. Off. Anal. Chem.* 60:957 (1977).
42. M. Jurzysta, A simplified method for quantification of saponins in alfalfa by use of *Trichoderma viride* fungus, *Hodowla Rośl. Aklim. Nasienn.* 1:16 (1979).
43. A. Scardavi and F.C. Elliot, A review of saponins in alfalfa and their bioassay utilizing *Trichoderma* spp., *Bull. Michigan Agric. Exp. St.*, East Lansing, 50:163 (1967).
44. M. Nonaka, Variable sensitivity of *Trichoderma viride* to *Medicago sativa* saponins, *Phytochemistry* 25:73 (1986).
45. S. Shany, Y. Birk, B. Gestetner, and A. Bondi, Lucerne saponins. III. Effect of lucerne saponins on larval growth and their detoxification by various sterols, *J. Sci. Food Agric.* 21:508 (1970).
46. Y. Assa, B. Gestetner, I. Chet, and Y. Henis, Fungistatic activity of lucerne saponins and digitonin as related to sterols, *Life Sci.* 11:637 (1972).
47. W. Oleszek, K.R. Price, and G.R. Fenwick, The sensitivity of *Trichoderma viride* to medicagenic acid, its natural glucosides (saponins) and derivatives, *Acta Soc. Bot. Pol.* 57:361 (1988).
48. M. Levy, U. Zehavi, M. Naim, I. Polacheck, and R. Evron, Structure--biological activityrelationship in alfalfa antimycotic saponins: The relative activity of medicagenic acid and synthetic derivatives thereof against plant pathogenic fungi, *J. Phytopathol.* 125:209 (1989).
49. W. Oleszek and M. Jurzysta, New saponins from alfalfa tops: the structure and biological activity, *Proc. X Int. Conf. EUCARPIA Medicago spp. Group*, pp. 315-316, Lodi (1992).
50. S. Martyniuk, B. Wroblewska, M. Jurzysta, and Z. Bialy, Saponins as inhibitors of cereal pathogens: *Gaeumannomyces graminis* v. *tritici* and *Cephalosporium gramineum*, *Proc. 11th Int. Symp. "Modern fungicides and antifungal compounds"* Reinhardsbrunn/Friedrichroda (1995).
51. W. Oleszek, J. Nowacka, J.M. Gee, G.M. Wortley, and I.T. Johnson, Effects of some purified alfalfa (*Medicago sativa*) saponins on transmural potential difference in mammalian small intestine, *J. Sci. Food Agric.* 65:35 (1994).
52. M.R. Malinow, W.P. McNulty, P. McLaughlin, C. Stafford, A.K. Burns, A.L. Livingston, and G.O. Kohler, The toxicity of alfalfa saponins in rats, *Food Cosmet. Toxicol.* 19:443 (1981).

53. M.R. Malinow, W.P. McNulty, D.C. Houghton, S. Kessler, P. Stenzel, S.H. Goodnight, E.J. Bardana, J.L. Palotay, P. McLaughlin, and A.L. Livingston, Lack of toxicity of alfalfa saponins in monkeys, *J. Med. Primatol.* 11:106-118 (1992).
54. G. Zbinden and M. Flury-Rovesi, Significance of the LD_{50} -- test for the toxicological evaluation of chemical substances, *Arch. Toxicol.* 47:77 (1981).
55. W. Majak, R.E. Wowarth, A.C. Fesser, B.P. Goplen, and M.W. Pedersen, Relationships between ruminant bloat and the composition of alfalfa herbage. II. Saponins, *Can. J. Anim. Sci.* 60:679 (1980).
56. A. Tava, D. Forti, and M. Odoardi, Alfalfa saponins: Isolation, chemical characterization and biological activity against insects, *Proc. X Int. Conf. EUCARPIA Medicago spp. Group*, pp. 283-288, Lodi (1992).
57. L. Puszkar, A. Jastrzebski, M. Jurzysta, and Z. Bialy, Alfalfa saponins as a chance in the integrated hop protection, *Proc. XXXIV Session IOR*, vol. II, pp. 255-259, Poznań (1994).
58. D. Waligora and J. Krzymanska, The influence of secondary plant substances: glucosinolates, alkaloids and saponins on the feeding of Colorado potato beetle (*Leptinotarsa decemlineata* Say), *Proc. XXXIV Session IOR*, vol. II, pp. 9-12, Poznań (1994).
59. B.N. Mishustin and A.N. Naumova, Secretion of toxic substances by alfalfa and their effect on cotton and soil microflora, *Izv. Akad. Nauk USSR, Ser. Biol.* 6:3 (1955).
60. P.M. Gorski, J. Mierch, and M. Ploszynski, Production and biological activity of saponins and canavanine in alfalfa seedlings, *J. Chem. Ecol.* 17:1135 (1991).
61. W. Oleszek and M. Jurzysta, The allelopathic potential of alfalfa root medicagenic acid glycosides and their fate in soil environments, *Plant and Soil* 98:67 (1987).
62. C.L. Wyman-Simpson, G.R. Waller, M. Jurzysta, J.K. McPherson, and C.C. Young, Biological activity and chemical isolation of root saponins of six cultivars of alfalfa (*Medicago sativa* L.), *Plant and Soil* 135:83 (1991).
63. W. Oleszek, M. Jurzysta, and P. Gorski, Alfalfa saponins -- the allelopathic agents, in: *Allelopathy, Basic and Applied Aspects*. S.I.W. Rizvi, V. Rizvi, eds., Chapman and Hall, London (1992).
64. C. Tarikov, A.E. Timbekova, M.K. Abubakirov, and R.K. Koblov, Growth regulating activity of triterpene glycosides isolated from alfalfa (*Medicago sativa*) roots, *Uzbek. Biol. J.* 6:24 (1988).
65. G.R. Waller, M. Jurzysta, and R.L.Z. Thorne, Allelopathic activity of root saponins from alfalfa (*Medicago sativa* L.) on weeds and wheat, *Bot. Bull. Acad. Sin.* 34:1 (1993).
66. G.R. Waller, M. Jurzysta, and R.L.Z. Thorne, Root saponins from alfalfa (*Medicago sativa* L.) and their allelopathic activity on weeds and wheat, *Allelopathy J.* 2:21-30 (1995).
67. W. Oleszek, Allelopathic potentials of alfalfa (*Medicago sativa*) saponins: their relation to antifungal and hemolytic activities, *J. Chem. Ecol.* 19:1063 (1993).
68. F. Macias, Allelopathy in the search for natural herbicide models, in: *Allelopathy - Organisms, Processes and Applications*. Inderjit, K.M.M.Dakshini, F.A. Einhellig, eds. American Chemical Society, Washington, DC (1995).
69. Y. Tencer, S. Shany, B. Gestetner, Y. Birk, and A. Bondi, Titrimetric method for determination of medicagenic acid in alfalfa (*Medicago sativa*), *J. Sci. Food Agric.* 20:1149 (1972).
70. M. Jurzysta and A. Jurzysta, Gas-liquid chromatography of trimethylsilyl ethers of soyasapogenols and medicagenic acid, *J. Chromatogr.* 148:517 (1978).
71. P.R. Brawn, N.M. Linder, J.M. Miller, and G.M. Telling, A gas chromatographic method for the determination of medicagenic acid in lucerne (alfalfa) leaf protein concentrate, *J. Sci. Food Agric.* 32:1157 (1981).
72. D. Rao and G. Bories, Simple gas chromatographic method for the determination of medicagenic acid in alfalfa (*Medicago sativa*), *J. Chromatogr.* 410:169 (1987).
73. W. Oleszek, M. Jurzysta, K.R. Price, and G.R. Fenwick, High performance liquid chromatography of alfalfa root saponins, *J. Chromatogr.* 519:109 (1990).
74. J. Nowacka and W. Oleszek, High performance liquid chromatography of zanhic acid glycoside in alfalfa (*Medicago sativa*), *Phytochem. Anal.* 40:227 (1992).
75. D.E. Fenwick and D. Oakenful, Saponin content of food plants and some prepared foods, *J. Sci. Food Agric.* 34:186 (1983).
76. S. Burda, M. Jurzysta, and W. Oleszek, Biologically active saponin content in eight cultivars of alfalfa (*Medicago sativa*), *Proc. EUCARPIA Lucern Section Meeting*, pp. 128-130, Lusignan (1994).
77. K. Hiller and G. Voigt, Neue Ergebnisse in der Erforschung der Triterpenesaponine, *Pharmazie* 32:365 (1977).
78. W. Oleszek, M. Jurzysta, S. Burda, and M. Ploszynski, *Acta Soc. Bot. Pol.* Triterpene saponins of *Trigonella monspeliaca* L., 56:281 (1987).
79. A. Marston, F. Gafner, S.F. Dossagi, and K. Hostettman, Fungicidal and molluscicidal saponins from *Dolichos kilimandscharicus*, *Phytochemistry* 27:1325 (1988).
80. M. Jurzysta and E. Nowacki, Saponins of the genus *Medicago*, *Acta Agrobot.* 32:127 (1979).
81. E. Small, P. Lassen, and B. Brookes, An expanded circumscription of *Medicago* (Leguminosae, Trifolieae) based on explosive flower tripping, *Wildenowia* 16:415 (1987).
82. M. Jurzysta, S. Burda, W. Oleszek, and M. Ploszynski, The chemotaxonomic significance of laricytrin and medicagenic acid in the tribe Trigonelleae, *Can. J. Bot.* 66:363 (1988).
83. M. Jurzysta, E. Small, and C. Nozzolillo, Hemolysis, a synapomorphic discriminator of an expanded genus *Medicago* (Leguminosae), *Taxon*, 37:354-363 (1988).
84. E. Small, M. Jurzysta, and C. Nozzolillo, The evolution of hemolytic saponin content in wild and cultivated alfalfa (*Medicago sativa*, Fabaceae), *Econ. Bot.* 44:226 (1990).
85. M. Jurzysta, S. Burda, W. Oleszek, M. Ploszynski, E. Small, and C. Nozzolillo, Chemical composition of seed saponins as a guide to the classification of *Medicago* species, *Can. J. Bot.* 70:1384 (1992).

CHEMISTRY AND BIOLOGICAL ACTIVITY OF TRITERPENOID GLYCOSIDES FROM *MEDICAGO SATIVA*

A.E. Timbekova, M.I. Isaev, and N.K. Abubakirov

Institute of Chemistry of Plant Substances, Uzbek AS
Kh. Abdullaev Ave. 77 Tashkent 170, Uzbekistan

INTRODUCTION

Alfalfa [*Medicago sativa* L.,(Leguminosae)] is of interest as a fodder plant and grown on large areas in many world regions. In Uzbekistan alfalfa is widely cultivated in crop rotation with cotton and this makes the plant an important raw material. Alfalfa saponin is well known to influence animal organisms[1,2], and to affect insect and fungus growth[3]. It was shown by Slotwinska[4] that the saponins induced the reduction of mitotic index, viability, and growth of cells in L-cells culture. Study of the growth-retarding activity of alfalfa saponins on higher plants was by Waller, Jurzysta and Thorne[5]. Owing to this we have studied triterpene glycosides of this plant of great national economic significance. Objects of study were alfalfa roots and leaves. As a result of thin layer chromatography (TLC) there have been discovered 13 glycosides in the roots and 10 in the leaves. Qualitative compositions of glycosides from plant organs are different, and do not have identical components. The structures were established for 10 root glycosides[6].

RESULTS AND DISCUSSION

Saponins were extracted from alfalfa grown in irrigation fields in Uzbekistan. Alfalfa roots were extracted in boiling methanol, and the extracts partitioned between 1-butanol and water. The procedures yielded 15 g/kg of crude saponin. The total root saponin was subjected to repeated column chromatography on silica gel in chloroform-methanol-water systems. Column chromatography gave the substances **A, B, C, E, F, G, H, I, J,** and **L,** named as medicosides (Tables 1,2).Treating alfalfa leaves with boiling water, partition of the extract between 1-butanol and water, and precipitation with acetone gave a crude saponin mixture (1.6 g/kg).

Medicagenic Acid Glycosides

Medicagenic acid is present in 5 glycosides (Table 1). Glycoside **4** was identified as the known glucopyranoside of medicagenic acid previously isolated by Morris[7]. The remaining four bisdesmoside glycosides, medicosides **G** (**6**), **H** (**8**), **J** (**10**), and **L** (**16**), can be considered as derivatives of medicagenic acid glucopyranoside.

The structure of glycoside **6** was elucidated on the basis of acid hydrolysis as a result of which medicagenic acid and the above-mentioned medicagenic acid 3-*O*-β-D-glucopyranoside have been obtained. Alkaline hydrolysis of **6** yielded compound **4**. Saponins **4** and **6** were methylated according to Hakomori[8] to obtain permethylates **5** and **7**. Acid hydrolysis of the permethylates afforded 2,23,28-tri-*O*-methylmedicagenic acid (**2**) and 2,23-di-*O*-methylmedicagenic acid (**3**) respectively. ^{1}H NMR spectra of the compounds **2,3** and medicagenic acid (**1**) revealed little difference between chemical shift values of H-3 signals, but H-2 signals suffered upfield 0.75-0.77 ppm shift as compared with the same signal of medicagenic acid. Therefore C(3)-OH group in compounds **2** and **3** were not methylated. The electron impact mass spectrum of compound **3** exhibited retro-Diels-Alder fragmentation, which produced ions characteristic of both ends of the molecule, at *m/z* 281 and 248. It was concluded that the C-28 carboxy group in compound **3** was free. Gas-liquid chromatography (GLC) analysis of the permethylates **5** and **7** revealed methyl 2,3,4,6-tetra-*O*-methyl-D-glucopyranoside in methanolysis products of the compounds. ^{1}H NMR spectra of medicoside **G** (**6**) and its permethylate **7** showed two doublet signals of anomeric protons at δ 4.91, 6.10 (2H, *d*, ^{3}J = 7.5 Hz) and 4.34, 5.65 (2H, *d*, ^{3}J = 7.0 Hz)(C_5D_5N) respectively. Spin-spin coupling constants of the signals were characteristic of $^{4}C_1$ conformations of the D-glucopyranosides and β-configurations of the glycoside bonds. Thus it was concluded that medicoside **G** had structure 3,28-di-*O*-β-D-glucopyranoside of medicagenic acid.

Both medicosides **H** (**8**) and **J** (**10**) after acid hydrolysis yielded medicagenic acid and 3-*O*-β-D-glucopyranoside of medicagenic acid. On the basis of GLC analysis it was shown that the compound **8** had consisted of D-glucose, L-arabinose, and L-rhamnose and the compound **10** consisted of the same sugars and D-xylose in addition. Methylation of medicosides **H** and **J** according to Hakomori afforded the permethylates **9** and **11**. Methanolysis of the compounds **9** and **11** gave methyl 2,3,4,6-tetra-*O*-methyl-D-glucopyranoside, methyl 2,3,4-tri-*O*-methyl-L-rhamnopyranoside, methyl 3,4-di-*O*-methyl-L-arabinopyranoside and methyl 2,3,4,6-tetra-*O*-methyl-D-glucopyranoside, methyl 2,3-di-*O*-methyl-L-rhamnopyranoside, methyl 3,4-di-*O*-methyl-L-arabinopyrano-side, methyl 2,3,4-tri-*O*-methyl-D-xylopyranoside, respectively. Acid hydrolysis of compounds **9** and **11** gave 2,23-di-*O*-methylmedicagenic acid. The above-mentioned carbohydrate compositions determined by GLC analysis led to a conclusion: the terminal sugars in medicoside **H** were D-glucose and L-rhamnose and those in medicoside **J** were D-glucose and D-xylose. That is why low-field signals in ^{1}H NMR spectra of the compounds **8-11** (δ 6.32-6.09, C_5D_5N) were assigned to anomeric protons of L-arabinopyranosyl units. It was concluded that medicoside **H** contained ester-linked chain Ara-Rha located at C-28 and medicoside **J** contained Ara-Rha-Xyl chain at the same position. In order to confirm the sequence of the sugar chain in medicoside **J** its permethyate **11** was subjected to $LiAlH_4$ decomposition. The following methyl glycosides were identified by GLC in the reaction products methyl 2,3,4,6-tetra-*O*- methyl-D-glucopyranoside, methyl 2,3-di-*O*-methyl-L-rhamnopyranoside, methyl 2,3,4-tri-*O*-methyl-D-xylopyranoside. The corresponding $^{3}J_{H\text{-}1,H\text{-}2}$ couplings observed in ^{1}H NMR spectra of compounds **8-11** were characteristic of β-D-glucopyranosyl, β-D-xylopyranosyl, and α-L-rhamnopyranosyl units. The $^{3}J_{H\text{-}1,H\text{-}2}$ couplings of arabinopyranosyl residues varied from 3.4 Hz for compound **10** to 4.3 Hz for its acetate **12**. Vicinal coupling constants of the rest arabinose proton signals were measured from increased scale ^{1}H NMR spectra of medicoside **J** and medicoside **J** acetate. The arabinose protons resonate at δ 5.68 (*d, J*=3.4 Hz, H-1), 3.82 (*dd, J*=7.7, 3.4 Hz, H-2), 3.90 (*dd, J*=10.2, 7.9 Hz, H-3), 3.46 (*td, J*=10.2 , 5.3 Hz, H-4) for compound **10** in deuteromethanol and at δ 5.74 (*d, J*=4.3 Hz, H-1), 3.89 (*dd, J*=6.2, 4.5 Hz, H-2), 5.16 (*dd, J*=10.9, 6.2 Hz, H-3), 5.23 (*ddd, J*=10.5, 6.8, 3.4 Hz, H-4), 3.93 (*dd, J*=12.0, 6.7 Hz, H-5), 3.72 (*dd, J*=12.8, 3.0 Hz, H-5') for compound **12** in deuterochloroform. The proton assignments showed close coincidence with the assignments made by the Massiot group

Table 1. Medicagenic acid glycosides and their derivatives.

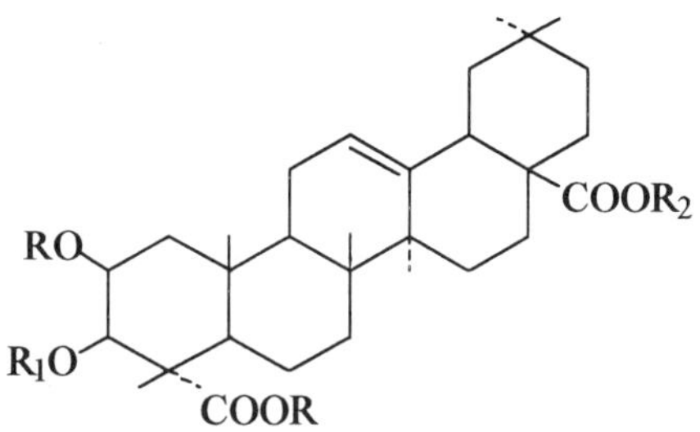

1-17

1. R= R_1 = R_2 = H Medicagenic acid
2. R = R_2 = Me, R_1 = H
3. R = Me, R_1 = R_2 = H

NAME	R	R_1	R_2
4*. GLUCOPYRANOSIDE OF MEDICAGENIC ACID	H	β-D-Glcp	H
5. PERMETHYLATE OF **4**			
6*. MEDICOSIDE **G**	H	β-D-Glcp	β-D-Glcp
7. PERMETHYLATE OF **6**			
8*. MEDICOSIDE **H**	H	β-D-Glcp	α-L-Rhap$\xrightarrow{2}$β-L-Arap
9. PERMETHYLATE OF **8**			
10.* MEDICOSIDE **J**	H	β-D-Glcp	β-D-Xylp$\xrightarrow{4}$α-L-Rhap$\xrightarrow{2}$β-L-Arap
11. PERMETHYLATE OF **10**			
12. ACETATE OF **10**			
13. DIGLUCOPYRANOSIDE OF MEDICAGENIC ACID	H	β-D-Glcp$\xrightarrow{2}$β-D-Glcp	H
14. PROGENIN OF **16**	H	β-D-Glcp ↑2 α-L-Rhap$\xrightarrow{3}$β-D-Glcp ↑2 β-D-Glcp	H
15. PERMETHYLATE OF**14**			
16*.MEDICOSIDE **L**	H	β-D-Glcp ↑2 α-L-Rhap$\xrightarrow{3}$β-D-Glcp ↑2 β-D-Glcp	β-D-Glcp
17. PERMETHYLATE OF **16**			

* The compound isolated from the plant

for identical compound with the help of modern techniques such as COSY, relayed COSY, and HOHAHA[9,10]. Medicoside **J** was also isolated from *Medicago sativa* roots by W. Oleszek with collaborators from Poland and England[11]. Identity of the compounds was proved on the basis of comparative analysis of the ^{13}C NMR spectra.

A structural peculiarity of medicosides **H** and **J** is the β-configuration of the L-arabinopyranoside residue. The configuration of the monosaccharide residue in medicoside **J** has been proposed on the basis of proton-proton coupling constants: $^3J_{H\text{-}1,H\text{-}2} = 3.4$ Hz, $^3J_{H\text{-}2,H\text{-}3} = 7.7$ Hz as well as carbon-proton coupling constant $^1J_{C\text{-}1,H\text{-}1} = 173.3$ Hz (off resonance ^{13}C NMR spectrum in C_5D_5N) indicating the equatorial orientation of anomeric proton and axial orientations of the second and third protons. We examined ^{1}H NMR spectra of medicoside **J** on an AM-300 Bruker spectrometer at 250 MHz using C_5D_5N-D_2O (7:1) solution at 85°C with TMS as an internal reference. The ^{1}H signals of the arabinose ring were assigned by homonuclear double resonance experiments difference spectra. The signals observed at δ 6.26 (*d*, $^3J_{H\text{-}1,H\text{-}2} = 2.5$ Hz, H-1), 4.31 (*dd*, $^3J_{H\text{-}2,H\text{-}3} = 4.1$ Hz, H-2), 3.82 (*dd*, $^3J_{H\text{-}3,H\text{-}4} = 9.7$ Hz, H-3), 3.99 (*ddd*, $^3J_{H\text{-}4,H\text{-}5ax} = 9.4$ Hz, H-4), 3.40 (*dd*, $^3J_{H\text{-}5ax,H\text{-}5eq} = 10.4$ Hz, H-5ax), 4.08 (*dd*, $^3J_{H\text{-}4,H\text{-}5eq} = 4.7$ Hz, H-5eq). Similar proton-proton coupling constants were measured for the arabinose signals in the above mentioned ^{1}H NMR spectra of the glycoside **10** in CD_3OD and its acetate **12** in $CDCl_3$. The glycosides **8** and **10** have been proposed to be medicagenic acid 3-*O*-β-D-glucopyranoside, 28-*O*-[α-L-rhamnopyranosyl(1→2)-β-L-arabinopyranoside] and medicagenic acid 3-*O*-β-D-glucopyranoside, 28-*O*-[β-D-xylopyranosyl(1→4)-α-L-rhamnopyranosyl(1→2)-β-L-arabinopyranoside] respectively.

Medicoside **L** (**16**) is the most polar compound isolated from alfalfa by us. Partial hydrolysis of compound **16** gave medicagenic acid, its 3-*O*-β-D-glucopyranoside, and medicagenic acid 3-*O*-β-D-glucopyranosyl(1→2)-β-D-glucopyranoside (**13**). The structure of the last compound was established on FAB mass and ^{13}C NMR spectral data. In the FAB mass spectrum there were observed cationated molecular ions (*m/z* 941, 903) and fragment ions as a result of two hexoses elimination.. In the ^{13}C NMR spectrum of the compound **13** anomeric carbon atom signal of D-hexose directly linked with the polycyclic nucleus was shifted upfield by δ 2.4 as compared with the spectrum of **4**. Moreover the C-2 signal of this D-glucose molecule was shifted from the position δ 75.1 to δ 83.6. This indicates 1→2 linkage of two hexoses in compound **13**. Carbohydrate chain substitution at the aglycone third position follows from the C- 3 chemical shift of medicagenic acid.

Medicoside **L** under alkali treatment gave progenin **14**. GLC analysis indicated the compound **14** contained D-glucose and L-rhamnose in proportion 3:1. In the FAB mass spectrum of this compound recorded with KCl addition there was observed the quasimolecular ion peak at *m/z* 1211 corresponding to a molecule containing three D-glucoses and L-rhamnose. Its decay proceeded on two directions $[M+2K-H]^+$-Rha and $[M+2K-H]^+$-Glc corresponding to mass values 1065 and 1049. The fragmentation indicates the availability of two terminal sugars and carbohydrate chain branching.

Glycoside **16** and its alkali progenin **14** was methylated by the method of Hakomori and the permethylates **17** and **15** were obtained. Both permethylates after methanolysis gave methyl 2,3,4,6-tetra-*O*-methyl-D-glucopyranoside, methyl 2,3,4-tri-*O*-methyl-L-rhamno-pyranoside, and methyl 3,4,6-tri-*O*-methyl-D-glucopyranoside. Attempts to identify methyl *O*-methyl-D-glucopyranoside corresponding to the branching centre were a failure. ^{1}H NMR spectra of permethylates **17** and **15** contain anomeric proton signals of four D-glucoses, one L-rhamnose and three D-glucoses, one L-rhamnose respectively. The downfield signal at δ 5.70 in the spectrum of **15** is absent.

Medicoside **L** subjected to degradation by the method of Smith[12] led to the receipt of medicagenic acid. Such result rules out branching on the D-glucose molecule directly linked to the genin and indicates branching at the second D-glucose molecule.

Anomeric carbon signals are observed at δ 95.7, 101.0, 103.0, 104.6, and 106.4 in the ^{13}C NMR spectrum of medicoside **L**. The first signal should be assigned to are acylglucose; the second one is due to L-rhamnose; the signal at δ 103.0 is characteristic of an anomeric carbon of D-glucose substituted on the second position. Of the remaining two signals the latter one has the value characteristic of terminal D-glucose. Therefore the signal at δ 104.6 belongs to D-glucose at the branching centre. Since the signal at δ 104.6 was shifted upfield δ 1.5 as compared with the corresponding signal of **13** it was concluded that the C-2 hydroxy group of **16** was substituted. Further the glycosylated carbon signals are observed at δ 84.3, 84.8 and 85.3. L-Rhamnose linked to the second or fourth position of D-glucose shifts the glycosylated carbon signal to δ 78.1-79.5[13], but L-rhamnose linked to the third position of D-glucose slifts the last-mentioned signal to δ 84.5[14]. Hence L-rhamnose more probably substitutes the third position. Therefore the second hydroxyl of the D-glucose molecule was glycosylated by D-glucose. A conclusion about the configuration of glycoside bonds was made on the base of chemical shifts of anomeric carbon atom signals in the ^{13}C NMR spectrum of medicoside **L** and the couplings of anomeric proton signals in ^{1}H NMR spectra of permethylates **15** and **17**. Medicoside **L** has the structure 3-*O*-[β-D-glucopyranosyl(1→2)],[α-L-rhamnopyranosyl(1→3)]-β-D-glucopyranosyl(1→2)-β-D-glucopyranoside, 28-*O*-β-D-glucopyranoside of medicagenic acid.

Hederagenin Glycosides

Hederagenin as well as medicagenic acid were found in the composition of 5 alfalfa glycosides (Table 2). Glycoside **20** proved to be a known bioside and was identified as caulosaponin **B**, previously isolated by Murakami[15]. Caulosaponin **B** forms the basis of the structure of three new hederagenin glycosides named as medicosides **C** (**21**), **F** (**26**), and **I** (**29**). It is significant that acyloside chains of hederagenin bisdesmosides unlike the glycosides of medicagenic acid are identical and include one D-glucose molecule only.

Medicoside **C** (**21**) and medicoside **I** (**29**) have related structures. Acid hydrolysis of both glycosides led to hederagenin (**18**). Partial acid hydrolysis of medicoside **C** gave the above-mentioned glycoside **20**. Alkali treatment of medicoside **I** gave medicoside **C**. Methanolysis of compounds **21** and **29** and following GLC analysis of carbohydrate silyl derivatives show D-glucose and L-arabinose presence in these compounds in the proportions 1:2 and 1:1 respectively. Permethylates **22** and **30** were subjected to GLC analysis. In methanolysis products of compound **22** there were methyl 2,3,4-tri-*O*-methyl-L-arabinopyranoside, methyl 3,4,6-tri-*O*-methyl-D-glucopyranoside, methyl 3,4-di-*O*-methyl-L-arabinopyranoside but in methanolysis products of **30** apart from the above-mentioned compounds there was also methyl 2,3,4,6-tetra-*O*-methyl-D-glucopyranoside. In the ^{1}H NMR spectra of permethylates **22** and **30** double signals of three and four anomeric protons were observed respectively. The downfield signal in the spectrum of **30** at 5.88 (*d*, ^{3}J = 6.0 Hz) belongs to D-glucose linked to hederagenin carboxy group. The couplings of the signals indicate β-configuration of D-glucoses and α-configuration of L-arabinoses. Consequently medicoside **C**(**21**) is 3-*O*-[α-L-arabinopyranosyl(1→2)-β-D-glucopyranosyl(1→2)-α-L-arabinopyranoside] of hederagenin and medicoside **I** (**29**) is 3-*O*-[α-L-arabinopyranosyl(1→2)-β-D-glucopyranosyl (1→2)-α-L-arabinopyranoside],28-*O*-β-D-glucopyranoside of hederagenin.

Table 2. Hederagenin glycosides and their derivatives.

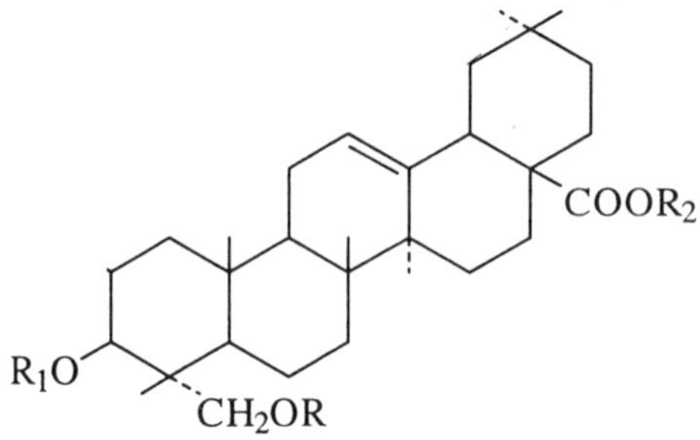

18. R= $R_1 = R_2$ = H Hederagenin
19. R = R_2 = Me, R_1 = H

18-30

NAME	R	R_1	R_2
20*. CAULOSAPONIN **B**	H	β-D-Glcp$\xrightarrow{2}$α-L-Arap	H
21*. MEDICOSIDE **C**	H	α-L-Arap$\xrightarrow{2}$β-D-Glcp$\xrightarrow{2}$α-L-Arap	H
22. PERMETHYLATE OF **21**			
23*. MEDICOSIDE **E**	H	β-D-Glcp$\xrightarrow{3}$β-D-Xylp	β-D-Glcp
24. PERMETHYLATE OF **23**			
25. PERACETATE OF **23**			
26*. MEDICOSIDE **F**	H	β-D-Glcp$\xrightarrow{2}$α-L-Arap	β-D-Glcp
27. PERMETHYLATE OF **26**			
28. PERACETATE OF **26**			
29*. MEDICOSIDE **I**	H	α-L-Arap$\xrightarrow{2}$β-D-Glcp$\xrightarrow{2}$α-L-Arap	β-D-Glcp
30. PERMETHYLATE OF **29**			

* The compound isolated from the plant

Two minor glycosides named by us as medicosides **E** (**23**) and **F** (**26**) include hedera-genin as aglycon. GLC analysis of the compounds shows the presence of D-glucose and D-xylose (2:1) in medicoside **E** and D-glucose and L-arabinose (2:1) in medicoside **F**. The permethylates **24** and **27** obtained by methylation of the compounds were subjected to methanolysis. By GLC analysis of the products we found that permethylate **24** contained 2,3,4,6-tetra-*O*-methyl-D-glucopyranose and 2,4-di-*O*-methyl-D-xylopyranose but per-methylate **27** contained 2,3,4,6-tetra-*O*-methyl-D-glucopyranose and 3,4-di-*O*-methyl-L-arabinopyranose. The compounds **25** and **28** were obtained by acetylation of medicosides **E** and **F** with acetic anhydride/pyridine. In ^{1}H NMR spectra of the acetates **25** and **28** there are observed proton signals of two fully acetylated D-glucopyranoses. Proton signals were assigned using double resonance experiments. Anomeric proton signals of one D-glucose in both spectra were shifted downfield (δ 5.54) indicating the aglycone carboxy group substitution by the sugar. In the spectrum of compound **25** the H-3 signal of D-xylopyranose is at δ 3.80 indicating the substitution at this position. Chemical shifts of L-arabinose proton signals observed in the spectrum of **28** have values corresponding to arabinopyranose substituted at the second position. Carbohydrate proton coupling in this spectra correspond to the $^4C_1$ conformation of the sugars. The signals at δ 3.37 and δ 3.40 in the ^1H NMR spectra of the compounds **25** and **28** were assigned to the third proton of hederagenin substituted by glycosidic linkage on C-3. The above given data for medicosides **E** (**23**) and **F** (**26**) are in conformity with those of the ^{13}C NMR spectra. *J*-modulation technique and literature data[16] were used for assigning the signals. Aglycone C-3 signals in both spectra were shifted downfield and C-28 signals were shifted upfield as compared with analogous signals in hederagenin spectrum. In the medicoside **E** spectrum there was observed a downfield signal at δ 88.4 assigned to C-3 of D-xylose but in medicoside **F** spectrum downfield signal was observed at δ 81.4 which was assigned to L-arabinose C-2. Hence medicosides **E** and **F** have the structures of hederagenin 3-*O*-[β-D-glucopyranosyl(1→3)-β-D-xylopyranoside], 28-*O*-β-D-glucopyranoside and hederagenin 3-*O*-[β-D-glucopyranosyl-(1→2)-α-L-arabinopyra-noside],28-*O*-β-D-glucopyranoside respectively.

Biological Activity

Alfalfa total saponins from aerial parts (top saponins) and roots (root saponins) as well as root saponin components: medicagenic acid 3-*O*-β-D-glucopyranoside (**4**), medicosides **G** (**6**), **H** (**8**), **J** (**10**), **C** (**21**), and **I** (**29**) were studied for antimicrobial and fungicidal action, growth, and seed germination[17,18].

Antimicrobial and fungicide properties were studied by the method of diffusion in agar. As test microorganisms there were used *Bacillus mesentericus*, yeast-like fungi *Candida albicans*, *Candida utilis*, *Candida tropicalis*, milk-acid bacteria *Agrobacterium tumefaciens*, *Corynebacterium insidiosum*, *Pseudomonas lachrymans*, and *Corynebacterium michiganense*. Data obtained are given in Table 3. All substances completely suppressed the growth of tomato wilt *C. michiganense* and vine root canker in *A. tumefaciens* with the exception of medicoside **C**. Total root saponin as well as medicagenic acid 3-*O*-β-D-glucopyranoside and medicoside **I** suppressed the growth of *C. insidiosum*, an alfalfa disease agent. No studied substance influenced agent growth of tomato wet rot *Erwinia carotovora.* Aerial parts saponin fully suppressed cotton homeosis *Xanthomonas campestris.*

We studied the effect of alfalfa root saponins and a few isolated components on the growth of wheat coleoptiles and cotton rootlets (Figure 1). It was noted that low concentrations of root saponins up to 150 μg/ml stimulated the growth of wheat coleoptiles but at higher concentrations inhibited their growth. Analogous concentration thresholds for medicagenic acid 3-*O*-β-D-glucopyranoside and medicosides **G** and **C** were at 50, 180, and 30 μg/ml. This shows that these monodesmosides **4** and **21** are stronger growth inhibitors than bisdesmoside **6**. Growth stimulation by alfalfa root saponins in given concentration intervals was produced apparently by more polar bisdesmosides of root saponins.

Alfalfa root saponins was found to be an inhibitor of cotton germination at all concentrations studied with the exception of 10 μg/ml. Medicagenic acid 3-*O*-β-D-gluco-pyranoside and medicoside **C** at concentrations over 160 and 275 μg/ml respectively were inhibitors of cotton germination. Medicoside **G** in all studied concentrations (10-500 μg/ml) was a cotton germination stimulator. Morover 100% germinating power was observed for this compound at the concentration of 50 μg/ml.

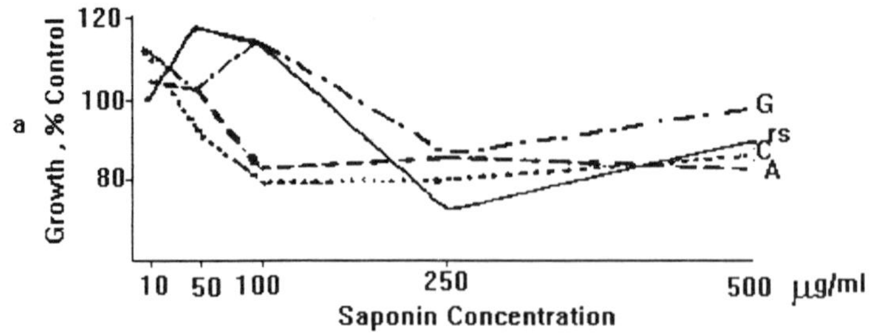

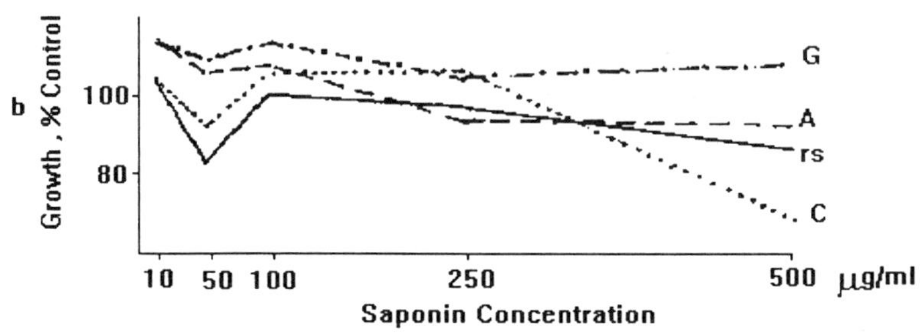

Figure 1. Growth of wheat coleoptiles (a) and cotton rootlets (b) as a percentage of the distilled water control in the presence of different concentrations of saponins: total root saponin (rs)(---), medicagenic acid 3-*O*-β-D glucopyranoside (**A**)(---), medicoside **G** (G)(.–. –.), medicoside **C** (C)(.. –..–..).

The determination of alfalfa saponins effect on cabbage seed germinating power and the sprout damage by phytopathogenic fungi and bacteria were conducted by the use of a moist chamber[19]. Obtained data are shown in Figure 2. We compared the actions of two total saponins from alfalfa aerial parts and roots, related medicagenic acid 3-*O*-β-D-glucopyranoside **(4)** and medicoside **G (6)**, medicosides **H (8)** and **J (10)**, medicosides **C (21)** and **I (29)**. Top and root saponins were weak shoot stimulators since during all periods there was increase of germinated seed percentage as compared with the control. Top saponins and root saponins effects on sprout damage were slightly different . Top saponins fully suppressed bacterium and stimulated fungus damage while root saponin suppressed both. In these cases the percentage of sound sprouts was higher than in the control but average sprout weight was lower than for the control.

Medicagenic acid 3-*O*-β-D-glucopyranoside **(4)** significantly reduced seed germination and average sprout weight. High germination percentage was observed and bacterium damage was absent in three- and seven-day experiments with medicoside **G (6)**. There were higher percentages of sound sprouts and slightly lower average sprout weights in the presence of medicoside **G**.

Medicosides **H (8)** and **J (10)** were found to act differently on seed germination, bacterium and fungus damage. Medicoside **J** in the two-day experiment sharply reduced cabbage seed germination and damage by bacteria and fungi; in the seven-day experiment seed germination reached the control and the damage proved lower than at the control level. Medicoside **H** gave a higher percentage of germinated seeds in the seven-day experiment but the damage by fungi was also slightly higher as compared with the control.

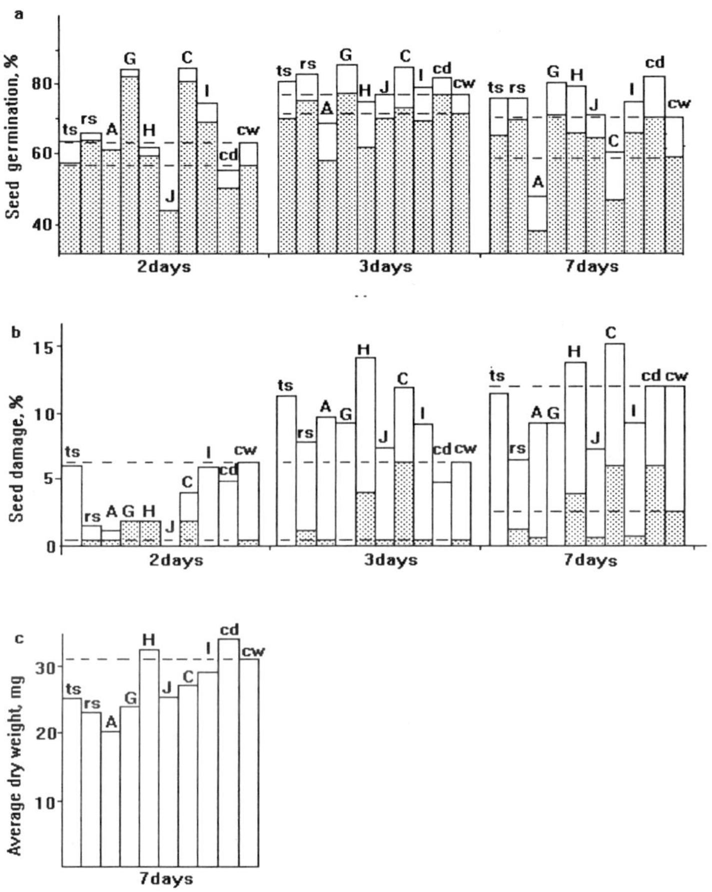

Figure 2. Cabbage seed germination in the presence of the saponins: a) seeds germinated as a percentage of the taken seeds (sum total damaged sprouts is not shaded), b) sprouts damaged by bacteria (shaded) and fungi (not shaded) as a percentage of the taken seeds, c) average dry weight of the cabbage sprouts. Abbreviations: top saponin (ts), root saponin (rs), medicagenic acid 3-*O*-β-D-glucopyranoside (**A**), medicosides **C**, **G**, **H**, **I**, **J** (**C**, **G**, **H**, **I**, **J** respectively), control dry seeds (cd), distilled water control (cw).

Table 3. Antimicrobial activity of alfalfa glycosides (growth suppression zone, mm).

Compound	Conc., μg/ml	*Bacillus mesentericus*	*Candida albicans*	*Candida utilis*	*Candida tropicalis*	*Saccharomyces cerevisiae*	*Agrobacterium tumefaciens*	*Corynebacterium insidiosum*	*Pseudomonas lachrymans*	*Corynebacterium michiganense*
Top saponin	1000	-	-	-	-	-	full	-	-	full
	500						suppr.			suppr.
Root saponin	1000	-	-	-	13.5	12.5	full	29.5	-	full
	500				10.5	9.5	suppr.	14.5		suppr.
Glucopyranoside of	1000	-	-	-	16.0	18.0	full	26.0	-	full
medicagenic acid (**4**)	500				11.0	16.0	suppr.	12.0		suppr.
Medicoside **G** (**6**)	1000	-	stim.	-	-	-	full	-	full	full
	500		11.0				suppr.		suppr.	suppr.
Medicoside **J** (**10**)	1000	-	24.5	-	-	11.0	full	-	full	full
	500		18.5			9.0	suppr.		suppr.	suppr.
Medicoside **C** (**21**)	1000	stim.	-	-	-	14.0	-	-	-	full
	500	22.0				10.0				suppr.
Medicoside **I** (**29**)	1000	-	19.0	stim.	stim.	11.0	full	14.5	full	full
	500		10.5	15.0	18.0	9.0	suppr.	10.5	suppr.	suppr.

Abbreviations: Conc.-concentration, full suppr.-full suppression, stim.-stimulation

* The experiments were carried out in the Microbiology Department of Moldova Academy of Sciences by V.M. Boguslavskii and S.A. Burceva (Kishinev)

Medicoside **C (21)** did not affect fungus damage; it stimulated seed germination and bacterial damage too. As a result in the seven-day experiment the percentage of sound sprouts was lower and average sprout weight did not essentially differ from the control level. In the presence of medicoside **I** the seed germination was slightly higher and the damage by fungi and bacteria was less than in the control. As a result sound sprout percentage was higher and average sprout weight was on the control level.

So to summarize the effects of medicagenic acid 3-*O*-β-D-glucopyranoside **(4)** and medicosides **G (6)**, **C (21)**, and **I (29)** on seed germination and bacterialnd fungus damage,it should be noted that the substitution of a carboxy group led to activity change. Medicagenic acid 3-*O*-β-D-glucopyranoside **(4)** is the inhibitor but medicagenic acid 3,28-di-*O*-β-D-glucopyranoside (medicoside **G**) is a stimulator of seed germination. In the cases of hederagenin monodesmoside (medicoside **C**) and hederagenin bisdesmoside (medicoside **I**) the sup-pression of bacterialamage increases from the former to the later compound.

In both medicagenic acid glycosides, medicosides **H (8)** and **J(10)** carboxy groups are substituted. This glycoside pair differs each from other by one molecule of xylose in the acyloside chain. Medicoside **J** suppressed the fungus and bacterial damage, and inhibited the seed germination , in contrast with medicoside **H**.

It was found that all compounds with the exception of medicoside **H** inhibited the growth of cabbage sprouts. Medicagenic acid glycosides that are medicagenic acid 3-*O*-β-D-glucopyranoside, medicosides **G** and **J**, were the most toxic but hederagenin glycosides medicosides **C** and **I** were less toxic. The seed germination in the presence of medicagenic acid glycosides is reduced as follows: medicoside **G** > medicoside **H** > medicoside **J** > medicagenic acid 3-*O*-β-D-glucopyranoside.

So the obtained data show that the components of alfalfa root saponin differ in their growth inhibitory action and antimicrobial activities. The effects depend primarily on the aglycone and the carbohydrate chain structure and position in the molecule secondarily.

SUMMARY

Ten triterpenoid glycosides have been isolated from the roots of *Medicago sativa* L. (Leguminosae) including eight new glycosides and two previously known, medicagenic acid 3-*O*-β-D-glucopyranoside and hederagenin 3-*O*-[β-D-glucopyranosyl(1→2)-α-L-arabinopyranoside]. Four new saponins were identified as glycosides of medicagenic acid, four from hederagenin. Their structures were elucidated by a combination of chemical and spectroscopic techniques (IR, EI/MS, NMR ^{1}H and ^{13}C). In some cases we used fast atom bombardment mass spectra (FAB/MS), NMR ^{1}H difference spectra (homonuclear double resonance, Overhauser effect), NMR ^{13}C partially decoupled spectra (off resonance), and two-dimensional ^{1}H-^{1}H correlating spectra (COSY ^{1}H-^{1}H). The inhibitory effects of these glycosides on growth and seed germinations and on some phytopathogenic bacteria have also been tested.

REFERENCES

1. G.Reshef, B.Gestetner, Y.Birk and A.Bondi, Effect of alfalfa saponins on the growth and lipid metabolism of mice and quails, *J. Sci. Food Agric.* 27:63 (1976).

2. M.R.Malinow, W.E.Connor, P.McLaughlin, C.Stafford, D.S.Lin, A.L.Livingston, G.O.Kohler and W.P.McNulty, Cholesterol and bile acid balance in *Macaca fascicularis*. Effect of alfalfa saponins, *J. Clin. Invest.* 67:156 (1981).

3. E.Horber, K.T.Leath, B.Berrang, V.Marcarian and C.H.Hanson, Biological activities of saponin components from DuPuits and Lahontan alfalfa, *Entomol. Exp. Appl.* 17:410 (1974).

4. M.Slotwinska, The effect of alfalfa saponins on L-cells culture in vitro, *Ann. Univ. Mariae Curie-Sklodowska, Sect. C* 38:177 (1983).

5. G. R.Waller, M.Jurzysta and R.L.Z.Thorne, Root saponins from alfalfa (*Medicago sativa* L.) and their allelopathic activity on weeds and wheat, *Allelopathy J.* 2:21 (1995).

6. A.E.Timbekova, A.S.Shashkov and N.K.Abubakirov, Triterpene glycosides of alfalfa. VII. Medicosides E and F, *Khim. Prirod. Soedin.* 701 (1993).

7. R.I.Morris, W.B.Dye and P.S.Gisler, Isolation, purification, and structural identity of an alfalfa root saponin, *J. Org. Chem.* 26:1241 (1960).

8. S.Hakomori, Rapid permethylation of glycolipids and polysaccharides, catalyzed by methylsulfinyl carbanion in dimethyl sulfoxide, *J. Biochem.* 55:205 (1964).

9. G.Massiot, C.Lavaud, L.Le Men-Olivier, G.Van Binst, S.P.F.Miller and H.M.Fales, Structural elucidation of alfalfa root saponins by mass spectrometry and nuclear magnetic resonance analysis, *J. Chem. Soc., Perkin Trans. I* 3071 (1988).

10. G.Massiot, C.Lavaud, V.Besson, L.Le Men-Olivier and G.Van Binst, Saponins from aerial parts of alfalfa (*Medicago sativa*), *J. Agric. Food Chem.* 39:78 (1991).

11. W.Oleszek, K.R.Price, I.J.Colquhoun, M.Jurzysta, M.Proszynski and G.R.Fenwick, Isolation and identification of alfalfa (*Medicago sativa* L.) root saponins: their activity in relation to a fungal bioassay, *J. Agric. Food Chem.* 38:1810 (1990).

12. M.Abdel-Akher, J.K.Hamilton, R.Montgomery and F.Smith, A new procedure for the determination of the fine structure of polysaccharides, *J. Am. Chem. Soc.* 74:4970 (1952).

13. Y.Imakura, S.Kobayashi and A.Mima, Bitter phenyl propanoid glycosides from *Campsis chinensis*, *Phytochemistry*.24:139 (1985).

14. M.M.Benidze, O.D.Dzhikiya, T.A.Pkheidze, E.P.Kemertelidze and A.S.Shashkov, Steroid glycosides of the leaves of *Jucca aloifolia*, *Khim. Prirod. Soedin.* 537 (1987).

15. T.Murakami, M.Nagasawa, S.Urayama and T.Satake, New triterpenoid saponins in the rhizome and roots of *Caulophyllum robustum*, *Yakugaku Zasshi* 88:321 (1968).

16. H.Kizu, S.Kitayma, F.Nakatani, T.Tomimori and T.Namba, Studies on Nepalese crude drugs. III. On the saponins of *Hedera nepalensis* K.Koch., *Chem. Pharm. Bull.* 33:3324 (1985).

17. S.Tarikov, A.E.Timbekova, N.K.Abubakirov and R.K.Koblov, Growth regulating activity of triterpene glycosides isolated from alfalfa roots (*Medicago sativa*), *Uzbek. Biol. Zh.* 6:24 (1988).

18. V.M.Boguslavskii, S.A.Burceva, A.E.Timbekova and N.K.Abubakirov, Antimicrobial activity of alfalfa root glycosides, in: *Microbiologiya v Sel'skom Khozyaistve*, G.V. Merenyuk, Ed., Shtiintsa, Kishinev (1991).

19. K.I.Bel'tyukova and M.S.Matyshevskaya, *Metody Issledovaniya Vozbuditelei Bakterial'nykh Boleznei Rastenii*, Naukova dumka, Kiev (1968).

SAPONINS AND ARTIFACTS

Georges Massiot, Marie-Geneviève Dijoux and Catherine Lavaud

Laboratoire de Pharmacognosie, URA CNRS 492
51, rue Cognacq-Jay
51096 Reims Cedex
France

INTRODUCTION

Saponins are complex multifunctional molecules made up of an aglycone (steroid or triterpene) and of one or several chains of sugars. They are polar and their isolation and separation require solvents and adsorbents of high polarities, sometimes associated with moderately high temperatures. A necessary step in their structural elucidation is the identification of their components, and this is often done by chemical or enzymic degradations. Under these conditions or during the isolation process, these fragile molecules lend themselves to structural changes and what is thought *bona fide* to be of natural origin, may be the artifactual product of a chemical rearrangement. Such transformations occur on the aglycone skeleton or on the sugar part, or consist in the loss of a labile substituent.

The purpose of this paper is to give a few examples of such transformations observed in the authors' laboratory or reported in the literature.

DEHYDRATION REACTIONS

Hydrolysis reactions, run under strongly acidic conditions (typically, 5N HCl at reflux temperature), are able to bring about dehydration of free alcohols. One of the best known examples concerns saponins with protobassic acid as aglycone. The dehydration product of protobassic acid, bassic acid, was known long before the isolation of the natural aglycone[1]. The natural triterpene escaped isolation after acidic hydrolysis, owing to the fragility of the 6β alcohol function. It was isolated only after enzymatic hydrolysis of the saponin by Kitagawa *et al.* in 1972[2]. We became acquainted with the problem when we worked on the saponins of *Tridesmostemon claessenssi* (Sapotaceae), whose hydrolysis product showed the two double bonds of Δ-5, Δ-12 triterpenes[3]. Of course, the ^{1}H NMR spectrum of the intact acetylated saponin did not display the expected signals for the diene system but instead an extra proton signal for H-6. It is worth noting here that the location of an OH group on a triterpene is no longer a problem difficult to solve owing to the powerful resolution of 1- and 2-D ^{1}H-^{1}H and ^{1}H-^{13}C NMR experiments.

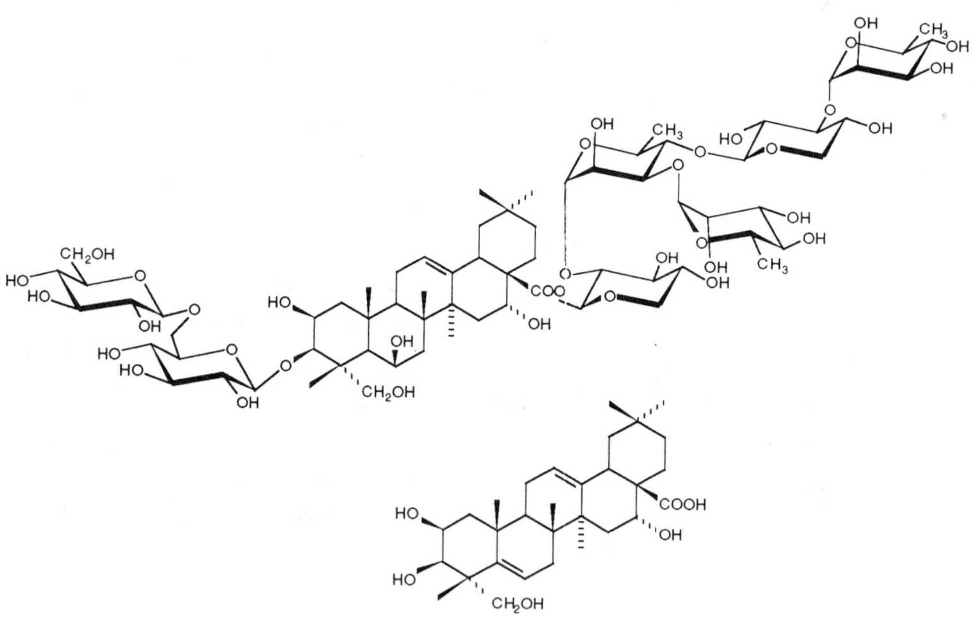

Figure 1. Saponin from *Tridesmostemon claessenssi* with 16-OH protobassic acid as aglycone; 16-OH bassic acid.

The second example of misbehavior of carbenium ions concerns the well known saponins of soybean. Hydrolysis of their complex mixtures yields a series of triterpenes named the soyasapogenols, among which are an alcohol (soyasapogenol B) and the corresponding olefin (soyasapogenol C) and *O*-methyl ether (soyasapogenol D). The tools available at the time of the first investigation favored location of the alcohol function at C-21 (although there was not much basis to discriminate it from C-22)[4].

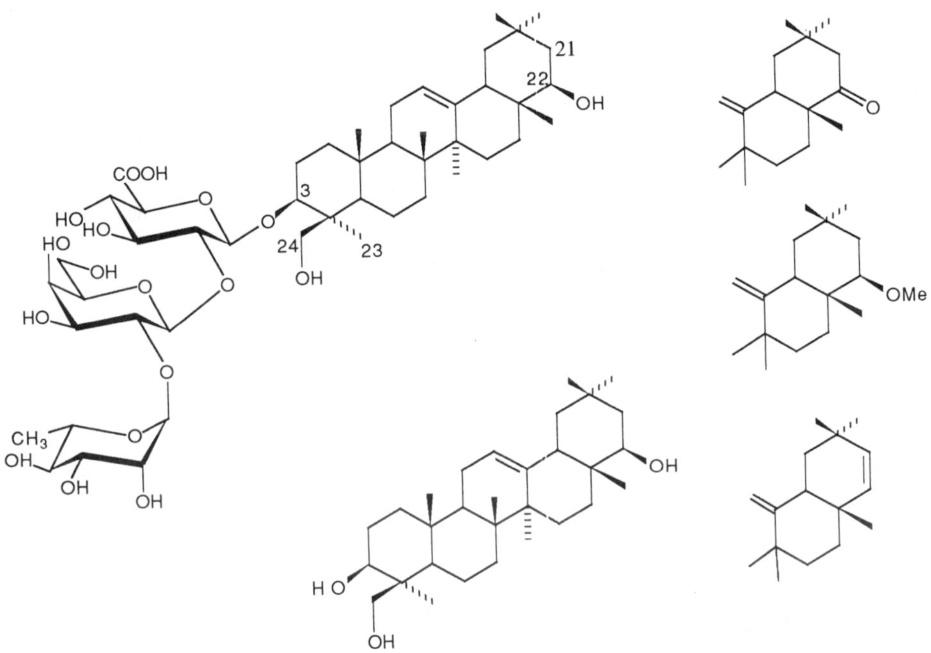

Figure 2. Soyasaponin I; soyasapogenols B, E, D and C (left to right; top to bottom).

Identification of the three aforementioned compounds as arising from a common alcohol through solvolytic reactions should have indicated C-22 instead. Protonation of the olefin may lead to its displacement between C-13 and C-18 and hence to activation of the C-22 alcohol through homoallylic activation, a situation quite similar to the one of the *i*-steroids. The problem was definitively settled later by X-ray analysis[5].

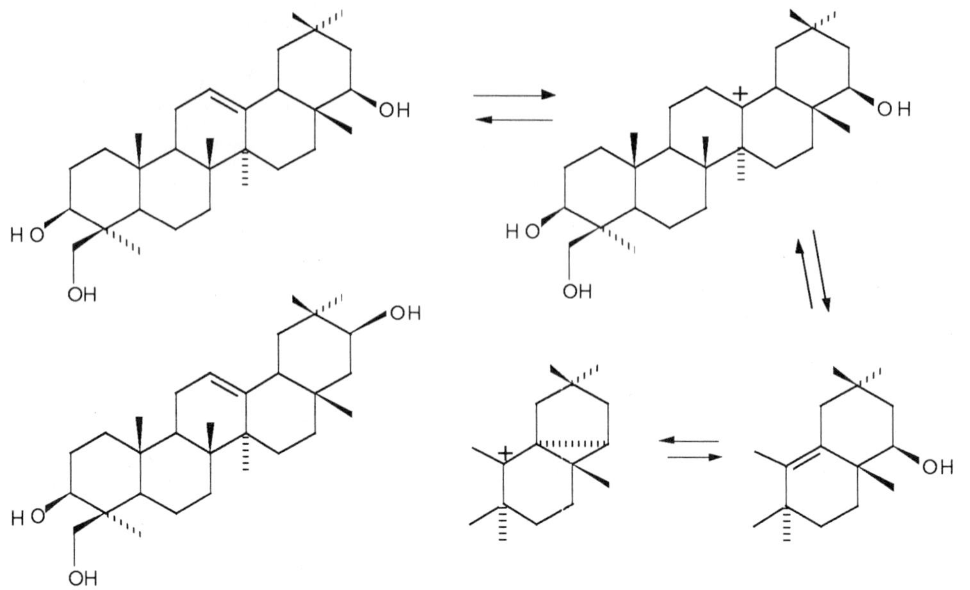

Figure 3. Two alternative structures for soyasapogenol B and formation of an *i*-steroid-like intermediate by protonation of soyasapogenol B.

LACTONIZATION REACTIONS

Alfalfa (*Medicago sativa*) is a rich source of proteins in temperate climates; its use as animal feed is however hampered by the presence of saponins with antifeedant properties. These saponins were the object of early work culminating in the structural elucidation of medicagenic acid by Djerassi *et al.* in 1957[6]. In 1959, Livingston published a note on another triterpene of alfalfa; the structure was not fully elucidated but the compound was characterized as a lactonic triterpene with three alcohol and one acid functions[7]. Lucernic acid, as it was named, was cited on several instances in the literature in reports on quality of alfalfa cultivars but its structural elucidation had to wait until 1988[8]. It turned out that lucernic acid is not a genuine natural compound but rather a lactone arising from 16-hydroxymedicagenic acid. For reasons which are not completely clear to us, it appears that 16α-hydroxy groups activate C-23 acids towards lactonization. The structure of highly complex saponins with this triterpene (also named zanhic acid) have recently been identified in *Medicago* species[9].

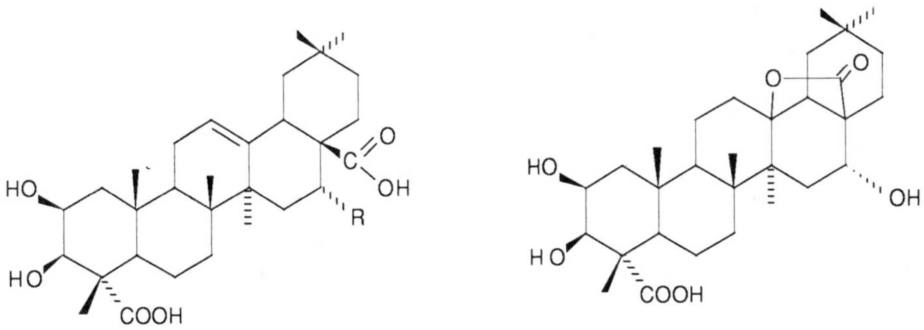

Figure 4. R = OH; 16-OH medicagenic acid (zanhic acid) and lucernic acid.

REACTIONS OCCURRING ON THE SUGAR PART

While working on alfalfa seed saponins, we isolated soyasaponin I, which we tried to characterize as a peracetate methyl ester. To our surprise the reaction turned out to be nonreproducible, giving sometimes as the main product a derivative in which the uronic acid had dehydrated into an α,β-unsaturated acid. One such compound was recently introduced into the literature as a natural product but in the absence of a characterization of the underivatized compound, this must await confirmation[10]. Further investigation of the soyasaponin problem showed that the dehydration yield depended much on the reaction

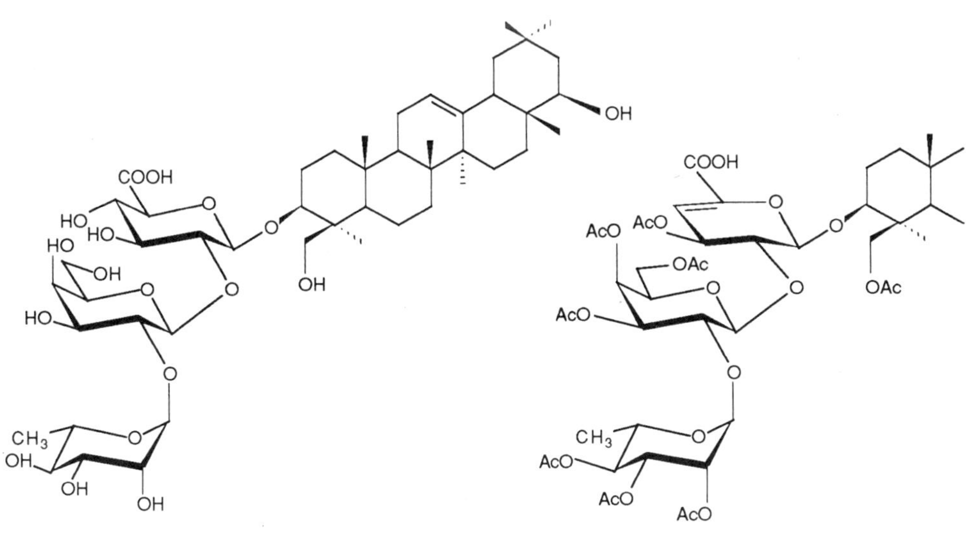

Figure 5. Soyasaponin I and rearranged acetyl derivative.

sequence and on the quality of the starting material. Depending on extraction conditions (1-butanol extract, precipitate in methanol/ether...), acids are either isolated as such or as salts and that makes a difference at the acylation stage. The proper means to obtain a clean derivative consists in first displacing salts by ion-exchange treatment, then treating with diazomethane, and finally performing acetylation (4-dimethylaminopyridine, pyridine).

LOSS OF ELEMENTS OF THE SAPONIN

Saponins are now described containing a large variety of radicals, most of them being acid residues. Some are labile and are only isolated through rapid processes; such is the case for formic or malonic residues, for instance. Some are reactive and give rise to side reactions. Some time ago, we isolated from an African Lecythidaceae plant, *Petersianthus macrocarpus*, a series of saponins containing tiglic and nilic acids[11]. This latter acid, 2-methyl-3-hydroxybutyric acid, is rare and seldom totally characterized. It is also fragile and may decompose by β elimination into tiglic acid. When one isolates tiglate derivatives of saponins, it may be advisable to look back at the genuine nature of the compounds and, for example, examine the ^{1}H NMR spectra of fractions for the characteristic signals of tiglate esters.

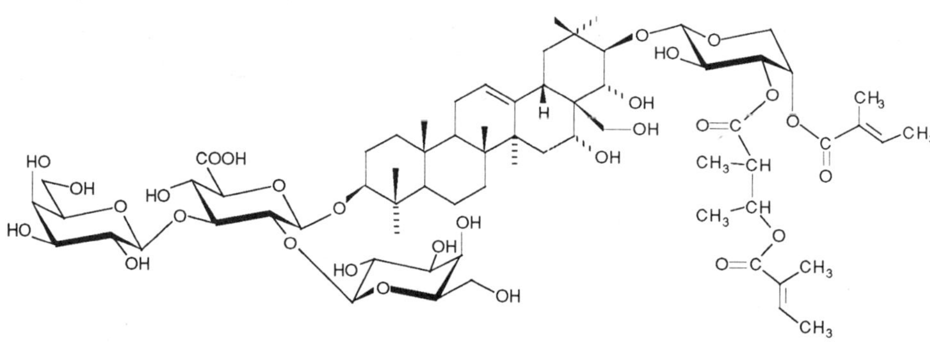

Figure 6. Saponin from *Petersianthus macrocarpus* containing tiglic and nilic acids.

The recording of spectra of crude fractions or extracts is very often outlawed by people in charge of instruments. We would like to describe at least one example where this proved to be of special importance. As said before, our laboratory spent a fair amount of energy on alfalfa seed saponins, looking at varieties of the plant and at their saponin contents. NMR of the crude saponin fractions repeatedly showed a sharp AB-like system at 2.5 and 2.9 ppm, which could not be assigned to any of the known alfalfa saponin structures. Careful purification was monitored by ^{1}H NMR and led to a pure compound with all characteristic features of soyasaponin I but whose ^{1}H and ^{13}C NMR contained a series of signals corresponding to six linked carbon and hydrogen atoms forming a dihydropyran ring. We named this compound soyasaponin VI[12] but its independent and almost simultaneous isolation in two Japanese laboratories introduced two other names for the same compound : chromosaponin I[13] and soyasaponin BeA[14] (soyasaponin βg[15]). A proposal for assignment of the anomeric configuration of this pseudosugar was made on the basis of an interpretation of a CD spectrum but definitive proof must await further confirmation.

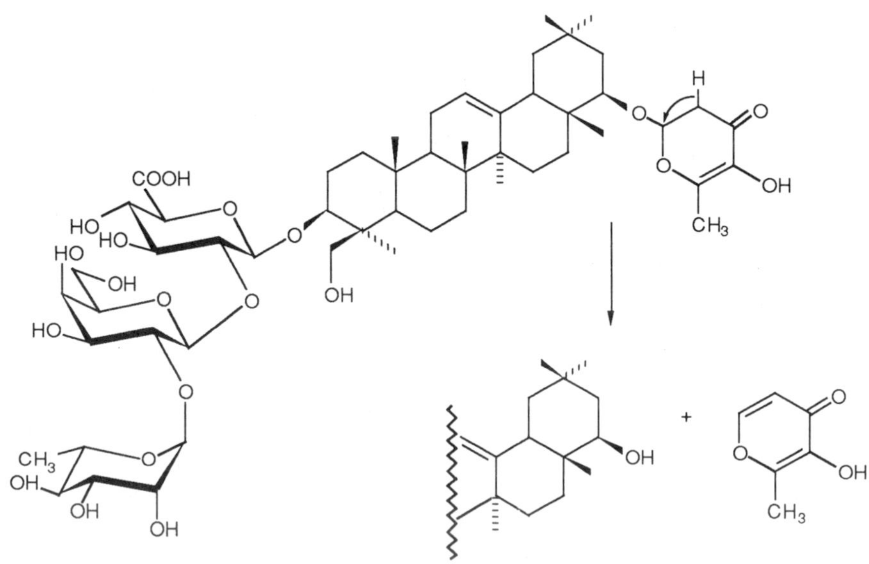

Figure 7. Soyasaponin VI and degradation products: soyasaponin I and maltol.

Examination of soya beans, peanuts, chickpeas, and several varieties of beans showed that soyasaponin VI was present in large quantities despite the fact that soyasaponin I was previously reported as the major compound. Soyasaponin VI readily decomposes into soyasaponin I in acidic or basic pH or simply on standing in alcoholic solutions. Soyasaponin I is thus, at least in part, an artifact made from soyasaponin VI.

Most surprisingly, another path of decomposition of soyasaponin VI was observed on samples kept in $CDCl_3$ for prolonged periods of time. The product of degradation obtained with up to 30% yield under these conditions was dehydrosoyasaponin I. The scheme below gives a tentative explanation for this decomposition. Dehydrosoyasaponin I may thus also be an artifact coming from soyasaponin VI.

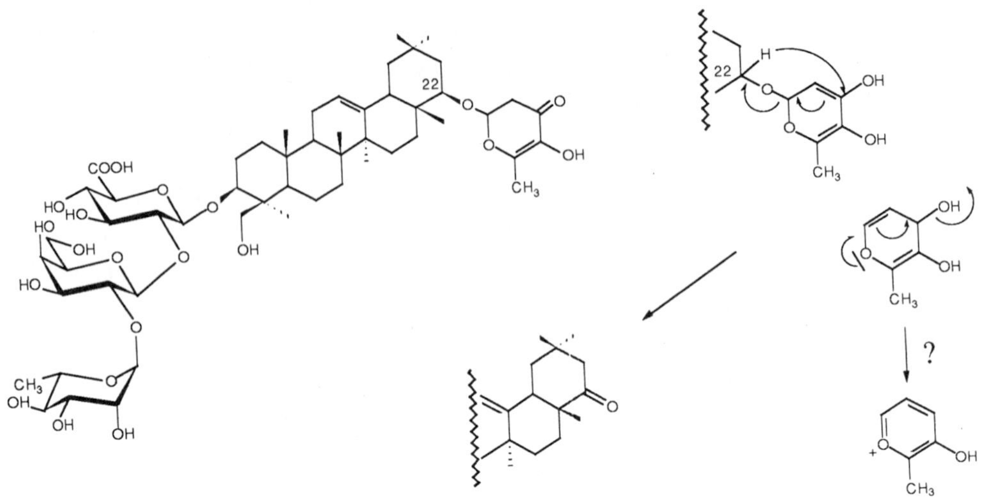

Figure 8. Decomposition of soyasaponin VI into dehydrosoyasaponin I.

WEIGHT MAY BE GAINED DURING ISOLATION !

On examination of crude sapogenin fractions from alfalfa seeds, we observed ^1H NMR signals that did not fit any of the structures of the known triterpenes. After tedious purification, we isolated a 24-*O*-linoleoyl derivative of soyasapogenol B. After going back to the saponins, a considerable amount of time was devoted to a search for saponins with fatty acid chains, substances heretofore unknown. That was in vain and the conclusion of the story is that these substances do not exist and that the 24-*O*-linoleoyl soyasapogenol B was an artifact produced during the hydrolysis stage in the presence of residual triglycerides. This reaction was eventually reproduced with soyasapogenol B, a triglyceride extract from alfalfa, and an acid catalyst!

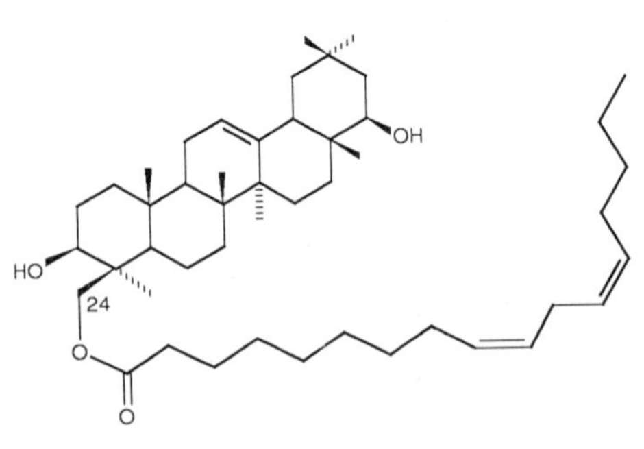

Figure 9. 24-*O*-linoleoyl soyasapogenol B.

A NOT-YET-SOLVED PROBLEM

Sugar beet (*Beta vulgaris*) contains a small amount of saponins which induce foam and turbidity (floc) during sugar processing. These saponins were studied a long time ago but only simple glucosides of oleanolic acid were identified. We were able to isolate these compounds anew, but besides these we found more polar compounds, di-acids arising from oxidative cleavage of pentoses[16]. We were not able to obtain these compounds in the pure state without derivatization and at the present time, it is not certain whether the natural compound is a diacid or a lactone.

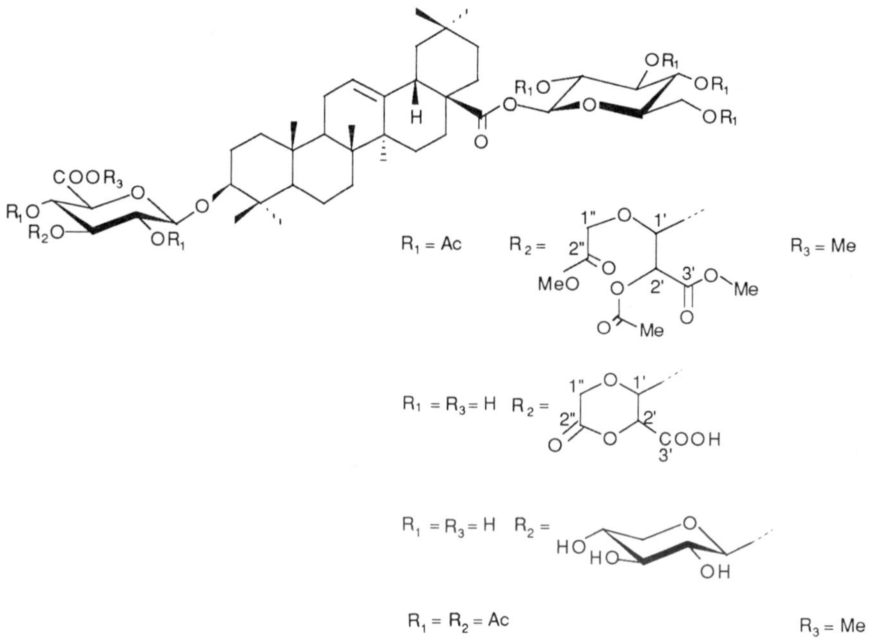

Figure 10. Saponins from *Beta vulgaris*.

CONCLUSION

As always, when dealing with natural products, one must ask oneself questions about the natural or artifactual nature of any isolate. Structural elucidation is difficult in the saponin or sapogenin domains and the pleasure of writing new structures should not make one forget this elemental interrogation.

Some of this work was done in the laboratory of Pharmacognosy of the Faculty of Pharmacy of the University of Reims and received the acknowledgeable support of MESR and CNRS. Fruitful discussion with the colleagues mentionned in the list of publication which follows is acknowledged with gratitude.

REFERENCES

1. T. J. King and J. P. Yardley, The structure of bassic acid, *J. Chem. Soc.* 4308 (1961).
2. I. Kitagawa, A. Inada, I. Yosioka, R. Somanathan, and M. U. S. Sultanbawa, Protobassic acid, a genuine sapogenol of seed kernels of *Madhuca longifolia* L., *Chem. Pharm. Bull.* 20:630 (1972).
3. G. Massiot, C. Lavaud, C. Delaude, G. Van Binst, S. P. F. Miller, and H. M. Fales, Saponins from *Tridesmostemon claessenssi*, *Phytochemistry* 29:3291 (1990).
4. I. Kitagawa, M. Yoshikawa, and J. Yosioka, Saponin and sapogenol XIII. Structures of three soybean saponins: soyasaponin I, soyasaponin II, and soyasaponin III, *Chem. Pharm. Bull.* 24:121 (1976).
5. I. Kitagawa, M. Yoshikawa, H. K. Wang, M. Saito, V. Tosirisuk, T. Fujiwara, and K. Tomita, Revised structures of soyasapogenols A, B and E, oleanene-sapogenols from soybean. Structures of soyasaponin I, II and III, *Chem. Pharm. Bull.* 30:2294 (1982).
6. C. Djerassi, D. B. Thomas, A. L. Livingston, and C. R. Thompson, The structure and stereochemistry of medicagenic acid, *J. Am. Chem. Soc.* 79:5292 (1957).
7. A. L. Livingston, Lucernic acid, a new triterpene from alfalfa, *J. Org. Chem.* 24:1567 (1959).
8. G. Massiot, C. Lavaud, D. Guillaume, and L. Le Men-Olivier, Reinvestigation of the sapogenins and prosapogenins from alfalfa (*Medicago sativa*), *J. Agric. Food Chem.* 36:902 (1988).
9. W. Oleszek, M. Jurzysta, M. Ploszynski, I. J. Colquhoun, K. R. Price, and G. R. Fenwick, Zahnic acid tridesmoside and other dominant saponins from alfalfa (*Medicago sativa* L.) aerial parts, *J. Agric. Food Chem.* 40:191 (1992).
10. M. Kapundu, A. Penders, R. Warin, C. Delaude, and R. Huls, Structure du kotschyoside, saponoside nouveau isolé de *Aspilia kotschyi* (Sch. Bip.) Oliv., *Bull. Soc. Chim. Belg.* 97:329 (1988).
11. G. Massiot, X-F. Chen, C. Lavaud, L. Le Men-Olivier, C. Delaude, A. Viari, P. Vigny, and J. Duval, Saponins from stem bark of *Petersianthus macrocarpus*, *Phytochemistry* 31:3571 (1992).
12. G. Massiot, C. Lavaud, M. Benkhaled, and L. Le Men-Olivier, Soyasaponin VI, a new maltol conjugate from alfalfa and soybean, *J. Nat. Prod.* 55:1339 (1992).
13. S. Tsurumi, T. Takagi, and T. Hashimoto, A γ-pyronyl-triterpenoid saponin from *Pisum sativum*, *Phytochemistry* 31:2435 (1992).
14. S. Kudou, M. Tonomura, C. Tsukamoto, M. Shimoyamada, T. Uchida, and K. Okubo, Isolation and structural elucidation of the major genuine soybean saponin, *Biosci. Biotech. Biochem.* 56:142 (1992).
15. S. Kudou, M. Tonomura, C. Tsukamoto, T. Uchida, and K. Okubo, Isolation and structural elucidation of DDMP-conjugated soyasaponins as genuine saponins from soybean seeds, *Biosci. Biotech. Biochem.* 57:546 (1993).
16. G. Massiot, M-G. Dijoux, C. Lavaud, L. Le Men-Olivier, J. D. Connolly, and D. M. Sheeley, *Seco*-glycosides of oleanolic acid from *Beta vulgaris*, *Phytochemistry* 37:1667 (1994).

TRITERPENOID SAPONINS FROM A SIBERIAN CHEMORACE OF THE *THALICTRUM* GENUS

A.A. Semenov, V.I. Lutsky, A.S. Gromova, T.V. Ganenko,
E.A. Khamidullina and N.N. Trofimova

Irkutsk Institute of Organic Chemistry, Siberian Division, Russian Academy of Sciences, Irkutsk, 664033, Russia

INTRODUCTION

Plants of the *Thalictrum* genus (*Ranunculaceae* family) grow in the temperate zones of the both hemispheres of the globe. Many of these species are met worldwide. *Thalictrum minus*, for example, is known in Europe, Asia, and North America. However, secondary metabolism in this species is variable. Thus, in the Bulgarian population of the plant a bisbenzylisoquinoline alkaloid, thalicarpin, has been isolated as a major secondary metabolite [1], whereas the Moldavian chemorace met hereabout produces a seco-protoberberine base, β-allocryptopine [2]. In general, all the *Thalictrum* species were thought to serve as producers of diverse isoquinoline and, on rare occasions, of diterpene alkaloids. *T. minus* widely spreads from east to west over the whole territory of Russia and is involved in the Siberian biocenoses. That is why 15 years ago we tried to assess its potential as a producer of alkaloids. However any tertiary bases have not been detected. Instead, we have found a fairly large amount of saponins. This seemed strange, since earlier in the literature there were only scarce and unreliable data on the presence of saponins of unidentified structure in the representatives of this genus. Therefore, we decided to concentrate all research efforts on the elucidation of the chemical nature of the glycosides found.

In East Siberia ten species of the *Thalictrum* genus are met. *T. minus*, *T. foetidum, T. squarrosum*, *T. appendiculatum* and *T. petaloideum* turned out to be effective producers of saponins. *T. simplex* does not produce surfactant glycosides. The rest of species have not been studied. For the isolation of saponins the plant material was thrice extracted with 80% methanol at 50° C. Methanol was distilled from the extract obtained and the aqueous residue was exhaustively extracted with chloroform and butanol. The saponin-containing butanolic extract was evaporated until dry and dissolved in methanol upon heating. When allowed to stay in a refrigerator, the solution gave a bulky precipitate. After filtering, the filtrate was evaporated to a small volume and diluted with acetone. In so doing a saponin-containing fraction precipitated. For the removal of low molecular carbohydrates the fraction was dissolved in water and the solution filtered repeatedly through G25 Sefadex. Phenolic substances were removed by passing the solution diluted with methanol (1:1) through aluminum oxide. Individual glycosides were isolated by multiple chromatography on silica gel using various systems of methanol-containing solvents.

THALICTRUM MINUS SAPONINS

T. minus was the object of the most detailed investigation. The saponin-bearing chemorace has been found to inhabit a vast territory in central Siberia between the 90-120° meridians and 50-64° parallels. During blooming in different places of this area *T. minus* produces from 1.4 to 2.3% of saponins (based on air dry weight).

The composition of the *T. minus* oligoside fraction is complex, but two glycosides called thalicosides A and B predominate in it. The first one is the major constituent. This substance is rather resistant to acid hydrolysis; thus the reaction can be carried out only by 6-h boiling with 10% sulfuric acid. In this case triterpenoid I (Fig. 1) and a 1:1 mixture of glucose and galactose are formed (GLC of aldononitrile acetates).

Substance I was purified by double column chromatography on silica gel in an ethyl acetate/methanol/water (10:3:2) solvent system and by subsequent double crystallization from ethyl alcohol. The chemical structure proposed on the basis of the results of NMR spectroscopy and mass spectrometry was confirmed by X-ray analysis [3].

The comparison of ^{1}H NMR spectra of the hydrolysis product and thalicoside A indicates that triterpene I is not a native sapogenol but is formed by the effect of the acid. In order to isolate the native genol, thalicoside A was first treated with 10% aqueous solution of sodium periodate for 24 h at room temperature and then, following filtration and evaporation, boiled for 1 h in a 3% water-ethanolic KOH solution. The reaction solution was neutralized on a cation exchanger, extracted with ethyl acetate, and the residue obtained by evaporation of the extractant was purified by chromatography on aluminum oxide using a gradient mixture of ethyl acetate and methanol. The triterpenoid, thalicogenin, obtained in this way has the structure III. The structure was proved by NMR spectroscopy and mass-spectrometry taking into consideration the structural data on artefact I. Substances of the cycloartane series are readily recognized by their

^{1}H NMR spectra, in which the cyclopropane ring protons resonate in a very high field as two doublets. In case of thalicogenin their chemical shifts are 0.44 and 0.62 ppm and the constants of spin-spin coupling are 4.5 Hz. For details on structure III see reference [4].

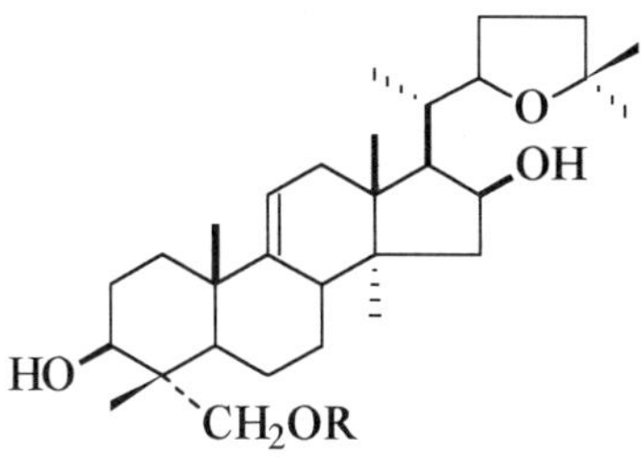

I. R = H

II. R = β-D-Glc

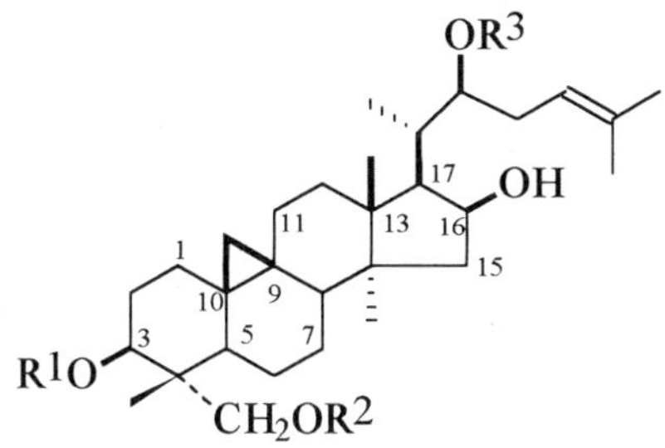

III. $R^1 = R^2 = R^3 = H$

IV. R^1 = β-D-Gal, R^2 = β-D-Glc, R^3 = H

V. R^1 = β-D-Gal, $R^2 = R^3$ = β-D-Glc

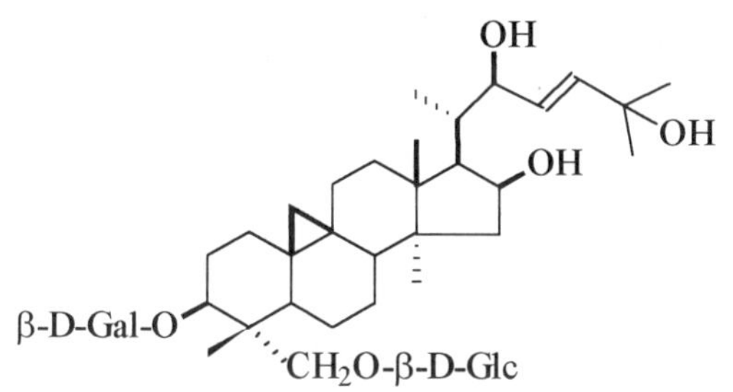

VI

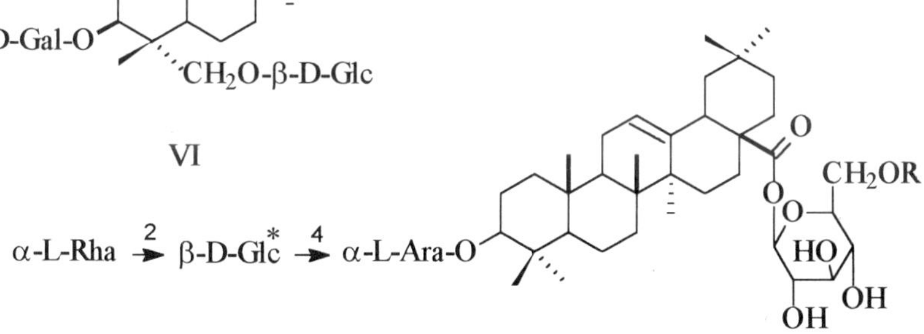

VII. R = H; VIII. R = β-D-Glc

Figure 1. *T. minus* saponins and sapogenols.

As already mentioned, the carbohydrate portion of the thalicoside A molecule consists of glucose and galactose. As shown by methylation with subsequent hydrolysis and GLC, the saponin contains a molecule of each of the above monosaccharides. The chemical shifts 3.63 (82.1), 3.52, and 3.37 (71.5) ppm in NMR spectra indicate points of their attachment to carbon atoms C-3 and C-29. Heating of thalicoside A with 80% acetic acid for 6 h afforded as the main product a monoglycoside containing glucose attached to C-29, which was evident from the observed chemical shifts in ^{13}C NMR spectrum:

72.7 (C-29) and 72.6 (C-3) ppm. On the basis of these and other results reported [5], the product of partial hydrolysis and thalicoside A have been assigned structures II and IV, respectively.

Two other glycosides of the cycloartane series have been isolated from the above-ground part of *T. minus*. Their portion in the total amount of saponins of meadow rue is very low. One of them, thalicoside C, is a trioside of the same thalicogenin aglycone. It has been obtained by multiple column chromatography of the saponin fraction followed by purification with DEAE-cellulose. Its approximate content in the saponin fraction is 1.5%.

The acid hydrolysis of saponin with 10% sulfuric acid yielded artefact I, and glucose and galactose in a 2:1 ratio. The treatment of the glucoside with an enzymic mixture from *Helix pomatia* stomach yielded thalicoside A (IV). The narrow region of 75.4 - 75.6 ppm in the ^{13}C NMR spectrum of the latter has a group of three signals arising from C-2 of glucose and galactose and a C-22 atom of the aglycone. One of these signals is displaced into the 85.3 ppm region due to glycosylation in thalicoside C. By heteronuclear correlation HETCOR this signal was found to correspond to a multiplet at 4.50 ppm in the proton spectrum. The TOCSY spectrum indicates that this multiplet belongs to the spin system of the aglycone side chain. Thus, in thalicoside C the second glucose molecule is attached to C-22 and it has a structure of the tris-desmoside V. This conclusion is supported by other data [6].

The third cycloartane glycoside of *T.minus* was isolated by multiple chromatography of the residues and chromatographic fractions formed during purification of other saponins. This glycoside was named thalicoside E and assigned chemical structure VI. The examination of two-dimensional NMR spectra of thalicoside E revealed numerous features similar to thalicoside A, e.g., the cyclic system of aglycone, the nature and attachment sites of carbohydrates to the aglycone. Differences are observed in the genol side chain. Given the H-17 proton chemical shift, it was possible to determine the spin system of the aliphatic fragment by the TOCSY method, establish the relations between all protons, and reconstruct the molecule side chain by the COSY experiment as shown in formula VI. Detailed structural studies are reported [7].

Thalicoside B, the second most abundant saponin of *T. minus*, is a glycoside of oleanolic acid. Its chemical structure was proved by classical methods. Acid hydrolysis with 5% sulfuric acid yielded oleanolic acid and the carbohydrates rhamnose, arabinose, and glucose (1:1:2), which was determined by GLC of aldononitrile or polyol acetates. One of the glucose molecules is attached to the carboxyl group since it is readily split off by alkaline hydrolysis. The sequence of sugar residues in the other saccharide chain was established by partial acid hydrolysis: heating with 0.75% sulfuric acid for 3 h formed a mixture of all possible progenols. The methylation of thalicoside B with methyl iodide in the presence of sodium hydride in dimethyl sulfoxide produced the permethylate. The mixture of methylated methyl glycosides obtained by permethylate methanolysis was acetylated and particular acetates of methyl sugars were identified in it by GLC-MS. Comparison of all the data obtained suggests thalicoside B to have structure VII. Previously [8] a structure with a 1-3-glycoside bond between glucose and arabinose has been suggested. Subsequently [9] this conclusion has been found invalid.

The structure of thalicoside B was revised in the structural determination of the minor oleanane saponin of *T. minus*, thalicoside D (VIII) by two-dimensional NMR spectroscopy and FAB mass spectrometry. The identification of all progenins of saponins VII and VIII as native substances in *T. minus* plants [10] alleviated the problem. The examination of NMR spectra of these progenins and oligosides VII and VIII allowed assignment of all carbon and most proton signals. Moreover, it has been found that in the ROESY and NOESY spectra the anomeric proton of Glc* has, apart from cross-peaks with H-2, H-3 and H-5 signals of its own molecule, an intense cross-peak at 4.28 ppm, where the arabinose H-4 proton resonates, thus indicating the position of the interglycoside bond.

THALICTRUM FOETIDUM L. SAPONINS

Like the *T. minus*, *Thalictrum foetidum* is a small herbaceous plant that grows in many temperate climatic zones of the earth. The presence of saponins among secondary metabolites of this plant has not been reported previously. But we have found that the chemorace of *T. foetidum* found in the vicinity of Lake Baikal produces triterpene saponins. They amount to 0.9% of the dry above-ground part collected during flowering. As in the preceding case, these saponins are chemically classified as oligosides of cycloartane genols and oleanolic acid. Altogether, three glycosides called cyclofoetosides A and B and foetoside C have been identified in this plant. The first two saponins are cycloartane derivatives.

The hydrolysis of saponin A with 0.5% sulfuric acid proceeded readily and yielded aglycone, cyclofoetigenin A, and a mixture of glucose, arabinose, and rhamnose in equal ratios. According to all spectral data the aglycone molecule was pentacyclic and contained an aliphatic side chain and four hydroxyl groups. One of the rings was three-membered, which is supported by the presence of two high-field signals (0.20 and 0.45 ppm) in the ^{1}H NMR spectra. Two of the hydroxyl groups are in the side chain and constitute α-glycol group since during oxidation with sodium periodate the three-carbon fragment is lost and cyclic hemiacetal XI is formed (Fig. 2). The latter evidence indicates at the same time the position of the third hydroxyl group at C-16. On the basis of these and other data [11] cyclofoetigenin A was assigned chemical structure IX.

The ^{13}CMR spectrum of cyclofoetoside A shows that its molecule contains two carbohydrate chains attached to the C-3 and C-16 atoms (glycosylation shifts are 78.0 → 88.7 and 72.0 → 82.7 ppm, respectively). Two progenols were isolated by partial acid hydrolysis with 0.25% sulfuric acid [12]. One of them contained β-D-glucopyranosylic residue at C-16 and the other had α-L-arabinose at C-3. Therefore, cyclofoetoside A is related to the bis-desmoside type. In the ^{13}C spectrum of the glucose-containing progenol the chemical shift of the glucose hydroxymethyl group was 63.1 ppm, while in the saponin it underwent a paramagnetic shift up to 68.4 ppm. Consequently, it is the glucose C-6 atom to which the rhamnose residue is attached and the molecule of cyclofoetoside A has the structure XII.

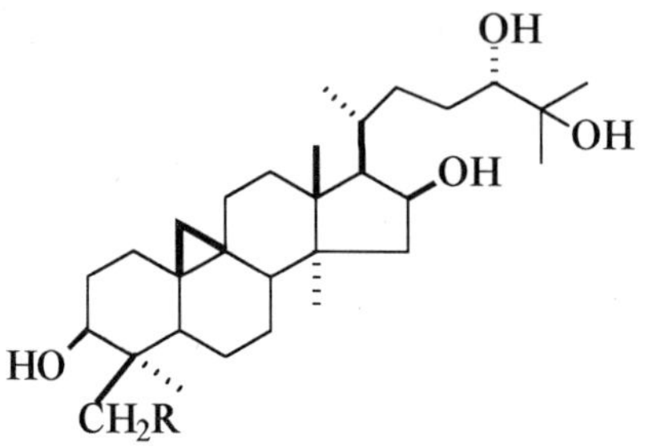

IX. R = H.

X. R = OH.

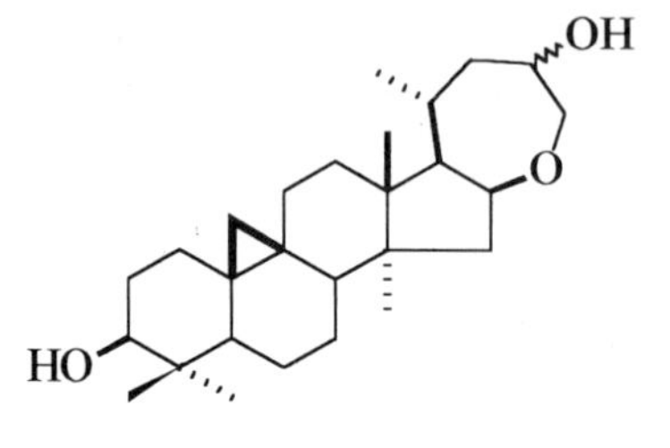

XI

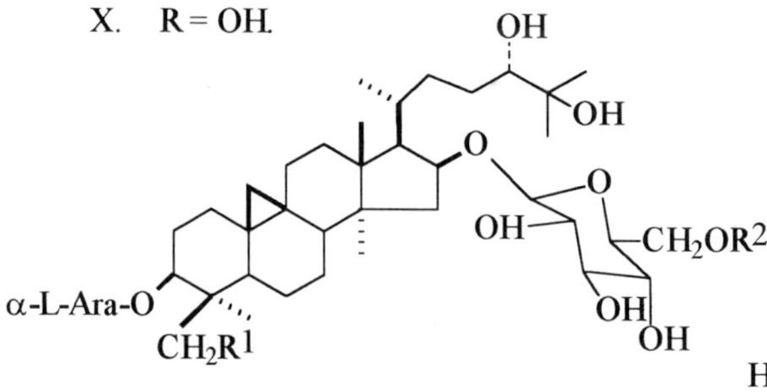

XII. R^1 = H; R^2 = α-L - Rha

XIII. R^1 = OH; R^2 = α-L - Rha

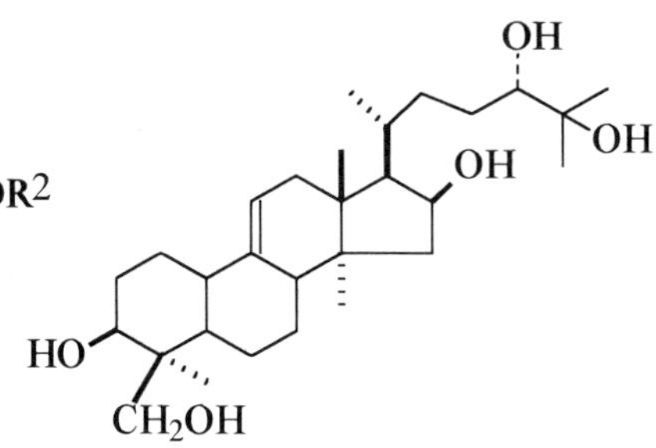

XIV

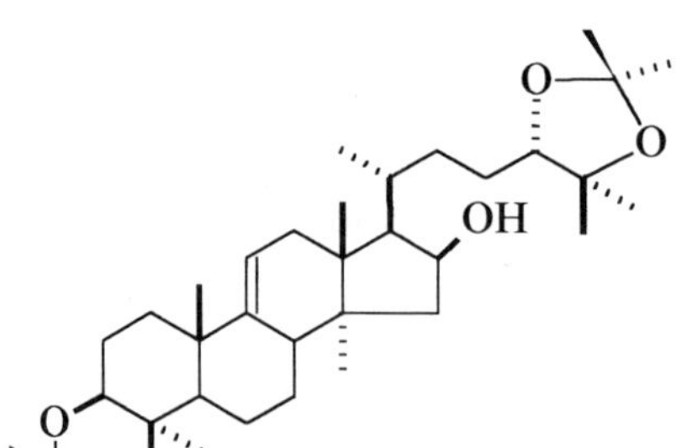

XVI

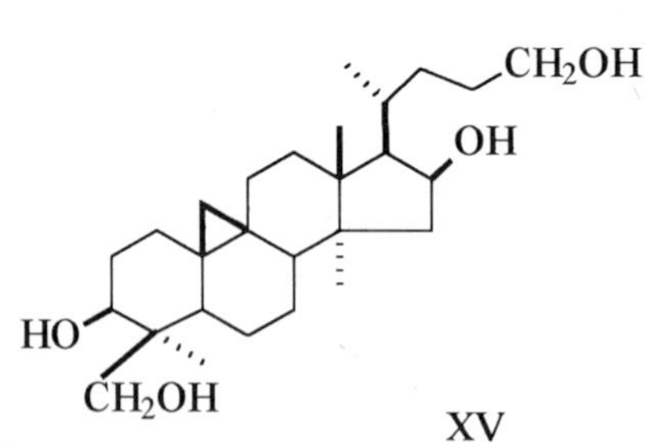

XV

Figure 2. *T. foetidum* saponins, sapogenols, and their derivatives.
(Continued)

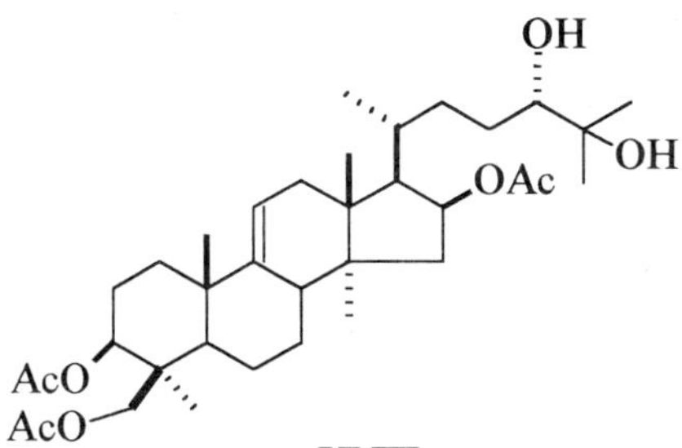

XVIII

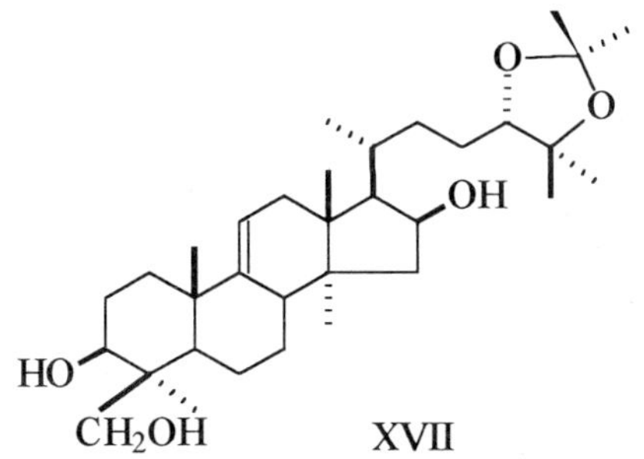

XVII

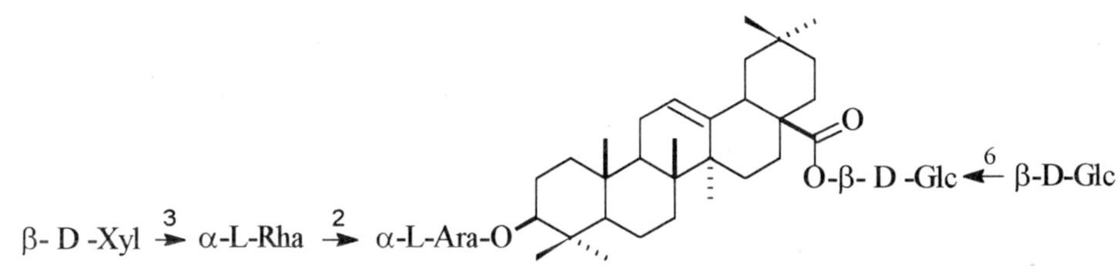

XIX

Figure 2. (*cont.*)

As distinguished from glycoside XII, the second cycloartane saponin of *T. foetidum* was resistant to mild acid hydrolysis and under rigid conditions produced the artefact XIV containing no cyclopropane ring. The oxidation of the glycoside with sodium periodate followed by borohydride reduction resulted in the trinortriterpenoid XV. In order to obtain the native sapogenol, cyclofoetoside B was transformed into peracetate. When held in a 0.5% KOH methanol solution at room temperature for 1 h the latter yielded a mixture of two diacetates. The same acetyl group in the two components of this mixture acylated the tertiary hydroxyl. Hence the side chain containing no α-glycol group, remained intact during periodate oxidation of diacetates followed by alkaline hydrolysis [13], and the native aglycone, cyclofoetigenin B was isolated. Its chemical structure X was proved by the methods and deductions similar to those used for the assessment of the genol IX structure.

Both substances contain chiral hydroxy- and hydroxymethyl groups. Axial orientation of the 3-OH group followed from the observed spin-spin coupling constants in the ^1H NMR spectrum of the aglycone and its derivatives. Diacetonide XVI and monoacetonide XVII were obtained in order to establish the configuration of the hydroxyl group at C-16 and the hydroxymethyl function at C-30. The monoacetonide is formed by keeping the triterpenoid X in anhydrous acetone containing 0.2% sulfuric acid. To obtain the diacetonide the aglycone X was treated with dimethoxypropane in dimethyl formamide in the presence of *p*-toluenesulfonic acid. Monoacetonide triacetate was hydrolyzed with 50% acetic acid to produce α-diol XVIII. The circular dichroism spectrum of the latter in the presence of Eu(fod$_3$) revealed the positive Cotton effect at 325 nm ($\Delta\varepsilon$ = + 1.97°) from which follows an S-configuration at C-24 [14].

Besides, the diacetate can be obtained from the acetonide XVIII. The hydroxyl group at C-16 remained free in the diacetate. The difference of molecular rotations between tri- and diacetate, $\Delta[M]_D$, was + 248°.According to the results reported in [15], β-orientation of the OH-16 group is suggested. As compared to cyclofoetigenin A (IX), the signal of the quaternary carbon C-4 of the hydroxymethyl analog X in the ^{13}CMR spectra experienced a paramagnetic shift of +2.2 ppm, whereas the chemical shift in the hydroxymethyl group C-atom was 64.6 ppm. This indicates an axial orientation of this group [4, 16].

Complete hydrolysis of cyclofoetoside B yielded a mixture of glucose, arabinose, and rhamnose (1:1:1). Treatment of an aqueous solution of saponin with gastric juice of the *Helix pomatia* resulted in the formation of aglycone X and two progenols containing one or two monosaccharide residues [17]. Examination of the ^{13}CMR spectra of the latter and their comparison with each other and with aglycone spectrum suggest that cyclofoetoside B should have the chemical structure of bis-desmoside XIII.

The third saponin isolated from *T. foetidum* and named foetoside C is an oleanolic acid derivative. Acid hydrolysis of foetoside C produced along with oleanolic acid a 2:1:1:1 mixture of glucose, xylose, arabinose, and rhamnose. Alkaline hydrolysis resulted in the formation of progenol containing no glucose. Therefore, its two glucose molecules are attached to the carboxylic group and the oxygen atom at C-3 bears a three-membered carbohydrate chain. This chain lacks any branching, which follows from the results of periodate oxidation. The structure of both carbohydrate chains was established by the method of methylation followed by hydrolysis and by the analysis of the partially methylated monosaccharides formed [18]. In accordance with these data foetoside C has the structure XIX.

Thus, there is much in common between the oligoside XIX and *T. minus* thalicoside D.

SAPONINS OF *THALICTRUM SQUARROSUM* ST. EX WILLD

The herbaceous plant *Thalictrum squarrosum* is not as abundant as other two species. However, in southern regions of Siberia eastward to Lake Baikal one can meet small steppe areas have dense stands of it.

We have found that the above-ground part of *T. squarrosum* contains a mixture of closely related saponins. Because of the complicated character of this mixture it was decided to start the investigation of aglycones with hydrolysis of the methanol extract of the plant. It was heated with 15% sulfuric acid for 20 h and the nonpolar fraction of the hydrolysate was chromatographed on a silica gel column and extracted by gradient elution with chloroform-methanol (0 → 25%). The structure of the only triterpenoid isolated was assessed [19] by a combination of chemical and physico-chemical methods and confirmed later by X-ray analysis [20]. This triterpenoid was called squarrofuric acid. Its chemical formula XX (Fig. 3) resembles structures of artefacts formed from cycloartane precursors under the action of mineral acids. However, no cycloartane aglycone corresponding to the acid XX has been found among *T. squarrosum* saponins. Presumably the precursor of squarrofuric acid exists in a non-glycoside form.

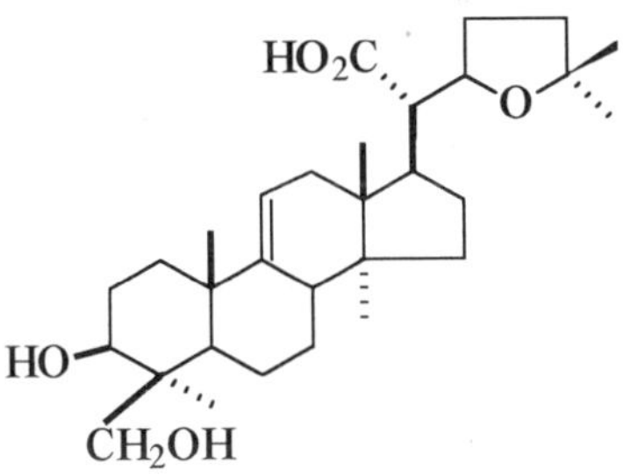

XX

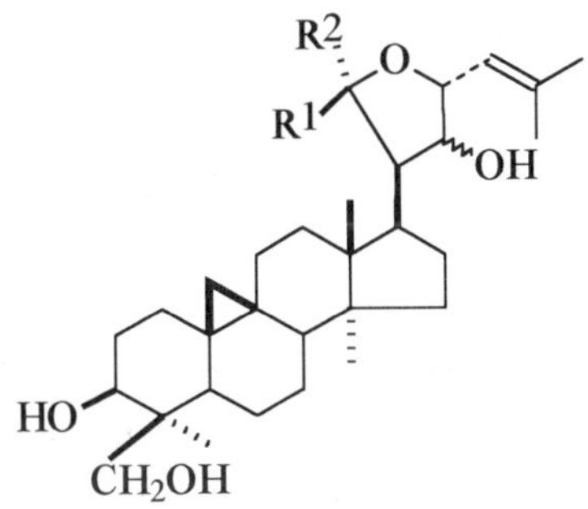

XXI. R^1 = OMe, R^2 = H

XXII.. R^1 = H, R^2 = OMe

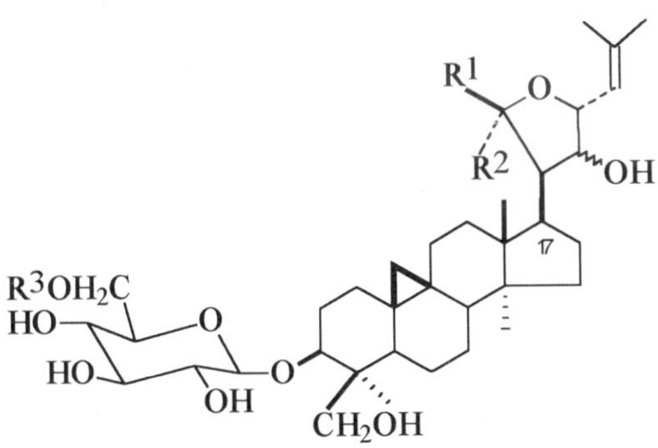

XXIII. R^1 = OMe, R^2 = R^3 = H.

XXIV. R^1 = R^3 = H, R^2 = OMe.

XXV. R^1 = OMe, R^2 = H, R^3 = α-L-Rha.

XXVI. R^1 = H, R^2 = OMe, R^3 = α-L-Rha.

XXVII. R^1 = OH, R^2 = H, R^3 = α-L-Rha.

XXVIII. R^1 = H, R^2 = OH, R^3 = α-L-Rha.

Figure 3. *T. squarrosum* saponins and sapogenols

It was possible to isolate *T. squarrosum* native aglycones by enzymatic hydrolysis of partially separated saponins. The total glycosides represent a complicated and hardly separable mixture of no less than eight components. We succeeded in obtaining a mixture of four quantitatively prevailing substances by silica gel column chromatography of the plant methanol extract. By combining countercurrent distribution (methanol-chloroform-water 20:10:7.5 and 5:6:4) and column chromatography on silica gel and polyamide the mixture could be separated into two fractions each containing two compounds. They were designated as squarrosides A1, A2 and B1, B2. The first pair consisted of monosides and the second, biosides. Saccharide components in each pair were identical. Glycosides were distinguished by aglycones denoted as squarrogenins 1 and 2. Individual saponins could not be obtained in the pure form.

The mixture of squarrosides A1 and A2 was enzymatically hydrolyzed with gastric juice of *Helix pomatia.* All the triterpenoids obtained by hydrolysis were separated on a silicagel column using for gradient elution a chloroform-methanol mixture ($0 \rightarrow 25\%$) and then hexane-acetone. Squarrogenins 1 and 2 were desorbed at the acetone content of 13 and 14%, respectively.

The chemical structures of these aglycones was mainly determined by NMR spectroscopy [21]. The pentacyclic moiety of the molecule is similar to the cyclic fragment of cyclofoetigenin B (X) though it lacks the hydroxyl group at C-16. It was immediately clarified that both substances contained the methyl acetal functional group but differed in its configuration. The acetal group is present in the side chain. The examination of spin-spin coupling in combination with mass-spectrometric data suggested that squarrogenins 1 and 2 should have the chemical structure XXI and XXII with the tetrahydrofuran ring in the side chain. The relative stereochemistry of this part of the molecule was determined by the Overhauser nuclear effect in uni- and two-dimensional variants. Nevertheless, while the orientation of the isobutyl group is beyond any doubt, contradictory results have been obtained for the hydroxyl group at C-22. Yoshimitsu et al. [22] described the isolation of saponins closely related to squarrosides from a non-identified *Thalictrum* species and attributed α-orientation to this hydroxyl group. As regards the acetal function in aglycones from *T. squarrosum*, it has β-orientation in squarrogenin 1 (XXI) and the opposite one in squarrogenin 2.

In this connection, the structure of glucosides became apparent, though, as mentioned above, we could not obtain them in the pure state. The first pair presents a mixture of glucosides, in which a β-D-glucoside residue is attached to the C-3 position of the aglycones [23]; that is, squarrosides A1 and A2 have the structure XXIII and XXIV, respectively.

As already mentioned, the group of saponins B is a mixture of biosides. It is interesting to note that this mixture has a clearly defined melting point (200 - 202°C, ethanol). The enzymatic hydrolysis of squarrosides B1 and B2 produced, apart from glucose and rhamnose, two genols and one progenol identical to squarrogenins 1 and 2 and squarroside A1, respectively. It is in this way that the proper glucoside XXIII could be obtained. The glucose C-6 position of attachment of the rhamnosyl residue was determined by a combination of FAB-mass spectroscopy, ^{13}CMR, and examination of the Smith oxidation products. These data indicate that squarrosides B1 and B2 have the chemical structure XXV and XXVI, respectively [23].

The *T. squarrosum* plant also contains a third more polar pair of squarrosides, B3 and B4. They were isolated similarly to squarrosides B1/B2. The examination of NMR spectra showed that this pair was structurally closely related to the latter. All four substances have similar carbohydrate components attached in the same way to similar aglycones. The difference is that saponins B1/B2 are methyl acetals and the B3/B4 pair is an anomeric mixture of hemiacetals XXVII and XXVIII. It is clear that this mixture is in principle inseparable. Nevertheless, in ^{1}H and ^{13}C NMR spectra the signals of each individual anomer can be assigned by methods of two-dimensional resonance. One anomer, squarroside B3, was isolated as the peracetate. It was obtained by acetylation of an anomeric mixture of saponins followed by column chromatyography [24].

It was possible to isolate from *T. squarrosum* squarrosides A1/A2 and B1/B2 (0.1 and 0.5%, respectively, on the basis of air dry plant material) and much less of B3/B4 squarrosides. However, visual examination of thin-layer chromatograms of fresh methanolic extracts indicates the latter pair to prevail in the saponin mixture. Therefore, some doubt was cast upon the native origin of A1/A2 and B1/B2 pairs since they could be formed during isolation and purification using methyl alcohol. To solve this problem, a special experiment in which all the saponins present were extracted and isolated only with ethyl alcohol was performed. The mass spectroscopy of the material obtained demonstrated that the mixture produced contained all the three pairs of squarrosides and in addition a set of ethyl acetals. This means that the *T. squarrosum* plant synthesizes both the anomeric pair of hemiacetal squarrosides B3/B4 and their methyl acetals A1/A2 and B1/B2. However, the real proportions are distorted as a consequence of the isolation and purification procedures.

PHYSIOLOGICAL ACTIVITY OF *THALICTRUM* SAPONINS

The physiological activity of saponins from the *Thalictrum* species has been investigated in mammals. Two types of properties have been clearly revealed. These are antitumor activity and the effect on the reproductive system. It is known that the ability to inhibit malignant growth is typical of many steroid and triterpene saponins. We have found [25] that all saponins of meadow rue inhibit tumor growth. However, it is only foetoside C (XIX), cyclofoetoside B (XIII), and thalicoside A (IV) that exhibit a rather high level of activity. Some results of our studies with experimental tumors in the rat are presented in the

Table 1 below. As seen from the table, some of these substances are worthy of a more detailed investigation, but it has not been carried out yet.

Table 1. Antitumor effect of the *Thalictrum* saponins

Saponin	Tumor	Dose mg/kg	Inhibition, %
Foetoside C	Sarcoma 45	30	91
	Breast cancer PMK-1	30	85
	Pliss lymphatic sarcoma	30	86
	Walker carcinosarcoma	30	85
	Alkylant-resistant sarcoma 45	30	90
Thalicoside A	Pliss lymphatic sarcoma	50	73
	Sarcoma 45	50	78
Cyclofoetoside B	Sarcoma 45	50	87

By using some *Thalictrum* saponins, such as thalicoside B and cyclofoetoside B (XIX and XIII), it is possible to reach 100% contraception in rats by subcutaneous injection of low doses (0.01 mg/kg). However, the effect of active saponins on the reproductive system is ambiguous. The result depends on the time of administration. The high contraceptive activity is evident when saponins are injected soon after *coitus*. Later (from 1 to 4 days) no effect on pregnancy is observed. On the contrary, the ovulation capacity and fertility of animals significantly increase when foetoside C (XIX) and cyclofoetoside B (XIII) are administered preventively. For instance, female and male rats were stressed by sexual limitation. After elimination of the stress factor the female fertility was 7-10 baby-rats, while in saponin-administrated prestressed animals it achieved 15-20 (two-fold increase) and was even higher than in animals not subjected to stress.

The mechanism of the effect of saponins on the reproductive system involves several components. *Thalictrum* glycosides interfere with the formation of endometrium, and decrease the rate of the ovum transport, but do not affect the process of implantation. Effects on the formation of gonadotrophic hormones of the hypophysis are likely to be the primary mechanism of contraceptive action. The administration of saponins results in an increase in the level of follicle-stimulating hormone in the blood serum and a decrease in the concentration of the luteinization hormone. The resulting hormonal imbalance may be the main reason for pregnancy failure [26].

CONCLUSION

Summarizing the results of our investigation, it has been found that populations of some *Thalictrum* species growing in central and north-eastern regions of Asia produce triterpene saponins. Oleanolic acid and oxidized cycloartane derivatives act as their aglycones. According to the general chemical structure the oleanane glycosides of meadow rue do not differ from a majority of analogs found in other plants. Most cycloartane saponins found in *Thalictrum* are of bis-desmoside type. Their carbohydrate chains consist of one or three monosaccharide units. Molecules with branched carbohydrate chains are found neither in oleanane nor in cycloartane series.

Cycloartane glycosides containing the hydroxymethyl group at C-4 are hydrolyzed with mineral acids only under severe conditions. In this case the cyclopropane ring is split and reactions can proceed in the side chain. But when only methyl groups are present at C-4, the hydrolysis proceeds readily and the native aglycones are preserved.

REFERENCES

1. A. Maleev, J. Stoichov, D. Todorov, K. Mikhaleva, N. Mollov, N.K. Duchevska. Antitumor effect of the alkaloid thaliblastine, *Dokl. Bulg. Akad. Nauk.* 25: 713 (1975).
2. K.I. Kuchkova and G.V. Lazur'évskii. Alkaloids from *Thalctrum minus* growing in Moldavia. I. The Identity of thalictrimine and β-allocryptopine, Izv. Akad. Nauk Mold.SSR, Ser. Khim. Biol. 11:43 (1965).
3. A.S. Gromova, V.I. Lutsky, A.A. Semenov, E.V. Slavyanov, V.K. Biyushkin, and G.I. Malinovsky. Triterpenoid saponins from *Thalictrum minus*. II. The structure of artefact, *Khim. Prirod. Soed.* 718 (1982).
4. A.S. Gromova, V.I. Lutsky, A.A. Semenov, V.A. Denisenko, and V.V. Isakov. Triterpenoid saponins from *Thalictrum minus*. III. The structure of thalicogenin, *Khim. Prirod. Soed.* 207 (1984).
5. A.S. Gromova, V.I. Lutsky, A.A. Semenov, V.A. Denisenko, and V.V. Isakov. Triterpenoid saponins from *Thalictrum minus*. IV. The structure of thalicoside A, *Khim. Prirod. Soed.* 213 (1984).
6. A.S. Gromova, V.I. Lutsky, S.V. Zinchenko, N.N. Trofimova, A.A. Semenov, and N.A. Nakhova. Triterpenoid saponins from *Thalictrum minus*. VI. The structure of thalicoside C, *Khim. Prirod. Soed.* 103 (1993).
7. A.S. Gromova, V.I. Lutsky, S.V. Zinchenko, T.V. Ganenko, and A.A. Semenov. Triterpenoid saponins from *Thalictrum minus*. VII. The structure of thalicoside E, *Khim. Prirod. Soed.* 567 (1993).

8. A.S. Gromova, V.I. Lutsky, A.A. Semenov, R.B. Valeev, G.A. Kalabin, and V.N. Elkin. Triterpenoid saponins from *Thalictrum minus*. V. The structure of thalicoside B, *Khim. Prirod. Soed.* 670 (1985).
9. A.S. Gromova, A.A. Semenov, V.I. Lutsky, S.V. Zinchenko, and N.N. Trofimova. Triterpenoid saponins from *Thalictrum minus*. VIII. The structure of thalicoside D, *Khim. Prirod. Soed.* 398 (1994).
10. A.S. Gromova, V.I. Lutsky, A.L. Vereshchagin, and A.A. Semenov. Application of ^{13}C NMR spectroscopy and FAB- mass-spectrometry for investigation of minor triterpenoids of *Thalictrum* minus, *Khim. Prirod. Soed.* 107 (1987).
11. T.V. Ganenko, M.I. Isaev, M.B. Gorovits, N.D. Abdulaev, V.I. Lutsky, A.A. Semenov, and N.K. Abubakirov. Triterpenoid glycosides and their genols from *Thalictrum foetidum*. II. The structure of cyclofoetigenin A, *Khim. Prirod. Soed.* 370 (1985).
12. T.V. Ganenko, M.I. Isaev, V.I. Lutsky, A.A. Semenov, N.D. Abdulaev, M.B. Gorovits, and N.K. Abubakirov. Triterpenoid glycosides and their genols from *Thalictrum foetidum*. III. The structure of cyclofoetoside A, *Khim. Prirod. Soed.* 66 (1986).
13. T.V. Ganenko, M.I. Isaev, A.S. Gromova, N.D. Abdulaev, V.I. Lutsky, M.F. Larin, A.A. Semenov, M.B. Gorovits, and N.K. Abubakirov. Triterpenoid glycosides and their genols from *Thalictrum foetidum*. IV. The structure of cyclofoetoside B, *Khim. Prirod. Soed.* 312 (1986).
14. J. Dillon and K. Nakanishi. Absolute configurational studies of vicinal glycols and amino alkohols. I. With bis(acetylacetonato) nickel, J. Amer. Chem. Soc. 97:5417 (1975).
15. L. Fieser and M. Fieser. Steroids, Reinhold Publishing Corporation, New York and Chapman & Hall, Ltd., London (1959).
16. W. Wehrli and T. Nishida. Progress in the Chemistry of Organic Natural Products, Springer-Verlag, Wien (1979).
17. T.V. Ganenko, M.I. Isaev, A.S. Gromova, N.D. Abdulaev, A.A. Semenov, M.B. Gorovits, and N.K. Abubakirov. Triterpenoid glycosides and their genols from *Thalictrum foetidum*. V. The structure of cyclofoetoside B, *Khim. Prirod. Soed.* 341 (1986).
18. T.V. Ganenko, M.I. Isaev, M.B. Gorovits, A.S. Gromova, V.I. Lutsky, A.A. Semenov, and N.K. Abubakirov. Triterpenoid glycosides and its genols from *Thalictrum foetidum*. I. The structure of foetoside C, *Khim. Prirod. Soed.* 458 (1984).
19. A.S. Gromova, V.I. Lutsky, A.A. Semenov, M.F. Larin, and R.B.Valeev. Triterpenoid glycosides of *Thalictrum squarrosum*. I. The structure of squarrofuric acid, *Khim. Prirod. Soed.* 376 (1987).

20. Yu.V. Gatilov, I.Yu. Bagryanskaya, V.I. Lutsky, A.S. Gromova, and A.A. Semenov. Triterpenoid glycosides of *Thalictrum squarrosum*. II. Molecular and crystal structure of squarrofuric acid, *Khim. Prirod. Soed.* 533 (1987).
21. V.I. Lutsky, E.A. Khamidullina, A.S. Gromova, and A.A. Semenov. Triterpenoid glycosides of *Thalictrum squarrosum*. III. The structure of squarrogenins 1 and 2, *Khim. Prirod. Soed.* 510 (1989).
22. H. Yoshimitsu, K. Hagoshi, M. Kumabe, and T. Nahara. Two new cycloartane glycosides from *Thalictrum Herba*, *Chem. Pharm. Bull.* 41:786 (1993).
23. E.A. Khamidullina, A.S. Gromova, V.I. Lutsky, A.L. Vereshchagin, A.A. Semenov, and M.F. Larin. Triterpenoid glycosides of *Thalictrum squarrosum*. IV. The structure of squarrosides A1, A2, B1, and B2, *Khim. Prirod. Soed.* 516 (1989).
24. E.A. Khamidullina, A.S. Gromova, V.I. Lutsky, A.A. Semenov, and S.V. Zinchenko. Triterpenoid glycosides of *Thalictrum squarrosum*. V. The structure of squarrosides B3 and B4, *Izv. Akad. Nauk, Ser. Khim.* in press.
25. K.D. Rakhimov, S.M. Vermenichev, V.I. Lutsky, A.S. Gromova, T.V. Ganenko, and A.A. Semenov. Triterpenoid glycosides of *Thalictrum foetidum, Thalictrum minus* and their antitumor activity, *Khim.-Farm. Zhurn.* 1434 (1987).
26. M.N. Matz, V.V. Korchov, V.I. Lutsky, A.S. Gromova, T.V. Ganenko, and A.A. Semenov. Triterpenoid glycosides of *Thalictrum minus, and Thalictrum foetidum,* and their contraceptive activity, *Rastitelnye Resursy*. 27:570 (1988).

NEW PREPARATION OF TRITERPENOID-3-SULFATES BY THE USE OF SO_3-DMSO COMPLEX

V.I. Grishkovets, A.E. Kondratenko, and V.Ya. Chirva

Department of Organic Chemistry
Simferopol State University
333036 Simferopol, UKRAINE

INTRODUCTION

It is known that steroidal 3-sulfates are widely spread in nature, while triterpenoid-3-sulfates have been found in natural sources only recently where they have been found in *Bupleurum rotundifolium*[1], *Patrinia scabiosaefolia*[2], *Schefflera octophylla*[3,4] and *Hedera taurica*[5,6].

The known reagents for sulfatation of 3-hydroxy groups in steroids and triterpenoids are concentrated sulfuric acid in the presence of dicyclohexylcarbodiimide (DCCD)[7], complexes of sulfur(VI) oxide with pyridine (SO_3-Py)[1], or sulfur oxide with trimethylamine (SO_3-Me_3N)[8]. Study of the sulfation properties of some sulfur trioxide complexes with 1,4-dioxane, tetrahydrofuran, and *N,N*-dimethyl- and diethylanilines does not show any advantages. These methods do not give high yields of desired products and require long reaction time with loss of heterogeneity of the reaction mixture[5]. It stimulated our search of suitable reagents and methods.

EXPERIMENTAL

Sulfur trioxide (10 ml) is added dropwise to 100 ml of dimethyl sulfoxide stirred at 10°-15 °C at the rate of not more than 1 ml per min. Approximately 100 ml of methylene chloride is then added, and the solid complex is collected by filtration. The complex is washed several times in methylene chloride and dried in a desiccator in vacuo. The complex SO_3-DMSO can be stored until needed. The amount of sulfur trioxide present is determined by alkalimetric titration to a phenolphthalein end-point.

RESULTS AND DISCUSSION

The use of sulfur trioxide-dimethyl sulfoxide (SO_3-DMSO) complex in dimethyl sulfoxide solution which was proposed for cellulose sulfation[9], gives excellent yields of 90-100% of desired products during a 15-20 min. reaction time at room temperature in homogeneous solution. The excess of the required reagent is no more than 30-50%. The work up of the reaction mixture is quite simple and consists of ice-water dilution and extraction of the triterpenoid 3-sulfate with benzene, chloroform, or 1-butanol. In that way

3-sulfates of oleanolic, betulinic and echinocystic acids have been prepared. The sulfation of echinocystic acid gives an example of high reagent selectivity for an equatorial 3-OH group in the presence of an axial 16-OH group (Fig. 1).

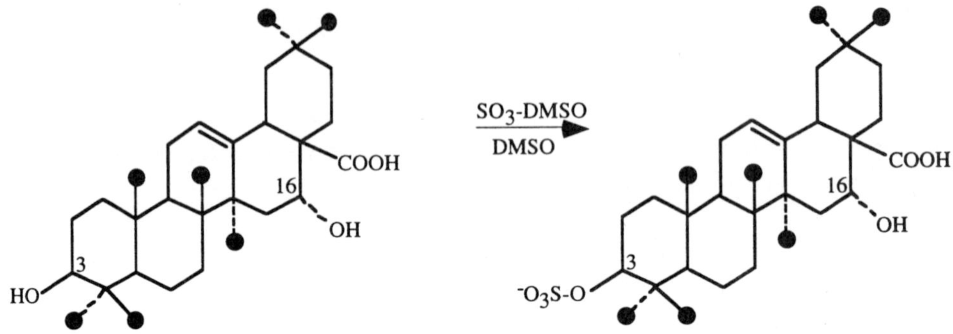

Fig.1. The formation of 3-sulfate of echinocystic acid.

SUMMARY

The use of sulfur trioxide-dimethyl sulfoxide (SO_3-DMSO) complex in dimethyl sulfoxide solution for the preparation of triterpenoid-3-sulfates with heavily quantitative yields has been described. The high reagent selectivity for equatorial 3-hydroxy groups in presence of an axial 16-hydroxy group was demonstrated.

REFERENCES

1. E. Akai, T. Takeda, Y. Kobayashi, and Y. Ogihara, Sulfated triterpenoid saponins from the leaves of *Bupleurum rotundifolium* L., *Chem. Pharm. Bull.* **33**: 3715 (1985).

2. A. Inada, M. Yamada, and H. Murata, Phytochemical Studies of Seeds of medicinal plants I. Two sulfated triterpenoid glycosides, sulfapatrinosides I and II, from seeds of *Patrinia scabiosaefolia* Fischer, *Chem. Pharm. Bull.* **36**: 4269 (1988).

3. J. Kitajima, M. Shindo, and Y. Tanaka, Two new triterpenoid sulfates from the leaves of *Schefflera octophylla*, *Chem. Pharm. Bull.* **38**: 714 (1990).

4. T.V. Sung and G. Adam, A sulphated triterpenoid saponin from *Schefflera octophylla, Phytochemistry*. **30**: 2717 (1991).

5. V.I. Grishkovets, N.V. Tolkacheva, A.S. Shashkov, and V.Ya. Chirva, Triterpenoid glycosides of *Hedera taurica* VIII, *Khim, Prir. Soedin.* 860 (1991).

6. V.I. Grishkovets, N.V. Tolkacheva, A.S. Shashkov, and V.Ya. Chirva, Triterpenoid glycosides of *Hedera taurica* X. *Khim, Prir. Soedin.* 683 (1992).

7. K. Shimada, Y. Fujii, and T. Nambara, Isolation of bufalin-3-sulfate from skin of *Bufo vulgaris formosa* Boulenger, *Tetrahedron Lett.* 2767 (1974).

8. L. Arnostova, I. Gerny, V. Pouzar, and P. Drasar, Synthesis of 5α-cholestane type glycoside sulfates, *Coll. Czech. Chem. Commun.* **85**: 1180 (1993).

9. R.L. Whistler, Sulfation of polysacchrides, *Meth. carbohydrate Chem.* **6**: 426 (1992).

TRITERPENE AND STEROID SAPONINS ISOLATED FROM TWO *MELILOTUS* SPECIES

G.V. Khodakov, Yu. A. Akimov, A.S. Shashkov, P.K. Kintia, and V.I. Grishkovets

Institute of Genetics
Academy of Science
Chisinau, 2002 Moldovia

INTRODUCTION

Of the eleven species comprising the genus *Melilotus*, four occur in Crimea. Among these, two species which are commonly grown in Moldovia are plaster clover (*Melilotus officinalis* L. Pall.) and common sweet clover (*M. albus* Medik.), the third has the range restricted to the Mediterranean region, *M. neapolitanus* Tep., and the fourth is an endemic species, *M. tauricus* (Bieb.) Ser.[1,2]. The best-studied species in terms of chemical composition are the plaster clover and the common sweet clover. In the aerial parts of these species, substances have been found such as coumarins (coumarin, dihydrocoumarin, dicoumarol)[3], acids (*o*-hydrocinnamic, melilotic, coumarinic, and *o*-coumaric)[4], coumarin glycosides[5], fatty acids, among which linoleic acid predominates[6], and flavonoids (quercetin, coumestrol)[7], as well as xylan[8]. Screenings of the Crimean representatives of the Graminoseae family revealed the presence of saponins in various sweet clover species. Evidence on these is limited to estimates of their total content in comparison with that present in other legumes[9].

The present paper is concerned with proving the structure of previously undescribed triterpenoid saponins that have been isolated from roots of common sweet clover growing in Crimea (village of Rybachje) and steroid saponins from *M. tauricus*. In a methanol extract of *M. albus* roots at least eight substances were discovered which were classified as triterpene glycosides. The structure of four of these, termed melilotosides **I, II, III,** and **IV** have been studied; for *M. tauricus* seeds the steroid structures of **V, VI, VII**, and **VIII** were found.

RESULTS AND DISCUSSION

Triterpenoid Saponins: Analysis of the products of complete acid hydrolysis of each compound in question, carried out using paper chromatography (PC) and TLC techniques (characterized by comparing their chromatographic mobilities with that of the known test substances, showed the presence of soyasapogenol B (3β,22β, 24-trihydroxyolean-12-ene) as the aglycone. Soyasapogenol B was found in a chloroform extract from the plant root extract. In melilotoside **I**, the carbohydrate component is arabinose, with the ^{13}C nuclear magnetic resonance (NMR) spectrum corresponding to a

Table 1. NMR chemical shifts of carbon atoms (δ, ppm, *O*-TMS, C_5D_5N) of melilotosides **I** and **III.**

Aglycone nuclei	I	III	Sugar residue nuclei	I	III
1	38.7	38.6	Ara		
2	26.8	26.6	1'	106.4	105.4
3	89.2	91.2	2'	73.6	77.6
4	44.6	44.4	3'	75.8	76.7
5	56.3	56.1	4'	75.5	74.1
6	19.0	19.0	5'	63.6	63.7
7	33.5	33.4			
8	40.2	40.0	Gal		
9	48.0	47.8	1"		101.8
10	36.8	36.6	2"		78.5
11	24.3	24.1	3"		75.7
12	122.7	122.5	4"		71.2
13	145.1	144.9	5"		76.5
14	42.6	42.4	6"		61.7
15	27.3	26.6			
16	28.7	28.8	Rha		
17	38.2	38.1	1"'		102.3
18	45.5	45.4	2"'		72.5
19	47.0	46.8	3"'		72.7
20	31.1	31.0	4"'		74.5
21	42.6	42.4	5"'		69.5
22	75.8	75.7	6"'		18.6
23	23.5	23.1			
24	63.6	63.7			
25	15.7	15.9			
26	17.3	17.1			
27	25.9	25.8			
28	28.7	28.8			
29	33.5	33.4			
30	21.4	21.3			

single monosaccharide residue (one signal for the anomeric carbon atom position) (Table 1). Comparisons of the magnitude of chemical shifts of the glycoside carbon atoms along with data in the literature for the unsubstituted aglycone[10,11] indicated that the sugar was attached at C-3 of soyasapogenol B (downfield shift from 80.2 ppm to 89.2 ppm).

In melilotoside **II**, the carbohydrate components are arabinose and galactose. The 'H NMR spectrum in conventional and differential double homonuclear resonance experiments confirms arabino- and galacto-configurations of the sugar residues in the pyranose form. Arabinose had the α-configuration and the galactose had the β-configuration from the spin-spin coupling constant of anomeric protons of the monosaccharides (7.5 Hz in both residues).

Melilotoside **III** has been found to contain galactose, rhamnose, and arabinose. Its ^{13}C NMR spectrum corresponds to a composition of sugar ratios of 1:1:1 [(three signals in the position of anomeric carbon atoms) (Table 1)]. On the basis of the PMR and ^{13}C spectra of the glycosides, the presence in their aglycone component of seven methyl groups was confirmed (Table 1). The support for their belonging to CH groups coming from the attached protons is generated from comparisons with data in the literature for soyasapogenol B. All of these are tertiary groups with the respective protons resonating as singlets. The presence of one double bond is evidenced by two resonance signals of carbons, respectively 122.7 and 145.1 ppm; the presence of one secondary CH group with its axial orientation follows from the chemical shift of a geminal proton (3.69 ppm) which has the characteristic doublet-doublet splitting. The resulting spectral characteristics are in good agreement with the data published in the literature [10,11].

To determine the sequence of carbohydrate residues within the molecule of saponin **III**, partial hydrolysis of the compound was carried out with chromatographic separation of the two resultant prosapogenins (**IIIb, IIIc**). They were subjected to TLC analysis, determination of physicochemical constants, and complete acid hydrolysis. This resulted in prosapogenins **IIIb** being identified as melilotoside **I** and **IIIc** as melilotoside **II**. Thus, the carbohydrate component of melilotoside **III** is a structure with the following arrangement: arabinose, galactose, and rhamnose.

The nature of splitting and the spin-spin coupling constants of all skeletal protons of saponin **III** sugar residues correspond perfectly to arabino-, galacto-, and rhamno-configurations of monosaccharides in pyranose form. What is more, the α-configuration of the glycoside center of the arabinose residue (7.5 Hz) and the β-configuration of the anomeric center of the galactose residue (7.5 Hz) (Table 2) follow from the *J* values. The nearly perfect agreement of chemical shifts in the ^{13}C spectrum of the rhamnose residue corresponds with data from the literature for terminal α-L-rhamnopyranose, and is consistent with the α-configuration of its glycoside center (Table 1)[4].

Comparisons of the chemical shifts of carbon atoms in the glycoside with data in the literature[12,13], of its component unsubstituted fragments, indicate an addition of a carbohydrate component at C-3 of the aglycone (downfield shift from 80.2 ppm to 91.2 ppm), and C-2 substitution in α-L-arabinopyranose and β-D-galactopyranose (respectively downfield shifts from 73.9 ppm to 77.5 ppm and from 72.7 ppm to 78.4 ppm) (Table 1). This is also borne out by the PMR spectrum data for the glycoside acetate, in which it was possible, using the technique of double homonuclear resonance, to identify signals of monosaccharide skeletal protons and to observe a relatively high position of signals for protons of the H-2 residue of arabinose and H-2 residue of galactose, suggesting the presence of 1→2 coupling bonds between the sugars.

Comparisons of the glycosylation effects in the ^{13}C NMR spectrum of glycoside C-1 with data in the literature are supported by nuclear Overhauser effect observations in a rotating frame experiment (two-dimensional ROESY)[14]. The observed cross-peaks indicate drawing together of the H-1 residue of α-L-arabinopyranose and H-3 atom of the aglycone, the H-1 of α-L-rhamnopyranose and H-2 atom of β-D-galactopyranose.

Another triterpene saponin termed melilotoside **IV** was isolated from roots of common sweet clover. Complete acid hydrolysis of **IV** allowed soyasapogenol B (3β, 22β, 24-trihydroxyolean-12-ene) to be identified as the aglycone. This is borne out by the coincidence of the 13C NMR and PMR spectra of compound **IV** (Table 2) with data in the literature[10,11].

Table 2. Chemical shifts of carbon atoms (δ, ppm, *O*-TMS, C_5D_5N) of melilotoside (**IV**).

C atom	δ	C atom	δ	C atom	δ
		Aglycone			
1	39.0	11	24.3	21	42.5
2	26.6	12	122.7	22	75.9
3	91.7	13	145.1	23	23.3
4	44.2	14	42.7	24	63.8
5	56.6	15	26.6	25	15.9
6	18.9	16	29.1	26	17.3
7	33.6	17	38.2	27	25.8
8	40.3	18	45.7	28	28.8
9	48.1	19	47.1	29	33.2
10	36.8	20	31.0	30	21.1
		Sugars			
	Arabinose		Galactose		Rhamnose
1'	105.3	1"	102.1	1" '	102.6
2'	77.5	2"	78.4	2" '	72.4
3'	76.6	3"	84.1	3" '	72.8
4'	72.4	4"	71.2	4" '	74.5
5'	63.8	5"	76.4	5" '	69.5
		6"	62.2	6" '	18.9
1" "	106.3				
2" "	73.9				
3" "	74.2				
4" "	69.				
5" "	66.4				

The locations and the order of monosaccharides within the molecule of saponin **IV** were determined by using the techniques of double homonuclear resonance (in conventional and differential versions) and nuclear Overhauser effect, as well as homonuclear (COSY) and heteronuclear (^{1}H-^{13}C) two-dimensional NMR spectroscopy. These techniques enabled unambiguous identification of all the glycoside signals in the ^{13}C NMR spectrum and of its sugar component in the ^{1}H NMR spectrum.

Spin-spin interaction constants and the nature of splitting of skeletal protons of all sugar residues correspond to galactorhamno- and two arabino-configurations in pyranose form, as well as to α-configurations of residues of two arabinoses ($J_{1,2}$ = 7.5 Hz) and β-configuration of the anomeric center of the galactose residue ($J_{1,2}$ = 7.5 Hz), with the α-configuration of the glycoside center of the rhamnose residue following from perfect coincidence of chemical shifts in the C spectrum of this residue with published data for terminal α-L-rhamnopyranose[13]. Comparisons of carbon atom spectral data for the aglycone component of saponin **IV** with those for unsubstituted soyasapogenol B revealed a downfield shift of the C-3 atom (from 80.2 ppm to 91.7 ppm), suggesting an addition to it of a sugar component. Comparisons of carbon atom chemical shifts for sugar residues with those for their analogues in related compounds revealed substitutions for C-2 atoms of α-L-arabinopyranose and β-D-galactopyranose, as well as for C-3 atom of β-D-galactopyranose (Table 2)[12,13,14].

The effects of glycosylation in the ^{13}C NMR spectrum of **IV** are compared with data in the literature and they are supported by nuclear Overhauser effect observations, which indicate that the H-1 of α-L-arabinopyranose and H-3 of the aglycone, H-1 of β-D-galactopyranose and H-2 of the same α-L-arabinopyranose, H-1 of α-L-rhamnopyranose and H-2 of β-D-galactopyranose, and H-1 of terminal α-L-arabinopyranose and H-3 of galactopyranose are an integral part of the saponins. The structures of **I, II, III**, and **IV** are shown in Fig. 1.

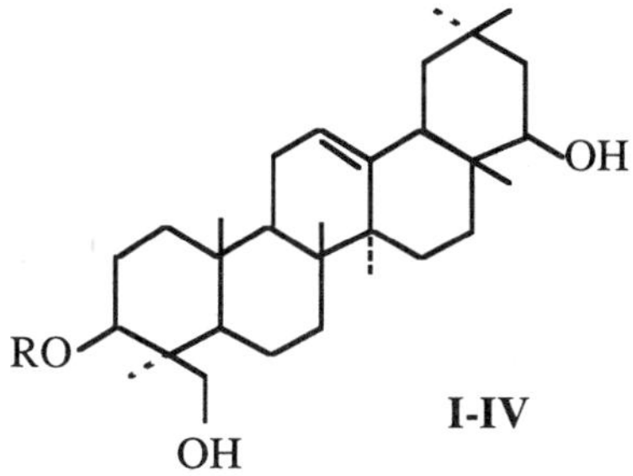

I-IV

I R = 3-*O*-[α-L-Arabinopyranosyl] soyasapogenol B
II R = 3-*O*-[β-D-Galactopyranosyl-(1→2)-*O*-α-L-arabinopyranosyl] soyasapogenol B
III R = 3-*O*-[α-L-Rhamnopyranosyl-[(1→2)-*O*-galactopyranosyl-(1→2)-*O*-L-arabinopyranosyl] soyasapogenol B
IV R = 3-*O*-[α-L-Rhamnopyranosyl-(1→2)-*O*-[α-L-arabinopyranosyl-(1→3)]-β-D-galactopyranosyl-(1→2)-*O*-α-L-arabinopyranosyl] soyasapogenol B

Fig. 1. Structures of triterpene saponins **I, II, III,** and **IV** isolated from roots of sweet clover (*Melilotus albus*) common in the Crimean region.

Steroid saponins: Saponins **V** and **VI** were detected from *M. tauricus* by TLC assays with vanillin-phosphoric acid[12] as yellow spots; these same spots could not be detected with Ehrlich reagent, allowing them to be represented as steroid glycosides of the spirostane series. This is shown also by diosgenin ((25*R*)-spirost-5-en-3β-ol) formation following acid hydrolysis of the saponins as well as by characteristics for steroid glycosides of the *R*-series, which have characteristic bands of absorption in the IR spectrum at 900 up to 920 cm.

In the carbohydrate component of acid hydrolysates, glucose was detected in saponins **V** by paper chromatography. TLC analysis of the glycoside showed the presence of a known test substance allowing trilline (3-*O*-[β-D-glucopyranosyl]diosgenin) to be identified[15]. By using paper chromatography (PC) and gas-liquid chromatography (GLC), glucose and rhamnose were detected in saponin **VI**, with monosaccharide ratio of 1:2. This finding is in good agreement with ^{13}C NMR spectra which clearly exhibit signals of three anomeric atoms of carbon at 100.8 ppm, 103.5 ppm, and 102.4 ppm (Table 3). TLC analysis of products of partial hydrolysis of glycoside **VI** allowed the prosapogenin to be identified as trilline.

Compound **VI** was methylated according to Hakomori[16]. An exhaustive methylation product that was isolated by column chromatography was subjected to methanolysis with subsequent analysis of the resulting products by GLC to yield methyl 3, 6-di-*O*-methyl-D-glucopyranoside and methyl 2,3,4-tri-*O*-methyl-L-rhamnopyranoside in a ratio of 0.95:2.07. Thus it was shown that the atoms that were affected by glycosylation were C-3 of the aglycone component, and C-2 and C-4 of the D-glucose residue were similar to those shown by ^{13}C NMR spectral data in the literature[4].

Table 3. NMR chemical shifts of carbon atoms (δ, ppm, *O*-TMS, C_5D_5N) of glycoside **VI.**

C atom	δ	C atom	δ	C atom	δ
		Aglycone			
1	37.3	11	20.8	21	16.5
2	30.4	12	40.4	22	110.9
3	78.4	13	40.6	23	37.4
4	39.0	14	56.8	24	28.8
5	141.0	15	32.0	25	34.6
6	121.7	16	81.5	26	75.8
7	31.8	17	64.3	27	17.9
8	31.5	18	16.7		
9	50.4	19	19.6		
10	37.0	20	40.9		
		Sugars			
Glucose		Rhamnose		Rhamnose	
1'	100.8	1"	103.5	1"'	102.4
2'	79.0	2"	72.8	2"'	72.4
3'	77.9	3"	72.1	3"'	72.5
4'	78.0	4"	73.7	4"'	73.8
5'	77.1	5"	69.2	5"'	70.5
6'	61.5	6"	18.5	6"'	18.7

Chemical shifts of anomeric carbon atoms of saponin **VI** were indicative of α-configuration of glycoside centers of both rhamnose molecules and β-configuration for those of glucose (Table 3)[17]. Chromatography of the extracts of the saponins yielded fractions rich in compounds **VII** and **VIII**. To improve purification from phenolic compounds, the latter were transformed into phenolates and were removed. During TLC, the saponins were detected with Ehrlich reagent as intensely light-red spots. In the IR spectrum, there was a low-intensity, broadened band at 900 cm. The data permits these saponins to be classified as steroid glycosides of the furostan series.

Analysis of products of complete hydrolysis of the saponins on the basis of physicochemical constants as well as by TLC and PC techniques resulted in identification of diosgenin and D-glucose in compound **VII**. Comparisons of physicochemical constants and chromatographic mobility of this aglycone with those of samples of known structure permits it to be interpreted as diosgenin or funkioside B or 26-*O*-[α-D-glucopyranoside](25*R*)-furost-5-ene-3β,22α,26-triol.

Identified in compound **VIII** were diosgenin, glucose, and rhamnose. The ^{13}C NMR spectrum of the saponin corresponds to a structure with a sugar residue ratio of 2:2 (four signals in the position of anomeric carbon atoms). Application of the GLC technique to products of methanolysis of permethylate of **VIII** which were produced by methylation according to Hakomori[16] allowed Me 2,3,4,6-tetra-*O*-methyl-D-glucopyranoside, Me 3,6-di-*O*-methyl-D-glucopyranoside and Me 2,3,4-tri-*O*-methyl-L-rhamnopyranoside to be identified in a ratio of 0.95:1.07:1.89. Comparison of spectral data of ^{13}C NMR-ART (attached proton test) with data in the literature[17] confirms that the aglycone structure is (25*R*)-furost-5-ene 3β, 22α, 26-triol and indicated substitution is for C-3 atom (downfield shift from 75.3 ppm to 78.2 ppm) and C-26 (the signal of a methylene atom of carbon at 75.2 ppm) (Table 4).

Table 4. NMR chemical shifts of carbon atoms (δ, ppm, *O*-TMS, C_5D_5N) of glycoside **VIII.**

C atom	δ	C atom	δ	C atom	δ
		Aglycone			
1	37.3	11	21.3	21	16.3
2	30.0	12	41.0	22	110.0
3	78.2	13	40.2	23	37.3
4	39.2	14	56.7	24	28.5
5	141.5	15	32.5	25	34.4
6	121.9	16	81.3	26	75.2
7	31.9	17	64.0	27	17.5
8	31.9	18	16.5		
9	50.7	19	19.5		
10	37.2	20	40.6		
		Sugars			
Glucose		Rhamnose		Rhamnose	
1'	100.5	1"	103.0	1"'	102.0
2'	79.8	2"	72.2	2"'	72.4
3'	78.0	3"	72.9	3"'	72.7
4'	78.2	4"	73.9	4"'	73.6
5'	76.8	5"	69.5	5"'	70.7
6'	61.8	6"	18.4	6"'	18.6
1""	104.8				
2""	75.2				
3""	78.6				
4""	72.1				
5""	78.6				
6""	63.1				

Determination of the configuration of **VIII**, sizes of oxygen rings, and confirmation of points of attachment of monosaccharides within the molecule were performed by PMR using double homonuclear resonance (in conventional and differential versions), nuclear Overhauser effect, and homonuclear (COSY) two-dimensional NMR spectroscopy. These signals enabled unambiguous interpretation of the PMR spectrum of the sugar component. The ^{13}C NMR spectrum of **VIII** was interpreted by comparing chemical shifts of carbon atoms with published data on structural analogues[18-26].

From constants of spin-spin interactions and characteristic splitting of skeletal protons of sugar residues, two gluco- and two rhamno-configurations of monosaccharides in pyranose form follow with the SSIC magnitudes corresponding to the α-configuration of anomeric centers of the two molecules of rhamnose (singlets), and the β-configuration of the two molecules of glucose (7.5 Hz).

Spatial proximity was proved of H-1 of β-D-glucopyranose and H-3 of the aglycone, H-1 of one α-L-rhamnopyranose and H-2 as well as H-1 of the other α-L-rhamnopyranose and H-4 of that same β-D-glucopyranose (data not shown). The structures of **V-VIII** are shown in Fig. 2.

Biological Activity: Preliminary results have been obtained on the antioxidant, antimicrobial and growth-regulating (allelochemical) activities determined for the total steroidal saponins; however, we are waiting for confirming results. No data were obtained for individual saponins

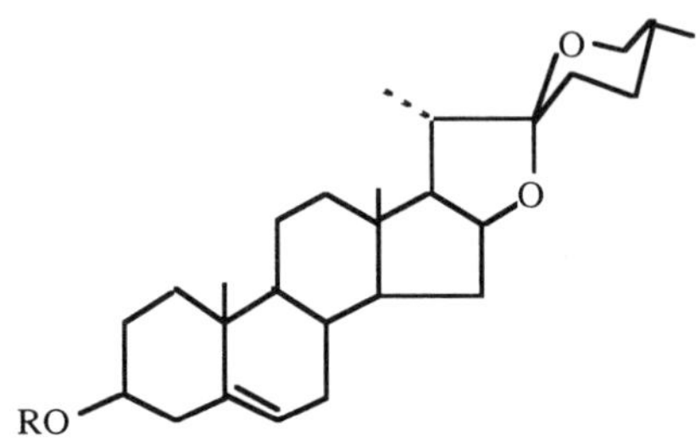

V. R = 3-*O*-β-D-Glucopyranoside-diosgenin

VI. R = 3-*O*-α-L-Rhamnopyranosyl (1→2)-[α-L-rhamnopyranosyl (1→3)-*O*-β-D-glucopyranoside-diosgenin

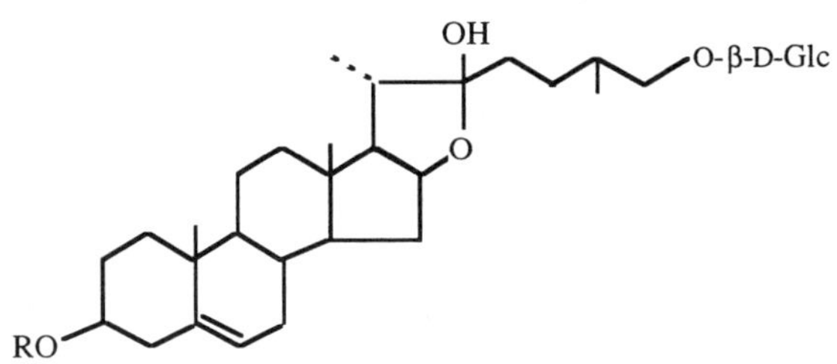

VII. R = H,26-*O*-β-D-Glucopyranoside(25*R*)-furost-5-ene 3β,22α,26-triol

VIII. R = 3-*O*-α-L-Rhamnopyranosyl (1→4)-α-*O*-β-D-glucopyranoside(25*R*)-furost-5-ene-3β,22α,26-triol-26-*O*-β-D-glucopyranoside

Fig. 2. Structures of steroid saponins **V, VI, VII,** and **VIII** isolated from *Melilotus tauricus* seeds from the Crimean region.

EXPERIMENTAL

Triterpenoid saponins: ***General remarks*****:** Paper chromatography of sugars was performed on FN-11 paper, using the solvent system 6, 1-butanol-benzene-pyridine-water (5:1:3:3). For sugar detection, aniline phthalate was used; for detection of glycosides, their acetates and aglycones, 25% phosphotungstic acid was employed with subsequent heating of the chromatogram. TLC was carried out on Silufol plates, and column chromatography on a 100-250-μm silica gel column, using the following solvent systems: 1. chloroform-methanol (4:1), 2. chloroform-methanol (7:3), 3. chloroform-methanol-water (65:35:7) for glycosides, 4. chloroform-methanol (9:1) for aglycone, and 5. benzene-acetone (4:1) for glycoside acetates.

Isolation of glycosides: Eight kilograms of ground air-dry roots of common sweet clover harvested at maturity were extracted with 70% aqueous ethanol. The resulting extract, concentrated by evaporation to 1/3 of the original volume, was successively extracted with chloroform and 1-butanol. The butanol extract was evaporated to give a total of 100 g of extracted substances. Chromatographic separation with successive application of systems 1, 2, and 3 yielded fractions rich in compounds **I, II, III,** and **IV**.

Formation of acetates of I, II, III, and IV: Saponins **I-IV** were acetylated with acetic anhydride in pyridine (1:1, 20 °C, 48 h) with subsequent dilution with water and chloroform extraction. The chloroform layer was washed and evaporated to dryness. The resulting material was loaded onto the column chromatography column and eluted with system 5 to yield: 108 mg of **Ia** ($C_{45}H_{68}O_{12}$), mp 136-138 °C; 72 mg of **IIa** ($C_{57}H_{84}O_{20}$), mp 137-140 °C; 1.9 g of **IIIa** ($C_{67}H_{98}O_{26}$), mp 148-151 °C. **IVa** [(90 mg), $C_{78}H_{110}O_{32}$] in the IR showed with ν^{KBr} discs: 1740 cm^{-1} (C=O) and absorption of OH-groups lacking.

The PMR spectrum of **IIIa** (δ, ppm, OTMS, C_5D_5N): 4.50 (*dd*, $J_{2,3}$ = 10.0 Hz, H-2'), 5.42 (*d*, $J_{1,2}$ = 7.5 Hz, H-1"), 4.42 (*dd*, $J_{2,3}$ = 10.0 Hz, H-2"), 5.55 (*dd*, $J_{3,4}$ = 3.0 Hz, H-3"), 5.86 (*d*, $J_{4,5}$ = 4.0 Hz, H-1"), 5.50 (*d*, $J_{1,2}$ = 2.0 Hz, H-1"), 5.58 (*dd*, $J_{2,3}$ = 3.0 Hz, H-2"), 5.79 (*dd*, $J_{3,4}$ = 10.0 Hz, H-3"), 5.63 (*t*, $J_{4,5}$ = 10.0 Hz, H-4"), 4.68 (*m*, H-5" '), 1.55 (H-6" '), 3.48 (H-3), 4.85 (*t*, $J_{22e,21a} = J_{22e,21e}$ = 4.0 Hz, H-22).

Deacetylation of Ia, IIa, IIIa, and IVa: 100 mg of **Ia**, 70 mg of **IIa**, 1.5 g of **IIIa**, and 1.5 g of **IVa** were dissolved respectively in 15, 10, 20, and 20 ml of 5% KOH solution of ethanol and the solutions boiled for 5 h on a water bath with a reflux condenser. The reaction mixture was diluted with water, and ethanol was evaporated. The aqueous solution was repeatedly extracted with 1-butanol saturated with water. The combined butanol extract was washed to neutral reaction, evaporated to dryness, and chromatographed on silica gel, eluting with the following systems: (1 for **II**, 2 for **IIa**, 3 for **IIIa** and 4 for **IVa**) yielding 80 mg (0.001% of the weight of air-dry raw material) of melilotoside **I** ($C_{35}H_{58}O_7$); 42 mg (0.0005% of the weight of air-dry raw material) of melilotoside **II** ($C_{41}H_{68}O_{12}$), decomposition temperature 219-220 °C (from system 2); 1.15 g (0.014% of the weight of air-dry raw material) of melilotoside **III** ($C_{47}H_{78}O_{16}$), decomp. temp. 239-240 °C (from system 3), all with ν^{KBr}, cm^{-1}: 3580-3200 (OH), 1630 (C=C); 1.15 g (0.014% of the weight of air-dry raw material) of melilotoside **IV** ($C_{52}H_{86}O_{20}$), decomp. temp. 239-240 °C, ν^{KBr}, cm^{-1}: 3580-3205 (OH),1620 (C=C).

Complete acid hydrolysis of saponins: Saponins **I, II, III,** and **IV** were dissolved in 10 ml of 5% aqueous solution of sulfuric acid and the solutions was boiled on a sand bath under a reflux condenser for 24 h. Then each reaction mixture was diluted with water and repeatedly extracted with chloroform. The combined extract was washed with water to a neutral reaction, evaporated to dryness, and chromatographed on silica gel with system 4. This yielded 10 mg of aglycone with mp 253-255 °C (from methanol). TLC analysis of the aglycone of the above compounds resulted in identification of soyasapogenol B.

The hydrolysate was neutralized with sodium carbonate, evaporated to dryness, and supplemented with a minimal volume of methanol. By using the technique of paper chromatography in system 5, the following sugars were identified: arabinose in glycoside **I**; arabinose and galactose in **II**; and arabinose, galactose, and rhamnose in **III** and **IV**.

Partial acid hydrolysis of triterpenoid saponins III and IV: 100 mg of **III** and **IV** were dissolved in 15 ml of 2.5% aqueous solution of sulfuric acid, and the solution boiled on a sand bath under a reflux condenser for 24 h. Then the reaction mixture was diluted with water and repeatedly extracted with water-saturated butanol. The combined butanol extract was washed to a neutral reaction, evaporated to dryness and chromatographed successively with systems 4, 1, and 2. This yielded 20 mg of prosapogenin (**IIIe**), and 27 mg of prosapogenin (**IIIf**), decomp. temp. 219-220 °C (from system 2).

Complete acid hydrolysis of prosapogenins: Solutions (5 mg each) of **IIIe** and **IIIf** dissolved in 5 ml of 5% aqueous sulfuric acid were boiled on a sand bath under a reflux condenser for 24 h. The reaction mixture was neutralized with barium carbonate, filtered, evaporated to dryness and the residue supplemented with a minimal volume of methanol. By using paper chromatography in system 5, the following sugars were identified in the residue: arabinose in **IIIe**, and arabinose and galactose in **IIIf**.

Steroid saponins: *General remarks*: The following solvent systems were used: 1. chloroform:methanol (8:2), 2. chloroform:methanol (7:3) for glycosides; 3. chloroform:methanol (20:1) for aglycones; 4. benzene:acetone (10:1) for methyl esters of glycosides; 5. butanol:benzene:pyridine:water (5:1:3:3) for sugars. TLC, CC (column chromatography), PC, GLC, and spectral analysis were run to identify the compounds unambiguously.

Isolation of glycosides. Combined extracts recovered by standard techniques from 800 g of cleaned and ground seeds of *M. tauricus* were chromatographed successively with systems 1 and 2 to yield fractions rich in compounds **V, VI, VII,** and **VIII**. The fractions were evaporated until dry residues were obtained which were dissolved in 5% aqueous solution of KOH and the mixture boiled for 2 h. Then the reaction mixture volume was increased to 100 ml and repeatedly extracted with butanol. The combined butanol extract was evaporated to dryness and chromatographed, elution being performed with the respective systems which yielded 21 mg of **V** (0.0029% of the weight of air-dry raw material), 58 mg of **VI** (0.0072%), 24 mg of **VII** (0.003%), and 150 mg of **VIII** (0.019%).

Complete acid hydrolysis: This was performed with 2.5% sulfuric acid for 5 h. TLC analysis of chloroform extracts of the hydrolysates allowed diosgenin to be identified. Detected in the hydrolysates, upon applying the PC technique, were glucose in compound **V**, glucose and rhamnose in **VI**, glucose in **VII**, and glucose and rhamnose in **VIII**.

Partial acid hydrolysis: This was carried out with 5% solution of sulfuric acid in methanol for 2.5 h. The reaction mixture was diluted with water, and methanol was evaporated. The aqueous residue was extracted with butanol. The extract was washed to a neutral reaction and evaporated until a dry residue was obtained which was diluted in a minimum quantity of methanol. TLC analysis enabled progenins to be identified.

Permethylation of VI and VIII. To glycosides **VI** and **VIII** (20 mg each) dissolved in 5 ml of absolutely dry DMSO, 35 mg of sodium hydride was added in small portions with constant stirring and the solution left for 1 h. Then 2 ml of methyl iodide was added drop-wise to the reaction mixture and stirring was continued for 5 h. The reaction mixture was poured into 25 ml of 2% aqueous solution of sodium hyposulfite and repeatedly extracted with chloroform. The extract was washed, evaporated to dryness and chromatographed on a column, using system 4 for elution. The recovered product was subjected to hydrolysis with 2.5% methanolic solution of sulfuric acid.

Application of the GLC technique allowed the following to be identified in the products of methanolysis: Me 3,6-di-*O*-methyl-D-glucopyranoside and 2,3,4-tri-*O*-methyl-L-rhamnopyranoside in a ratio of 0.95:2.07 for **VI** and, in addition to these two, also Me 2,3,4,6-tetra-*O*-methyl-D-glucopyranoside in a ratio of 0.95:1.06:1.89, respectively, for **VIII**.

SUMMARY

Four new triterpene saponins of the oleanan series – melilotosides **I, II, III, IV** and nonglycosylated soyasapogenol B have been isolated from roots of *Melilotus albus* Medik. (Leguminosae) and four steroid saponins were isolated from *M. tauricus* (Bieb.) ser. seeds. The chemical structure of the saponins and of their aglycone was proved based on chemical transformations and spectral data. Melilotoside **I** has the structure of 3-*O*-[a-L-arabinopyranosyl] soyasapogenol B, melilotoside **II** that of 3-*O*-[β-D-galactopyranosyl-(1→2)-*O*-α-L-arabinopyranosyl]soyasapogenol B, melilotoside **III** that of 3-*O*-[α-L-rhamnopyranosyl-(1→2)-*O*-β-D-galactopyranosyl-(1→2)-*O*-α-L-arabinopyranosyl] soyasapogenol B, and melilotoside **IV** that of 3-*O*-[α-L-rhamnopyranosyl-(1→2)-[α-L-arabinopyranosyl-(1→3)]-*O*-β-D-galactopyranosyl-(1→2)-*O*-α-L-arabinopyranosyl soyasapogenol B. The steroid saponins have been identified as 3-*O*-β-D-glucopyranoside-diosgenin (**V**); 3-*O*-α-L-rhamnopyranosyl (1→2)-[α-L-rhamnopyranosyl(1→3)-*O*-β-D-glucopyranoside-diosgenin (**VI**), 26-*O*-β-D-glucopyranoside (25*R*)-furost-5-ene 3β,22α,26-triol (**VII**), and 3-*O*-α-L-rhamnopyranosyl(1→4)-α-*O*-β-D-glucopyranoside (25*R*)-furost-5-ene 3β,22α,26-triol-26-*O*-β-D-glucopyranoside (**VIII**).

REFERENCES

1. N.I. Rubtsov, *Index of higher plants for the Crimea.* Nauka, Leningrad, p. 548 (1972) (In Russian).

2. P.D. Sokolov, *Plant resources of the USSR. The Hydrangeaceae-Haloragaceae Families.* Nauka, Leningrad, p. 159-161 (1987).

3. G.A. Kuznetsova, *Natural Coumarins and Furocoumarins.* Nauka, Leningrad, p. 246 (1967) (In Russian).

4. W.R. Akeson, H.J. Gorz, and F.A. Haskins, Effect of genotype and growth stage on distribution of melilotic acid, o-coumaric acid, and coumarinic acid in *Melilotus alba* Desr. *Crop Sci.* 3:167 (1963).

5. J.B. Harborne, *Biochemistry of Phenolic* Compounds, Translated into Russian, Moscow, p. 618 (1968).

6. A.S. Akramov, A.N. Umarov, and A.L. Markman, Oil of *Robinia pseudacacia* and *Melilotus officinalis* seeds, *Khim. Prirod. Soed.* 314 (1968) (In Russian).

7. M. Torck, L. Bezanger-Beauquesne, and M. Pinkas, Flavanoids from flowers of the legumes, *Cercia siliquastrum, Onobrychis viciaefolia*, and *Melilotus alba, Ann. Pharm. Fr.* 27:419(1969);*Chem. Abstr.* 72, 63577 (1970).

8. E.I. Kozarets, M.S. Dudkin, and I.S. Kazanskaya, Structure of *Melilotus alba* xylan,. *Khim. Prirod. Soed.* 249 (1980) (In Russian).

9. P.K. Kintia, I.M. Fadeev, and Yu.A. Akimov, *Terpenoids from plants*, Shiintsia, Chisinau, Moldovia, p. 150 (In Russian).

10. A.I. Kalinovsky, ^{13}C NMR Spectroscopy of oleanane derivatives, *Khim. Prirod. Soed.* 3 (1992) (In Russian).

11. T. Fukunaga, K. Nishiya, K. Takeya, and H. Itokawa, Studies on the constituents of goats rue (*Galega officinalis* L.) *Chem. Pharm. Bull.* 35:1610 (1987).

12. G.V. Shevchuk, Yu.S. Vollerner, A.S. Shashkov, and V.Ya. Chirva, Spirostanol and furostanol steroids from *Nolina microcarpa*. II. Structure of nolinospiroside D and nolinofurasides D, E, F. *Khim. Prirod. Soed.* 678 (1991) (In Russian).

13. V.I. Grishkovets, A.A. Loloiko, A.S. Shashkov, and V.Ya Chirva, Triterpene glycosides from *Hedera taurica*. **VI** Structure of hederosides G, H_1, H_2 and I from berries of Crimean ivy, *Khim. Prirod. Soed.* 686 (1990) (In Russian).

14. A.A. Bothner-By, R.L. Stephans, J. Lee, C.D. Warren, and R.W. Jeanlos, Structure determination of a tetrasaccharide: Transient nuclear Overhauser effects in the rotating frame, *J. Am. Chem. Soc.* 106:811 (1984).

15. R.E. Marker and J.J. Krueger, Sterols CXII. Sapogenins. XLI. The preparation of trilline and its conversion to progesterone, *J. Am. Chem. Soc.* 62:349 (1940).

16. S. Hakomori, Permethylation of glycolipids and polysaccharides catalyzed by methylsulfonyl carbanion in dimethyl sulfoxide. A rapid permethylation of glycolipids and polysaccharides catalyzed by methyl sulfonyl carbonion dimethylsulfoxide, *J. Biochem.* (Tokyo) 55:205 (1964).

17. L.I. Strigina and V.V. Isakov, ^{13}C NMR spectra of steroidal glycosides. glycosides of pennogenin, *Khim. Prirod. Soed.* 474 (1982).

A NOVEL TRITERPENOID SAPONIN FROM THE LEAVES OF *MAJORANA HORTENSIS* (MOENCH)

R.N. Yadava

Natural Products Laboratory
Department of Chemistry
Dr. H.S. Gour University
SAGAR 470 003 (M.P.) India

INRODUCTION

Majorana hortensis Moench (Labiatae, common name : 'murwa' in Hindi), commonly grown in southern Europe, North Africa, and India, is widely used as an indigenous drug for treating asthma, hysteria, and paralysis[1]. A literature survey reveals the presence of different flavones in the leaves of the plant[2,3]. The present paper deals with the isolation and structural elucidation of a new triterpenoid saponin, provisionally named murwaoside (**1**).

RESULTS AND DISCUSSION

The chloroform extract of the leaves of *Majorana hortensis* gave compound **1**, $C_{46}H_{74}O_{17}$, mp 226°. The ^{1}H and ^{13}C-NMR spectra (Table 1) showed signals for seven methyl groups : ^{1}H-NMR signals at δ 0.94 (6H, s,2 x CH_3), 0.96 (6H, s, 2 x CH_3), 1.05, 1.20, 1.23 (each 3H, s, 3 x CH_3) and ^{13}C-NMR signals at 33.4, 27.1, 29.64, 23.8, 18.46, 17.4 and 16.2 ppm. The spectra also exhibit signals for a carboxylic group (ν^{max}_{KBr} cm^{-1} : 1715, 1792), a double bond (ν^{max}_{KBr} cm^{-1} :1240, a proton signal at δ 5.32, dd and δ 122.6) secondary alcoholic groups (ν^{max}_{KBr} cm^{-1} 3445, δ 69.52 and 80.41) and downfield shift at δ 2.04 (2H, dd, J = 4.0 Hz, H-11) and 2.51 (1H, t, J = 4.9, 13.5 Hz, H-18β) characteristic of an olean - 12-en-28-oic acid type triterpene. Three anomeric protons observed at δ 4.29, 4.32, and 4.35 indicated the presence of three sugar units.

Compound **1** on acetylation yielded a decaacetate (**3**). On hydrolysis with EtOH/ H_2SO_4, compound **1** furnished aglycone (**2**) and sugar moieties as *D*-arabinose, *D*-glucose, and *L*-arabinose.

Compound **2** by various chemical degradations yielded **2a**, **2b** and **2c**; their structure showed the presence of two OH groups. The ^{1}H-NMR spectrum of **2c** exhibited signals at δ 5.12 (1H, dd, J = 10.0 Hz, H-2β), 4.74 (1H, d, J = 10.0 Hz, H-3α), 2.05 (3H, s, α-OAc), and 1.96 (3H, s, β-OAc). These values are in very close agreement with those expected for a 2α, 3β- diol system in the aglycone[4-6]. Observation of fragment ions at *m/z* 765 and 603 in the positive FABMS of compound **1** indicated the presence of a terminal *L*-arabinose unit, and fragment ions of **2** from retro-Diels-Alder cleavage at *m/z* 248 and 203 were assigned as typical of Δ^{12}-pentacyclic triterpenes. Aglycone **2** was identified as 2α, 3β–dihydroxyolean-12-en-28-oic acid by direct comparison (mp, mmp, IR and ^{1}H-NMR and ^{13}C-NMR spectra) with the literature[7-9]. Compound **2** gave a positive test for acid but **1** did not, strongly indicating that the series of sugar untis were linked to the 28-carboxylic group in ester. form[10-11].

Compound **1** on partial hydrolysis with Kiliani mixture afforded two saponins (**4** and **5**). Acid hydrolysis of compound **4** yielded **2**, *D*-arabinose, and *D*-glucose while compound **5** yielded **2** and *D*-arabinose. Permethylation of **1**,**4** and **5** yielded **6**,**7** and **8**; this followed by hydrolysis liberated [3,4-di-*O*-methyl-*D*-arabinose, 2,3,4-tri-*O*-methyl-*D*-glucose, and 2,3,4-tri-*O*-methyl-*L*-arabinose], [3,4-di-*O*-methyl-*D*-arabinose and 2,3,4,6-tetra-*O*-methyl-*D*-glucose] and [2,3,4-tri-*O*-methyl-*D*-arabinose] respectively, indicating that C-1 and C-2 of *D*-arabinose, C-1 and C-6 of *D*-glucose, and C-1 of *L*-arabinose are involved in the glycosidic linkage and were in pyranose form [12,13].

Compound **1** on hydrolysis with the enzyme Taka-diastase liberated *L*-arabinose, confirming the presence of the α-configuration between *L*-arabinose and *D*-glucose; further enzymatic hydrolysis of **1** with almond emulsion produced *D*-glucose and *D*-arabinose, confirming the presence of the β-configuration between *D*-glucose and *D*-arabinose as well as between *D*-arabinose and the aglycone in the glycosidic linkage.

Thus, the compound **1** was identified as 2α, 3β-dihydroxyolean-12-en-28-oic-acid 28-*O*-[α-*L*-arabinopyranosyl-(1→6)-*O*-β-*D*-glucopyranosyl-(1→2)-*O*-β-*D*-arabinopyranosyl] ester.

EXPERIMENTAL

General Experimental Procedures

All melting points were uncorrected. Optical rotations were measured on a Perkin-Elmer 241 polarimeter. The ^{1}H-NMR spectrum was measured on a Varian EM-360 L instrument at 100 MHz and ^{13}C-NMR was measured on a Varian CVT-20 at 25 MHz with TMS as internal standard. The IR spectra were recorded on a Perkin Elmer 577 instrument and EIMS (70 eV) were taken with a direct inlet on JEOL D-300 spectrometer. Column chromatography was performed on silica gel (Merck, 60-120 mesh), and tlc on kieselgel (Merck, 60 G); spots on tlc were visualized with 7% H_2SO_4 followed by heating. Paper chromatography was carried out on Whatman No 1 filter paper using the descending mode and visualized with anline hydrogen phthalate.

Plant Material

The plant material of *Majorana hortensis* was procured from M/s United Chemical and Allied products, Calcutta, identified by the Department of Botany of this University, and deposited in the herbarium of the Chemistry Department.

Extraction and Isolation of Compound 1

The defatted leaf powder (2 kg) was extracted exhaustively with $CHCl_3$ in a Soxhlet appartus, the extract evaporated, and the residue (58.3g) chromatographed through silica gel (120-200 mesh, 800g) column. Eution with $CHCl_3$: EtOAc 1:1(2 x 500 ml) afforded 10.05g of product. On tlc Examination it was found to be a homogeneous, light brown crystalline solid compound **1**.

Compound 1

Compound **1** gave a positive test for triterpenes; it was soluble in ether and pyridine, mp 226°, anal. calcd. $C_{46}H_{74}O_{17}$, C=62.25, H = 8.25%; found C = 62.27, H = 8.23%; $[\alpha]_D^{25}$ + 15.8° (c = 1.0, C_5H_5N); IR $\nu_{KBr}^{max}cm^{-1}$: 3492, 3025, 2962, 2902, 2842, 1720, 1662, 1420, 1378, 1240 and 803; ^{13}C-NMR (pyridine-d_5,25MHz) (see Table 1); FABMS *m/z* 898, 765, 749, 603, 587, 455, 439, 411, 248, 223, 206, 203, 192, 189, and 133.

Acetylation of 1

Compound **1** (50mg) dissolved in 50% MeOH (10-15 ml) was treated with Ac_2O (4ml) in C_5H_5N (4ml) at room temperature for 12h to furnish the decaacetate **3**, $C_{86}H_{114}O_{37}$, mp 287-289° $[\alpha]_D^{25}$ - 3.6° (c = 3.98, MeOH), ^1H-NMR spectrum : δ 0.95 (6H, s, 2 x CH_3), 0.97 (6H, s, 2 x CH_3), 1.05, 1.20, 1.23 (each 3H, s, 3 x CH_3), 1.96 (3H, s, β-OAc, C-3), 2.05 (3H, s, α-OAc, C-2), 1.27-2.01 (2OH, m, polymethylene envelop CH_2 and CH), 2.04 (2H, dd, J = 4 Hz, C-11), 2.51 (1H, t, J = 4.0 Hz, C-18), 4.74 (1H, d, J = 10 Mz, C_3-αH), 5.12 (1H, dd, J = 10 Hz, C_2-βH), 5.32 (1H, dd,C-12-vinylic protons), 4.29, 4.32, 4.35 (each 1H, d, J = 7.5, 7.8, 7.9Hz, 3 anomeric protons - C_1', C_1'', C_1'''), 4.67-4.82 (5H, m, C_2'-H, C_3'-H, C_4'-H, C'_5-2H), 4.69-4.84 (6H, m, C_2''-H, C_3''-H, C_4''-H, C''_5-H, C_6''-2H), 4.71-4.87 (5H, m, C_2'''-H, C_3'''-H, C_4'''-H, C'''_5-2H), 2.10, 2.14 (each 3H, s, C_3'-OAc, C_4'-OAc), 2.04, 2.02, 2.09 (each 3H, s, C_2''-OAc, C_3''-OAc, C_4''-OAc), 2.08, 2.13, 2.16 (each 3H, s, C_2'''-OAc, C_3'''-OAc, C_4'''-OAc).

Acid Hydrolysis of 1

Compound **1** (100mg) was dissolved in MeOH (25ml) / 7% H_2SO_4 (0.5ml) and the solution refluxed for 3 h on a water bath. The mixture was diluted with ice cold H_2O and extracted with $CHCl_3$. The $CHCl_3$ layer was evaporated to dryness and the residue chromatographed over silica gel with $CHCl_3$: MeOH (2:1) as solvent to give compound **2**.

Indentification of Sugars

The aqueous layer from the hydrolysis was filtered, neutralized with $BaCO_3$, freed of $BaCO_3$ and $BaSO_4$ by filtration, concentrated, and subjected to pc examination using BuOH : AcOH : H_2O (4:1:5 v/v); this showed *L*-arabinose, *D*-glucose, and *D*-arabinose (R_f values 0.21, 0.17, and 0.23).

Compound 2

This compound gave positive test for triterpene (i.e. the Lieberman-Burchard test and the Salkowski test), mp 290°, $[\alpha]_D^{25}$+32° (c = 1.0, $CHCl_3$), anal. calcd. $C_{30}H_{48}O_4$, C = 76.23, H = 10.24; found C = 76.16, H = 10.21 %; M^+ at *m/z* 472; UV (MeOH) λ^{max} 226 nm; IR $\nu_{KBr}^{max}cm^{-1}$; : 3445, 2950, 2892, 1715, 1628, 1340, 1240, 1030; ^{13}C-NMR see Table 1.

Compound 2a

Compound **2** was dissolved in Et_2O and treated with diazomethane; the solid obtained after the usual workup procedure was chromatographed on silica gel. Elution with AcOH : MeOH (1:1) gave a solid which on recrystallization from $CHCl_3$ furnished **2a,** mp

Table : 1. ^{13}C-NMR chemical shifts (25 MHz, pyridine-d_5, TMS as internal standard)

No. of carbon	Compound 1	Compound 2	Compound 2-C	No. of carbon	Compound 1	Compound 2	Compound 2-C	Sugar residue No.of carbon	Sugar residue 1
C-1	41.6	41.62	43.25	C-16	23.6*	23.5*	23.6*	C'-1	101.8
C-2	69.23	69.3	69.54	C-17	45.9	45.9	46.8	C'-2	78.9
C-3	79.10	79.10	80.42	C-18	42.8	42.8	42.9	C'-3	72.4
C-4	39.24	39.25	39.25	C-19	46.5	46.5	43.3	C'-4	67.4
C-5	54.35	54.35	55.34	C-20	31.02	31.02	31.10	C'-5	65.5
C-6	18.54	18.57	18.67	C-21	38.6	33.7	33.8	C''-1	105.6
C-7	35.36	35.37	35.36	C-22	32.6*	32.6*	32.7*	C''-2	75.6
C-8	38.4	39.6	39.6	C-23	29.64	29.64	29.66	C''-3	78.1
C-9	47.5	47.6	47.6	C-24	17.46	18.51	17.46	C''-4	71.6
C-10	36.7	36.6	36.7	C-25	17.40	17.21	16.47	C''-5	76.0
C-11	23.7*	23.2*	23.6*	C-26	16.20	16.20	17.36	C''-6	68.5
C-12	122.6	122.4	122.5	C-27	27.10	27.10	26.12	C'''-1	105.2
C-13	145.6	145.7	145.7	C-28	179.2	179.3	180.1	C'''-2	72.4
C-14	42.2	42.2	44.1	C-29	33.4	33.41	33.22	C'''-3	74.2
C-15	28.5	28.5	28.4	C-30	23.80*	23.80*	23.8*	C'''-4	66.9
				C-2αOAc			170.59	C'''-5	66.4
				C-3βOAc			170.33		

*The assignment in any vertical column may be interchanged

Table : 2. ^{13}C-NMR chemical shifts (25 MHz, pyridine-d_5, TMS as internal standard)

No. of	Compound		No. of	Compound		No.of	Sugar residue	
carbon	**4**	**5**	carbon	**4**	**5**	carbon	**4**	**5**
C-1	41.6	41.6	C-16	23.7*	23.5*	C'-1	101.7	101.6
C-2	69.12	69.11	C-17	45.8	45.7	C'-2	78.9	78.8
C-3	79.11	79.12	C-18	42.8	42.9	C'-3	72.4	73.3
C-4	39.24	39.24	C-19	46.5	46.51	C'-4	67.4	67.5
C-5	54.35	54.35	C-20	31.03	31.05	C'-5	65.49	65.5
C-6	18.55	18.55	C-21	38.6	38.60	C"-1	105.7	
C-7	35.36	35.34	C-22	32.7*	32.69	C"-2	75.6	
C-8	38.38	38.37	C-23	29.63	29.64	C"-3	78.11	
C-9	47.49	47.49	C-24	17.47	17.43	C"-4	71.6	
C-10	36.9	36.9	C-25	17.38	17.39	C"-5	76.01	
C-11	23.6*	23.4*	C-26	16.21	16.22	C"-6	68.52	
C-12	122.6	122.7	C-27	27.09	27.09			
C-13	145.6	145.59	C-28	179.2	179.2			
C-14	42.3	42.31	C-29	33.4	33.4			
C-15	28.4	28.4	C-30	23.81*	23.83*			

* The assignment in any vertical column may be interchanged.

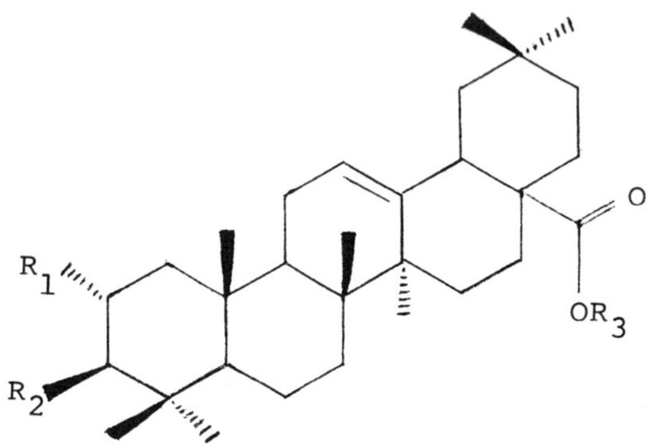

	R1	R2	R3
(1)	OH	OH	*D*-Arab (2→1)-*D*-Glu (6→1)-*L*-Arab
(2)	OH	OH	H
(2a)	OH	OH	Me
(2b)	O	O	Me
(2c)	OAc	OAc	Me
(3)	OAc	OAc	*D*-Arab (2→1)-*D*-Glu (6→1)-*L*-Arab octaacetate
(4)	OH	OH	*D*-Arab (2→1)-*D*-Glu
(5)	OH	OH	*D*-Arab
(6)	OMe	OMe	3,4-di-*O*-Me-*D*-Arab(2→1)-2,3,4-tri-*O*-Me-*D*-Glu (6→1)-2,3,4-tri-*O*-Me-*L*-Arab
(7)	OMe	OMe	3,4-di-*O*-Me-*D*-Arab (2→1)-2,3,4,6-tetra-*O*-Me-*D*-Glu
(8)	OMe	OMe	2,3,4-tri-*O*-Me-*D*-Arab

253°, $[\alpha]_D^{25}$ + 14° (*c* = 0.98, $CHCl_3$), M^+ at *m/z* 486, and calcd. $C_{31}H_{50}O_4$, C = 76.50, H = 10.35; found C = 76.48, H = 10.32%, identical with an authentic sample (mmp, co-tlc, co-IR).

Compound 2b

Compound **2a** (60mg) in C_5H_5N (1ml) was treated with Cro_3 / C_5H_5N (10ml) below 5° for 12h. After the usual workup the solid was chromatographed on silica gel [eluants C_6H_6 : petroleum ether (40-60); 2:1], furnishing diketone **2b,** mp 179-181°, M^+at *m/z* 482, anal. calcd. $C_{31}H_{46}O_4$, C = 77.14, H = 9.61; found C = 77.11, H=9.64%; $[\alpha]_D^{25}$ + 16° (*c* = 0.64, $CHCl_3$); UV (EtOH) λ^{max} 274 nm; IR ν_{KBr}^{max} cm^{-1} : 3442 (OH), 1732 (COOMe), 1662, 1642 (α, β-unsaturated ketone), 1240 (>C=C<), identical with an authentic sample by mmp and co-tlc.

Compound 2c

Compound **2a** (20mg) was treated with Ac_2O/C_5H_5N 1:1 (20.0 ml) at room temperature for 5h. The reaction mixture was diluted Et_2O (30ml), washed successively with 5% H_2SO_4 (x2), 5% $NaHCO_3$ (x2) and brine (x2) and evaporated. The residue chromatographed on silica gel [elution with EtOAc : petroleum ether (40-60); 1:1] furnished diacetate **2c,** mp 166°; anal. calcd. $C_{35}H_{54}O_6$, C = 73.65, H = 9.54; found C = 73.61, H = 9.57%; $[\alpha]_D^{25}$ + 13° (*c* = 0.92, $CHCl_3$); M^+ at *m/z* 570; ^{13}C-NMR see Table 1.

Partial Hydrolysis of 1

Compound **1** (100mg) was treated with Kiliani mixture (75 ml, HCl : AcOH : H_2O; 15:35:50) at room temperature for 12h; extraction with BuOH yielded **4** and **5** (approx. 40 mg each).

Compound 4

mp 214°, anal. calcd. $C_{41}H_{66}O_{13}$, C = 64.09, H = 8.62; found C = 63.95. H = 8.63%, M^+ at *m/z* 766; ^{13}C-NMR see Table 2.

Compound 5

mp 199.4°, anal. calcd. $C_{35}H_{56}O_7$, C = 71.73, H = 9.65; found C = 71.67, H = 9.62%; M^+ at *m/z* 588; ^{13}C-NMR see Table 2.

Permethylation and Hydrolysis of 1,4 and 5

Compounds **1,4** and **5** (30 mg each separately) in MeI (2ml) / Ag_2O (50 mg) / DMF (4 ml) were held at room temperature for 24h, diluted with H_2O, and extracted with $CHCl_3$ (25ml) to yield **6,7** and **8** respectively.

Compound **6, 7** and **8** were hydrolyzed with Kiliani mixture separately; the reaction mixtures were filtered, the filtrates neutralized with $BaCO_3$, freed of $BaCO_3$ and $BaSO_4$ by filtration, evaporated, and the residues chromatographed on silica gel with BuOH : EtOH : H_2O (4:1:5) as eluants. The methylated sugars were identified by co-pc and co-tlc[14-16]

Enzymatic Hydrolysis of 1

Compound **1** (50 mg) was suspended in enzyme Taka-diastase at 35° for 24h. The hydrolysate on paper chromatography using BuOH : AcOH : H_2O (4:1:5) as eluants and aniline hydrogen phthalate as spraying reagent showed one spot corresponding to *L*-arabinose (R_f 0.21). The residue treated with almond emulsion (30 ml) as described above showed two spots, corresponding to *D*-glucose and *D*-arabinose (R_f 0.17 and 0.23).

Acknowledgements

Thanks are due to the Director, C.D.R.I., Lucknow, for spectral analysis and Prof. S.P. Banerjee, Head, Department of Chemistry, Dr.H.S.Gour University, Sagar (M.P.) India, for providing the necessary laboratory facilities. Thanks are also due to Prof. V.K. Saxena, Department of Chemistry, Dr. H.S.Gour University, Sagar (M.P.) India, for his keen interest and encouragement during this investigation.

REFERENCES

1. R.N. Chopra, S.L. Nayar, and I.C. Chopra, *Glossary of Indian Medicinal Plants*, CSIR, New Delhi, p. 242 (1985).
2. S.S. Subramanian, A.G.R. Nayar, E. Rodriguez, and T.J. Mabry, Polyphenols of the leaves of Majorana hortensis, *Curr. Sci.* 41.202 (1972)
3. B. Voirin, J. Favre-Bonvin, V. Indra, and A.G.R. Nayar, Structural revision of the flavone majoranin from Majorana hortensis, *Phytochemistry* 23:2973 (1984).
4. G.K. Narayan, I.R. Row, and C. Suryaprakash Shastri, Chemical examination of the leaves of Barringtonia aintangula Gaerth, *Curr.Sci.* 45:518 (1976).
5. R. Higuchi, T. Kawasaki, M. Bishwas, V.B. Pandey, and B. Dasgupta, Triterpenoid saponins from the stem bark of Symplocos spicata, *Phytochemistry* 21: 907 (1982).
6. B. Saha, D.B. Naskaer, D.R. Mishra, B.P. Pradhan, and H.N. Khastigir, Baccatin, a noble nor-triterpene peroxide isolated from Sapium baccatum Roxb., *Tetrahedron Lett.* 3095 (1977)
7. E.P.Kemertelidze, L.N. Gvazava, M.D. Alania, V.S. Kikoladze, and V.G. Tsitsishvili, Digitoside, a novel triterpene glycoside from Digitalis ciliata., *J. Nat. Prod.* 55 : 217 (1991)
8. T.K. Chen, D.C. Ales, N.C. Baenziger, and D.F. Wiemer, Ant repellant triterpenoids from Cordia alliodora, *J.Org.Chem.* 48 : 3535 (1983).
9. T.N. Misra, R.S. Singh, T. Upadhyay, and R. Shrivastava, Chemical constituents of Vernonia cinerea Part I Isolation and spectral studies of triterpenes, *J.Nat.Prod.* 47:368 (1984).
10. Y.Oshima, T. Ohsava, and H. Hikino, Part 55 in the series on the validity of the oriental medicines. Structure of dianosides C, D, E and F triterpenoid saponins of Dianthus superbus var. longicalycinus herb., *Planta Med.*, 50:43 (1984)
11. J.S. Choi and D.S. Woo, Triterpenoid glycosides from the roots of Patrinia scabiosaefolia, *Planta Med.* 53:62 (1987).
12. O.P.Sati, U. Rana, D.C. Chaukiyal, and M. Sholichin, A new spirostanol glycoside from Agave cantala, *J. Nat. Prod.* 50:263 (1987).
13. O.P.Sati and G.Pant, Spirostanol glycosides from Asparagus plumosus, *Phytochemistry* 24:123 (1985).
14. S. Seo, Y. Tomita, K. Tori, and Y. Yoshimura, Determination of the absolute configuration of a secondary hydroxy group in a chiral secondary alcohol using glycosidation shifts in carbon-13 Nuclear Magnetic Resonance spectroscopy, *J.Am.Chem.* Soc. 100:3331 (1978).
15. S.K.Uniyal, V. Badoni, and O.P.Sati, A new triterpenoidal saponin from Acacia auriculoformis, *J.Nat.Prod.* 55 : 500 (1992).
16. D.R. Daryl and H.N. Roger, Norolenane saponins from Celmisia petriei, *Phytochemistry* 23 : 639 (1984).

CHEMILUMINESCENCE OF OXYGEN RADICAL SCAVENGERS SUCH AS DDMP SAPONINS IN THE PRESENCE OF RADICALS AND ALDEHYDE

Yumiko Yoshiki[1], Kazuyoshi Okubo[1], and Kiharu Igarashi[2]

[1]Faculty of Agriculture, Tohoku University
1-1, Tsutsumidori, Amamiyamachi, Aoba-ku, Sendai-shi, Japan 981
[2]Faculty of Agriculture, Yamagata University
1-23, Wakabamachi, Tsuruoka-shi, Japan 997

INTRODUCTION

Many kinds of saponins have been isolated from soybean seeds, and can be divided into two groups, group A[1] and DDMP (2,3-dihydro-2,5-dihydroxy-6-methyl-4*H*-pyran-4-one) saponins[2] (Fig.1). Group A saponins are present only in the soybean hypocotyl, whilst DDMP saponins are widely distributed in leguminous seeds and are present in both hypocotyl and cotyledon, especially in the case of soybean. From the pharmacological or health point of view, it has been reported that soybean saponins show various properties such as hypolipidemic and antioxidative activity[3] , and HIV-infection-inhibitory properties[4]. Although both group A and DDMP saponins are present in the hypocotyl of soybean seeds at high concentration, the interrelationship concerning their physiological role is not clear.

Recently, we reported that the phenolic compounds such as flavonoids and anthocyanins, which are known natural radical scavengers, exhibit weak light emission (chemiluminescence; CL) via a non-enzymatic reaction at room temp. and neutral pH value in the presence of active oxygen species and acetaldehyde (MeCHO)[5,6]. CL intensity of these compounds was in good agreement with the radical scavenging activites determined by using an enzymatic method[7] or an HPLC-electrochemical method[8]. In the present study, we examined the active oxygen- or radical-scavenging activity, and photon constants of eight kinds of catechins and gallic acid by measuring their CL intensity in the presence of MeCHO and hydrogen peroxide (H_2O_2), hydroxyl radical (HO·), *tert*-butyl hydroperoxide (*t*-BuOOH) or it's derived radical (*t*-BuO·). In addition, this paper describes interrelationships between group A saponins and DDMP saponins based on the CL exhibited in the presence of active oxygen species.

group A saponins

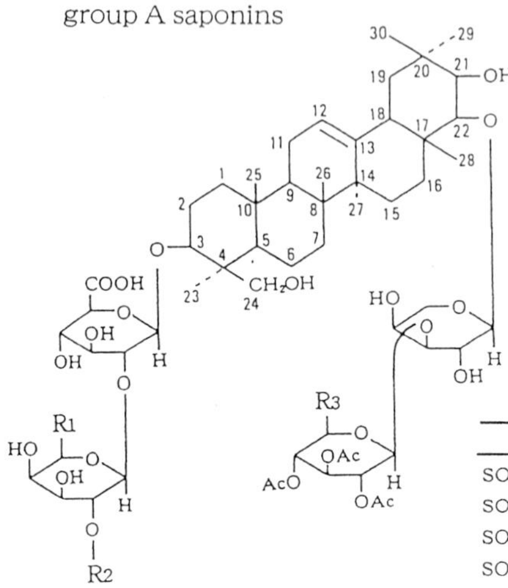

	R1	R2	R3
soyasaponin Aa	CH_2OH	β-D-Glc	H
soyasaponin Ab	CH_2OH	β-D-Glc	CH_2OAc
soyasaponin Ac	CH_2OH	α-L-Rha	CH_2OAc
soyasaponin Ad	H	β-D-Glc	CH_2OAc
soyasaponin Ae	CH_2OH	H	H
soyasaponin Af	CH_2OH	H	CH_2OH
soyasaponin Ag	H	H	H
soyasaponin Ah	H	H	CH_2OH

DDMP saponins

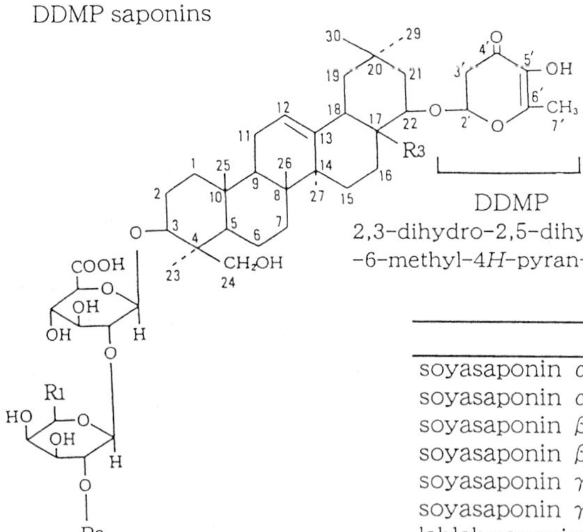

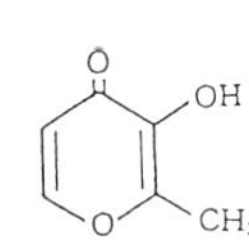

maltol

	R1	R2	R3
soyasaponin αg	CH_2OH	β-D-Glc	CH_3
soyasaponin αa	H	β-D-Glc	CH_3
soyasaponin βg	CH_2OH	α-L-Rha	CH_3
soyasaponin βa	H	α-L-Rha	CH_3
soyasaponin γg	CH_2OH	H	CH_3
soyasaponin γa	H	H	CH_3
lablab saponin I	CH_2OH	β-D-Rha	CHO

Fig.1 Structures of soybean saponins and DDMP saponins

MATERIALS AND METHODS

Chemicals and reagents

Gallic acid was purchased from Nakarai Co.Ltd. (Kyoto, Japan), and *tert*-butyl hydroperoxide and hydrogen peroxide from Katayama Chem. (Tokyo, Japan) and Santoku Chem. (Tokyo, Japan), respectively. (-)-Catechin (C), (-)-epicatechin (EC), (-)-gallocatechin (GC), (-)-epigallocatechin (EGC), (-)-catechin gallate (CG), (-)-epicatechin gallate (ECG), (-)-gallocatechin gallate (GCG) and (-)-epigallocatechin gallate (EGCG) of *Camellia sinensis* were purchased from Kurita Co.Ltd. (Tokyo, Japan). Gallic acid and catechins were solubilized in 50% MeOH by sonification just before use.

Soyasaponin Ab and soyasaponin βg were extracted from hypocotyls (500g) of soybean with 70% ethanol, containing 0.01% EDTA and centrifuged at 8×10^3 rpm for 15 min. The extract was evaporated and dissolved in H_2O : 1-BuOH (1:1, v/v), the 1-BuOH layer evaporated, and the soyasaponin Ab and DDMP saponin fraction was isolated as described previously[1,9]. Soyasaponin Ab and soyasaponin βg were solubilized in 50% MeOH by sonification just before use.

Measurement of chemiluminescence

The chemiluminescence (CL) was measured with a filter-equipped photon-counting-type spectrophotometer (CLD-110, Tohoku Electronic Ind.), connected to a Waters Model 510 pump and U6K injector. The dispersed light at the grating was simultaneously detected on the photocathode with the image sensor. The number of photons counted in the wavelength between 300 and 650 nm was computed as total spectral intensities.

Measurement of antioxidative activity

To investigate antioxidative activity in the XYZ system, gallic acid and MeCHO were used as Y and Z, respectively. The reaction mixture contained 2.51% linoleic acid, 50mM phosphate buffer (pH 7.0), and various concentrations of gallic acid and MeCHO (total 5ml) reacted at 37 ℃. The oxidation ratio of linoleic acid was determined by ferric thiocyanate method[10].

RESULTS AND DISCUSSION

The chemiluminescence (CL) of catechins was measured in the presence of active oxygen species and 2% MeCHO. (-)-Epigallocatechin (EGC) exhibited the strongest CL among the eight kinds of catechins (EGC > EGCG > GC > GCG > EC > C > ECG > CG > gallic acid). When the reaction mixture composed of EGC, MeCHO and H_2O_2 was subjected to HPLC and LC-MS, it was comfirmed that EGC was almost not lost during the reaction, EGC itself did not show CL. Therefore, this reaction may proceed via a route similar to that described Gotoh and Niki[11].

active oxygen + phenolic compound → phonolic compounds· or -OOH 1)
phenolic compound· + acetaldehyde → phenolic compound + products +light 2)

CL intensity of the reaction mixture was dependent on the concentrations of H_2O_2 (X; active oxygen species), phenolic compound (Y; catalytic species) and MeCHO (Z; receptive species) (Fig.2).

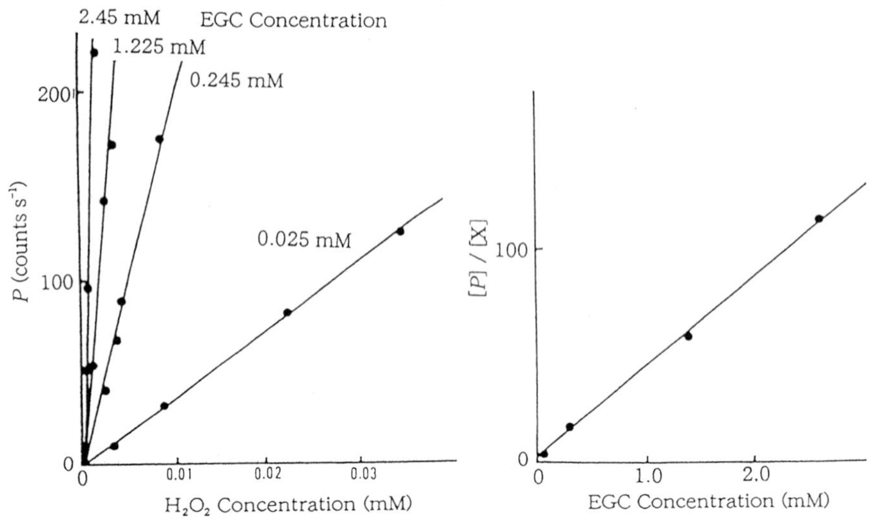

Fig.2 Effect of EGC (Y; catalytic species) and H_2O_2 (X; active oxygen species) concentration on photon intensity (*P*) in the presence of MeCHO (Z; receptive species)

EGC; (-)-epigallocatechin, MeCHO; acetaldehyde

We propose that the CL intensity, *P*, in the presence of H_2O_2 and MeCHO is given by [*P*] = *k*[X][Y] f(z). The calculated photon constants (log *k*) are summarized in Table 1. Log *k* in the presence of HO· and *t*-BuO· was calculated from H_2O_2 and *t*-BuOOH concentrations. In the presence of H_2O_2 as X, photon constants of catechins were closely related to the characteristics of their chemical structures.

Table 1. Photon constants catechins or gallic acid (Y; catalytic species) in the presence of acetaldehyde (Z; receptive species) and active oxygen species (X), in $M^{-2}s^{-1}$ counts at pH 7.0 and 23℃

	Substituent at		log *k* (*k* ; photon constant)			
Name	C-3	C-5	HOOH	HO·	*t*-BuOOH	*t*-BuO·
(-)-Catechin	OH	H	6.52	6.38	5.98	5.72
(-)-Epicatechin	OH	H	7.42	6.50	6.25	6.20
(-)-Gallocatechin	OH	OH	7.99	6.39	6.93	7.46
(-)-Epigallocatechin	OH	OH	10.64	7.12	6.90	7.73
(-)-Catechin gallate	O-Ga	H	6.46	5.85	5.80	6.23
(-)-Epicatechin gallate	O-Ga	H	6.85	5.69	5.79	6.26
(-)-Gallocatechin gallate	O-Ga	OH	7.39	5.98	6.72	6.27
(-)-Epigallocatechin gallate	O-Ga	OH	9.41	5.51	6.42	6.77
Gallic acid			6.38	6.84	5.46	6.71

Ga; galloyl group

C-3 and C-5' are substituted as follow

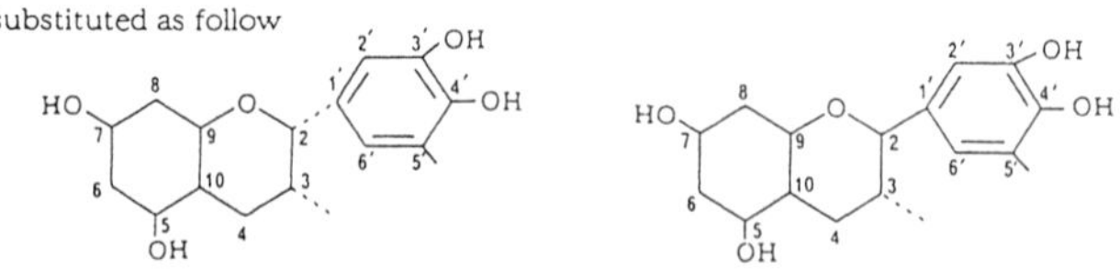

Photon constants were increased by the presence of following partial structures: (a) the pyrogallol structure rather than the catechol structure in the B-ring as described by Bors *et al*[12], (b) a free hydroxyl group at C-3, rather than its esterified form, (c) the stereoscopic structure between the C-3 hydroxyl group and B-ring, which is epimeric with respect to carbon atom 2. From the CL obtained with catechins, it is suggested that the active oxygen species were scavenged by the X (active oxygen species); Y (catalytic species); Z (receptive species) system with a relationships described by the equation; $[P] = k[X][Y]f(z)$ (k : photon constant). The present study suggested that photon constants of catechins, which represent radical scavenging activity of catechin, can be calculated from, [X] (concentration of active oxygen or radical species), [Y] (concentration of catalytic species) and [Z] (concentration of receptive species).

On the basis of the CL exhibited in the presence of active oxygen species, interrelationships between group A saponins and DDMP saponins were investigated. The polymorphisms of group A saponin composition existing in soybean seeds are controlled by the same gene and the presence of both soyasaponin Aa and Ab as the main saponins of group A can be explained by co-dominant acting genes[13]. Soybean hypocotyl contains soyasaponin Aa or Ab and soyasaponin βg at relatively high concentration of 1-5%[14].

H_2O_2, soyasaponin Ab and soyasaponin βg in 50mM phosphate buffer (pH 7.0) did not exhibit CL. The latter two compounds as a mixture also did not exhibit CL. To study the role of group A and DDMP saponins on CL, CL of soyasaponin Ab and soyasaponin βg was measured in the presence of H_2O_2 (X) and MeCHO (Z) in which phenolic compounds such as flavonoids, anthocyanins and catechins showed the strongest CL as factor Y. Although soyasaponin βg did not exhibit significant CL, soyasaponin Ab showed CL of about one tenth when compared with that of EGC, which exhibited the highest CL among phenolic compounds. CL of soyasaponin Ab and soyasaponin βg was also measured in the presence of H_2O_2 (X) and EGC (Y). Although soyasaponin βg gave a stronger CL than MeCHO, soyasaponin Ab only exhibited a slight CL. These results indicate that soyasaponin Ab and soyasaponin βg act as factors Y and Z, respectively. Soyasaponin βg is unstable on heating and in basic media is easily converted to soyasaponin I and maltol. Maltol exhibited CL as factor Z in the presence of H_2O_2 and EGC, but the CL intensity of 1mM maltol (0.5 counts s^{-1}) was low when compared with that of 1mM soyasaponin βg (1×10^3 counts s^{-1}).

CL of soyasaponin βg was higher than that of MeCHO by 10 fold as factor Z in the presence of the EGC and H_2O_2. This result suggests that CL is influened by the kind of factor Z and that the combination of Y and Z factors is important for CL. CL of soyasaponin Ab and soyasaponin βg in the presence of H_2O_2 was measured to clarify the interrelationships between group A and DDMP saponins (Fig.3). When CL of soyasaponin Ab [Y; catalytic species] was measured in the presence of H_2O_2 [X; active oxygen species] and soyasaponin βg [Z; receptive species], photon intensity [*P*] increased linearly depending on the concentrations of H_2O_2 and of soyasaponin Ab. This result indicates that CL of soybean saponins was emitted from XYZ system; $[P] = k[X][Y]f(z)$. Photon intensity of soyasaponin βg in the presence of soyasaponin Ab against H_2O_2 was about 100 times as much as MeCHO.

CL occurs in the presence of X (active oxygen species), Y (catalytic species), and Z (receptive species). We found that soybean saponin played a role as Y or Z. It has been reported that group A saponins of soybean have antioxidative activity and may act as catalytic factors Y, because phenolic compounds such as flavonoids, anthocyanins and catechins which set as factor Y in the presence of X and Z also have antioxidative activity. Since a mixture of H_2O_2 and soyasaponin Ab does not exhibit CL, it is thought that CL occurs in two stages; (1) electron exchange to produce an excited state of soyasaponin Ab (Y' ·), since H_2O_2 is not a radical, and CL increased rapidly in the presence of the hydroxyl radical (HO·) by the Fenton reaction, and (2) electron exchange to ground state (Y) in the presence of Z. A clear difference in the role for CL reaction of H_2O_2 between soyasaponin Ab and soyasaponin βg seems to be explained by the structural difference found in those saponins because the sugar chain attached to C-22 of the aglycone occurs in saponins of soyasaponin Ab, while it is absent in soyasaponin βg (Fig.1). Nishida *et al.* found that group A saponin strongly inhibits the

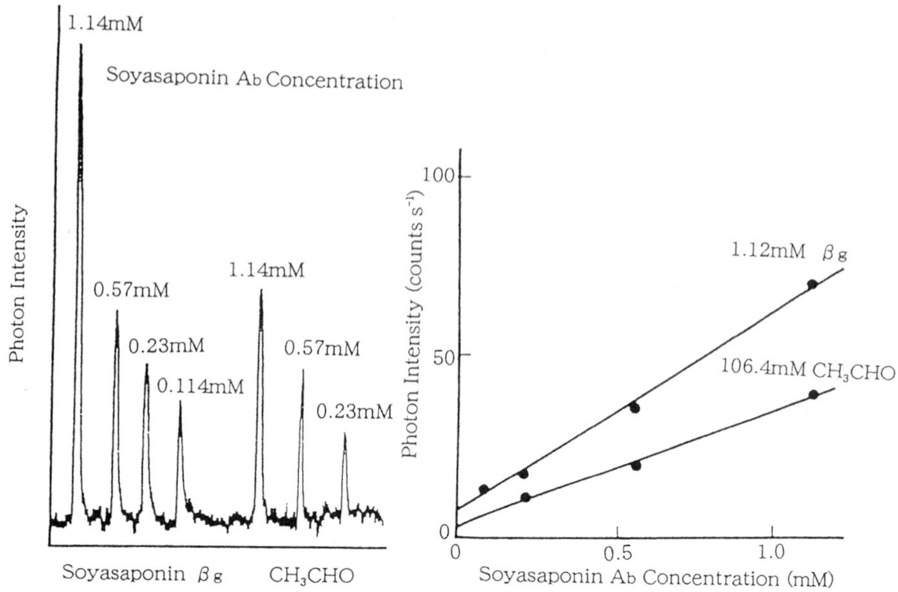

Fig.3 Chemiluminescence of soyasaponin Ab (Y; catalytic species) in the presence of hydrogen peroxide (X; active oxygen species) and soyasaponin βg or acetaldehyde (Z; receptive species)

CCl_4-induced lipid peroxidation in mouse liver microsomes and suggested that the sugar chain attached to C-22 of the aglycone of group A saponins may play a similar role to the phytyl side chain of α-tocopherol which is known to react directly with peroxy radicals[15]. Although another saponin glycyrrhizin, which has a similar chemical structure except for substitution at C-10, C-26 and C-30, and sugar chain at C-3 of the aglycone, also exhibited CL as Y in the presence of H_2O_2 and MeCHO, glycyrrhetinic acid did not (data not shown). These results also suggest that the presence of a sugar chain is related to the occurrence of CL. The sugar chain may also play an important role in making the structure suitable for the electron exchange, especially the structural complexity found in saponins because CL was affected by species and combination of Y and Z.

Spin density distribution theory is becoming increasingly popular as a tool for calculating the ground-state properties of molecules[16]. Thus mojor contributions to spin density distribution of DDMP saponin were calculated (Fig.4). These results suggested that C-4, C-5 and C-6 of DDMP moiety are related to radical reactions of DDMP saponins.

The antioxidative effect of gallic acid and MeCHO or soyasaponin βg was examined by measuring the inhibition for autoxidation of linoleic acid. Fig.5 shows typical antioxidative activity in the XYZ system. MeCHO, which was Z in the XYZ system acted as proxidant. However, when MeCHO was added to linoleic acid solution containing 1mM gallic acid, the MeCHO showed stronger antioxidative activity, compared with antioxidative activity of 1mM gallic acid for oxidation of linoleic acid. Although soyasaponin βg also showed strong oxidative activity, a similar result was obtained when soyasaponin βg was added to gallic acid.

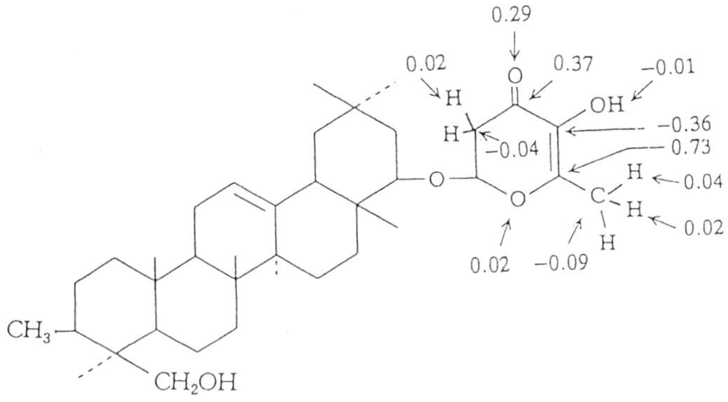

Fig.4 Major contributions to spin density distribution (| density | > 0.01)

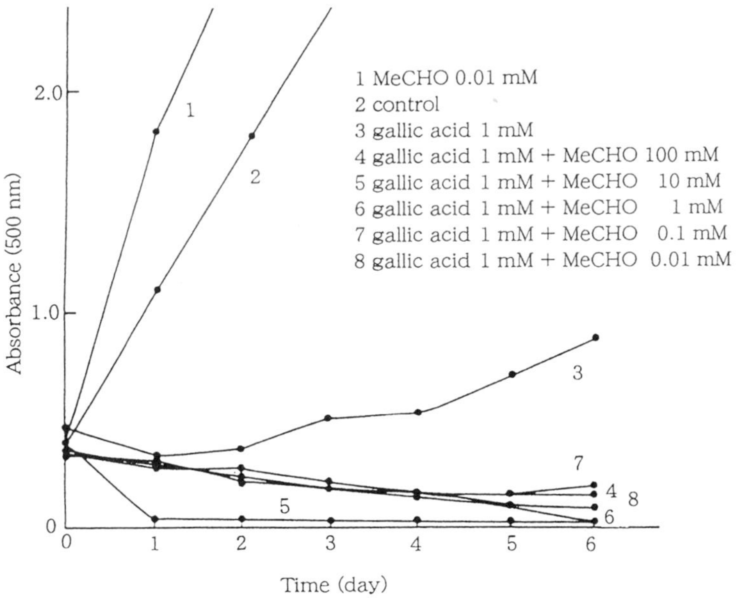

Fig.5 Effect of acetaldehyde on antioxidative activity of gallic acid (1mM)

MeCHO; acetaldehyde

Radical scavenging activity of antioxidants is generally espressed to represent how rapidly they react with active oxygen to scavenge this; however, only the secondarily formed oxidative substances or radicals have attracted pharmacological attention in relation to their cytotoxic and mutagenic/antimutagenic activities. It may be also important to investigate how to be eliminated high potential energy which may be produced in the presence of active oxygens and antioxidants such as phenolic compounds and soyasaponins in biological systems.

REFERENCES

1. M.Shiraiwa, S.Kudou, M.Shimoyamada, K.Harada, and K.Okubo, Composition and structure of group A saponin in soybean seed. *Agric.Biol.Chem.*, 55:315(1991)
2. S.Kudou, M.Tonomura, C.Tsukamoto, T.Uchida, T.Sakabe, N.Tamura, and K.Okubo, Isolation and structural elucidation of DDMP-conjugated soyasaponins as genuine saponins from soybean seeds. *Biosci.Biotech.Biochem.*, 57:546(1993).
3. H.Ohminami, Y.Kimura, H.Okuda, S.Arich, M.Yoshikawa, and I.Kitagawa, Effects of soyasaponins on liver injury induced by highly peroxidized fat in rats. *Planta Medica*, 46:440(1984).
4. H.Nakashima, K.Okubo, Y.Honda, T.Tamura, S.Matsuda, and N.Yamamoto, Soybean saponin and isoflavonoids against HIV in vitro. *AIDS*, 3:655(1989).
5. Y.Yoshiki, K.Okubo, M.Onuma, and K.Igarashi, Chemiluminescence of benzoic and cinnamic acids, and flavonoids in the presence of aldehyde and hydrogen peroxide or hydroxyl radical by Fenton reaction, *Phytochemistry*, 39:225(1995).
6. Y.Yoshiki, K.Okubo, and K.Igarashi, Chemiluminescence of anthocyanins in the presence of aldehyde and *tert*-butyl hydroperoxide. *J.Biolumin.Chemilumin.*, 10;335(1995)
7. A.Puppo, Effect of flavonoids on hydroxyl radical formation by Fenton-type reactions; influence of the iron chelator, *Phytochemistry*, 31:85(1992).
8. S.R.Husain, J.Chillard, and P.Chillard, Hydroxyl radical scavenging activity of flavonoids, *Phytochemistry*, 26:2489(1987).
9. Y.Yoshiki, J.H.Kim, K.Okubo, I.Nagaoya, T.Sakabe, and N.Tamura, A saponin conjugated with 2,3-dihydro-2,5-dihydroxy-6-methyl-4*H*-pyran-4-one from *Dolichos lablab*, *Phytochemistry*, 38:229(1995)
10. N.Nakatani, and H.Kikuzaki, A new antioxidative glucoside isolated from oregano (*Oreganum vulgare* L.), *Agric.Biol.Chem.*, 51:2727(1987).
11. N.Gotoh, and E.Niki, Rate of interactions of superoxide with vitamin E, vitamin C and related compounds as measured by chemiluminescence, *Biochim.Biophys.Acta*, 1115:201(1992).
12. W.Bors, W.Heller, C.Michel, and M.Saran, Flavonoids as antioxidants: Determination of radical-scavenging efficiencies, *Methods in Enzymology*, 186:343(1990).
13. C.Tsukamoto, A.Kikuchi, K.Harada, K.Kitamura, and K.Okubo, Genetic and chemical polymorphisms of saponins in soybean seeds, *Phytochemistry*, 34:1351(1993).
14. M.Shiraiwa, K.Harada, and K.Okubo, Composition and contents of saponins in soybean seed according to variety, cultivation year and maturity, *Agric.Biol.Chem.*, 55:323(1991).

15. K.Nishida, Y.Araki, Y.Nagamura, Y.Ohta, M.Ito, and I.Ishiguro, Preventive effect of saponin fractions separated from hypocotyl of soybean on CCl_4-induced injury in isolated rat hepatocytes. *Med,Biol.*, 126:87(1993).
16. Y.Qin, and W.A.Ralph, Density-functional methods give accurate vibration frequencies and spin densities for phenoxyl radical, *J.Chem.Phys.*, 102;1689(1995).

HERNIARIA SAPONIN B, A NOVEL TRITERPENOID SAPONIN FROM *HERNIARIA FONTANESII*

Z. Charrouf[1], A. Nait-Mbark[1], D. Guillaume[2], Y. Leroy[3], and O. Kol[3]

[1] Laboratoire de Chimie des Plantes et de Synthèse Organique et Bioorganique, Faculté des Sciences. B. P. 1014 Rabat, Morocco

[2] Laboratoire de Chimie Thérapeutique, URA 1310 du CNRS, Faculté des Sciences Pharmaceutiques et Biologiques, 75270 Paris Cedex 06 France

[3] Laboratoire de Chimie Biologique et UMR n° 111. Université des Sciences et Technologie de Lille, 59655 Villeneuve d'Ascq Cedex, France

ABSTRACT

The aerial parts of *Herniaria fontanesii* have yielded a new saponin named herniaria saponin B. Its structure was characterized by means of mass spectrometry, NMR techniques, and chemical analysis gave 28-*O*-{α-L-rhamnopyranosyl-(1→2)-[(α-L-rhamnopyranosyl-(1→3)]-β-D-xylopyranosyl (1→2)-β-D-fucopyranosyl} ester of 3-*O*-[β-D-glucopyranosyl uronic-(1→4)-α-L-rhamnopyranosyl]-2β, 3β-16α-trihydroxy-olean-12-ene-23, 24-dioic acid.

INTRODUCTION

Herniaria fontanesii Gay (caryophyllaceae) is a widely distributed plant in the Mediterranean area. Its aerial parts are used in folk medicine as a diuretic and for the treatment of lithiasis. We have recently reported the presence of glycosides in the title plant aerial parts (Nait-Mbark et al., 1995; 1996) and now report the isolation and structure elucidation of a new oleanane saponin : herniaria saponin B (HSB, **1**).

RESULTS AND DISCUSSION

The dried aerial parts of *H. fontanesii* were extracted with CH_2Cl_2, then MeOH. The alcoholic extract was dissolved in water and successively extracted with CH_2Cl_2, EtOAc, and n-BuOH. The latter extract was dissolved in MeOH and the saponin content precipitated with Et_2O. The crude saponin fraction (CSF) was purified by a combination of gel filtration on Sephadex LH 20 and HPLC to yield HSB (0.9% of the CSF).

The negative-mode FAB/MS of **1** displayed at *m/z* 1409 $[(M-H)^-]$ a molecular ion, indicating the molecular formula $C_{65}H_{102}O_{33}$, together with significant peaks at *m/z* 1263, 1087, and 839 suggesting successive losses of deoxyhexose (d), d + glucuronic acid, and 3d + pentose (p). Additional fragment peaks at *m/z* 483 and 517 were respectively attributed to $(3d+p-104-H)^-$ (Fraisse et al., 1986) and $(aglycon-H)^-$.

From its ^1H and ^{13}C NMR data, the genin of HSB ($C_{30}H_{46}O_7$) was rapidly identified as zanhic acid (Klein et al., 1982; Dimbi et al., 1984).

Methanolysis of HSB yielded glucuronic acid (glcA), rhamnose (rhm), fucose (fuc), and xylose (xyl) derivatives identified by GC of their trimethylsilylated methylglycosides in the respective molecular ratio 1/3/1/1.

Alkaline hydrolysis of HSB, under reductive conditions, allowed the isolation of an oligosaccharide residue composed of fucitol, rhamnose, and xylose in the molar ratio 1/2/1 respectively (GC analysis of the trimethylsilylated methylglycosides). Thus, the fucose residue was identified as the sugar moiety esterifying directly the aglycone at the C-28 position. Hydrolysis of the permethylated derivative of **1** followed by reduction and acetylation of the sugar derivatives afforded a mixture of alditol acetates whose GC/MS analysis suggested the presence of two terminal rhamnose (rhmt) units, one 2,3-disubstituted xylose, one 4-substituted rhamnose (rhmi), and one 2-substituted fucose residues.

Interglycosidic linkages were definitely established by using ^1H, ^{13}C and 2D homo and heteronuclear COSY and finally comparison with the NMR data of herniaria saponin A (Nait-Mbark et al., 1995). The oligosaccharide esterifying position C-28 of the aglycone was identified as rhm-(1→2)-[rhm-(1→3)]-xyl-(1→2)-fuc. The disaccharide substituting position C-3 was identified as glcA-(1→4)-rhm. Thus the structure of HSB was established as 28-*O*-{α-L-rhamnopyranosyl-(1→2)-[(α-L-rhamnopyranosyl-(1→3)]-β-D-xylopyranosyl (1→2)-β-D-fucopyranosyl} ester of 3-*O*-[β-D-glucopyranosyl uronic-(1→4)-α-L-rhamnopyranosyl]-2β, 3β-16α-trihydroxy-olean-12-ene-23, 24-dioic acid.

In order to confirmed the claimed biological activity of *H. fontanesii*, we studied its litholytic properties. Hence, human bile stones were transplanted into the bile canal of dogs to which a solution of herniaria saponins was then administrated, three times a day, during two months. This treatment induced a reduction in the biliary secretion of cholesterol and the rate of dissolution of the stones was significantly accelerated.

Table 1: Selected ^{1}H NMR data (δ ppm, *J* Hz) of HSB in CD_3OD at 300 MHz

aglycone

H-2	4.51 *m*	H-12	5.49 *t*-like
H-3	4.28 (*d*, 3.2)	H-16	4.63 *m*

3-O-sugar

glcA-1	4.47(*d* ,7.6)	rhmi-1	5.73(*d*, 1.5)
glcA-2	3.47	rhmi-2	4.13
glcA-3	3.63	rhmi-3	3.98
glcA-4	3.73	rhmi-4	3.67
glcA-5	4.06	rhmi-5	4.02
		rhmi-6	1.57

28-O-sugar

fuc-1	5.46(*d* ,8.2)	xyl-1	4.52(*d*, 7.8)	rhmt-1	5.41(*d*, 1.5)	rhmt-1	5.23(*d*, 1.5)
fuc-2	4.06	xyl-2	3.52	rhmt-2	4.06	rhmt-2	4.26
fuc-3	3.87	xyl-3	3.69	rhmt-3	3.90	rhmt-3	3.96
fuc-4	3.79	xyl-4	3.70	rhmt-4	3.60	rhmt-4	3.64
fuc-5	3.80	xyl-5	3.42/3.90	rhmt-5	4.02	rhmt-5	4.06
fuc-6	1.37			rhmt-6	1.43	rhmt-6	1.45

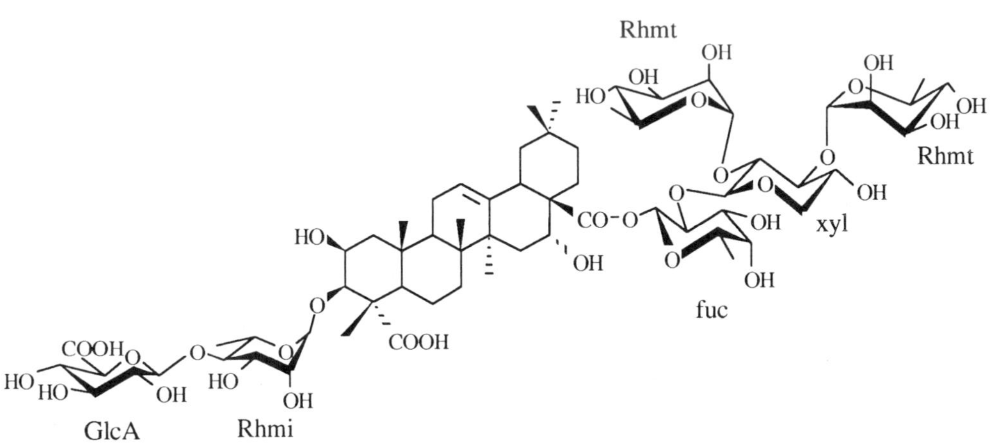

1

Table 2: ^{13}C NMR data (δ ppm) of HSB in CD_3OD at 75 MHz

aglycone					
C-1	45.5	C-11	25.0	C-21	37.0
C-2	72.0	C-12	123.7	C-22	33.8
C-3	88.7	C-13	144.9	C-23	186.4
C-4	54.7	C-14	43.4	C-24	15.2
C-5	53.4	C-15	31.6	C-25	17.7
C-6	22.1	C-16	76.7	C-26	18.0
C-7	37.0	C-17	49.2	C-27	27.2
C-8	41.4	C-18	42.7	C-28	177.2[b]
C -9	48.8[a]	C-19	48.7[a]	C-29	34.2
C-10	37.9	C-20	32.6	C-30	25.0

3-O-sugar			
glcA-1	108.1	rhmi-1	100.9
glcA-2	77.2	rhmi-2	71.9
glcA-3	74.4	rhmi-3	72.5
glcA-4	74.4	rhmi-4	84.5
glcA-5	74.0	rhmi-5	68.8
glcA-6	177.4[b]	rhmi-6	18.5[c]

28-O-sugar							
fuc-1	95.3	xyl-1	104.9	rhmt-1	102.9	rhmt-1	102.2
fuc-2	72.6	xyl-2	76.7	rhmt-2	72.5	rhmt-2	72.5
fuc-3	77.8	xyl-3	85.5	rhmt-3	72.6	rhmt-3	72.6
fuc-4	72.6	xyl-4	70.6	rhmt-4	74.2	rhmt-4	74.4
fuc-5	72.5	xyl-5	67.5	rhmt-5	69.7[d]	rhmt-5	69.9[d]
fuc-6	16.8			rhmt-6	18.2[c]	rhmt-6	18.2[c]

[a-d]: interchangeable values

EXPERIMENTAL

Plant material: Aerial parts of *Herniaria fontanesii* were collected in Oujda (Morocco) in spring 1993 and authenticated by Drs. Kahouadji and Bentatou (Department of Botany, University Mohammed V Rabat, Morocco). A voucher specimen (Ch. Sauvage 8918) is preserved in the RAB Herbarium of Scientific Institute of Rabat.

Extraction and isolation: Dried plant material (250 g) was extracted at room temperature with CH_2Cl_2, then with MeOH. The dried MeOH extract (15 g) suspended in water was extracted successively by CH_2Cl_2, EtOAc, and *n*-BuOH. The *n*-BuOH extract (2 g) was dissolved in MeOH and the saponin content precipitated with Et_2O. The precipitate was subjected to chromatography with Sephadex LH 20 and column chromatography, using MeOH as solvent, another chromatography over an ODS Zorbax column (25 x 0.94 cm ID) eluted with MeCN-H_2O afforded **1** (18 mg).

Methods and apparatus: Methods used for the determination of the molar carbohydrate composition, for alkaline hydrolysis, methylation, GC/MS analysis, and for fast atom bombardment mass spectrometry (FAB/MS) are described in ref 1. ^{1}H and ^{13}C spectra were recorded in CD_3OD with a Bruker AC-300 spectrometer.

Herniaria saponin B: $[\alpha]_D$= -29 (*c* = 0.8, MeOH); FABMS *m/z* 1409 [(M-H)$^-$], 1263, 1087, 839, 517 , 483. ^{1}H-NMR (CD_3OD) δ : 1.43/2.28 (2 H-1), 4.51 (1H, H-2), 4.28 (1H, d, *J* = 3.2, H-3), 2.11/2.20 (2 H-11), 5.49 (1H, t-like, H-12), 1.62/1.71 (2 H-15), 4.63 (1H, H-16), 3.10 (1H, H-18), 1.25/2.50 (2 H-19), 1.58 (3H, s, H-24), 1.54 (3H, s, H25), 1.10 (3H, s, H-26), 1.43 (3H, s, H-27), 0.90 (3H, s, H-29), 1.05 (3H, s, H-30). For ^{13}C NMR see Table 2 .

ACKNOWLEDGMENTS

We thank UNESCO for financial support to A. N-M.

REFERENCES

Dimbi, M. Z., Warin, R., Delaude, C., Huls, R., Kapundu, M., and Lami, N., 1984, Structure de l'acide zanhique et l'acide zanhique γ-lactone, triterpenoides nouveaux isolés de *Zanha golugensis* et de *Ganophyllum giganteum., Bull. Soc. Chim. Belg.*, **93**, 323-328.

Fraisse, D., Tabet, J. C., Becchi, M., and Raynaud, J., 1986, Fast atom bombardment mass spectrometry of triterpene saponins, *Biomed. Environ. Mass Spectrometry*, 13, 1-14.

Klein, G., Jurenitsh, J., and Kubelka, W., 1982, Struktur der sapogenin von herba herniariae (*Herniaria glabra* L. und *Herniaria hirsuta* L.), *Sci. Pharm.*, **50**, 216-233.

Nait-Mbark, A., Charrouf, Z., Wieruszeski, J.-M., Leroy Y., and Kol, O., 1995, Herniaria saponin A, a novel saponin from *Herniaria fontanesii*, *Nat. Prod. Lett.*, **6**, 233-240.

Nait-Mbark, A., Guillaume, D., Cherrah, Y., Wieruszeski, J.-M., Ricart, G., and Charrouf, Z, 1996, An acylated isorhamnetin glycoside from *Herniaria fontanesii*, *Phytochemistry*, in press.

STEROIDAL GLYCOSIDES FROM *NICOTIANA TABACUM* L. SEEDS AND THEIR BIOLOGICAL ACTIVITY

S.A. Shvets, O.N. Gutsu, and P.K. Kintia

Institute of Genetics
Academy of Science
Chisinau 2002 Moldovia

INTRODUCTION

During recent years steroidal spirostanol and furostanol glycosides have attracted the attention of researchers as biologically active material having an independent importance due, first of all, to the diversity of their biological action[1]. The present work reports the isolation of spirostanol and furostanol glycosides from *Nicotiana tabacum* L. seeds, elucidation of their chemical structure, and establishment of some of their biological activities for the first time.

RESULTS AND DISCUSSION

Gel filtration of Sephadex and silica gel chromatography of the total methanolic extract in solvent systems differing in polarity yielded individual glycosides, tentatively named nicotianosides A (**1**), B (**2**), C (**3**), D (**4**), E (**5**), F (**6**), and G (**7**). Compounds **1**, **2**, **3**, and **4** were assigned to derivatives of the (25*S*)-spirostan series according to their positive reaction with the Sannié reagent[2] and negative reaction with the Ehrlich reagent[3], as well as IR spectra, in which absorption bands were present at 960, 920, 900, 862, and 835 cm^{-1} [4]. Compounds **5**,**6**, and **7** reacted positively with the Sannié and Ehrlich reagents and their IR spectra contained a low-intensity broadened absorption band at 900 cm^{-1}, which allowed their assignment to glycosides of the furostane series.

Acid hydrolysis of nicotianosides **1-7** resulted in a genin, identified by physicochemical constants and R_f mobility in a silica gel thin layer impregnated with 2% $AgNO_3$, in the presence of neotigogenin, which served as authentic sample. ^{13}C NMR spectra show (ppm): 37.3, 30.0, 70.6, 38.2, 44.7, 29.0, 32.2, 35.4, 54.6, 36.0, 21.4, 40.3, 40.9, 56.6, 31.7, 81.3, 62.9, 16.5, 12.5, 42.3, 15.0, 109.9, 27.6, 26.3, 26.5, 65.2, and 16.4 (C_1 - C_{27}). They are consistent with the literature data[5].

Since **5-7** have been assigned to glycosides of the furostane series, (25*S*)-5α-furostane-3β,22α,26-triol is their natural aglycone.

Paper chromatography was used to determine the qualitative monosaccharide composition of the oligosaccharide part of each nicotianoside, and GLC of preisolated acetates of aldonitryl sugar derivatives[6] resulted in the establishment of quantitative monosaccharide ratios. Glucose was identified for **1**, glucose and rhamnose in the ratio of 1:1 for **2**, glucose and rhamnose (1:2) for **3**, glucose and rhamnose (1:3) for **4**, glucose and rhamnose (2:1) for **5**, glucose and rhamnose (2:2) for **6**, and glucose and rhamnose (2:3) for **7**.

Methylation of nicotianosides **1-7** according to Hakomori's method[7], followed by methanolysis of permethylated glycosides and analysis of the products isolated by GLC, allowed the identification of methyl 2,3,4,6-tetra-*O*-methyl-D-glucopyranoside (**a**) for **1**, methyl 2,3,4-tri-*O*-methyl-L-rhamnopyranoside (**b**) and methyl 3,4,6-tri-*O*-methyl-D-glucopyranoside (**c**) for **2**; **b** and methyl 3,6-di-*O*-methyl-D-glucopyranoside (**d**) for **3**; **b** and methyl 6-mono-*O*-methyl-D-glucopyranoside (**e**) for **4**; **a**, **b**, and **c** for **5**; **a**, **b**, and **d** for **6**; and **a**, **b**, and **e** for **7**.

Periodate oxidation of the nicotianosides followed by acid hydrolysis of the oxidized compounds showed glucose only in the hydrolyzates of saponins **3**, **4**, **6**, and **7**, which confirmed the methylation data. Thus, the methylated monosaccharides were identified as nicotianoside E (**5**) as in spirostanol **2**, as nicotianoside F (**6**) as in **3**, for nicotianoside G (**7**) as in **4**, and additionally a completely methylated glucopyranoside was identified in each case.

Mild acid decomposition of nicotianosides **2**, **3**, and **4** showed the sequence of carbohydrate fragment linkage in the glycoside oligosaccharide chain. Afterwards, a progenin was isolated from nicotianoside B (**2**), identical to **1** by its physicochemical constants. Its hydrolysis led to the isolation of neotigogenin and D-glucose. Nicotianoside C (**3**) provided three progenins, one of which, neotigogenin monoside, is identical to **1**; the other, neotigogenin rhamnopyranoside, coincides with **2** by its constants. The third one was neotigogenin rhamnoglucopyranoside, methylation and methanolysis of which showed that the terminal rhamnose is attached to glucose linked with the aglycone by the C-1 → C-4 bond. Under mild reactive conditions, nicotianoside D (**4**) yielded one progenin, neotigogenin monoside, identical to **1**.

Mild acid decomposition of nicotianoside E (**5**) provided two progenins, namely monoside, identical to **1** by physicochemical constants, and a bioside, identical to **2**. Nicotianoside F (**6**) afforded the same progenins as **3** and an additional progenin, identical to nicotianoside C (**3**) by physicochemical characteristics. Decomposition of nicotianoside G (**7**) resulted in similar fragments, which characterize nicotianoside D (**4**) as saponin **4** itself.

The analysis of the data obtained in the course of acid hydrolysis and methylation led us to the conclusion that glycosides from **5-7** differ from **2-4** by the structure of the carbohydrate part in that they have an additional molecule of glucose. Taking into consideration the nature of the glycosides under study, it is natural to suggest that this molecule of glucose in each glycoside is attached to the aglycone at its C-26 position. By enzymatic splitting of nicotianosides **5-7** with a complex enzyme from the gastric juice of the grape snail (*Helix pomatia*), the active agent of which is β-glucosidase, glucose attached at the C-26 position of the genin is readily disconnected and glucosides of the furostane series are converted into their spirostanol analogs. Thus, enzymatic splitting of nicotianosides E (**5**), F (**6**), and G (**7**) afforded spirostanol analogs, identical to nicotianosides B (**2**), C (**3**), and D (**4**) respectively.

The structure of nicotianosides of the spirostane and furostane series is confirmed by ^{1}H and ^{13}C NMR spectral data. Thus, for example, in the PMR spectrum of nicotianosides **3** (Table 1) signals of three anomeric protons are present (characteristic doublet splitting) in the region of 4.8-6.5 ppm. Signals with chemical shifts of 6.12 ppm and 5.65 ppm based on a low value of spin-spin interaction constants (SSIC) ($J_{1,2}$ = 1.5 Hz) are assigned to anomeric protons of two rhamnose residues (Rha' and Rha"). Two doublet signals of methyl groups in the spectral high-field region with the chemical shifts of 1.64 ppm and 1.52 ppm and SSI_C of 6.5-7 Hz correspond to them, as well. The positions of the signals of the remaining protons of two of these rhamnose residues were found beginning from protons H-1 and H-1" by using the method of double homonuclear resonance in a difference variant[8]. The pattern of their splitting and SSIC are in complete agreement with the L-rhamnose having the pyranose structure of these residues. The signal of the anomeric proton with a chemical shift of 4.87 ppm and a higher value of SSIC ($J_{1,2}$ = 7.5 Hz) undoubtedly belongs to the remaining residue of glucose and determines the configuration of its glycoside bond. The splitting pattern of the other protons of this residue also confirms D-glucose in the pyranose structure.

Table 1. Chemical shift data (δ, ppm, *O*-TMS, C_5D_5N) and spin-spin interacting constants (*J*, Hz) of H protons of nicotianoside C (**3**).

Protons of aglycone and sugars III	δ, ppm, *J*, Hz		Protons of sugars III	δ, ppm, *J*, Hz	
H-3	3.82 *m*			L-Rha'	
CH_3-18	0.71 *s*		H-1	6.12 *d*	$J_{1,2}$ = 1.5
CH_3-19	0.76 *s*		H-2	4.62 *dd*	$J_{2,3}$ = 3.5
CH_3-21	1.06 *d*	$J_{20,21}$ = 7.0	H-3	4.42 *dd*	$J_{3,4}$ = 9.5
CH_3-27	1.03 *d*	$J_{25,27}$ = 7.0	H-4	4.18 *t*	$J_{4,5}$ = 9.5
H-26A	3.62 *m*	$J_{26A,26B}$ = 11.0	H-5	4.72 *d*	$J_{5,6}$ = 6.5
H-26B	3.30 *d*		H-6	1.64 *d*	$J_{6,5}$ = 6.5
	D-Glc			L-Rha"	
H-1	4.82 *d*	$J_{1,2}$ = 7.5	H-1"	5.65	$J_{1,2}$ = 1.5
H-2	4.0 *t*	$J_{2,3}$ = 7.3	H-2"	4.50	$J_{2,3}$ = 3.5
H-3	4.06 *dd*	$J_{3,4}$ = 9.8	H-3"	4.40	$J_{3,4}$ = 9.5
H-4	4.38 *t*	$J_{4,5}$ = 10.0	H-4"	4.19	$J_{4,5}$ = 9.5
H-5	3.5 *m*		H-5"	4.73	$J_{5,6}$ = 7.0
H-6A	3.9 *dd*	$J_{6A,AB}$ = 12.5	H-6"	1.52	$J_{6,5}$ = 7.0

The remaining signals in the high-field spectral region (0.5-2.0 ppm) are assigned to methyl groups at C-18 and C-19 (two singlets) and C-21 and C-27 (both with a characteristic doublet splitting with $J = 7$ Hz having a proton at the adjacent C-20 or C-25 atom). The signals of H-3 protons (3.82 ppm, a characteristic doublet-doublet splitting) and H-26, A,B (3.62 and 3.30 ppm, $J_{26A,26B}$ = 11 Hz) have also been found. All these confirm the steroid nature of the aglycone belonging to the spiro series. Close values of the chemical shifts of protons C-21-CH_3 and C-27-CH_3 prove that the steroid belongs to the derivatives of the (25*S*)-neo series[9].

The types of glycosidic linkages in the carbohydrate part of the glycoside are found from the two-dimensional ROESY experiments[10]. Thus, the presence of cross peaks at 4.87 and 3.83 ppm in the ROESY spectrum proves the spatial closeness of proton H-1 Glc and H-3 of the aglycone and unequivocally determines the monosaccharide bound to the aglycone. Signals at 6.16 and 4.05 ppm show the bonding of the residue Rha' with the C-2-OH group of the residue Glc at 5.65 and 4.42 ppm, proving that the C-1 → C-4 bond is between Rha" and Glc.

The ^{13}C NMR spectrum of nicotianoside C (**3**) (Table 2) also contains the signals of three anomeric C atoms in the region of 100-105 ppm. The assignment of the C atom signals of rhamnose terminal residues was carried out using literature data for unsubstituted methyl α-L-rhamnopyranoside[11] and literature data for chemical shifts of terminal rhamnoses in different glycosides[8]. The C atom signals of 2,4-di-substituted β-glucose were assigned by using data on α- and β-effects of glycosylation. Signals of C atoms of the glycosidic aglycone part are in good agreement with the literature data for neotigogenin[5]. The positive effect on C-3 (6.7 ppm) in the aglycone and negative β-effect on C-2 and C-4 confirm the location of the carbohydrate chain attachment to the aglycone.

The ^{13}C NMR spectrum of nicotianoside F (**6**) (Table 2) in many respects is similar to that of nicotianoside C (**3**). However, there is another signal at 105.2 ppm in the region of anomeric C atoms. Besides, the spectral range of 60-80 ppm contains five additional signals as compared with the spectrum of saponin **3**. They are unequivocally assigned to the terminal residue of β-glucose. The remaining signals in the region of 60-80 ppm are in complete agreement with those of **3**.

Table 2. Chemical shift data of ^{13}C carbohydrate atoms of nicotianoside C (**3**) and nicotianoside F (**6**) (δ, ppm, *O*-TMS, C_5D_5N).

	Compounds			Compounds	
C atom	**3**	**6**	C atom	**3**	**6**
1	37.1	37.0	Glc		
2	29.9	30.2	1	100.0	100.0
3	78.0	78.6	2	79.0	78.4
4	34.3	33.9	3	77.4	77.1
5	44.8	44.5	4	78.4	78.1
6	28.9	28.6	5	77.1	76.9
7	31.9	32.1	6	61.6	61.6
8	34.8	35.1	Rha'		
9	54.1	54.3	1'	102.3	102.4
10	35.3	35.3	2'	72.6	72.9
11	21.2	21.0	3'	72.8	73.0
12	40.4	40.1	4'	74.2	74.2
13	40.6	40.3	5'	70.6	70.6
14	56.5	56.8	6'	18.6	18.7
15	32.1	32.0	Rha"		
16	80.6	81.0	1"	103.0	103.0
17	61.9	62.2	2"	72.6	72.7
18	16.4	16.5	3"	72.8	72.9
19	12.0	12.2	4"	74.0	74.0
20	42.3	42.1	5"	69.7	69.7
21	14.6	14.3	6"	18.8	18.8
22	109.2	112.6	Glc"		
23	27.1	35.4	1"		105.2
24	25.9	28.3	2"		75.4
25	26.1	36.1	3"		78.7
26	65.1	71.8	4"		71.9
27	16.2	17.2	5"		78.6
O-CH_3	–	49.8	6"		63.0

In the low-field region of the spectrum for saponin **6** compared with the spectrum of saponin **3**, the chemical shifts of the signals of atoms C-23, C-24, C-25, and C-27 are moved to the weak field, which is characteristic of furostanol steroids. The chemical shift of the C-26 atom signal (weak field shift from 65.1 ppm to 71.8 ppm) corresponds to the glycosylated form of furostane. Thus, nicotianoside F (**6**) contains two carbohydrate chains at the C-3 atom of the aglycone, as in **3**, and a residue of one glucose at the C-26 atom.

Summing up the above, the following structures in Figure 1 are suggested for nicotianosides A (**1**), B (**2**), C (**3**), D (**4**), E (**5**), F (**6**), and G (**7**):

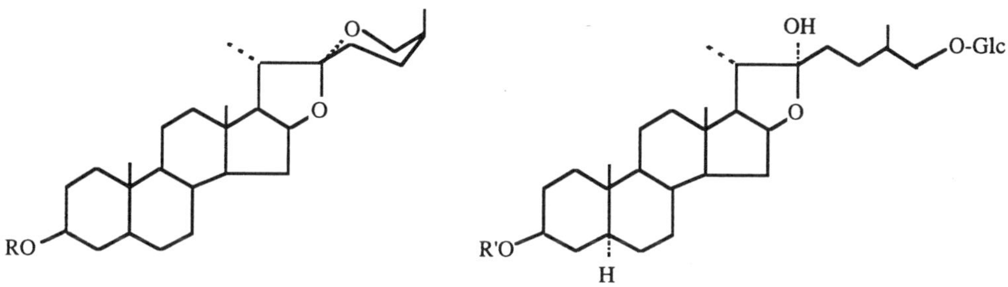

1. R = β-D-Glc (3-*O*-β-D-Glucopyranoside neotigenin)

2. **5**, R = R' = α-L-Rha (1→2) → β-D-Glc (3-*O*-α-L-rhamnopyranosyl(1→2)-*O*-β-D-glucopyranoside neotigenin)

3. **6**, R = R' = α-L-Rha 1→4 → β-D-Glc (3-*O*-{[α-L-rhamnopyranosyl(1→2)-[α-L-rhamnopyranosyl(1→4)]-β-D-glucopyranoside}neotigenin)
α-L-Rha 1→2 ↗

4, **7**, R = R' = α-L-Rha 1→2 ↘
α-L-Rha 1→3 → β-D-Glc (3-*O*-{[α-L-rhamnopyranosyl(1→2)[α-L-Rhamnopyranosyl(1→3)]-α-L-rhamnopyranosyl(1→4)}-β-D-glucopyranoside neotigenin)
α-L-Rha 1→4 ↗

Fig. 1. Structures of nicotianosides A (**1**), B (**2**), C (**3**), D (**4**), E (**5**), F (**6**), and G (**7**) with the scientific names being shown in the parentheses.

BIOLOGICAL ACTIVITY

Effect on Tobacco Seed Presowing: The glycosides isolated have been tested for phytostimulating activity. We studied the influence of presowing treatment of tobacco seed with an aqueous solution of total steroidal glycosides of the furostane series from tobacco seeds (nicotianosides) on transplant seedling development. Tobacco (cultivars Ostrolist 89 and Virginia 115) seeds were soaked in the nicotianoside solution for 24 h.[13] After germinating in a thermostat the seeds were sown in a hotbed. Seedling development was observed until maturation of the seedling, which was about 1 month. For each cultivar the glycoside solution was used in the concentrations of 0.1, 0.08, 0.01,

0.005, and 0.001%. The experiment was replicated four times, and 25 plants were studied in each treatment. In the control experiments seeds were soaked in a 2% solution of formalin, used in practice, and in water.

Observations of the seedling development showed that seed presowing treatment with glycosides reduced the period of their germination by 2 days, and transplant seedling maturation by 5-6 days; the seedlings grown from glycoside-treated seeds were more viable and better developed as compared with the control.[13]

With the aim of establishing the impact of nicotianosides on tobacco resistance to viral diseases, the transplanted seedlings grown from glycoside-treated seeds and control plants were field planted and at the 5-6-leaf stage the plants were exposed to a rigid infectious background of tobacco mosaic virus (TMV), tomato bronzetop virus (TBV), and potato Y-virus (TYV). The phytopathological assessment was performed according to conventional methods. The data expressed are an average degree of tobacco plant infection, and are presented in Table 3.

Table 3. Influence of tobacco seed presowing treatment with the nicotianoside solution on the infection rate in tobacco plants (cvs Ostrolist-89 and Virginia 115) with viral diseases (on a rigid artificial infectious background).

Treatment	Conc. %	Infection rate as related to the control, %		
		TMV	TBV	TYV
		Ostrolist-89		
Water (control)	–	100.0	100.0	100.0
Formalin (prototype)	2.0	100.0	96.0	92.0
Nicotianosides (**1-7**)	0.1	12.6	18.0	6.0
	0.08	0.0	8.0	0.0
	0.01	0.0	4.1	0.0
	0.005	0.0	0.0	0.0
	0.001	17.4	19.2	15.6
		Virginia 115		
Water (control)	–	100.0	100.0	100.0
Formalin (prototype)	2.0	100.0	100.0	100.0
Nicotianosides (**1-7**)	0.1	12.6	0.0	0.0
	0.08	0.0	0.0	0.0
	0.01	0.0	0.0	0.0
	0.005	4.3	13.5	4.7
	0.001	19.2	40.3	27.6

The results show that presowing nicotianoside tobacco seed treatment has a favorable effect upon the plant resistance to the virus action, and that it depends on the concentration of the products used to treat the seeds.

Protection Against Viral Infection: A four-year study showed the effects on tobacco plants exposed to the highly concentrated infections of TMV, TBV, and PVY. Table 4 shows the effects of nicotianosides **4, 5, 6**, and **7** lowering the tobacco plant sensitivity to TMV 10- to 20-fold, to TBV 3- to 10-fold, and to PVY 20-fold. Nicotianoside **4** reduces plant sensitivity to the combination of viruses 2.4-fold. Tobacco plants grown from nicotianoside-treated seeds were not affected by viral diseases, their sensitivity being, in some cases, 2 to 3 times lower than that of the control, suggesting a possibility for using steroid glycosides to improve tobacco plant resistance to a combination (complex) of viral diseases. Chemical analysis of the leaves showed that the quality of raw material is not reduced, and that in some cases it was improved. The dry leaf yield in the treatment variant was 0.1-0.15 ton/ha higher than in the control. Improved resistance of

Table 4. Tobacco plants were affected (%) by viral diseases Ostrolist 89(cv) following treatment with nicotianoside **4, 5, 6** and **7**.

Nicotianoside	TMV'	TBV'	PVY'	Viral Complex
		% Live Plants		
4	5.2		5.2	–
5	0.0	7.6	5.8	–
6	10.0	7.1	4.7	–
7	–	25.0	–	41.6
Control	100.0	–	100.0	100.0
		85.0		

'Artifically infected.

tobacco plants to viral diseases resulting from treatment with nicotianosides provides a change from the currently used tobacco-growing technology of 1 to 2 chemical treatments of the tobacco plantation against insect vectors transmitting viral diseases of plants, which is very important from the economic and social standpoint.

Proposed Mechanism of Action. In order to elucidate the mechanisms of action of steroid glycosides, an electron microscopic evaluation was made of the chloroplast structure of the tobacco plants exposed to tobacco mosaic infection and treated with nicotianoside[12,13]. Leaves for electron microscopic studies were fixed in 6.5% glutaraldehyde in a phosphate buffer (pH 7.4) for 4 h, refixed in 2% osmic acid, dehydrated in increased concentrations of a series with alcohol and acetone, and placed in Epon. Thin sections were examined under an JEM-100CX electron microscope with an operative multiplication of 14×10^3. Quantitative estimation of organelle structure parameters was made using the method described elsewhere.[14]

Quantitative estimation of chloroplast structure parameters in TMB-resistant and TMB-sensitive tobacco cultivars exposed to infection and treated with steroid glycosides has shown that the glycoside treatment to normalize the chloroplast ultrastructural organization in the sensitive cultivar resulted in a more complicated chloroplast membrane system of the cell in the resistant cultivar.

The evidence presented indicates that differences in chloroplast structure between sensitive and resistant cultivars are detectable at the level of the control plants, without infection and without nicotianoside treatment (Table 5). Thus, six out of the eleven chloroplast characters examined showed significant differences in the cultivar studied, with two parameters (the number of lipid globules and form factor) in the resistant cultivar being significantly higher and the other parts lower than in the sensitive cultivar. In general, the chloroplast structure parameters were observed to decrease in the resistant cultivar relative to those of the sensitive cultivar.

Inoculation of tobacco plants of both cultivars with TMV resulted in changes in the majority of quantitative characters of these organelles. Thus, in plants of the resistant cultivar Yubileyny, the perimeter and area of the organelle section, the chloroplast length and width, the number of lipid globules, and the grana number and area per section are reduced relative to the control. An appreciable increase occurs in the number of starch grains, area of a single grana, and form factor. Plants of the resistant cultivar Trapezond 689 exhibited similar changes, but changes in the ultrastructure parameters were less pronounced than in the sensitive cultivar, suggesting differential responses of the sensitive and resistant tobacco cultivars to viral infection. The results agree with the findings of Ladygina et al.[15]. Thus, changes in chloroplast structure following nicotianoside treatment have been found to be similar in the sensitive and in the resistant tobacco cultivars, suggesting very similar cell metabolism changes in the tobacco plants studied

Table 5. Electron microscopic examination of the structure of leaf mesophyll protoplasts of a TMB-sensitive (Yubileyny) and a TMV-resistant (Trapezond 689) tobacco cultivars.

	Yubileyny			Trapezond 689		
Microscopic Examination (conventional units)	Control	TMV-Infec.	Steroid Glycoside Treatment	Control	TMV-Infec.	Steroid Glycoside Treatment
Section perimeter	47**	38.77*	39.41*	44.4	41.4*	42.19*
Chloroplast length	21**	17.76*	17.08*	19.3	18*	18.73
Chloroplast width	6.7	5.6*	6.65*	7	6*	7.03
Lipid globule number	0.0**	6.1	7.06	11	6.9	7.45
Starch grain number	0.3	.61	0.07	0.10	0	0.06
Section area	99.5	65*	82.4*	92.2	81*	84.39*
Grana per section	30.2	15*	17.15*	30.1	19*	16.89*
Total grana area	27**	19*	16.64*	23.4	15*	7.25*
Single grana area	0.90**	1.25	1.00	0.79	.87	1.03*
Total grana area to chlorplast area ratio	0.28	0.21	0.21*	0.26	.20*	0.20*
Formative Factor	0.33**	0.39	0.40*	0.36	.37	0.37

* Difference from the control is significant at $P<0.05$.
** Difference between the controls is significant at $P<0.05$.

following treatment with steroid glycosides. Preventive steroid glycoside treatment repairs the effect of viral infection, resulting in chloroplast structural organization changes.

EXPERIMENTAL

General remarks. Silica gel (100/160 and 40/100 μm), "Silufol" plates, FN-3 paper, and solvent systems: (1) chloroform/methanol (4:1), (2) chloroform/ethanol/water (65:35:3), (3) benzene/ethanol (9:1), (4) benzene/diethyl ether (7:3), (5) 1-butanol/benzene/pyridine/water (5:1:3:3, upper layer) were used for chromatography. The Sannié reagent (1% alcoholic solution of vanillin), Ehrlich reagent (1% alcoholic solution of *p*-dimethylaminobenzaldehyde), and aniline phthalate solution, and concentrated sulfuric acid served as developers.

GLC was carried out using a "Chrome-5" unit; a 2.4-m column loaded with 5% XE-60 on a N-AW-HMDS Chromaton was used for sugar derivatives. Helium served as gas carrier. Chromatography of aldononitryl sugar derivatives was performed at 180-230 °C with a temperature program of 3°/min; the temperature for chromatography of methylated methylglycosides and for sugars was 140 °C. The rate of flow of the gas carrier was 45 ml/min.

Melting points were measured on a Boetius table, and specific rotations on a polarimeter of the firm Zeiss. IR spectra were recorded with a "Specord-71-IR" spectrophotometer, mass spectra with a MX-1303 unit, and PMR spectra on AM-300 and WM-250 (Bruker) instruments for solutions in d_5-pyridine at 70 °C. ^{13}C NMR spectra were recorded on a AM-300 unit, the working frequency being 75 MHz under similar conditions.

Isolation of steroidal glycosides. 1 kg of air-dry tobacco seeds were ground and extracted with a chloroform/methanol mixture in the ratio of 2:1. The chloroform/methanolic extract was diluted with water and separated with a separatory funnel. The aqueous methanolic extract was evaporated to dryness, followed by repeated chromatography on a silica gel column using successively solvent systems 1 and 2 for elution. These yielded 150 mg of **1**, mp 276°, $[\alpha]_D^{20}$ –65° (c 1.0; CH_3OH); 390 mg of **2**, mp 241-242°, $[\alpha]_D^{20}$ –56° (c 1.0; CH_3OH); 820 mg of **3**, mp 268-269°, $[\alpha]_D^{20}$ –73° (c 1.0; CH_3OH), 120 mg of **4**, mp 283-284°, $[\alpha]_D^{20}$ –75° (c 1.0; Py), 160 mg of **5**, mp 178°, $[\alpha]_D^{20}$ –78° (c 1.0; CH_3OH), 1300 mg of **6**, mp 181-182°, $[\alpha]_D^{20}$ –75° (c 1.0; H_2O), and 290 mg of **7**, mp 189-192°, $[\alpha]_D^{20}$ –86° (c 1.0; H_2O).

Acid hydrolysis of nicotianosides 1-7. 100 mg of each glycoside were hydrolyzed with 2.5% sulfuric acid at 110° for 8 h in soldered ampules, followed by water dilution of the reaction mixture. The hydrolyzate solution was extracted (4 x 20 ml) with diethyl ether. The ether extract was concentrated and chromatographed on a column loaded with silica gel in system 4. For each nicotianoside neotigogenin was identified as the aglycone: mp 203°, $[\alpha]_D^{20}$ –75° (c 1.0; $CHCl_3$), $[M]^{+\cdot}$ 416, IR spectrum: 3360, 960, 918, 985, 860, and 835 cm^{-1}. The ^{13}C NMR spectrum of neotigogenin agrees with the literature data[11].

Methylation of 1-7 and methanolysis of their permethylated derivatives. 50 mg of each glycoside were dissolved in 15 ml of DMSO containing methylsulfonyl anion (prepared from 700 mg of NaH and 30 ml of DMSO) and mixed for 1 h at 50° in an argon atmosphere. 20 ml of CH_3I were added to the reaction mixture, which was left at room temperature for 12 h in darkness; the mixture was thereafter diluted with water and extracted with chloroform. The resulting chloroform extract was washed with a saturated solution of $Na_2S_2O_3$, then water, and concentrated in vacuo. The resulting permethylated glycoside derivatives were chromatographed on a column with silica gel using system 3.

Each of the permethylated glycosides was exposed to methanolysis with 72% $HClO_4$ in methanol (1:10) for 6 h at 110°. The precipitates were filtered out, and the solution was neutralized with ion-exchange resin, evaporated, and analyzed by using TLC and GLC to identify the methylglycosides described in the discussion.

Periodate oxidation of 2-7. 15 ml of 2% aqueous solution of $NaIO_4$ were added to the nicotianoside solution (30 mg) in 10 ml of methanol. The mixture was left to stand at room temperature for 48 h, followed by the addition of several drops of ethylene glycol and after 1 h it was extracted with 1-butanol. Butanolic extracts were evaporated and hydrolyzed with 2.5% H_2SO_4 for 8 h at 110 °C. Glucose was identified by using paper chromatography (system 5) in the hydrolyzates of **3**, **4**, **6**, and **7**.

Enzymatic hydrolysis of 5-7. 100 mg of **5**, 150 mg of **6**, and 180 mg of **7** were separately dissolved in 100 ml of water. β-Glucosidase was added, and the mixtures were incubated for 24 h at 37 °C. The reaction products were extracted with *n*-butyl alcohol, and the butanolic extract was concentrated and chromatographed on silica gel column with system 1. The enzymatic hydrolysis of **5** yielded progenin (72 mg) identical to **2** by physicochemical constants, 105 mg of **5** identical to **3**, and 113 mg of **7** identical to **4**.

Inoculation with the tobacco mosaic virus, tomato bronze top virus, and potato Y-virus. For inoculation of cultivars involved in the four-year study, TMV was isolated from dry leaves of mosaic-affected tobacco plants, and TBV was isolated from tobacco plants that had been exposed to natural infection during the early stages of development. The plants were inoculated with the virus, isolated at 90-100 days following seed treatment of the nicotianosides **1**, **2**, **3**, and **4** by rubbing into healthy leaves the sap of the diseased plants or liquid prepared from the virus leaves. The plant development was at 5-6 true leaf stage. The maximum rate of dilution of the infected sap was 1:50.

SUMMARY

Steroidal glycosides of the spirostane and furostane series have been isolated from the methanolic extract of *Nicotiana tabacum* L. seeds by using gel filtration on silica gel and chromatography on a silica gel column. They were tentatively named nicotianosides A

(**1**), B (**2**), C (**3**), D (**4**), E (**5**), F (**6**), and G (**7**). It has been shown that nicotianosides A (**1**), B (**2**), C (**3**), and D (**4**) are 3-*O*-β-D-glucopyranoside-; 3-*O*-α-L-rhamnopyranosyl (1→2)-*O*-β-D-glucopyranoside-; 3-*O*-{[α-L-rhamnopyranosyl (1→2)] - [α-L-rhamnopyranosyl (1→4)]-β-D-glucopyranoside}: 3-*O*-{[α-L-rhamnopyranosyl (1→2)-α-L-rhamnopyranosyl (1→3)]-α-L-rhamnopyranosyl (1→4)]-β-D-glucopyranoside}: neotigogenin, respectively.

The structures of nicotianosides E (**5**), F (**6**), and G (**7**) have been established as: 3-*O*-α-L-rhamnopyranosyl (1→2)-*O*-β-D-glucopyranoside; 3-*O*-{[α-L-rhamnopyranosyl-(1→2)]–[α-L-rhamnopyranosyl (1→4)]-β-D-glucopyranoside}: 3-*O*-{[α-L-rhamnopyranosyl (1→2)-[α-L-rhamnopyranosyl (1→3)]-α-L-rhamnopyranosyl (1→4)]-β-D-glucopyranoside}: (25*S*)-5-α-furostane-3β,22α,26-triol [-26-*O*-β-D-glucopyranoside].

Growth-stimulating activities, prevention of virus attacks on tobacco, and a possible mechanism of action were established for the nicotianosides.

REFERENCE

1. P.K. Kintia, G.V. Lazuryevskiy, N.N. Balashova, I.T. Balashova, A.I. Suruzhiu, and V.A. Lyakh, *Structure and Biological Activity of Steroidal Glycosides Of The Spirostan And Furostane Series*, Shtiintsa, Kishinev (1987) (In Russian).

2. C. Sannié and H. Lapin, Recherches sur les sapogenines a noyau sterolique. Identification les genines sur de petites quantites de plantes. *Bull. Soc. Chim. Fr.* **19**:1080 (1952).

3. S. Kiyosawa, M. Hutoh, T. Komori, T. Nohara, I. Hosokawa, and T. Kawasaki, Detection of prototype compounds of diosgenin and other spirostanol glycosides. *Chem. Pharm. Bull.* **16**: 1162 (1968).

4. M.E. Wall, C.R. Eddy, M.L. McClennan, and M.E. Klumpp, Detection and estimation of steroidal sapogenins in plant tissues. *Anal. Chem.* **24**: 1337 (1952).

5. P.K. Agrawal, D.C. Jain, R.K. Gupta, and R.S. Thakur, Carbon 13-NMR spectroscopy of steroidal sapogenins and steroidal saponins. *Phytochemistry* **24**: 2479 (1985).

6. V.V. Krokhmalyuk, P.K. Kintia, and V.Ya. Chirva, Gas-liquid chromatography of monosaccharides of triterpene glycosides. *Izv. AN MSSR* **1**: 103 (1975) (In Russian).

7. S. Hakomori, A rapid permethylation of glycolipids and polysaccharides catalyzed by methylsulfonyl carbanion in dimethyl sulfoxide. *J. Biochem.* (Tokyo) **55**: 205 (1964).

8. C. S. Xing and J.K. Snyder, Diosgenin-bearing molluscicidal saponins from *Allium vineale*: An NMR approach for the structural assignment of oligosaccharide units. *J. Org. Chem.* **54**: 3679 (1989).

9. A.V. Kamernitskiy, N.K. Abubakirov, M.V. Gorovits, Yu, E. Vollerner, N.E. Voyshvillo, I.G. Reshetova, and V.A. Paseshnichenko, *Chemistry of Spirostanols*, Nauka, Moscow (1986) (In Russian).

10. A. Aksel, B.Y. Bothner, R.L. Stephens, J.M. Lee, C.D. Warren, and R.W. Jeanloz. Structure determination of a tetrasaccharide: Transient nuclear Overhauser effects in the rotating frame. *J. Am. Chem. Soc.* **106**: 811 (1984).

11. S. Seo, Y. Tomita, K. Tori, and Y. Yoshimura, Determination of the absolute configuration of a secondary hydroxy group in a chiral secondary alcohol using glycosidation shifts in carbon-13 nuclear magnetic resonance spectroscopy. *J. Am. Chem. Soc.* **100**: 3331 (1978).

12. N.N. Balashova, *Plant Breeding: New Genetic Approaches and Solutions.* Shtiintsa, Chisinau, 340 (1991) (In Russian).

13. E.D. Scherban and A.P. Grosu, *On the techniques of estimating tobacco resistance to major diseases and pests.* Moldagroninformreklama, Chisinau, Moldovia (1992) (In Russian).

14. K. Tashke, *Introduction into quantitative cryohistological morphology*, Acad. Sci., Bucharest, Romania (1980).

15. M.E. Ladygina, V.P. Grishkova, and L.N. Parshakova, Photoinduced proton absorption in chloroplasts of tobacco species showing differential TMV resistance on exposure to viral infection. *Agr. Biol.* **2**:81 (1983) (In Russian).

NOVEL MICROBIAL TRANSFORMATIONS OF STEROIDS

K.M. Madyastha

Department of Organic Chemistry
Indian Institute of Science
Bangalore 560 012, India

INTRODUCTION

In recent years, reactions catalyzed by microbes or their enzymes have been extensively evaluated from the viewpoint of synthetic organic chemistry. These biocatalysts offer greater specificity than conventional organic reactions and hence used as reagents to carry out certain specific chemical reactions. Microbes find use in the synthesis and production of pharmacologically active steroids. Microbial transformation of steroids was recognised after the successful 11α-hydroxylation of progesterone by the fungus, *Rhizopus arrhizus*.[1] Microorganisms can be efficiently used to carry out hydroxylation of a steroid molecule at specific positions. A significant amount of work has already been carried out on the microbial steroid transformations.[2-4]. We isolated two versatile microorganisms identified as *Mucor piriformis* and *Moraxella* sp. that carry out novel and preparatively useful transformations of some steroids. The unique properties of these two microorganisms as biocatalyst are presented.

NEW PATHWAY IN THE MICROBIAL CONVERSION OF 3ß-ACETOXY-19-HYDROXYCHOLEST-5-ENE INTO ESTRONE

The steroid molecule contains numerous asymmetric centres and so the total synthesis of pharmacologically active steroids is a difficult task. Hence, partial synthesis of steroids from compounds of steroidal origin is favored. Since a cyclopentanoperhydrophenanthrene structure is necessary for all steroids, the best source for partial synthesis would be the abundantly available natural sterols such as cholesterol and phytosterols. Removal of the alkyl side chain from the C_{17} position in sterols leaves behind the cyclopentanoperhydrophenanthrene structure. Considerable work has been carried out on the microbial removal of the C_{17} hydrocarbon side chain of various sterols.[5-13] These studies have clearly demonstrated that the hydrocarbon side chain of sterols is degraded via C_{26}, C_{24}, and C_{22}

acids to C_{19} androstane derivatives. It is generally known that sterol-degrading microorganisms possess both side chain and steroid nucleus-degrading activities. The nucleus-degrading activity can be prevented either by using a specific inhibitor or by chemically modifying the substrate.[7] Modified sterols such as cholest-4-en-19-ol-3-one and 3ß-acetoxy-19-hydroxycholest-5-ene (19-HCA) readily undergo microbiological conversion into estrone.[5-7] These studies have demonstrated that the degradation of the C_{17} side chain proceeds via C_{22} phenolic acid intermediates. Further degradation involves cleavage of the three-carbon side chain of the C_{22} phenolic acid to estrone and propionic acid suggesting that complete removal of C_{17} side chain takes place only after the aromatization of the A ring.

During our search for microbial process to selectively eliminate the C_{17} side chain of cholesterol or its derivatives, a soil bacterium belonging to the genus *Moraxella* was isolated by an enrichment culture technique using the model compound isooctyl-cyclopentane as the sole carbon source.[11] The model compound resembles the D ring and side chain of cholesterol. The organism also accepts 3β-acetoxy-19-hydroxy-cholest-5-ene

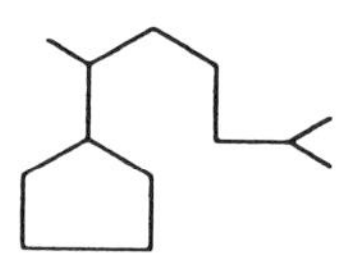

ISOOCTYLCYCLOPENTANE

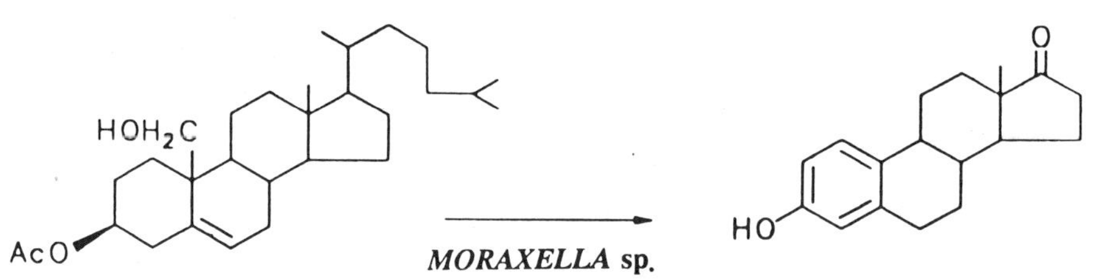

1 2

(19-HCA, 1) as a carbon source and readily transforms 19-HCA (1) into estrone (2).[11] Detailed studies were initiated to establish the mechanism involved in the cleavage of the C_{17} side chain and the aromatization of the A ring during the formation of estrone (2).

Moraxella sp. can be maintained on Seubert's medium[14] containing 0.05% cholesterol and 2% agar (pH 7.2) or in a liquid mineral salts medium[14] containing 0.05% cholesterol. Large-scale fermentation of 19-HCA (1) with *Moraxella* has been carried out to establish the degradative sequence in the conversion of 19-HCA to estrone (2). The procedures followed for fermentation, isolation, and characterization of metabolites have been reported earlier in great detail.[13,15] Fermentation of 19-HCA(1) with *Moraxella* resulted in the formation of estrone (2) as the major metabolite.[11,13,15] The minor metabolites formed (Fig.1) were 5α-androst-1-en-19-ol-3,17-dione (3), androst-4-en-19-ol-3,17-dione (4), androst-4-en-9α, 19-diol-3,17-dione (5) and androstan-19-ol-3,17-dione (6). The minor metabolite 5α-androst-1-en-19-ol-3,17-dione (3) appears to be hitherto unknown and its structure has been conclusively established by X-ray analysis.[13] Acidic metabolites were not formed during the transformation.

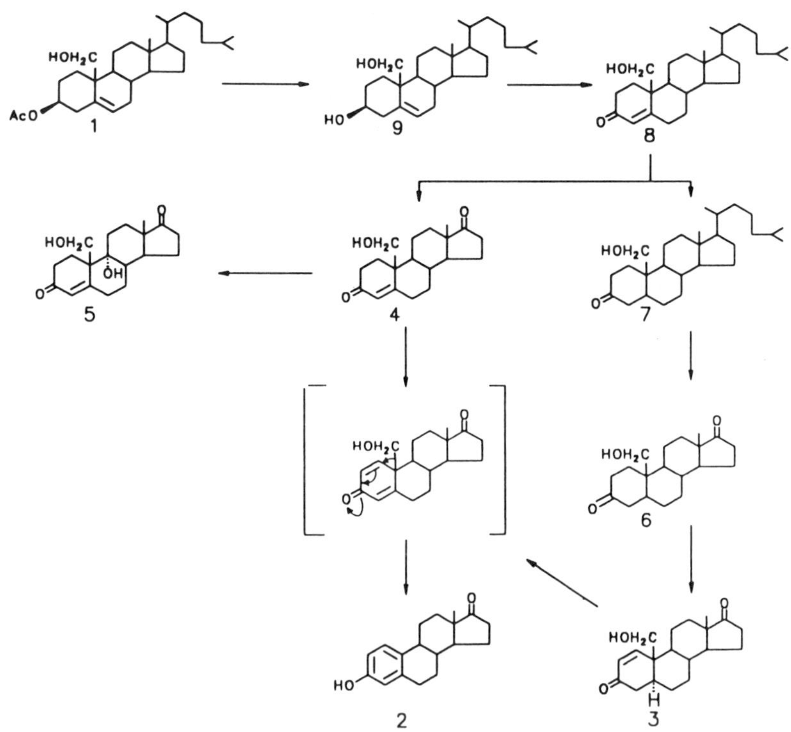

Fig. 1 Proposed pathway for the metabolism of 3β-acetoxy-19-hydroxycholest-5-ene (1) by *Muroxella* sp

The time course plot (Fig.2) of a 7 day fermentation of 19-HCA (1) indicated that during the early stages (2 days), nearly 10% of 19-HCA was metabolized and androst-4-en-19-ol-3,17-dione (4) was formed in small quantities. Prolonging the incubation period to 3 days resulted in the accumulation of 4 and low levels of estrone (2). When the fermentation was continued up to 7 days, the level of 4 dropped with a concomitant increase in estrone (2). The experimental details of these studies have been reported earlier.[15]

Other researchers reported that steroid nucleus-degrading activity by microorganisms can be prevented by specific inhibitors to enable detection of intermediary metabolites.[7,10,12] The addition of chelating agents such as α,α'-bipyridyl (α,α'-D) to cholesterol-decomposing microorganisms not only inhibits the 9α-hydroxylation reaction, but also induces the accumulation of acidic metabolites. To ascertain whether acidic metabolites are formed during the microbial conversion of 19-HCA (1) into estrone (2), we carried out fermentation of 19-HCA (1) with *Moraxella* in the presence of α,α'-D or 1-propanol. In these experiments, the organism was grown in mineral salts medium containing 0.2% glucose and 0.05% 19-HCA for 36 h. After this growth period, α,α'-D (0.6 mM) or 1-propanol (2%) was added and incubated for an additional 48 h. Metabolites formed were isolated and characterized as described earlier.[15] We observed that incubation in the presence of α,α'-D or 1-propanol did not result in the formation of acidic metabolites, even during the early stages of fermentation. However, these experiments revealed the formation of three additional neutral metabolites which were not seen when the incubation was carried out in the absence of inhibitors. In order to isolate and characterize these additional neutral metabolites, a large-scale resting-cell experiment was carried out in the presence of α,α'-D. In these experiments cells grown in mineral salts medium containing

0.2% glucose and 0.05% 19-HCA for 48 h were harvested, washed well with phosphate buffer (0.03M, pH 7.2), and then suspended in sterile phosphate buffer (0.03M, pH 7.2). To this cell suspension, 19-HCA and α,α'-D (0.6 mM) were added and the mixture incubated at 30 °C for 48 h. From the culture medium metabolites were isolated and characterized as described earlier.[15] These experiments led to the isolation and characterization of three additional metabolites, viz. cholestan-19-ol-3-one (7), cholest-4-en-19-ol-3-one (8) and cholest-5-ene-3ß,19-diol (9) (Fig.1). Similar results were also obtained when 1-propanol (2%) was used instead of α,α'-D.[15] These studies also indicated that α,α'-D completely inhibits 9α-hydroxylation.

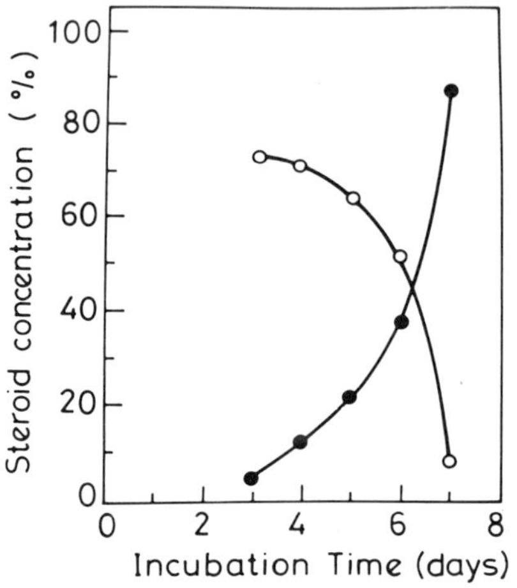

Fig. 2 Precursor - Product relationship between 4 and 2
o - o 19-Hydroxyandrost-4-ene-3,17-dione (4), ● - ● Estrone (2)

Transformation of the Intermediates

Resting cells grown on 19-HCA (1) readily converted both 3 and 4 into estrone (2).[15] However, the rate of conversion of 3 to 2 was considerably lower than the rate of conversion of 4 to 2. It was also noticed that incubation of 8 with resting cells resulted in the formation of 4 and 2. These studies provided information regarding the sequence of reactions taking place during the formation of estrone (2) from 19-HCA. To understand the mode of degradation of the sterol side chain and aromatization of A ring in estrone, studies were carried out at the cell-free level.

Cell-free extract was prepared from 19-HCA-induced *Moraxella* cells as reported earlier.[15] The cell-free extract converted both 3 and 4 into 2 in the presence of phenazine methosulfate, an artificial electron acceptor.[15] This indicated the presence of both steroid 1,2- and 4,5-dehydrogenases in the cell extract. However, the cell extract contained higher levels of steroid 1,2-dehydrogenase than steroid 4,5-dehydrogenase. The specific activities of the 1,2-dehydrogenase and 4,5-dehydrogenase were 3.1 and 0.7 nmol/min/mg, respectively.[15] This observation amply explains why the conversion of 3 to 2 was considerably slower than the conversion of 4 to 2. The available evidences also suggest that the major metabolite 2 is formed mostly from androst-4-en-19-ol-3,17-dione (4) through the action of 1,2-dehydrogenase (Fig.3). Introduction of 1,2-double bond in 4 facilitates a spontaneous

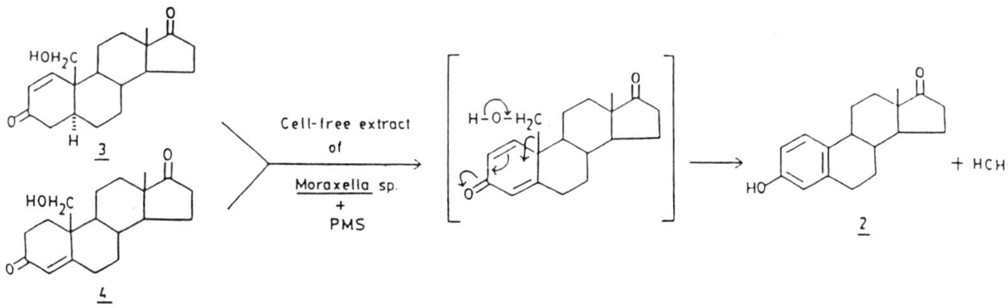

Fig.3 Steroid 1,2- and 4,5-dehydrogenation using cell-free extract of *Moraxella* sp.

nonenzymatic retro aldol-type reaction to yield estrone (2) and formaldehyde (Fig.3). Partially purified 1,2-dehydrogenase[15] converted 4 into 2 and formaldehyde in the presence of phenazine methosulphate. One would also expect a similar rearrangement to take place by introducing a 4,5-double bond in metabolite 3. The presence of steroid 4,5-dehydrogenase in the cell-free extract has also been demonstrated.[15] In addition, the resting cells transform metabolite 3 into estrone (2).

Our results suggest that the degradation of the hydrocarbon side chain of 19-HCA (1) does not proceed via C_{22} phenolic acid intermediates and complete removal of C_{17} side chain takes place prior to the aromatization of the A ring in estrone (2). The mode of degradation of the sterol side chain appears to be through the fission of the C_{17}-C_{20} bond. Efforts to isolate the hydrocarbon side chain of 19-HCA (1) as 6-methylheptan-2-one were not successful. However, we have noticed that cells adapted to 19-HCA (1) readily metabolize 6-methylheptan-2-one.[15] The data presented herein suggest that *Moraxella* sp. transforms 19-HCA (1) into estrone (2) by following a pathway (Fig.1) different from that established earlier.[5,6,7] Our results also appear to be different from the mammalian systems wherein sterols are converted into C_{17} steroids via cleavage of the C_{20}-C_{22} bond, followed by the cleavage of the C_{17}-C_{20} bond. Further conversion of 17-keto steroids to estrone is catalysed by the microsomal cytochrome P-450 enzyme aromatase by utilizing 3 moles each of NADPH and O_2.[16,17] Thus, we have established a new pathway (Fig.1) for the conversion of 19-HCA (1) into estrone (2), where neutral metabolites play a key role. The study also for the first time demonstrated the transformation of the hitherto unknown compound 5α-androst-1-en-19-ol-3,17-dione (3) into estrone (2).

The introduction of the 1,2-double bond into steroids represent one of the most valuable uses of microorganisms in steroid chemistry. The 1,2-dehydro derivatives of cortisone and cortisol namely, prednisone and prednisolone, showed increased antirheumatic and antiallergic activities and produced less undesirable side effects. It has also been demonstrated that cortico hormone analogues with an additional 1,2-double bond in ring A possess enhanced glucocorticoid and reduced mineralocorticoid activities.

Our studies on the microbial conversion of 19-HCA(1) into estrone (2) clearly established that steroid 1,2-dehydrogenase plays an important role in the aromatization of the A ring in estrone (2). Although steroid 1,2-dehydrogenase is very well documented in the literature, such activity has not been noticed in *Moraxella* sp. Hence a detailed study was carried out on the structure-activity relationship of the 1,2-dehydrogenation reaction using various steroids. These studies have demonstrated that *Moraxella* has the ability to carry out the 1,2-dehydrogenation of various C_{19}, 19-nor and C_{21} steroids with a 3-keto-4-en system. However, it appears that when the A-ring in a C_{21} or C_{19} steroid molecule is modified with a hydroxyl at C-3 without the 4,5- double bond or with a 5,6-double bond, then the organism initiates the reaction by converting such a system to a 3-keto-4-ene system before carrying out the 1,2-dehydrogenation reaction. The 3-keto-4-ene system or at least a 3-keto group appears to be one of the required structural feature in a C_{21} or C_{19} steroid molecule for the 1,2-dehydrogenase to accept as the substrate.

The 1,2-dehydrogenase from *Moraxella* has been shown to be inducible in nature. The enzyme is localized in the cytosol and does not require oxidized pyridine nucleotides (NAD^+, $NADP^+$) for activity. The enzyme is selective to steroids as it fails to convert various monocyclic, bicyclic, and acyclic ketones as substrates for 1,2-dehydrogenation reaction.

TRANSFORMATIONS OF STEROIDS BY *Mucor piriformis*

Mucor belongs to the family of Mucoraceae of the order *Mucorales*, under the sub-class *Zygomycetes*, which is a lower form of fungi. The *Mucorales* are a large and diverse group of fungi. The majority of these fungi are saprophytes that feed on vegetative debris or, to a lesser extent, on animal debris in the soil. These fungi have an extensive branched vegetative mycelium. Many of the fungi in this group are known to transform organic compounds of diverse structures.

Mucorales exhibit a variety of steroid-transforming activities.[18] These organisms mainly effect hydroxylation at 14α,11α,7α and 6ß positions of various C_{19} and C_{21} steroids.[19-24] Details of some of the transformations mediated by Mucorales are covered by patents.[25,26] We isolated a fungal strain identified as *Mucor piriformis*.[22] Two other reports appeared in the literature on the transformations of steroids by *Mucor piriformis*,[19,20] indicated the hydroxylation at 9,11 and 6-positions. The *Mucor piriformis* isolated in our laboratory has been shown to carry out preparatively useful transformations of some of the steroids and morphine alkaloids.[22,27-29] We present here some of our studies on steroid transformations carried out using *Mucor piriformis* and the scope of its synthetic applicability.

Mucor piriformis used in these studies was maintained and propagated as reported earlier.[22] All the steroid transformations studied using *Mucor piriformis* were carried out in sterile modified Czapek Dox medium.[30] Flasks containing 100 ml of sterile medium were inoculated with 1 ml spore suspension (2.2 x 10^8 counts/ml) from a 5-day-old culture and incubated at 29-30 °C on a rotary shaker (220 rpm) for 48 h. After this growth period, 50 mg of substrate (steroid) in acetone (0.5 ml) was added to each flask and the incubation continued for 48 h. At the end of the incubation period, the contents of the flasks were pooled and extracted with chloroform, metabolites were separated by column chromatography on silica gel and the purified metabolites were subjected to various spectral analysis.[22,27,28]

Transformations of Progesterone(10)

Mucor piriformis transforms progesterone (10) predominantly into 14α-hydroxyprogesterone, which further gets hydroxylated at 6ß,7α, or 7ß position (Fig.4).[22]

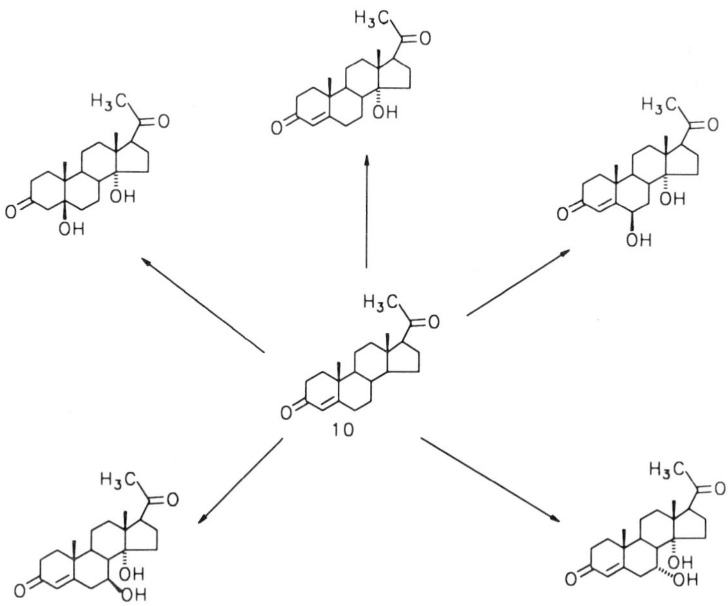

Fig.4 Transformations of progesterone (10) by *M. piriformis*

Mucor piriformis also produces the 5ß,14α-dihydroxy compound as a minor metabolite. Time course experiments have clearly demonstrated that during the early stages of fermentation (12h), nearly 75% of the progesterone (10) gets converted mostly into 14α-hydroxylated product whereas by 48 h dihydroxyprogesterones are formed.[22] Earlier studies have indicated that *Mucor* species have rigid stereoselectivity in their ability to hydroxylate various C_{19} and C_{21} steroids.[19] However, *Mucor piriformis* hydroxylates 14α-hydroxprogesterone at both 7α and 7ß positions.

Transformations of 17α-Hydroxyprogesterone (11) and 16-Dehydroprogesterone (12)

The organism transforms 17α-hydroxyprogesterone (11) into 17α,20α-dihydroxypregn-4-en-3-one, 7α,17α-dihydroxypregn-4-ene-3,20-dione,6ß,17α,20α-trihydroxypregn-4-en-3-oneand 11α,17α,20α-trihydroxypregn-4-en-3-one(Fig.5).[27] Time-course studies indicated that 17α,20α-dihydroxypregn-4-en-3-one is the initial product, which is further hydroxylated either at the 6ß or 11α position. Mucorales are known to hydroxylate various C_{19} and C_{21} steroids at the 14α-position.[18,19,22,24] However, *Mucor piriformis* failed to carry out the 14α-hydroxylation of 11. It is quite possible that the presence of a hydroxyl at the 17α-position would prevent hydroxylation at the 14α-position. It has been shown that an alkyl substitution at the 17α-position in a steriod molecule would prevent Mucorales from carrying out hydroxylation at the 14α-position.[31] However, steroids with a hydroxyl group at the 17ß-position, as in testosterone, are readily transformed to their 14α-hydroxyl derivatives.[24] So it appears that the bulky substitution at the 17α-position sterically hinders hydroxylation at the 14α-position and directs it to the 7α-position. The organism can also reduce the 20-keto group in 17α-hydroxyprogesterone (11,Fig.5), a reaction never reported by organisms of the order Mucorales.[27] Surprisingly, the reduction of the 20-keto group has not been observed in the case of progesterone, 16-dehydroprogesterone, and pregnenolone. A time-course experiment carried out with compound 11 indicated that this compound metabolized at a comparatively lower rate than

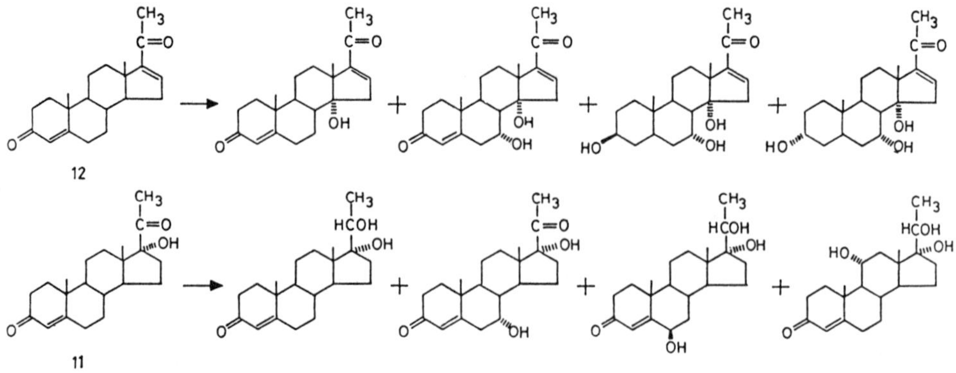

Fig.5 Transformations of 16-dehydroprogesterone (12) and 17α-hydroxyprogesterone (11) by *Mucor piriformis*

compound 12. The compound with the reduced 20-keto group (Fig.5) was formed in considerable amount during the early stages of incubation(12h).

Various fungal species transform 16-dehydroprogesterone (12) into different metabolites involving hydroxylations at the 11α and 21-positions, dehydrogenation at the C-1 position, and reduction of the C-16(17) double bond.[32] In contrast, *Mucor piriformis* efficiently hydroxylates at the 14α-position of 16-dehydroprogesterone[27] (12). Further transformation involves hydroxylation at the 7α-position. At the end of 48 h, nearly 80% of 12 gets transformed to 7α,14α-dihydroxypregna-4,16-diene-3,20-dione (Fig.5) and this metabolite accumulates in the culture medium in significant levels. The reduction of the 4-en-3-one group is not a major activity of this organism since metabolites resulting from this activity are formed in small quantities. Reduction of the 3-keto group is not stereoselective since metabolites with both 3α and 3β-hydroxyl function have been isolated (Fig.5) and these metabolites appear to be hitherto unkown. Reduction of the 4-en-3-one group in a steroid molecule by any *Mucor* sp. has not been reported so far.

To our knowledge, *Mucor piriformis* has not been used to carry out transformations in compounds 11 and 12. It appears that this organism does not produce any side-chain-cleaved product from either compound 11 or 12. Microorganisms appear to have altered specificity with regard to the C_{17} side chain. In cubation of 17α-hydroxyprogesterone (11) with a *Bacillus* sp was shown to yield a side chain cleaved product.[33]

Transformations of Androstenedione (13)

Androstenedione (13) appears to be a good substrate for *Mucor piriformis*. The organism transforms 13 into testosterone, 14α-hydroxyandrostenedione, 7α-hydroxy-androstenedione, 14α-hydroxytestosterone and 7α,14α,17β-trihydroxyandrost-4-en-3-one (Fig.6). The major metabolite formed at the end of 24 h has been shown to be 14α-hydroxytestosterone which further gets hydroxylated at the 7α-position resulting in the formation of a trihydroxy compound (Fig.6). 14α-Hydroxytestosterone could have been formed from 14α-hydroxyandrostenedione in the presence of a 17-keto oxidoreductase or from testosterone by 14α-hydroxylation. The formation of testosterone and 14α-hydroxytestosterone from androstenedione has been reported earlier using *Phycomyces blakesleeanus*.[34] Although the reduction of a 17-keto group has been reported in several microbial systems, it has never been shown in *Mucor piriformis*.

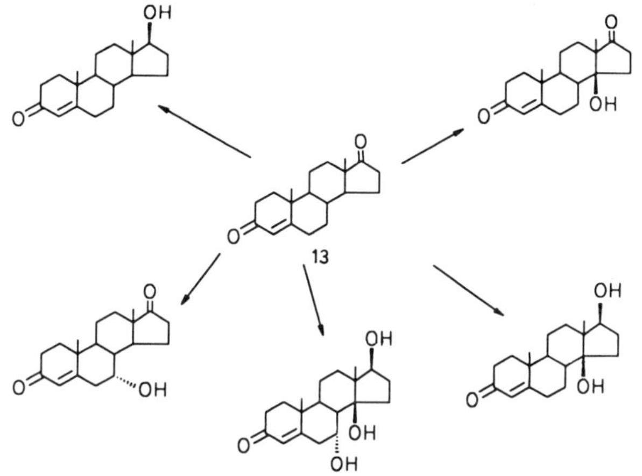

Fig.6 Transformations of androstenedione (13) by *M.piriformis*

Transformations of Dehydroepiandrosterone (14) and Pregnenolone(15)

To find out whether the organism has the ability to isomerize a 5-en-3-ol to a 4-en-3-one system in C_{19} and C_{21} steroids, representative steroids belonging to this category, viz. dehydroepiandrosterone (14) and pregnenolone (15) were used as substrates. These substrates have also provided an opportunity to find out the effect of a 5-en-3ß-ol system on the mode of transformation by *Mucor piriformis*. Incubation of 14 with this organism yielded metabolites resulting from hydroxylation at the 7α-position as well as reduction of the 17-keto group (Fig.7)[28]. *Mucor piriformis* fails

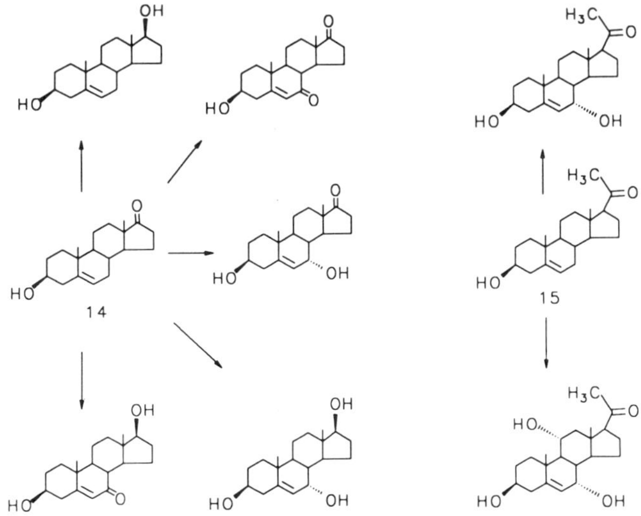

Fig.7 Transformations of dehydroepiandrosterone(14) and pregnenolone(15) by *M. piriformis*

to carry out 14α-hydroxylation of 14 and 15 . Both these compounds (14 and 15) contain a 5-en-3ß-ol system in their structure and it appears that one of the structural requirements needed for the 14α-hydroxylase to accept a steroid molecule as a substrate is the presence of a 4-en-3-one group in its molecule. Reduction of the 17-keto group to 17ß-hydroxyl as seen in the case of dehydroepiandrosterone (14) and androstenedione (!3) has not been reported in fungi of the order Mucorales.

The organism transforms pregnenolone (15) into 3ß,7α-dihydroxypregn-5-en-20-one and 3ß,7α,11α-trihydroxypregn-5-en-20-one (Fig.7).[28] Microbial conversion of 15 into progesterone as a result of the isomerization of the 3ß-hydroxy-5-ene to the 4-en-3-one system has been reported.[35] There are also reports on the further conversion of progesterone to hydroxylated progesterones with a hydroxyl function at 7ß, 11α, 12ß, or 15α-position.[36,37] So it appears that *Mucor piriformis* is not capable of isomerizing 5-en-3ß-ol to 4-en-3-one, a reaction many organisms are capable of performing.

Cursory examination of the metabolites formed from different steroids tested reveals that *Mucor piriformis* readily converts steroids with a 4-en-3-one group and without bulky sustitution at the 17α-position to their respective 14α-hydroxy derivatives, and also carries out 7α-hydroxylation very efficiently. Although organisms belonging to the order *Mucorales* are known for their unique ability to hydroxylate various C_{19} and C_{21} steroids at the 14α-position, very little is known regarding the enzyme system involved in this hydroxylation reaction. This hydroxylase system has never been fully characterized. So far only two fungal cell-free steroid hydroxylation systems have been studied - the 11α-hydroxylase from *Aspergillus* and *Rhizopus* species[38-41] and a 7α-hydrolase of *P.blakesleeanus*.[42]

Cell-free extracts with high 14α-hydroxylase activity have been prepared from induced vegetative cell cultures of *Mucor piriformis* by grinding in potassium phosphate buffer (0.05M, pH 8.0) containing glucose (0.25 M), KCl (1 mM), glutathione (1.0 mM) and glycerol (10%).[43] Differential centrifugation studies with the active cell-free extract have indicated that most of the 14α-hydroxylase activity was associated with the membrane fraction sedimenting at 105 x 10^3 g (microsomes). It was also observed that microsomes prepared from uninduced cells did not contain any hydroxylase activity. Both NADPH and O_2 are necessary for the hydroxylase activity. Microsomal 14α-hydroxylase activity can be inhibited to a significant extent by CO and p-CMB whereas moderate inhibition was seen with SKF-525A, metyrapone, and N-methylmaleimide. The inhibition studies indicate the possible involvement of cytochrome P-450 in the hydroxylation reaction. Microsomes isolated from progesterone-induced mycelia have been shown to carry out the 14α-hydroxylation of various C_{19} and C_{21} steroids.[43] These studies have also supported the earlier findings that a bulky substitution at a 17α-position would prevent hydroxylation at the 14α-position by *Mucorales*. Further studies are required to characterize this membrane-bound hydroxylase system.

REFERENCES

1. D.H.Peterson and H.C.Murry, Microbiological oxygenation of steroids at carbon-11, *J.Am.Chem.Soc.*74: 1871 (1952)
2. W.Charney and H.L.Herzog, *Microbial transformations of steroids: a hand book*, Academic Press, New York (1967).
3. S.B.Mahato and S.Banerjee, Steroid transformations by microorganisms-II, *Phytochemistry* 24: 1403 (1985).

4. S.B.Mahato, S.Banerjee and S.Podder, Steroid transformations by microorganisms-III, *Phytochemistry* 28: 7 (1989).
5. C.J.Sih, K.C.Wang and H.H.Tai, C_{22} Acid intermediates in the microbiological cleavage of cholesterol side chain, *J.Am.Chem.Soc.* 89: 1956 (1967).
6. C.J.Sih, K.C.Wang and H.H.Tai, Mechanism of steroid oxidation by microorganisms, XIII C_{22} acid intermediates in the degradation of the cholesterol side chain, *Biochemistry* 7: 796 (1968).
7. C.K.A.Martin, Microbiological cleavage of sterol side chain, *Adv.Appl.Microbiol.* 22: 29 (1977).
8. K.Arima, T.Nakamatsu and T.Beppu, Microbial production of 3-oxobisnorchola-1,4-dien-22-oic acid, *Agric.Biol.Chem.* 42: 411 (1978).
9. Y.Fujimoto, C.S.Chen, Z.Szeleczky, D.Ditullio and C.J.Sih, Microbial degradation of the phytosterol side chain.I. Enzymatic conversion of 3-oxo-24-ethylcholest-4-en-26-oic acid into 3-oxo-cholest-4-en-24-oic acid and androst-4-en-3,17-dione, *J.Am.Chem.Soc.* 104: 4718 (1982).
10. R.W.Owen, A.N.Mason and R.F.Bilton, The degradation of cholesterol by Pseudomonas sp. NC1B 10590 under aerobic conditions, *J. Lipid Res.* 24: 1500 (1983).
11. P.K.Bhattacharyya, M.K.Rao, N.R.Devi, M.Ramgopal, P.Madyastha and K.M.Madyastha, Microbial oxidation of sterol side chain, *J.Indian Chem.Soc.* 61: 1 (1984).
12. R.W.Owen, A.N.Mason and R.F.Bilton, The degradation of ß-sitosterol by Pseudomonas sp. NC1B 10590 under aerobic conditions, *J. Steroid Biochem.* 23: 327 (1985).
13. V.N.Shankar, T.N.Guru Rao and K.M. Madyastha, Evidence for a new pathway in the microbial conversion of 3ß-acetoxycholest-5-en-19-ol into estrone, *J.Chem.Soc.,Perkin Trans.I*: 2233 (1993).
14. W.Seubert, Degradation of isoprenoid compounds by microorganisms, *J. Bacteriol.* 79:426 (1960).
15. K.M.Madyastha and V.N.Shankar, Role of neutral metabolites in microbial conversion of 3ß-acetoxy19-hydroxy cholest-5-ene into estrone, *Appl.Environ.Microbiol.* 60: 1512 (1994).
16. S.Burstein and M.Gut, Intermediates in the conversion of cholesterol to pregnenolone: Kinetics and mechanism, *Steroids* 28: 115 (1976).
17. E.A.Thompson Jr. and P.K.Siiteri, The involvement of human placental microsomal cytochrome P-450 in aromatization, *J.Biol.Chem.* 249: 5373 (1974).
18. C.Vezina and K.Singh, Transformation of organic compounds by fungal spores, in: *Filamentous Fungi*, Vol. I, pp 158-192, J.E.Smith and D.R.Berry, eds. Edward Arnold, London (1975).
19. H.C.Murray and D.H.Peterson, 8-Hydroxy-11-deoxycorticosterones, U.S. Patent 2,800,490 (1957) CA 52: 1300 g (1958).
20. V.K.Eroshin, Capacity of *Mucorales* fungi to oxidize steroid S, *Mikrobiologiya* 31: 608 (1962).
21. H.L.Holland and E.Riemland, Microbial hydroxylation of steroids. Rearrangements during epoxidation and hydroxylation and the stepwise nature of these enzymic reactions, *Can.J.Chem.* 63: 1121 (1985).
22. K.M.Madyastha and J.Srivatsan, Novel transformation of progesterone by a *Mucor* sp. *Can.J.Microbiol.* 33: 361 (1987).
23. S.B.Mahato, S.Banerjee and S.Podder, Steroid transformation by microorganisms-III, *Phytochemistry* 28: 7 (1989).

24. R.Krishnan, K.M.Madyastha and M.A.Viswamitra, The crystal structure of 14α, 17ß-dihydroxyandrost-4-en-3-onemonohydrateand 14α, 17ß-dihydroxyandrost-1,4-dien-3-one monohydrate, *Steroids* 56: 440 (1991).
25. H.C.Murray and D.H.Paterson, US Patent 2,602,769 (1952) CA 46: 8331 f (1952).
26. R.M.Dodson and R.C.Tweitt, US Patent 2,924,611 (1960) CA 54: 11086 h (1960).
27. K.M.Madyastha and T.Joseph, Transformation of 16-dehydroprogesterone and 17α-hydroxyprogesterone by *Mucor pirifirmis*, *Appl.Microbiol.Biotechnol.* 41: 170 (1994).
28. K.M.Madyastha and T.Joseph, Transformation of dehydroepiandrosterone and pregnenolone by *Mucor piriformis*, *Appl. Microbiol. Biotechnol.* 43: 000 (1995).
29. K.M.Madyastha and G.V.B.Reddy, *Mucor piriformis*, an efficient N-dealkylating reagent for thebaine and its N-variants, *J. Chem. Soc., Perkin. Trans. I*, 911 (1994).
30. B.R.Prema and P.K.Bhattacharyya, Microbiological transformations of terpenes II. Transformations of α-pinene, *Appl. Microbiol.* 10: 524 (1962).
31. K.Singh, S.N.Sehgal and C.Vezina, Transformation of steroids by *Mucor griseocyanus*, *Can. J. Microbiol.* 13: 1271 (1967).
32. C.Vezina, S.N.Sehgal and K.Singh, Transformation of steroids by spores of microorganisms. I. Hydroxylation of progesterone by conidia of *Apergillus ochraceus*, *Appl. Microbiol.* 11: 50 (1963).
33. S.B.Mahato and S.Banerjee, Metabolism of 17α-hydroxyprogesterone by a *Bacillus* species, *Biochem. J.* 239: 473 (1986).
34. K.E.Smith, S.latif and D.N.Kirk, Microbial transformation of steroids. II. Transformation of progesterone, testosterone and androstenedione by *P.blakesleeanus*, *J.Steroid Biochem.* 32: 445 (1989).
35. D.Perlman, Microbiological conversion of pregnenolone to progesterone, *Science* 115: 529 (1952).
36. L.Tan and L.L.Smith, Microbial hydroxylations.III. 11α-Hydroxylation of some 3α,5α-cyclopregnane derivatives, *Biochem. Biophys. Acta* 164: 389 (1968).
37. K.Namboori, L.Pereira and J.R.Merchant, Fungal transformation of pregnenolone and progesterone with the marine fungus *Cladosporium herbarum*, *Indian J. Biochem. Biophys.* 17: 149 (1980).
38. M.Sibahara, J.A.Moody and L.L.Smith, Microbial hydroxylations. V. 11α-Hydroxylation of progesterone by cell-free preparation of *Aspergillus ochraceus*, *Biochim. Biophys. Acta* 202: 172 (1970).
39. K.Breskvar and T.Hudnik-Plevnik, Inducibility of cytochrome P-450 and NADPH-cytochrome C reductase in progesterone-treated filamentous fungi *R. nigricans*, *J. Steroid Biochem.* 14: 395 (1981).
40. C.R.Jayanthi, P.Madyastha and K.M.Madyastha, Microsomal 11α-hydroxylation of progesterone in *A. ochraceus* : part I. Characterization of the hydroxylase system, *Biochem.Biophys.Res.Commun.* 106: 1262 (1982).
41. K.M.Madyastha, C.R.Jayanthi, P.Madyastha and D.Sumathi, Studies on the microsomal 11α-hydroxylation of progesterone in *A.ochraceus*. Characterization and solubilisation of the hydroxylase system, *Can. J. Biochem. Cell. Biol.* 62: 100 (1984).
42. K.E.Smith, S.A.Latif and D.N.Kirk, Microbial transformation of steroids.VII. Hydroxylation of progesterone by extracts of *Phycomyces blakesleeanus*, *J.Steroid Biochem. Molec. Biol.* 38: 249 (1991).
43. K.M.Madyastha and T.Joseph, Studies on the 14α-hydroxylation of progesterone in *Mucor piriformis*, *J. Steroid Biochem. Molec. Biol.* 45: 563 (1993).

THERMAL BEHAVIOR OF STEROIDAL GLYCOSIDES

V.A. Bobeyko and P.K. Kintia

Institute of Genetics
Academy of Sciences
Chisinau 2002, Moldova

INTRODUCTION

At present, more than half a thousand C-27 steroidal glycosides of the furostane, spirostane, and spirosolane series, most of which posses diverse biological activity, have been isolated from various representatives of earth flora, and their chemical structures have been elucidated. Methods to establish and modify molecular structures of naturally occurring glycosides include a wide range of chemical techniques. However, the potential of changing their structures under the action of physical factors, particularly high temperatures, have not been studied sufficiently well. Investigations on the thermal behavior of steroidal saponins and their sapogenins are necessitated both by theoretical interest and a need in searching for the ways to obtain new substances from native compounds. The prospects of creating medicines and other biologically active preparations from them demand additional information on their properties. These considerations, in our opinion, justify the investigation of the thermal behavior of steroidal glycosides. A part of the findings in this investigation are presented in this work.

RESULTS AND DISCUSSION

Studies on the thermal properties of C-27 steroidal glycosides and sapogenins have been conducted in two steps. The first one included thermal analysis (TA) of the substances, resulting in thermal curves (DTG, DTA, and TG); the second one included identification of intermediates produced in the course of the probe heating in accordance with the effects detected on thermal curves, and determination of the paths of thermolysis of glycosides differing in structure.

TA data were compared within the group of biogenetically related compounds (furostanol glycoside, its spirostanol analog, and their sapogenins) and among representatives of a chemical group (individual furostanol, spirostanol, and sapogenin compounds).

Weight loss in the range of 2-9% is the first effect detected on the thermal curves of most samples studied, regardless of the structural group, in the temperature interval of 70-130 °C. The loss is recorded on the TG curve having a corresponding endothermic effect

Saponins Used in Food and Agriculture
Edited by Waller and Yamasaki, Plenum Press, New York, 1996

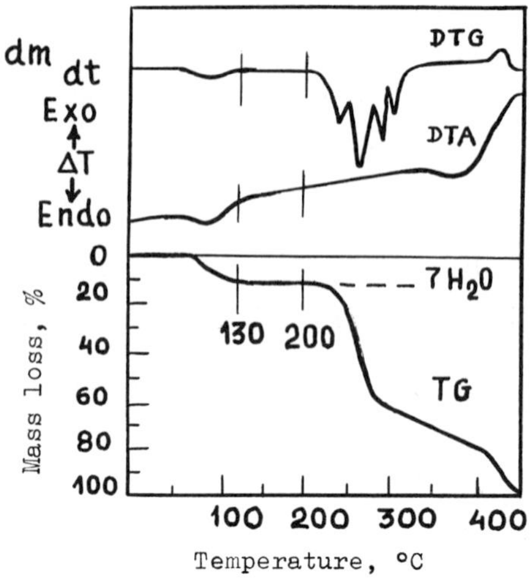

Fig. 1. Thermal curves of tomatoside.

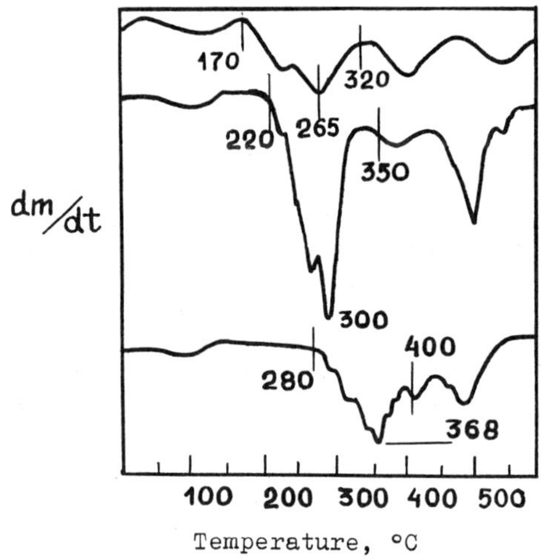

Fig. 3. DTA curves for: 1 - neotigogenin; 2 - gekogenin; 3 - tomatidin.

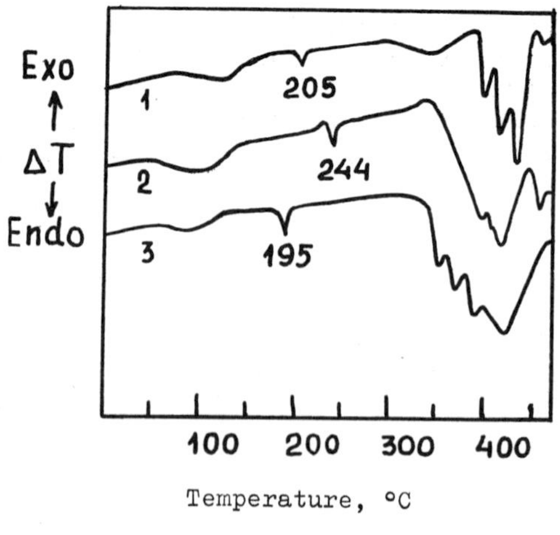

Fig. 2. DTG thermal curves for: 1 - deltoside; 2 - deltonin; 3 - diosgenin.

expressed by a wide endothermic peak on the DTA curve as is observed with particular reference to the thermal curves of tomatoside (Fig. 1), a furostanol glycoside from tomato seeds isolated and described for the first time by Sato and Sakamura[1].

Thin layer chromatography (TLC) and IR spectrometry of the samples of tomatoside, deltoside, a furostanol glycoside isolated by Paseshnichenko and Guseva[2] from *Dioscorea deltoidea* roots, their spirostanol analogs, tomatonin and deltonin, a spirostanol glycoside agavoside C, isolated by use from *Agava americana* L. leaves[3], and spirosolane α-tomatin[4] and their sapogenins of neotigogenin, diosgenin, gekogenin, and tomatidin heated at 130 °C showed that the substances remained unchanged in this temperature interval. They maintain their initial R_f values and specific coloration characteristic of each glycoside, when developed by sulfuric acid in TLC, and characteristic absorption bands in the IR spectrum (furostanol glycosides at 900 cm^{-1}, spirostanol glycosides and sapogenins at 850, 890, 920, and 980 cm^{-1}). This indicates that the first weight loss at the temperature under 130 °C is related to water release.

Further comparative studies of the thermal curves of a furostanol glycoside, deltoside, and its spirostanol analog, deltonin, exhibited great difference in the thermal behavior of three distinctive though biogenetically related compounds. DTG curves turned out to be most informative (Fig. 2).

Thus, on the DTG curve of a deltoside, the initiation of the second thermal effect is recorded at 170 °C with a minimum at 265 °C, after water loss, and in the case of deltonin, the corresponding effect starts at 220 °C and has a minimum at 300 °C. As regards to the sapogenin diosgenin, the effect following water loss begins at 280 °C and has a minimum at 368 °C. At these temperatures weight loss related to thermodegradation of the substances under study begins. These effects are completed at 320 °C, 350 °C, and 400 °C for the furostanol deltoside, the spirostanol deltonin, and the diosgenin sapogenin, respectively. The same regularity in the thermal behavior was observed in the group of biogenetically related furostanol tomatoside, its spirostanol analog tomatonin, and neotigogenin sapogenin, corresponding to them. The initiation, following water loss, of weight loss was recorded on the DTG and TG curves according to the above pattern at 200 °C (Fig. 1), 230 °C (Fig. 4A), and 290 °C (Fig. 4B).

It should be noted that in the case of sapogenins, a wide endothermic peak related to water loss (where it occurs) is followed on the DTA curve (Fig. 3) by another low peak, to which no weight loss corresponds and which is not expressed on the DTG curve.

The minimum of this peak in each case coincides with the melting point of the sample analyzed. These facts indicate that the appearance of the second endothermic peak on the DTA curves of sapogenin is associated with the solid-liquid transition. A similar peak was recorded in the course of thermal analysis of sterols, which indicates possibly more frequent display (than in the case of sapogenins) of steroid melting point on the DTA curve in thermal analysis. The effect makes it possible to additionally estimate the identity of degree and purity of the sample under study.

Nevertheless, in the case of the diosgenin sample studied, the peak of the transition has not been displayed on the DTA curve. However, a relatively low, sharp exothermic peak with a maximum at 190 °C was recorded with no visible effects on other curves (DTG and TG). The appearance of this peak is likely to be caused by a conformational change, but at present the explanation is not complete.

The temperature of initiation of the thermal decomposition (after water loss) varied in the representatives of the remaining groups of the substances studied: agavoside C (190 °C), its sapogenin, gekogenin (290 °C), tomatin (180 °C), and its sapogenin, tomatidin (280 °C) (Figs. 4A and B).

The thermal TG curves in Figs. 4A and 4B describing weight loss in spirostanol glycosides and their sapogenins in the course of heating show clearly that the temperatures of initiation of weight loss differ greatly in glycosides and sapogenins. It should be noted

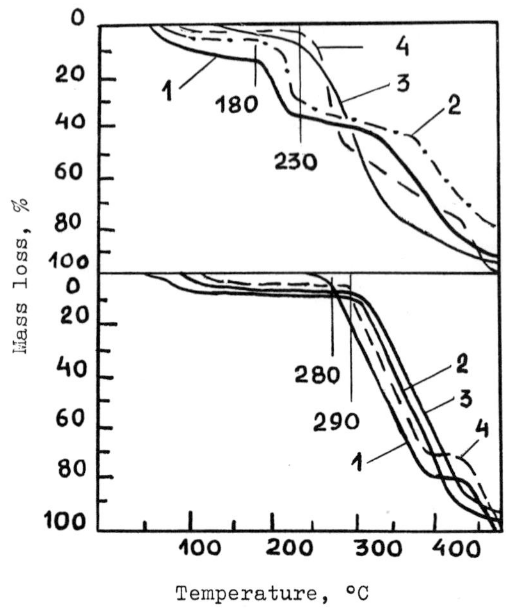

Fig. 4. TG thermal curves for: top; 1 - tomatin; 2 - tomatonin; 3 - agavoside C; 4 - deltonin (A); bottom; 1 - tomatidin, 2 - neotigogenin, 3 - gekogenin, 4 diosgenin (sapogenin B).

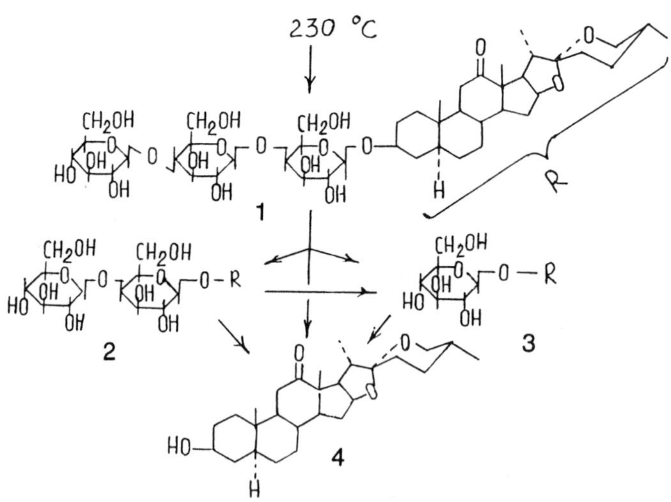

Fig. 5. The paths and main products of agavoside C thermolysis: 1 - agavoside C, 2 - gluco-galactoside of gekogenin (bioside), 3 - gekogenin galactoside (monoside), 4 - gekogenin.

at the same time that the difference between the temperatures of weight loss occurs in glycosides (though three of them, tomatonin, deltonin, and agavoside C are triosides quite similar in structure) is much higher than that in sapogenins and ranges from 30 to 50 °C. All samples studied started to lose weight in the temperature range of 280-290 °C. However, we have no sufficient evidence to what extent these temperature differences between glycosides, even of one chemical group, are associated with structural features of glycosides, and to what extent with other factors. In other experiments we recorded different temperatures of weight loss initiation in samples (purified by different techniques) of the same glycoside. Probably this index depends on TA conditions, degree of sample amorphism, thickness of its layer, and other characteristics.

The appearance of several minima in one peak on the DTG curves, with overlapping, especially expressed on the DTG curves indicates that, in fact, several simultaneous processes are occurring in the samples studied upon heating, one of which dominates at a certain temperature. The difference between a relatively low temperature of weight loss initiation in glycosides and a relatively high temperature of initiation of weight loss in sapogenins proves a more thermoresistant character of sapogenins in comparison to glycosides. These allowed us to propose and verify a process destroyed at the initial stage of steroidal glycoside thermodegradation.

Actually, heating of the samples of deltonin, tomatonin, agavoside C, and α-tomatin at 225-250 °C led to the formation of new, less polar substances, which retained their specific coloration for the original substance when developed with sulfuric acid after TLC and the coloration with specific reagents, that of Sannié in the case of spirostanol glycosides and Dragendorf in the case of a spirosolan glycoalkaloid, α-tomatin. IR spectrometry of spirostanol glycosides manifested that the absorption bands at 850, 890, 920, and 980 cm^{-1} characteristic of spirostanols are maintained after heating at the temperature indicated. These prove that the thermal degradation of C-27 steroidal glycosides begins at the carbohydrate moiety. The isolation of several less polar intermediate products from the reaction mixture of each glycoside made it possible to judge the pattern of the carbohydrate chain degradation (whether the entire chain is destroyed simultaneously, by one monosaccharide at a time, or some other way). Diosgenin, neotigogenin, gekogenin, and tomatidin, all sapogenins in each case, turned out to be the least polar products extracted with sulfuric acid from the thermolytic mixture and purified on a chromatographic column. In addition to physicochemical and spectral characteristics, the first three compounds were identified by using ^{13}C NMR spectrometry. The spectra obtained were consistent with the literature data[5].

Further chromatographic separation of the thermolytic products of spirostanol triosides (deltonin, tomatonin, and agavoside C) allowed us to isolate in each case two compounds identical to monosides and biosides in TLC produced from partial acid hydrolysis of the original triosides. Full acid hydrolysis of the intermediate products followed by identification of the aglycones and monosaccharides isolated confirmed their identity to be progenins resulting from partial hydrolysis of the original saponins. Tomatonin neotigogenin galactoside and neotigogenin glucogalactoside were isolated directly whereas diosgenin glucoside and diosgenin glucogalactoside were produced from deltonin. Agavoside C, (Fig. 5) gekogenin, galactoside and gekogenin glucogalactoside were also identified as reaction products.

A tetraoside, α-tomatin, a steroidal glycoalkaloid of the spirostane series differing from the above trioside in the size of the carbohydrate chain and the presence of amine nitrogen in the F ring of the aglycone, has also broken up into the aglycone, tomatidin, and a number of progenins (Fig. 6). Their composition was established by the identification data of full acid hydrolysis products and comparison of their chromatographic and physicochemical characteristics with those produced after partial hydrolysis of progenin α-tomatin (Fig. 6). Hence, the presence of progenins in glycoside thermolysis products proves that the carbohydrate moiety passes through a stepped degradation.

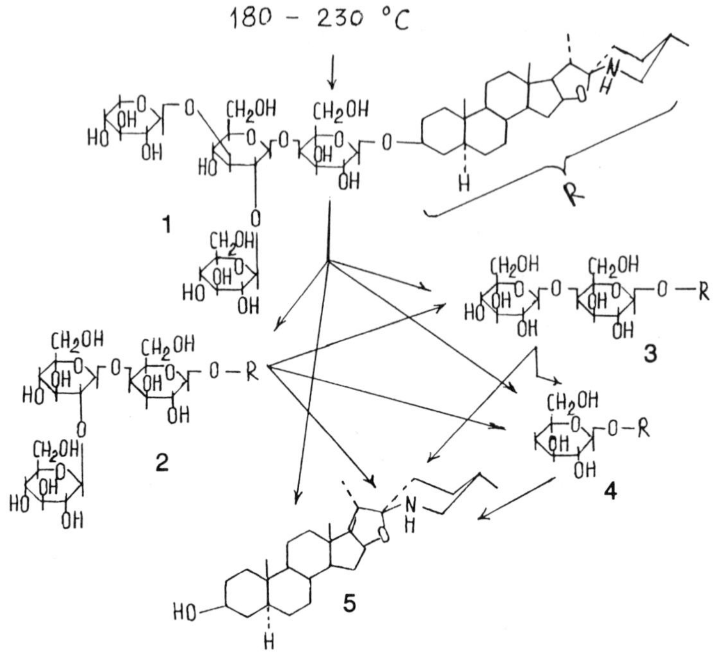

Fig. 6. The paths and main products of α-tomatin thermolysis: 1 - α-tomatin, 2 - tomatidin trioside, 3 - tomatidin bioside, 4 - tomatidin monoside, 5 - tomatidin.

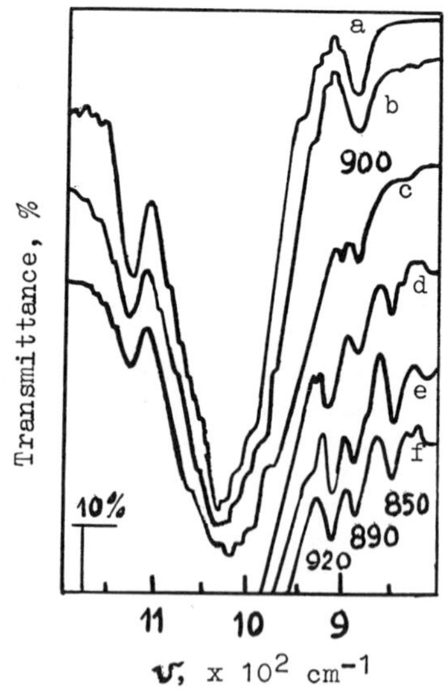

Fig. 7. IR spectra of tomatoside (a) native, (b) heated up to 120 °C, (c) heated up to 165 °C, (d) heated up to 250 °C, and of tomatonin (e) native, (f) heated up to 200 °C.

Heating of the furostanol glycosides, deltoside and tomatoside, to the temperature of the second (after water loss) thermoeffect produced much higher numbers of intermediates than in the case of spirostanol and spirosolane glycosides. TLC of the product mixture formed in the course of heating established the presence of compounds reacting with the Ehrlich reagent, specific for furostanol glycosides, less polar than the original substance, and compounds reacting with the Sannié reagent, specific for spirostanols, also less polar than the original substance. IR spectrometric monitoring of tomatoside thermolysis (Fig. 7) showed that during heating the furostanol aglycone is converted into a spirostanol one.

The absorption band in the IR spectrum with a minimum at 900 cm^{-1}, characteristic of furostanol glycosides, is maintained at 120 °C, is deformed at above 150 °C, and disappears completely at above 200 °C. Instead, the absorption bands with minima at 850, 920, and 980 cm^{-1}, characteristic of spirostanol compounds, appear in the spectrum, the peak at 920 cm-1 being higher than that at 890 cm^{-1}. This example is characteristic of spirostanols of the 25*R*-configuration, to which neotigogenin, a sapogenin produced after tomatosin acid hydrolysis, belongs. Diosgenin and neotigogenin have been isolated and identified from the diethyl ether extracts of the thermolyzed mixture of deltoside tomatoside after chromatographic purification on columns. Chromatographic separation of all the steroidal products of tomatoside decomposition led to two compounds of the furostane series and three of the spirostane (except neotigogenin). After the elucidation of the structure, spirostanol compounds were found to be galactoside (monoside), glucogalactoside (bioside), and gluco(bis)galactoside (trioside) named tomatonin. As regards the furostanol compounds formed during heating in the presence of 2% sulfuric acid, the less polar compound yielded neotigogenin and its galactoside, and the more polar one afforded neotigogenin and its gluco-galactoside[6]. Similar results were recorded in the case of deltonin thermolysis (Fig. 8).

In this case, deltonin, its progenins (monoside and diosgenin bioside) and the corresponding furostanol monoside and bioside were isolated from the thermolytic mixture and identified in addition to the aglycone. more prolonged heating (a few hours) of all the glycosides at 200-500 °C led to increase in the number of sapogenins in the mixture. A maximum output of free sapogenins was produced during chromatographically followed heating of glycosides.

TLC and IR spectrometric analyses of the glycosidic samples heated under conditions of subsequent thermoeffects, recorded on DTG curves, at temperatures over 300 °C, showed that the intermediates produced under these conditions lose their spirostanol features in addition to the glycosidic and furostanol characteristics. In this connection the investigation of these processes is a subject of another work.

EXPERIMENTAL

Samples of chromatographically pure substances (one spot in TLC) isolated and identified by the techniques described elsewhere[1-4] were used in the analyses.

The thermal behavior of compounds was studied using a derivatograph (Hungarian Optical Works MOM, Budapest), belonging to the F. Paulik, I. Paulic, and L. Erdey's system[7] in an air atmosphere, and in the temperature range of 20-500 °C. The sample size was 100 mg, DTA sensitivity 1/5, DTG sensitivity 1/10; a platinum crucible was used. Calcined aluminum oxide was applied as reference.

IR spectra of glycosides, sapogenins, and intermediates were recorded on a "Specord-75-IR" spectrophotometer in the region of 1300-800 cm^{-1}, in the case of gekogenin and its glycosides in the region of 1800-800 cm^{-1}. Readings were taken from a thin triturated sample paste in vaseline oil.

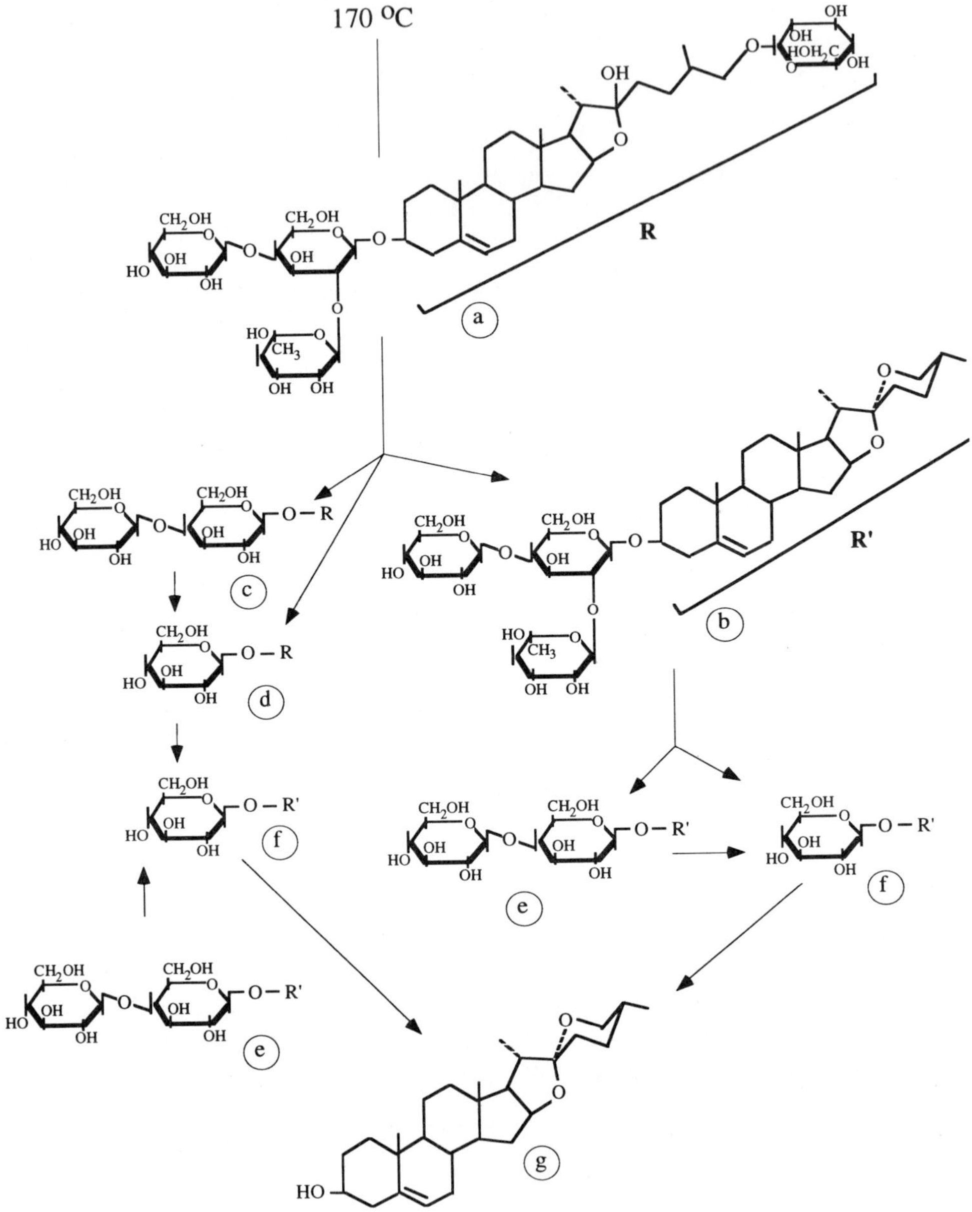

Fig. 8. The paths and main products of the thermolysis of a furostanol glycoside, deltoside: a - deltoside, b - deltonin, c - furostanol bioside, d - furostanol monoside, e - spirostanol bioside, f - spirostanol monoside, g - diosgenin.

Mass spectra were recorded on a MX-1303 mass spectrometer. TLC of glycosides and sapogenins was carried out on "Silufol" plates (firm "Chemopol") in the presence of authentic samples. A system of solvents, chloroform/ethanol/water (10:9:1, v/v), was used to elute glycosides, and benzene/diethyl ester (7:3) was used for sapogenins. Substances in TLC were detected by spraying plates with concentrated sulfuric acid or Ehrlich or Sannié reagents and heating up to 100-105 °C for 1-2 min.

Sapogenins (neotigogenin, diosgenin, gekogenin, tomatidin) were extracted from the products of thermolysis by diethyl ether extraction, chromatography on a column loaded with silica gel L 40/100 in the system benzene/diethyl ether (7:3). Monosides, biosides, and triosides (spirostanol and furostanol) produced in the course of thermolysis were isolated by chromatography of the thermolytic product mixture on columns filled with silica gel L 40/100 with elution with gradients of mixtures of chloroform/methanol (from 1 to 20% of methanol, v/v).

Intermediate glycosides were hydrolyzed in 5% sulfuric acid at 105 °C for 5 h in sealed glass ampules. Partial hydrolysis was performed in 2% sulfuric acid at 100 °C for 4 h in flasks fitted with reflux condensers.

Monosaccharides resulting from hydrolysis were identified using descending paper chromatography (PC) on FN-3 paper in a solvent system 1-butanol/benzene/pyridine/water (5:1:3:3, v/v, upper layer) in the presence of authentic samples of expected monosaccharides. The chromatographic paper strip was 40 cm. Monosaccharides were detected in PC by spraying the paper with aniline phthalate and heating at 105 °C for 3-4 min.

CONCLUSION

DTG, DTA, and TG thermal curves recorded up to 500 °C, major intermediate products, and the paths of the thermal degradation up to 250 °C of a number of glycosides belonging to the furostanol, spirostanol, and spirosolane class and their sapogenins are described. All the classes of steroidal glycosides and sapogenins under study are thermostable up to 150-170 °C. At higher temperature, furostanol glycosides are the first to modify their structure, glucose is split off from the C-26 position to form the F ring in the aglycone, and furostanol glycosides are converted into their spirostanol analogs. Within the temperature range of 190-250 °C the carbohydrate moiety of the aglycone C-3 is fragmented, and at more than 270 °C saponins are destroyed. It is possible to obtain a spirostanol glycoside from its native furostanol analog, and progenins and a free aglycone from spirostanol glycoside, by controlling the temperature and monitoring the duration of heating.

REFERENCES

1. H. Sato and S. Sakamura, Isolation and structure of a new furostanol saponin, *Agric. Biol. Chem.* 37:225 (1973).

2. V.A. Paseshnichenko and A.P. Guseva, Isolation and properties of saponins from *Dioscorea deltoidea* roots, *Prikl. Biokhim. Biol.* 21:94 (1975) (In Russian).

3. P.K. Kintia, W.A. Bobeyko, W.W. Krochmaliyk, and W.I. Tschirva, Über die Saponine in den Blättern von *Agava americana* L., *Pharmazie* 30:396 (1975).

4. T.D. Fontaine, G.W. Irving, R.Ma, J.B. Poole, and S.P. Doolittle, Isolation and partial characterization of crystalline tomatine, an antibiotic agent from the tomato plant, *Arch. Biochem.* 18:467 (1948).

5. P.K. Kintia, D.C. Jain, R.K. Gupta, and R.S. Thakur, Carbon-13 NMR spectroscopy of steroidal sapogenins and steroidal saponins, *Phytochemistry* 24:2479 (1985).

6. V.A. Bobeyko, P.K. Kintia, and I.V. Dranka, Thermal decomposition of furostanol glycoside - tomatoside, *J. Therm. Anal.* 36: 1307 (1990).

7. F. Paulik, I. Paulic, and L. Erdey, Derivatography, a complex method of thermal analysis, *Talanta* 13:1405 (1966).

PROTON AND CARBON-13 NMR STUDIES OF STEROIDS AND TRITERPENES

Geoffrey A. Cordell, Long-Ze Lin and Roberto R. Gil

Program for Collaborative Research in the Pharmaceutical Sciences
Department of Medicinal Chemistry and Pharmacognosy
College of Pharmacy
University of Illinois at Chicago
Chicago, IL 60612

INTRODUCTION

Steroids and triterpenes are common constituents of higher and lower plants, as well as marine flora and fauna. Derived from mevalonic acid with squalene as an intermediate, their skeletal diversity is essentially unmatched. Some of these compounds are important as mammalian hormones (testosterone, estrone), while others are potent toxic agents either as molluscicides, insecticides or as cytotoxic agents. Thus, as well as chemical diversity, these compounds also represent biological diversity in their mode of action. Some of the compounds, for example beta-sitosterol, are probably almost ubiquitous in the plant kingdom, indicating that while there may be many selective biosynthetic pathways depending on a particular taxonomic and ecologic niche, there is also a pathway which may be primordial in nature remaining unchanged and common to all plants.

Although the number of skeleta represented by the triterpenes and the steroids is substantial, one of the common structural features is that functional groups can sometimes be rather sparse in the molecule, perhaps only a single hydroxy or keto moiety, possibly with a quaternary carboxylic acid moiety. On the other hand some degraded triterpenoids, such as the limonoids and the quassinoids, present a wealth of functional groups and, as we shall see, more complex triterpene and steroid saponins with numerous sugar moieties are now routinely being examined, both through hydrolytic techniques and through direct analysis of the parent saponin.

A prototypical triterpene, such as oleanolic acid, contains 30 carbons, 48 hydrogens, and three oxygens (alcohol and carboxylic acid), in five rings with one trisubstitued double bond, 11 methylene groups, and seven quaternary methyl groups. and thus presents a formidable challenge for proton and carbon-13 NMR spectral assignment. Until recently, accepting such a challenge of spectral assignment even for more highly functionalized systems would have not have been considered. Recent progress in high-field NMR instrumentation, as well as the very substantial advances in pulse programming for pulse generation and to a series of experiments which can now begin to be used to make unambiguous proton and carbon-13 assignments in many types of triterpene and steroid a reality.

In this short survey, following a brief review of the history of proton and carbon-13 NMR studies in this field with reference made to some pertinent reviews, a strategic approach will be indicated together with some recent examples of our own work which has permitted the assignment of several sets of proton and carbon-13 data unambiguously. Finally, the biological significance of some of these studied compounds will be mentioned.

History of NMR Studies of Steroids and Triterpenes

Ironically, some of the earliest proton nmr studies of natural products were conducted on steroids when, in 1958, almost forty years ago, Shoolery and Rogers presented their seminal work on the NMR spectra of the steroids (Shoolery and Rogers, 1958). This was followed three years later by work by Zurcher (Zurcher, 1961) on a variety of steroids. The next year studies on the proton nmr spectra of triterpenes began to appear (Lehn, 1962; Jacquesy and Levisalles, 1962). These early studies on steroids were gathered into several compilations thus rendering them useful for correlation studies and for looking for substituent effects (Bhacca and Williams, 1964; Cohen and Rock, 1964 and Smith, 1964). For many years these substituent effects (Bhacca and Williams, 1964) were used extensively to make cursory spectral assignments and establish the structures of new derivatives. Such studies were also extended to cover specific groups of steroids, such as the substituted androstanes (Schneider *et al.*, 1985).

When commercial carbon-13 nmr instruments became available steroids were once again some of the earliest natural products examined, and for many years the original work of Roberts (Roberts *et al.*, 1969) on the carbon-13 nmr spectra of 28 closely related steroids was the basis of numerous subsequent assignments and structures. Subsequent studies on steroidal saponins and sapogenins were reviewed (Blunt and Stothers, 1977; Agrawal *et al.*, 1985), and the carbon-13 assignments of the pentacyclic triterpenes were reviewed (Agrawal and Jain, 1992; Mahato and Kundu, 1994), as were the data for the cardenolides and the bufadienolides (Robien *et al.*, 1987) and the withanolides (Gottlieb and Kirson, 1981). Some strategies for the structure determination f saponins have also been described (Chen and Snyder, 1993).

Although these data were certainly extremely useful and continue to provide a basis from which to examine both the proton and carbon-13 spectra of steroids and triterpenes, it was apparent that unlike studies on many other groups of natural products, the data for essentially all of the compounds were continuously inferred, rather than being independently established. Consequently, there arose some opportunities for improving the quality of the spectral data for steroids and triterpenes, namely: i) the unambiguous assignment of the carbon-13 spectra, ii) the stereotopical assignment of the proton spectra, iii) conformational studies of the compounds in solution in comparison with gas phase (molecular modeling) and solid phase (X-ray) data, and iv) the complete assignment of the spectra of saponins, including the sugar linkages, without degradation.

Our early structure elucidation studies at the University of Illinois were made possible through several programs aimed at the isolation and structure elucidation of anticancer agents from plants sponsored by the National Cancer Institute. In the course of this effort many different classes of natural product were found to display activity, including several types of steroid and triterpene derivative. Subsequent studies continued to examine plant anticancer agents, and were also expanded to embrace other biological activities involving many diverse groups of natural products; some of which were steroids and triterpenes. With enhanced NMR capabilities, it became possible to address the challenges indicated above and most of the remainder of this chapter will be devoted to a brief discussion of the most salient features of this work, recognizing that substantial additional detail, including representative spectra in some instances, can be found in the relevant primary literature associated with these compounds. The isolates that we have studied over the years that are members of the steroid and triterpene groups of natural products are shown in Table 1.

Table 1. Triterpenes and Steroids Isolated and Characterized

Plant (Family)	Compound	Reference
Ailanthus excelsa (Simaroubaceae)		
	Ailanthinone	Ogura *et al.*, 1977
	Glaucarubinone	Ogura *et al.*, 1977
	Glaucarubol 15-isovalerate	Ogura *et al.*, 1977
	13,18-Dehydroglaucarubol 15-isovalerate	Ogura *et al.*, 1977
	3*S*,24*S*,25-Trihydroxytirucall-7-ene	Sherman *et al.*, 1980
Ailanthus integrifolia subsp. *calycina* (Simaroubaceae)		
	Chaparrinone	Seida *et al.*, 1978
	6α-Tigloyloxychaparrinone	Seida *et al.*, 1978
Asclepias albicans (Asclepiadaceae)		
	Uzarigenin 3-*O*-β-D-glucopyranosyl(1→4)-β-D-glucopyranoside	Koike *et al.*, 1980
Aster batangensis (Asteraceae)		
	Asterbatanoside F⁺	Shao *et al.*, 1994a, 1995a
	Asterbatanoside G⁺	Shao *et al.*, 1994a, 1995a
	Asterbatanoside H⁺	Shao *et al.*, 1995a
	Asterbatanoside I⁺	Shao *et al.*, 1995a
	Asterbatanoside J⁺	Shao *et al.*, 1995b
Aster yunnanensis (Asteraceae)		
	Asteryunnanoside A⁺	Shao *et al.*, 1995c
	Asteryunnanoside B⁺	Shao *et al.*, 1995c
	Asteryunnanoside C⁺	Shao *et al.*, 1995c
	Asteryunnanoside D⁺	Shao *et al.*, 1995c
	Asteryunnanoside E⁺	Shao *et al.*, 1994b
	Asteryunnanoside F⁺	Shao *et al.*, 1994c
	Asteryunnanoside G⁺	Shao *et al.*, 1994c
Azadirachta indica (Meliaceae)		
	28-Deoxonimbolide*	Kigodi *et al.*, 1989
	Nimbolide*	Kigodi *et al.*, 1989
Brucea javanica (Simaroubaceae)		
	Javanicin	Lin *et al.*, 1990
Bupleurum longicaule var. *franchitii* (Umbelliferae)		
	2"-*O*-β-D-Glucopyranosylsaikosaponin b2*	Lin *et al.*, 1992a
Bupleurum wenchuanense (Umbelliferae)		
	2"-*O*-β-D-Xylopyranosylsaikosaponin b2*	Luo *et al.*, 1993c
	3",6"-*O,O*-diacetylsaikosaponin b2*	Luo *et al.*, 1993c
	Prosaikogenin G*	Luo *et al.*, 1993c
Caraipa densiflora (Clusiaceae)		
	3β-Hydroxy-28-*p*-coumaroyloxy-lup-20(29)-en-27-oic acid	Gunasekera *et al.*, 1983
Caralluma umbellata (Asclepiadaceae)		
	Carumbelloside I*	Lin *et al.*, 1994
	Carumbelloside II*	Lin *et al.*, 1994
Davidsonia pruriens (Davidsoniaceae)		
	Methyl 3-oxo-bauer-7-en-28-oate	Meksuriyen *et al.*, 1986
	3β-Hydroxy-bauer-7-en-28-oic acid	Meksuriyen *et al.*, 1986
Dianthus barbatus (Caryophyllaceae)		
	Barbatoside A	Cordell *et al.*, 1977
	Barbatoside B	Cordell *et al.*, 1977

Euphorbia nicaeensis subsp. *glareosa* (Euphorbiaceae)	
Cycloartanol derivatives	Öksüz *et al.*, 1993
Euphorbia cyparissias (Euphorbiaceae)	
Glut-5-ene-3α-methylbutyrate[*]	Öksüz *et al.*, 1994
Gelsemium sempervirens (Loganiaceae)	
12β-Hydroxy-5α-pregn-16-ene-3,20-dione[+]	Schun and Cordell, 1987
Glyptopetalum sclerocarpum (Celastraceae)	
20-Hydroxy-20-epi-tingenone[*]	Likhitwitayawuid *et al.*, 1993
22β-Hydroxytingenone[*]	Bavovada *et al.*, 1990; Likhitwitayawuid *et al.*, 1993
Gyrinops walla (Thymelaeaceae)	
Wallenone	Schun *et al.*, 1986
Jacaranda caucana (Bignoniaceae)	
Jacoumaric acid	Ogura *et al.*, 1977a
Jacarandic acid	Ogura *et al.*, 1977b
Larrea tridentata (Zygophyllaceae)	
3β-(3,4-Dihydroxycinnamoylcinnamoyl)-erythrodiol	Konno *et al.*, 1988
3β-(4-Hydroxycinnamoylcinnamoyl)-erythrodiol	Konno *et al.*, 1988
Marsdenia koi (Asclepiadaceae)	
Marsdekoiside A[+]	Yuan *et al.*, 1992
Marsdenia tenacissima (Asclepiadaceae)	
12β-*O*-Acetyltenacigenin A[*]	Luo *et al.*, 1993a
11α-*O*-Tigloyl-12β-*O*-acetyltenacigenin B[*]	Luo *et al.*, 1993b
11α-*O*-Benzoyl-12β-*O*-acetyltenacigenin B[*]	Luo *et al.*, 1993b
11α-*O*-2-Methylbutyryl-12β-*O*-acetyltenacigenin B[*]	Luo *et al.*, 1993b
11α-*O*-2-Methylbutyryl-12β-*O*-tigloyltenacigenin B[*]	Luo *et al.*, 1993b
11α-*O*-2-Methylbutyryl-12β-*O*-benzoyltenacigenin B[*]	Luo *et al.*, 1993b
11α-*O,O*-Ditigloyl-17β-tenacigenin B[*]	Luo *et al.*, 1993b
Marstenacigenin A[+]	Luo *et al.*, in press
Marstenacigenin B[+]	Luo *et al.*, in press
Nicandra physaloides (Solanaceae)	
Nicandrenone	Gunasekera *et al.*, 1981
Nierembergia aristata (Solanaceae)	
17-Epi-11α-hydroxy-6,7-dehydrostrophanthidin-3-*O*-β-boivinopyranoside[*]	Gil *et al.*, 1995
6,7-Dehydrostrophanthidin-3-*O*-β-boivinopyranoside[*]	Gil *et al.*, 1995
6,7-Dehydrostrophanthidin-3-*O*-β-oleandropyranoside[*]	Gil *et al.*, 1995
Simaba multiflora (Simaroubaceae)	
Chaparrinone	Arisawa *et al.*, 1983
6α-Senecioyloxychaparrin	Arisawa *et al.*, 1983
6α-Senecioyloxychaparrinone	Arisawa *et al.*, 1983
Simalikalactone D	Arisawa *et al.*, 1985
Hispidol B	Arisawa *et al.*, 1987
3*S*,23*R*,25-Trihdroxytirucall-7-en-24-one	Arisawa *et al.*, 1987
Soulamea soulameoides (Simaroubaceae)	
Glaucarubulone	Handa *et al.*, 1983
Holacanthone	Handa *et al.*, 1983
Isobrucein A	Handa *et al.*, 1983
Picrasin B	Handa *et al.*, 1983
Terminalia chebula (Combretaceae)	
Arjungenin[+]	Reddy *et al.*, 1994
Arjunglucoside-I[+]	Reddy *et al.*, 1994
Xylosma velutina (Flacourtiaceae)	
Velutinic Acid	Chang *et al.*, 1977

[*] Indicates complete ^{13}C data

[+] Indicates complete ^{1}H- and ^{13}C-data

Strategies for Unambiguous Proton and Carbon-13 Assignment

There are now a plethora of NMR techniques available to the natural product chemist for the structure elucidation of new triterpenes and steroids. Several books discussing these techniques are available [Atta-ur-Rahman, 1986; Derome, 1987; Martin and Zektzer, 1988], and some of the drawbacks in terms of resolution, time, sample size, computer capabilities, etc. have been discussed [Derome, 1989; Kriwacki and Pitner, 1989]. The list of techniques which are most useful in structure elucidation and unambiguous assignment include: COSY [Aue *et al.*, 1976; Bax and Freeman, 1981], DQF-COSY [Piantini *et al.*, 1982; Rance *et al.*, 1983], HOHAHA [Braunschweiler and Ernst, 1983; Bax and Davis, 1985; Davis and Bax, 1985; Edwards and Bax, 1986], NOESY [Jeener *et al.*, 1979], ROESY [Bothner-By *et al.*, 1984; Summers et al., 1986; Greisinger and Ernst, 1987], HETCOR [Bodenhausen and Freeman, 1978; Bax and Morris, 1981; Bax, 1983; Bax and Summers, 1986], HMQC [Bax and Subramanian, 1986], HMBC [Bax and Summers, 1986; Bax *et al.*, 1986], FLOCK [Reynolds *et al.*, 1989; Carpenter *et al.*, 1992], Selective INEPT [Bax, 1984], COLOC [Kessler *et al.*, 1984], and INADEQUATE [Levitt and Ernst, 1981; Bax *et al.*, 1981].

A recent chapter (Byrne, 1993) has indicated some of the key techniques and how they can be applied for structure elucidation, including determination of relative stereochemistry. Crucial, when high-field instrument time is charged, is that one needs to consider the optimum techniques in terms of available sample size, time and cost to gain a particular result. Potentially, very similar data can often be obtained from the long-range HETCOR, FLOCK, selective INEPT and HMBC spectra. If only limited ^{1}H-^{13}C correlations are required, a series of selective INEPT spectra rather than a HMBC or FLOCK spectrum has, in our hands, proved to be most useful. In subsequent sections we will summarize some of the techniques and strategies that we have used for steroid and triterpene saponins.

It is worth mentioning some aspects of the techniques that in our quite limited experience have proven to have use in addressing our goal of unambiguous resonance assignment. These techniques represent an integration of one- and two-dimensional NMR techniques coupled with molecular modeling as follows. The one-dimensional techniques of importance, after the acquisition of a high field NMR spectrum in an appropriate solvent, may include a difference nOe spectrum, a broad-band decoupled carbon-13 spectrum, a DEPT or APT spectrum, and spectra which offer the one-bond and three-bond correlations of protons and carbons. The most important spectrum is the high-field proton NMR spectrum, because the conduct of many other experiments and the strategies to be used for structure elucidation often depend directly on this spectrum. Many investigators focus on higher field to enhance the signal dispersion, in addition, we have focussed on optimizing signal dispersion through changing solvents. Effort expended at this time reaps substantial benefits in the analysis of most two-dimensional spectra. The two-dimensional techniques that we use include DQF-COSY, HOHAHA, NOESY, ROESY, HETCOR or HMQC and a variety of long-range HETCOR techniques, including COLOC, FLOCK and HMBC.

For molecular modeling, we have used MacroModel software on a Silicon Graphics Iris Computer with energy minimized with MM2 and more recently, PC Model 486 software, which operates on a desk-top computer system. It is the judicious interplay of these techniques which has proved to be effective in spectral assignment.

One particularly powerful pulse program, whose applications in our programs we have previously reviewed (Cordell, 1991; Cordell and Kinghorn, 1991) is the selective INEPT technique (Bax, 1984). The technique involves a soft proton pulse irradiation of preselected proton resonances which is J-modulated. This permits the quite selective determination of three-bond correlations, and typically with somewhat smaller J-values, the very useful two-bond correlations.

Selected Examples of Spectral Assignment and Structure Elucidation

Our first foray into the application of the selective INEPT technique to a steroid system was with 12β-hydroxy-5α-pregn-16-ene-3,20-dione from *Gelsemium sempervirens* (Loganiaceae) (Schun and Cordell, 1987) where it became possible to assign unambiguously the carbon-13 spectrum. Particularly, it was possible to make the distinction between the methylene carbons C-6 (30.77 ppm), C-7 (29.26 ppm) and C-15 (32.00 ppm) through the irradiation of H_3-19 and H-16, which resulted in revised assignments for C-6 and C-15. In Figure 1, the corresponding data for the co-occurring pregna-4,16-diene-3,20-dione (**1**) are displayed. We also used this technique to generate the first unambiguous assignments for the quassinoids chaparrinone, holacanthone, picrasin B, isobrucein A and bruceantin (154), thereby permitting the reassignment of several carbons.

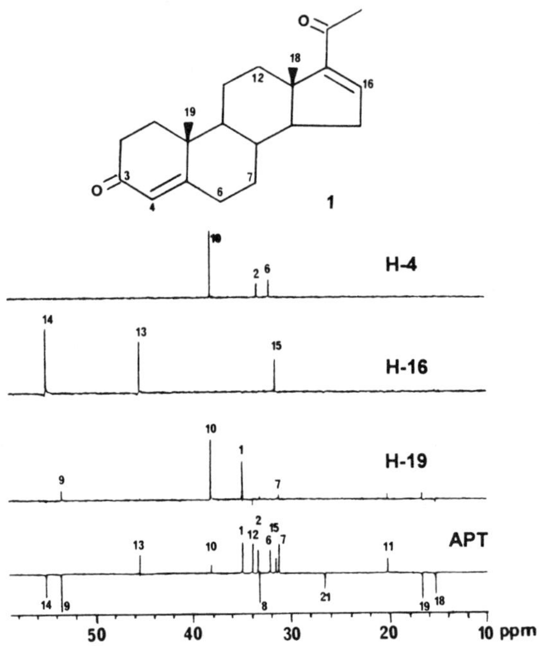

Figure 1. Selective INEPT spectra of pregna-4,16-diene-3,20-dione (**1**)

Another example of the power of the selective INEPT technique occurred in the nimbolide series (Kigodi *et al.*, 1989). Nimbolide (**2**) is a meliacin derivative from *Azadirachta indica* (Meliaceae), originally isolated in 1967, whose structure was based almost solely on hydrolysis to a known derivative (Ekong, 1967). When a new, closely-related derivative **3** was isolated, it became necessary to assign the proton and carbon-13 spectra (Kigodi *et al.*, 1989). The four methyl groups, at 1.22, 1.37, 1.48 and 1.70 ppm, in **2** were assigned through a combination of HETCOR and selective INEPT techniques. With the latter technique (Figure 2), irradiation of the resonance at 1.22 ppm was the only one which enhanced the carbonyl carbon at 200.55 ppm and must therefore be the 10-Me. This irradiation also enhanced the signal at 40.89 ppm which was enhanced when the resonance at 1.37 ppm was irradiated, indicating that the latter was the 8-Me. The signal at 49.23 ppm was enhanced from the irradiation of H-15 at 5.53 ppm and by the irradiation of the methyl resonance at 1.70 ppm, indicating the latter to be the 13-Me. Finally, the resonance at 174.8

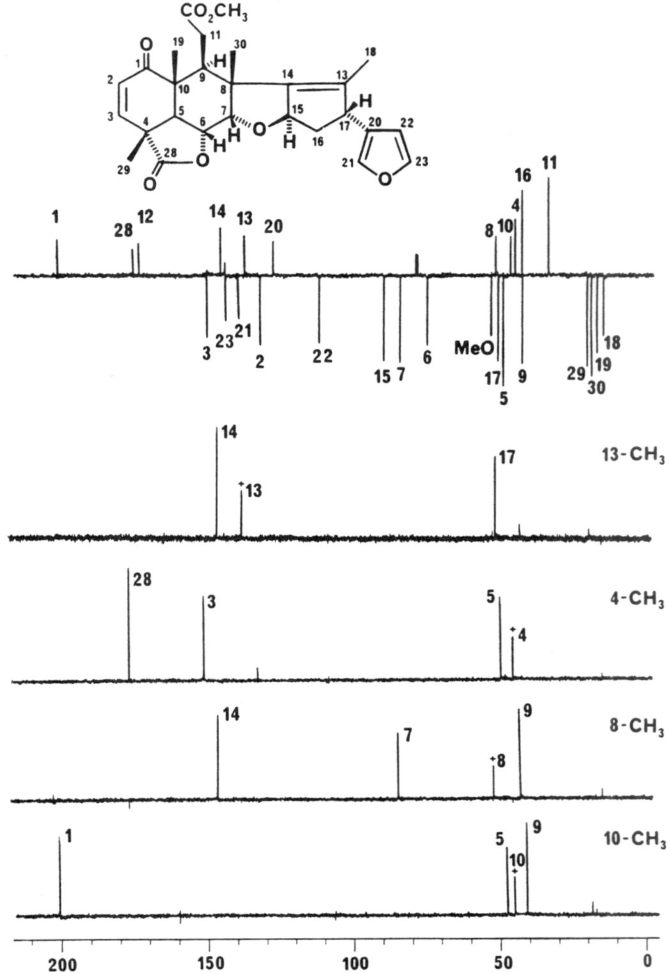

Fig. 2 Selective INEPT Spectra of Nimbolide (2)

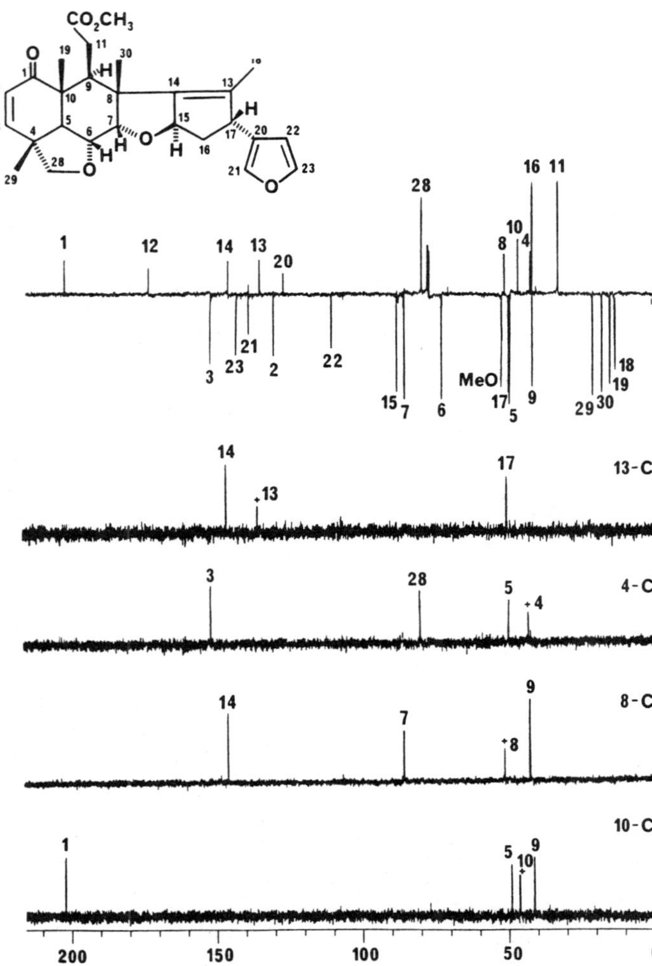

Fig. 3 Selective INEPT Spectra of 28-Deoxonimbolide (3)

ppm, the C-28 lactone carbonyl carbon, was enhanced when the methyl resonance at 1.48 ppm was irradiated, indicating it to be the 4-Me.

When it came to the assignment of the structure of 28-deoxonimbolide (**3**) it was not assumed that the sequence of the chemical shifts of the methyl resonances in the new isolate was the same as that in nimbolide which had been 10-Me (1.22 ppm), 8-Me (1.37 ppm), 4-Me (1.48 ppm), and 13-Me (1.70 ppm); rather the sequence was independently established for the four resonances at 1.17 ppm, 1.31 ppm, 1.35 ppm, and 1.67 ppm. Very similar data (Figure 3) were obtained for the irradiation of the resonances at 1.17 ppm (10-Me), 1.31 ppm (8-Me), and at 1.67 (13-Me) as had been obtained for **2**. When the signal at 1.35 ppm was irradiated the olefinic carbon at 152.0 ppm and a methylene carbon resonance at 79.4 ppm were enhanced indicating that this was the 4-Me and that the lactone carbonyl at C-28 in **3** was replaced by a methylene in **3** (Kigodi *et al.*, 1989).

In order to illustrate our strategy for structure elucidation, some recent work on the cardenolides from *Nierembergia aristata* (Gil *et al.*, 1995), a rearrangement in the tenacigenin series (Luo *et al.*, 1993a, 1993b; Qiu *et al.*, in press) and the spectral assignment of some saikosaponins (Lin *et al.*, 1992a; Luo *et al.*, 1993c) will be presented.

We recently reported on the first isolation of the cardenolides **4-6** from a plant in the family Solanaceae as a result of bioactivity-directed fractionation of a cytotoxic extract (Gil *et al.*, 1995). *Nierembergia aristata* is native to Argentina, and structure elucidation of the three active isolates was completed through a combination of DQF-COSY, HOHAHA, ROESY, HETCOR and selective INEPT experiments. Four features are of special note, i) the location and nature of the sugar unit, ii) the location of the disubstituted double bond, iii) the stereochemistry at C-17, and iv) the location of the secondary hydroxy group. Selective INEPT irradiation at H_3-18 did not enhance an olefinic carbon hence the double bond could not be at C-11,12, whereas irradiation of H-6 did enhance C-10 (54.7 ppm) which was also enhanced through the irradiation of H-19 (10.46 ppm). Irradiation of H-7 enhanced C-14 (83.19 ppm), C-5 (73.91 ppm), both of which have hydroxy groups attached and C-9 (43.36 ppm).

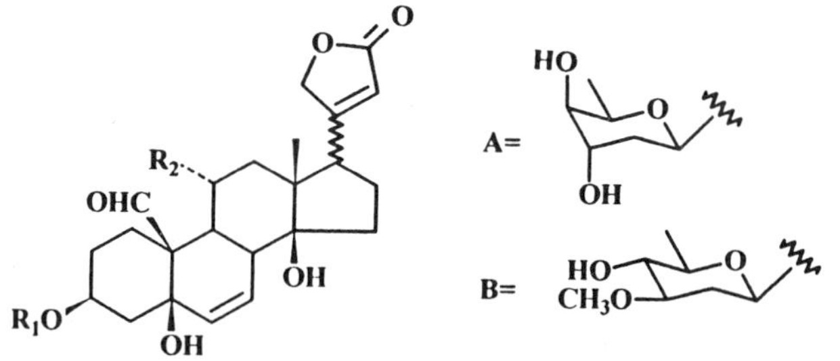

4 17-Epi-11*α*-Hydroxy-6,7-dehydrostrophanthidin-3-*O*-*β*-boivinopyranoside: R_1=A, R_2=OH, H-17*β*
5 6,7-Dehydrostrophanthidin-3-*O*-*β*-boivinopyranoside: R_1=A, R_2=OH, H-17*α*
6 6,7-Dehydrostrophanthidin-3-*O*-*β*-oleandropyranoside: R_1=B, R_2=OH, H-17*α*

Following the correlations in the DQF-COSY spectrum, the H-17 resonance could be used to assign the signals of H-16a, H-16b, H-15a and H-15b at 2.02, 1.80, 2.19 and 1.94 ppm, respectively. A cross correlation was also observed in the HOHAHA spectrum (Figure 4) between H-17 and H-15a,b, which eliminates the location of a double bond at C-15,16. The anomeric carbon at 98.68 ppm was cross-correlated with the doublet of doublets (*J* =

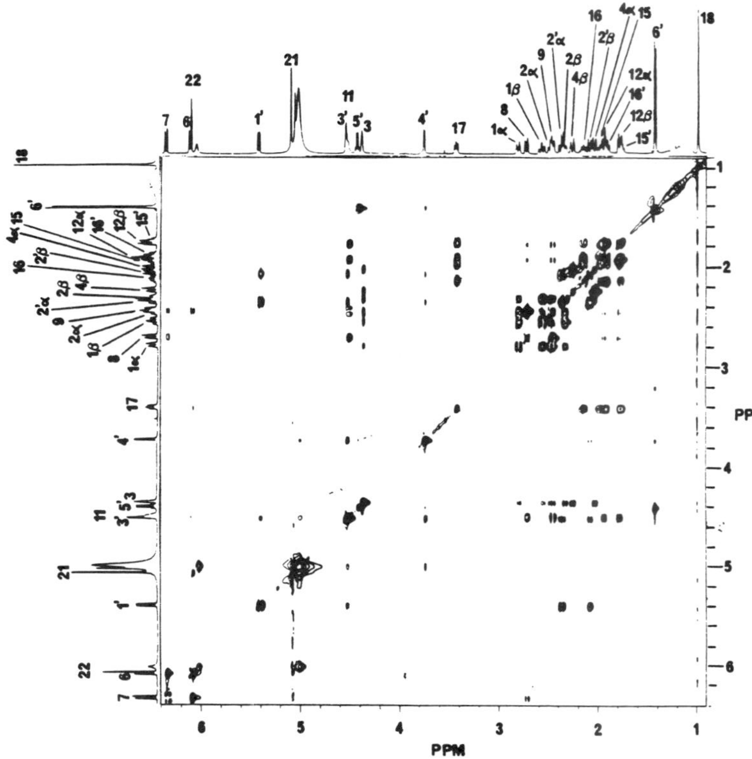

Fig. 4 HOHAHA spectrum of 4.

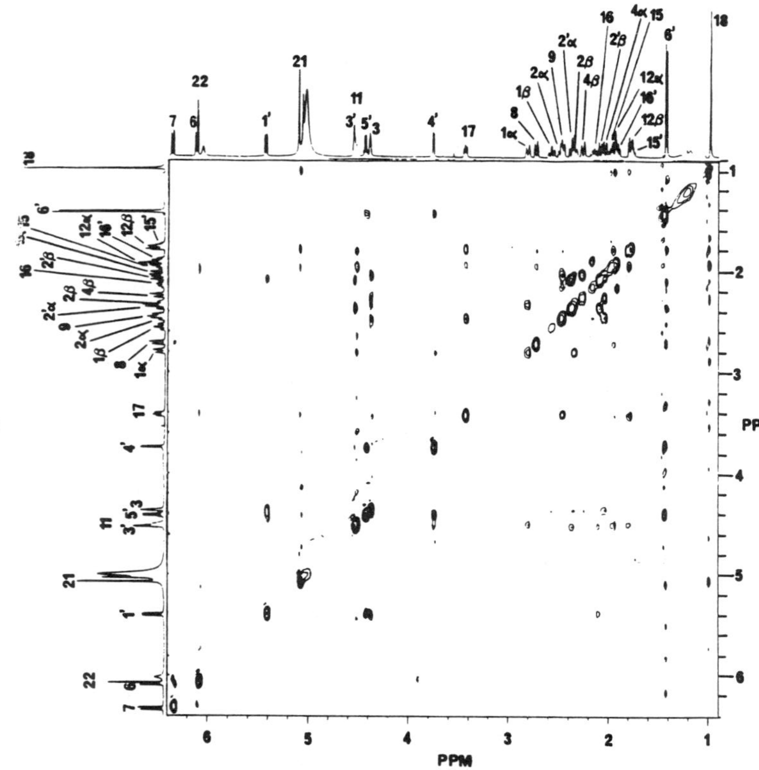

Fig, 5 ROESY spectrum of 4.

3,10 Hz) at 5.45 ppm in the HETCOR spectrum indicating that C-2' was unsubstituted and the DQF-COSY spectrum led to the assignment of H-2', H-3', H-4' and H-5'. The latter resonance also showed a correlation peak with the doublet methyl group (C-6') at 1.46 ppm. Selective INEPT irradiation of the anomeric proton enhanced C-3 (74.9 ppm) confirming the location of the sugar moiety, and the H-3 and H-1' resonances also showed a correlation in the ROESY spectrum (Figure 5). The DQF-COSY experiment permitted correlations to be established from H-6 through H-7, H-8, H-9 and a resonance at 4.55 ppm (66.6 ppm in the HETCOR). The H-9/H-11 coupling constant of 8 Hz indicates a *trans* diaxial relationship. In the selective INEPT experiment, irradiation of H-11 enhanced C-13 (49.0 ppm), as well as C-10 (54.7 ppm). Through DQF-COSY and HETCOR experiments the resonance at 19.7 ppm was unambiguously assigned to C-1, and not to either C-7 or C-11, so that this assignment may have to be reversed for some strophanthidin derivatives, as suggested previously (Pauli *et al.*, 1993).

A change in the configuration at C-17 from a 17β- to a 17α-cardenolide causes a shift of ca. -8 ppm in the shift of C-12 (Kawaguchi *et al.*, 1993), but this is when C-11 is unsubstituted. The chemical shift of H-17 has also been suggested as a measure of the stereochemistry (17α-H at *ca.* 2.8 ppm, and 17β-H at *ca.* 3.40 ppm. In the isolate, the chemical shift of H-17 was 3.46 ppm suggesting a 17β-H configuration and in support of this no nOe was observed with H_2-12 or between H_3-18 and H-17, although H-17 and H_3-18 were correlated in the ROESY spectrum with H_2-21 and H-22. In the second isolate **5**, the corresponding proton appeared at 2.77 ppm with no 11-OH group, and H-12a,b showed no ROESY cross peaks with H-17. Only the sugar unit was different between compounds **5** and **6**. The cardenolide **5** showed an ED_{50} of 0.03-0.09 mcg/ml in a broad range of human cancer cell bioassays.

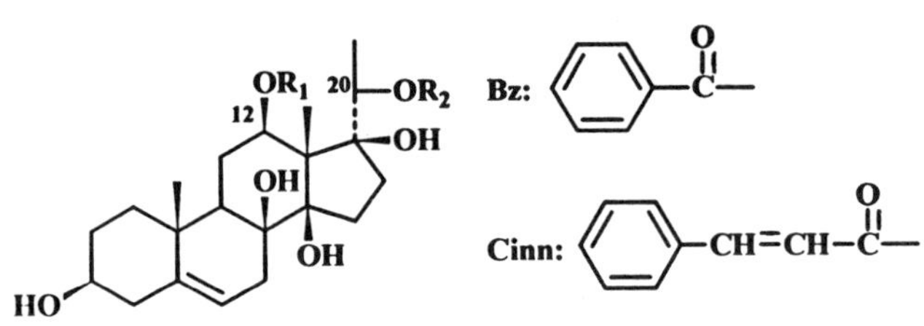

7 **Gymnemarsgenin: R_1 = Cinnamoyl, R_2 = Benzoyl**
Sarcostin: R_1 = R_2 = H
8 **20-*O*-Cinnamoylsarcostin: R_1 = H, R_2 = Cinnamoyl**

Marsdenia tenacissima (Asclepiadaceae) has been used in China for the treatment of cancer. We have isolated several new tenacigenin derivatives from this plant (Luo *et al.*, 1993a, 1993b, in press) and during the course of this work observed a phenomenon in which mild transesterification occurs under very mild conditions (Qiu *et al.*, in press). Thus when we reinvestigated the structure of gymnemarsgenin, an isolate of *Gymnema yunnanense* (Chen *et al.*, 1989) and *Marsdenia globifera* (Qiu *et al.*, 1990) using the selective INEPT technique it was found that rather than having the structure 12β-*O*-benzoyl-20-*O*-cinnamoylsarcostin, the structure was deduced to be 20-*O*-benzoyl-12β-*O*-cinnamoylsarcostin (7).

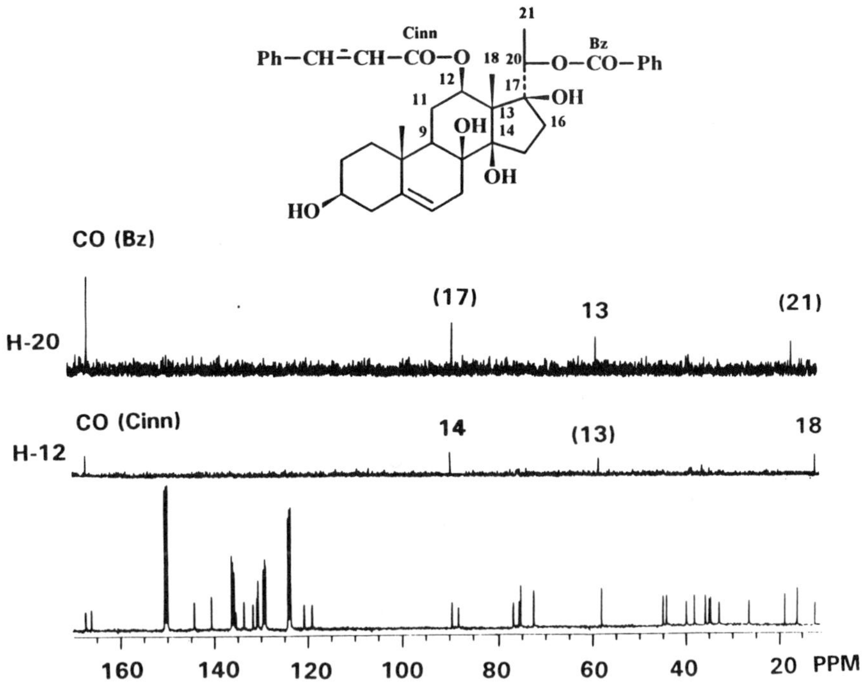

Figure 6. Selective INEPT spectra of gymnemarsgenin (7)

When the parent compound, rather than the hydrolysis product 20-*O*-cinnamoylsarcostin (**8**), was examined selective irradiation of H-12 (5.28 ppm) enhanced the carbon resonance at 167.2 ppm, which was also enhanced on irradiation of the α-proton of the cinnamoyl group (Figure 6). Corresponding data were obtained when irradiation of H-20 (5.25 ppm) led to the enhancement of the ester carbonyl at 166.0 ppm. A mechanism was proposed for the internal acyl migration of the 12-*O*-cinnamoyl group to C-20 during the mild alkaline hydrolysis (Qiu *et al.*, in press).

Bupleurum species are well-known in China for their use as anti-inflammatory agents and for the treatment of common colds (Hikino and Kiso, 1988). The active ingredients responsible for the former activity are the saikosaponins and more than 60 derivatives in this series have been isolated (Jia and Zhang, 1989). However, none of these isolations had been accompanied with complete spectral assignments. Our work has focussed on the derivatives of two *Bupleurum* species, *B. longicaule* var. *franchitii* (Lin *et al.*, 1992) and *B. wenchuanense* (Luo *et al.*, 1993c). From *B. longicaule* var. *franchitii* we isolated 2"-*O*-β-D-glucopyranosyl-saikosaponin b2 (**9**) and completed its spectral assignments using a combination of DQF-COSY, HOHAHA, ROESY, HETCOR and HMBC techniques. All of the appropriate spectra have been published and will not be reproduced.

The APT spectrum was used to characterize the carbon peaks into methyl, methylene, methine and quaternary carbons and this was followed by a HETCOR spectrum to reveal some of the direct proton-carbon attachments. The long range couplings and therefore the quaternary carbons and the skeleton were revealed through the HMBC experiment, and a DQF-COSY experiment yielded information regarding the coupling correlations of geminal and vicinal protons and to begin to assign the protons stereotopically. The direct connectivity and the relayed connectivity was revealed with a HOHAHA experiment, which since some of the couplings were small in magnitude were not apparent from the DQF-COSY spectrum.

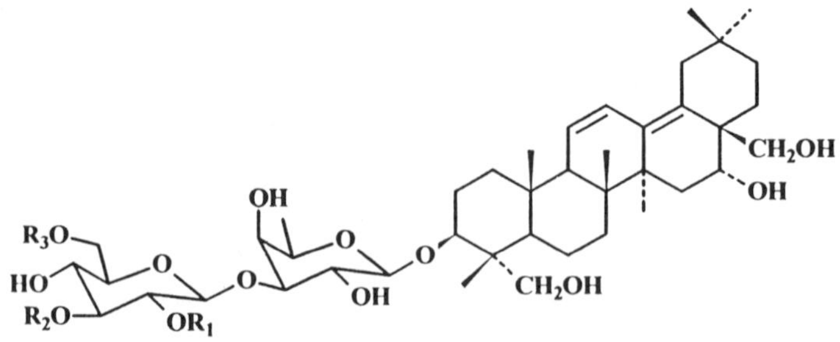

9 Saikosaponin B2: $R_1 = R_2 = R_3 = H$
2"-*O*-*β*-D-Glucopyranosylsaikosaponin B2:
R_1 = *β*-D-glucopyranosyl, $R_2 = R_3 = H$
2"-*O*-*β*-D-Xylopyranosylsaikosaponin B2:
R_1 = *β*-D-xylopyranosyl, $R_2 = R_3 = H$
2"-*O*-Acetylsaikosaponin B2: R_1 = Ac, $R_2 = R_3 = H$
3",6"-*O*, *O*-Diacetylsaikosaponin B2: R_1 = H, $R_2 = R_3$ = Ac
3"-*O*-Acetylsaikosaponin B2: R_2 = Ac, $R_1 = R_3 = H$

The ROESY spectrum indicated the proximity of some of the protons in the molecule and comparison could be made with a model structure and calculated distances.

The only assumption that was made in starting the analysis was that the signal at 4.20 ppm could be assigned to H-3α. This assumption was borne out by all of the subsequent analyses. Thus the A-ring protons could be readily assigned and beginning with H-5α the B-ring protons were determined. The methine protons at C-9 (2.18 ppm), C-11 (5.66 ppm), and C-12 (6.65 ppm) were recognized by their shifts in the HETCOR spectrum and H-11 and H-12 were attributed by their ROESY contours with the H-1β and the H-19 protons. A strong correlation was observed between H-16 and C-14 in the HMBC spectrum, and H-14 also generated weak couplings with H_2-15, one proton of which (β) showed a contour with the β-26-Me signal at 0.82 ppm and one of the β-28-H_2-protons centered at 3.70 ppm.

The three sets of methylene protons in the E-ring were assigned with the aid of the ROESY and DQF-COSY experiments. For example the C-19 methylene protons were assigned through the ROESY correlations with the β-28 methylene and β-30 methyl signals, which also allowed the distinction to be made between the 29- (0.98 ppm) and 30-methyl (0.93 ppm) groups. The former signal also showed a correlation contour with a signal at 1.45 ppm which is the α-proton at C-21 (β- at 1.79 ppm correlates with β-H_2-28). ROESY correlations of H-2β and H-6β with methyl resonances at 0.94 ppm and 0.82 ppm, respectively allowed the assignment of the 25- and 26-methyl groups.

Finally, the ROESY spectrum revealed decisive correlations for the sugar linkages, for example H-3α and H-F_1, H-F_3 and H-G_1, and H-G_2 and H-G'_1. Critically, these correlations were also paralleled in the HMBC spectrum and allowed all of the sugar linkages to be directly determined. The remaining sugar protons were quite overlapped and the assignments were clarified through the HOHAHA experiments using a spin lock time of 30-40 ms for the geminal, vicinal and protons three bonds away, and a time of 55 ms for the contours of protons separated by four bonds. Thus by working with the DQF-COSY and the HOHAHA spectra all of the sugar protons could be assigned.

The HMBC spectrum was then used to establish the various methylene and quaternary carbons as well as the six methyl resonances which had been tentatively assigned by other techniques. For example C-3 and the H_3-24 and H_2-23, C-5 and H_3-25, C-10 and H_3-26, H-7α and H-7β with C-26, etc. Once again, the internal consistency between the various sets of data from the different experiments was absolutely crucial (Lin *et al.*, 1992a). With these data in hand it was relatively straightforward to determine the structures of a substantial series of related compounds from *Bupleurum wenchuanense* (Luo *et al.*, 1993c).

Biological Activity of Isolated Steroids and Triterpenes

The rationale for essentially all of these studies was that the biological activity of each of the various compound groups was of interest and these data have been discussed in detail in the respective publications. For example, the quassinoids showed cytotoxic and antitumor activities (Ogura *et al.*, 1977; Seida *et al.*, 1978; Arisawa *et al.*, 1983; Arisawa *et al.*, 1985). The cardenolides of *Nierembergia aristata* (Gil *et al.*, 1995), the nimbolide derivatives (Kigodi *et al.*, 1989), the saikosaponin b2 derivatives (Luo *et al.*, 1993c), the tenacigenin derivatives (Luo *et al.*, 1993b) and the tingenone derivatives (Likhitwitayawuid *et al.*, 1993) also displayed potent cytotoxic activity. Very potent anti-inflammatory activity was noted for several of the saikosaponin derivatives (Brewer and Cordell, unpublished results) and nimbolide was established to be very potent as an antimalarial agent in our assays.

Conclusions

The aforementioned studies have demonstrated that nmr techniques are now routinely available for the unambiguous assignment of the proton and carbon spectra of steroids and triterpenes, their esters, and their glycosylated derivatives directly without hydrolysis, which in some cases may lead to rearrangements occurring. Strategies involving the judicious interweaving of various one- and two-dimensional nmr techniques, together with molecular modeling to establish preferred conformation and interproton distances have also been described.

Acknowledgements

This work was made possible by collaboration with several outstanding colleagues in the PCRPS at UIC and the dedication of a number of other postdoctoral fellows, visiting scholars and graduate students whose names are acknowledged on the various manuscripts. This work has also involved collaborators located in several countries of the world including Drs. A.V.N. Appa Rao (Kakatiya University, Warangal, India), Rapepol Bavovada and Kittisak Likhitwitayawuid (Faculty of Pharmaceutical Sciences, Chulalongkorn University, Bangkok, Thailand), Phillip G.K. Kigodi (Muhimbili Medical Center, Dar es Salaam, Tanzania), Si-Qi Luo (Shanghai Institute of Pharmaceutical Industry, Shanghai, People's Republic of China) and Yodhathai Thetaranonth (Mahidol University, Bangkok, Thailand). The biological aspects of this work were conducted in the Bioassay Research Laboratory in the Department under the direction of Dr. John M. Pezzuto. Dr. Michael E. Johnson in this department and his postdoctoral fellow Dr. Liang Xue collaborated on the molecular modeling aspects of the tenacigenin derivatives and Dr. George Doss originally established the selective INEPT technique at UIC. We also thank the Research Resources Center of the University of Illinois at Chicago for the provision of NMR spectroscopic facilities. This work was supported by a grant from the Division of Cancer Treatment, National Cancer Institute, Bethesda, MD.

REFERENCES

Agrawal, P.K.; Jain, D.C. ^{13}C NMR spectroscopy of oleanane triterpenoids. *Progr. NMR Spectr.* **1992**, *24*, 1-90.

Agrawal, P.K.; Jain, D.C.; Gupta, R.K.; Thakur, R.S. Carbon-13 nmr spectroscopy of steroidal sapogenin and steroidal saponins. *Phytochemistry* **1985**, *24*, 2479-2496.

Arisawa, M.; Cordell, G.A.; Kinghorn, A.D.; Farnsworth. N.R. Plant anticancer agents XXIII. 6α-Senecioyl-oxychaparrin, a new antileukemic quassinoid of *Simaba multiflora. J. Nat. Prod.* **1983**, *46*, 218-221.

Arisawa, M.; Fujita, A.; Morita, N.; Cox, P.J.; Howie, R.A.; Cordell, G.A.; Farnsworth, N.R. Triterpenes from *Simaba multiflora. Phytochemistry*, **1987**, *26*, 3301-3303.

Arisawa, M.; Fujita, A.; Morita, N.; Kinghorn, A.D.; Cordell, G.A.; Farnsworth, N.R. Plant anticancer agents XXXV. Further constituents of *Simaba multiflora* (Simaroubaceae). *Planta Med.* **1985**, *50*, 348-349.

Atta-ur-Rahman.*One and Two Dimensional Nuclear Magnetic Resonance Techniques*. Elsevier, Amsterdam, Holland, 1989; p. 578.

Aue, W.P.; Bartholdi, E.; Ernst, R.R. Two-dimensional spectroscopy. Application to nuclear magnetic resonance. *J. Chem. Phys.* **1976**, *64*, 2229-2246.

Bavovada, R.; Blaskó, G.; Shieh, H.-L.; Pezzuto, J.M.; Cordell, G.A. Spectral Assignment and Cytotoxicity of 22-Hydroxytingenone from *Glyptopetalum sclerocarpum. Planta Med.* **1990**, *56*, 380-382..

Bax., A. Broadband homonuclear decoupling in heteronuclear shift correlation nmr spectroscopy. *J. Magn. Reson.* **1983**, *53*, 517-520.

Bax, A. Structure determination and spectral assignment by pulsed polarization transfer via long-range ^{1}H-^{13}C couplings. *J. Magn. Reson.* **1984**, *57*, 314-318.

Bax, A.; Aszalos, A.; Dinya, Z.; Sudo, K. Structure elucidation of the antibiotic desertomycin through the use of new two-dimensional NMR techniques. *J. Am. Chem. Soc.* **1986**, *108*, 8056-8063.

Bax, A.; Davis, D.G. MLEV-17-Based two-dimensional homonuclear magnetization transfer spectroscopy. *J. Magn. Reson.* **1985**, *65*, 355-360.

Bax, A.; Freeman, R. Investigation of complex networks of spin-spin coupling by two-dimensional nmr. *J. Magn. Reson.* **1981**, *44*, 542-561.

Bax, A.; Freeman, R.; Frenkeil, T.A. An NMR technique for tracing out the carbon skeleton of an organic molecule. *J. Am. Chem. Soc.* **1981**, *103*, 2102-2104.

Bax, A.; Morris, G.A. An improved method for heteronuclear chemical shift correlation by two-dimensional nmr. *J. Magn. Reson.* **1981**, *42*, 501-505.

Bax, A.; Subramanian, S. Sensitivity-enhanced two-dimensional heteronuclear shift correlation nmr spectroscopy. *J. Magn. Reson.* **1986**, *67*, 565-569.

Bax, A.; Summers, M.F. ^{1}H and ^{13}C Assignments from sensitivity-enhanced detection of heteronuclear multiple-bond connectivity by 2D multiple quantum NMR. *J. Am. Chem. Soc.* **1986**, 108, 2093-2094.

Bhacca, N.S; Williams, D.H. *Application of NMR spectroscopy in organic chemistry, illustration for the steroid field.* Holden-Day, San Francisco, CA 1964; pp 198.

Blunt, J.W.; Stothers, J.D. ^{13}C N.m.r. spectra of steroids - a survey and commentary. *Org. Magn. Reson.* **1977**, *9*, 439-464.

Bodenhausen, G.; Freeman, R. Correlation of chemical shifts of potons and carbon-13. *J. Am. Chem. Soc.* **1978**, *100*, 320-321.

Bothner-By, A.A.; Stephens, R.L.; Lee, J.; Warren, C.D.; Jeanloz, R.W. Structure determination of a tetrasaccharide: transient nuclear Overhauser effects in the rotating frame. *J. Am. Chem. Soc.* **1984**, *106*, 811-813.

Braunschweiler, L.; Ernst, R.R. Coherence transfer by isotopic mixing: application to proton correlation spectroscopy. *J. Magn. Reson.* **1983**, *53*, 521-528.

Byrne, L.T. Nuclear magnetic resnance spectroscopy strategies for structural determination. In *Bioactive Natural Products. Detection, Isolation and Structure Determination*; Colegate, S.M.; Molyneux, R.J., Eds.; CRC Press, Boca Raton, FL, 1993; pp 75-104.

Carpenter, K.A.; Reynolds, W.F.; Yang, J.-P.; Enriquez, R.G. Further improvements in the FLOCK sequence. *Magn. Reson. Chem.* **1992**, 30, S35-S41.

Chang, P.T.O.; Cordell, G.A.; Fong, H.H.S.; Farnsworth, N.R. Velutinic acid, a new freidelane derivative from *Xylosma velutina. Phytochemistry*, **1977**, *16*, 1443-1445.

Chen, J.-J.; Qiu, S.-X.; Zhang, Z.-X.: Zhou, J. The chemical constituents of *Gymnema yunnanense. Acta Bot. Yunnan.* **1989**, *11*, 203-208.

Chen, S.-X.; Snyder, J.K. General strategy for the structure determination of saponins: molluscicidal saponins from *Allium vineale*. In *Bioactive Natural Products. Detection, Isolation and Structure Determination*; Colegate, S.M.; Molyneux, R.J., Eds.; CRC Press, Boca Raton, FL, 1993; pp 349-403.

Cohen, A.I.; Rock, S.; Anisotropy of substituents on the proton resonance of C-18 and C-19 methyl groups. Evaluation by computer regression. *Steroids* **1964**, *3*, 243-257.

Cordell, G.A. Selective INEPT spectroscopy - a powerful tool for the spectral assignment and structure elucidation of natural products. *Phytochem. Anal.* **1991**, *2*, 49-59.

Cordell, G.A.; Kinghorn, A.D. One-dimensional proton-carbon correlations for the structure determination of natural products. *Tetrahedron* **1991**, *47*, 3521-3534.
Cordell, G.A.; Lyons, R.L.; Fong, H.H.S.; Benoit, P.S.; Farnsworth, N.R. Biological and phytochemical investigations of *Dianthus barbatus* cv. "China Doll" (Caryophyllaceae). *Lloydia* **1977**, *40*, 361-363.
Davis, D.G.; Bax, A. Assignment of complex ^{1}H nmr spectra via two-dimensional homonuclear Hartmann-Hahn Spectroscopy. *J. Am. Chem. Soc.* **1985**, *107*, 2820-2821.
Derome, A.E. *Modern NMR Techniques for Chemistry Research*. Pergamon Press, Oxford, England, 1987; pp 280.
Derome, A.E. The use of N.M.R. spectroscopy in the structure determination of natural products: two-dimensional methods. *Nat. Prod. Repts.* **1989**, 6, 111-141.
Edwards, M.W.; Bax, A. Complete proton and carbon-13 nmr assignment of the alkaloid gephyrotoxin through the use of homonuclear Hartmann-Hahn and two-dimensional nmr spectrosscopy. *J. Am. Chem. Soc.* **1986**, *108*, 918-923.
Ekong, D.E.U. Chemistry of the meliacins (limonoids). The structure of nimbolide, a new meliacin from *Azadirachta indica*. *Chem. Commun.* **1967**, 808.
Gil, R.R.; Lin, L.-Z.; Chai, H.-B.; Pezzuto, J.M.; Cordell, G.A. Cardenolides from *Nierembergia stricta* (Solanaceae). *J. Nat. Prod.* In press.
Gottlieb, H.; Kirson, I. ^{13}C NMR spectroscopy of the withanolides and other highly oxygenated C_{28} steroids. *Org. Magn. Reson.* **1981**, *16*, 20-25.
Griesinger, C.; Ernst, R.R. Frequency offset effects and their elimination in nmr rotating frame cross-relaxation spectroscopy. *J. Magn. Reson.* **1987**, *75*, 261-271.
Gunasekera, S.P.; Cordell, G.A.; Farnsworth, N.R. Plant anticancer agents XX. constituents of *Nicandra physaloides* (Solanaceae). *Planta Med.* **1981**, *43*, 389-391.
Gunasekera, S.P.; Cordell, G.A.; Farnsworth, N.R. 3β-Hydroxy-28-*p*-coumaroyloxy-lup-20(29)-en-27-oic acid from *Caraipa densifolia*. *J. Nat. Prod.* **1983**, *46*, 118-122.
Hamburger, M.O.; Cordell, G.A. ^{1}H- and ^{13}C-NMR Studies of Selected Quassinoids. *Planta Med.* **1988**, *52*, 352-355.
Handa, S.S.; Kinghorn, A.D.; Cordell, G.A.; Farnsworth, N.R. Plant anticancer agents XXV. Constituents of *Soulamea soulameoides* (Simaroubaceae). *J. Nat. Prod.* **1983**, *46*, 359-364.
Hikino, H.; Kiso, Y. Natural products for liver diseases. *Economic and Medicinal Plant Research*; Hikino, H.; Wagner, H.; Farnsworth, N.R., Eds; Vol: 2, Academic Press, London, 1988; pp 39-72.
Jacquesy, J.C.; Levisalles, J. Stéréochemie de l'halogenation des céto-3(5α) stéroïdes. *Bull. Soc. Chim. Fr.* **1962**, 1866-1874.
Jeener, J.; Meier, B.H.; Bachman, P.; Ernst, R.R. Investigation of exchange processes by two-dimensional nmr spectroscopy. *J. Chem. Phys.* **1979**, *71*, 4546-4553.
Jia, Q.; Zhang, R.-Y. Progress in the research on the chemistry of the saponins from the genus *Bupleurum*. *Acta Pharm. Sinica* **1989**, *24*, 961-971.
Kawaguchi, K.; Asaka, I.; Hirotani, M.; Furuya, T.; Katsuki, S. Cardenoildes in the regenerated plants obtained from *Strophanthus divaricatus* calli. *Phytochemistry* **1993**, *34*, 1317-1321.
Kessler, H.; Griesinger, C.; Zarbock, J.; Loosli, H.R. Assignment of carbonyl carbons and sequence analysis in peptides by heteronuclear shift correlation via small coupling constants with broadband decoupling in t_1 (COLOC). *J. Magn. Reson.* **1984**, *57*, 331-336.
Kigodi, P.G.K.; Blaskó, G.; Thebtaranonth, Y.; Cordell, G.A. Spectroscopic and biological investigation of nimbolide and 28-deoxynimbolide from *Azadirachta indica*. *J. Nat. Prod.* **1989**, *52*, 1246-1251.
Koike, K.; Bevelle, C.; Talapatra, S.K.; Cordell, G.A.; Farnsworth, N.R. Plant anticancer agents V. Cardiac glycosides of *Asclepias albicans*. *Chem. Pharm. Bull. Japan.* **1980**, *28*, 401-405.
Konno, C.; Xue, H.-Z.; Lu, Z.-Z.; Soejarto, D.D.; Cordell, G.A.; Fong, H.H.S.; Hodgson, W. 3β-(3,4-Dihydroxy-cinnamoyl)-erythrodiol and 3β-(4-hydroxy-cinnamoyl) erythrodiol from *Larrea tridentata*. *Phytochemistry* **1988**, *27*, 233-235.
Kriwacki, R.W.; Pitner, T.P. Current aspects of practical two-dimensional (2D) nuclear magnetic resonance (NMR) spectroscopy: applications to structure elucidation. *Pharmaceut. Res.* **1989**, *6*, 531-553.
Lehn, J.M. Résonance magnétique nucléaire de produits naturels - II. Triterpènes de la série du dammarane: les groupes méthyles. *Bull. Soc. Chim. Fr.* **1962**, 1832-1837.
Levitt, M.H.; Ernst, R.R. Improvement of pulse performance in nmr coherence transfer experiments. A compensated INADEQUATE experiment. *Mol. Phys.* **1981**, *50*, 1109-1124.
Likhitwitayawuid, K.; Bavovada, R.; Lin, L.-Z.; Cordell, G.A. Revised structure of 20-hydroxytingenone and ^{13}C nmr assignments of 22β-hydroxytingenone. *Phytochemistry* **1993**, *34*, 759-763.
Lin, L.-J.; Lin, L.-Z.; Gil, R.R.; Cordell, G.A.; Ramesh, M.; Srilatha, B.; Reddy, B.; Appa Rao, A.V.N. New pregnane glycosides from *Caralluma umbellata*. *Phytochemistry* **1994**, *35*, 1549-1553.

Lin, L.-Z.; Cordell, G.A.; Ni, C.-Z.; Clardy, J. A quassinoid from *Brucea javanica*. *Phytochemistry* **1990**, *29*, 2720-2722.
Lin, L.-Z.; Lin, L.-J.; Cordell, G.A.; Luo, S.-Q.; Jiang, H.-F. Spectral assignment of 2-O-β-D-glucopyranosylsaikosaponin B2 by two-dimensional nmr techniques. *Magn. Reson. Chem.* **1992**, *30*, 1097-1103.
Luo, S.-Q.; Lin, L.-Z.; Cordell, G.A. Saikosaponins from *Bupleurum wenchuanense*. *Phytochemistry* **1993**, *33*, 1197-1205.
Luo, S.-Q.; Lin, L.-Z.; Cordell, G.A.; Xue, L.; Johnson, M.E. Assignment of the ^{1}H- and ^{13}C nmr spectra of the C_{21} steroids 12β-*O*-acetyltenacigenin A and tenacigenin A by two-dimensional nmr techniques and computer modeling. *Magn. Reson. Chem.* **1993a**, *34*, 215-221.
Luo, S.-Q.; Lin, L.-Z.; Cordell, G.A.; Xue, L.; Johnson, M.E. Polyoxypregnanes from *Marsdenia tenacissima*. *Phytochemistry* **1993b**, *34*, 1615-1620.
Luo, S.-Q.; Qiu, S.-X.; Lin, L.-Z.; Cordell, G.A. Minor polyoxypregnanes from *Marsdenia tenacissima*. *Phytochemistry*, In press.
Mahato, S.B.; Kindu, A.P. ^{13}C NMR spectra of pentacyclic triterpenoids - a compilation and some salient features. *Phytochemistry* **1994**, *37*, 1517-1575.
Martin, G.E. and Zektzer, A.S. (1988) *Two-Dimensional NMR Methods for Establishing Molecular Connectivity*. VCH Publishers, Inc., New York, NY, pp. 508.
Meksuriyen, D.; Nanayakkara, N.P.D.; Phoebe, Jr., C.H.; Cordell, G.A.; Farnsworth, N.R. Two triterpenoids from *Davidsonia pruriens*. *Phytochemistry* **1986**, *25*, 1685-1689.
Ogura, M.; Cordell, G.A.; Farnsworth, N.R. Jacoumaric acid, a new triterpene acid from *Jacaranda caucana* (Bignoniaceae). *Phytochemistry* **1977a**, *16*, 286-288.
Ogura, M.; Cordell, G.A.; Farnsworth, N.R. Plant anticancer agents IV. Constituents of *Jacaranda caucana*. *Lloydia* **1977b**, *40*, 157-168.
Ogura, M.; G.A. Cordell, A.D. Kinghorn and N.R. Farnsworth. Plant anticancer agents VI. Constituents of *Ailanthus excelsa*. *Lloydia* **1977**, *40*, 579-584.
Oksüz, S.; Shieh, H.-L.; Pezzuto, J.M.; Ozhatay, N.; Cordell, G.A. Biologically active compounds from the Euphorbiaceae. Part 1. Triterpenoids of *Euphorbia nicaeensis* subsp. *glabra*. *Planta Med.* **1993**, *59*, 472-473.
Oksüz, S.; Gil, R.R.; Chai, H.-B.; Pezzuto, J.M.; Cordell, G.A.; Ulubelen, A. Biologically active compounds from the Euphorbiaceae. Part 2. Triterpenoids of *Euphorbia cyparissias*. *Planta Med.* **1994**, *60*, 594-596.
Pauli, G.F.; Junior, P.; Berger, S.; Matthiesen, U. Alepposides, cardenolide oligoglycosides from *Adonis aleppica*. *J. Nat. Prod.* **1993**, *56*, 67-75.
Piantini, U; Sørenson, O.W.;Ernst, R.R. Multiple quantum filters for elucidating nmr coupling networks. *J. Am. Chem. Soc.* **1982**, *104*, 6800-6801.
Qiu, S.-X.; Lin, L.-Z.; Nan, Y.; Lin, P.; Chen, J.-J.; Zhang, Z.-X.; Zhou, J.; Cordell, G.A. Internal acyl migration of gymnemarsgenin. *Phytochemistry*, In press.
Qiu, S.-X.; Zhang, Z.-X.; Zhou, J. Studies on the constituents of *Marsdenia globifera*. *Acta Bot. Sinica* **1990**, *32*, 936-942.
Rance, M.; Sørenson, O.W.; Bodenhausen, G.; Wagner, G.; Ernst, R.R.; Wuthrich, K. Improved spectral resolution in COSY ^{1}H nmr spectra of proteins by double quantum filtering. *Biochem. Biophys. Res. Commun.* **1983**, *117*, 479-485.
Reddy, B.M.; Rao, N.K.; Ramesh, M.; Appa Rao, A.V.N.; Lin, L.-J.; Cordell, G.A. Chemical investigation of the fruits of *Terminalia chebula*. *Int. J. Pharmacog.* **1994**, *32*, 352-356.
Reich, H.J.; Jautelat, M.; Messe, M.T.; Weigert, F.J.; Roberts, J.D. Nuclear magnetic resonance spectroscopy. Carbon-13 spectra of steroids. *J. Am. Chem. Soc.* **1969**, *91*, 7445-7454.
Reynolds, W.F.; McLean, S.; Perpick-Dumont, M.; Enriguez, R.G. Improved ^{13}C-^{1}H shift correlation spectra for indirectly bonded carbons and hydrogens: the FLOCK sequence. *Magn. Reson. Chem.* **1989**, 27, 162-169.
Robien, W.; Kopp, B.; Schabl, D.; Schwarz, H. Carbon-13 nmr spectroscopy of cardenolides and bufadienolides. *Progr. NMR Spectr.* **1987**, *19*, 131-181.
Schneider, H.-J.; Buchheit, U.; Becker, N.; Schmidt, G.; Siehl, U. ^{1}H NMR analyses, shielding mechanisms, coupling constants, and conformations in steroids bearing halogen, hydroxy, oxo groups, and double bonds. *J. Am. Chem. Soc.* **1985**, *107*, 7027-7039.
Schun, Y.; Cordell, G.A.-Cytotoxic steroids of *Gelsemium sempervirens* (Loganiaceae). *J. Nat. Prod.* **1987**, *50*, 195-198.
Schun, Y.; Cordell, G.A.; Cox, P.J.; Howie, R.A. Wallenone, a new C-32 triterpenoid from the leaves of *Gyrinops walla*. *Phytochemistry* **1986**, *25*, 753-755.
Seida, A.A.; Kinghorn, A.D.; Cordell, G.A.; Farnsworth, N.R. Potential anticancer agents IX. The isolation of a new simaroubolide 6α-tigloyloxy-chaparrinone from *Ailanthus integrifolia* subsp. *calycina*. *Lloydia* **1978**, *41*, 584-587.
Shao, Y.; Zhou, B.-N.; Lin, L.-Z.; Cordell, G.A. Two new triterpenoid saponins, asterbatanoside F and G, from *Aster batangensis*. *Nat. Prod. Letts.* **1994a**, *5*, 233-240.

Shao, Y.; Zhou, B.-N.; Lin, L.-Z.; Cordell, G.A. Asteryunnanoside E, A new triterpenoid saponin from *Aster yunnanensis*. *Chin. Chem. Letts.* **1994b**, *5*, 761-764.
Shao, Y.; Zhou, B.-N.; Lin, L.-Z.; Cordell, G.A. Two oleanolic acid saponins from *Aster yunnanensis*. *Chin. Chem. Letts.* **1994c**, *5*, 843-846.
Shao, Y.; Zhou, B.-N.; Lin, L.-Z.; Cordell, G.A. Triterpenoid saponins from *Aster batangensis*. *Phytochemistry* **1995a**, *38*, 927-933.
Shao, Y.; Zhou, B.-N.; Lin, L.-Z.; Cordell, G.A. The structure of asterbatanoside J from *Aster batangensis*. *Chin. Chem. Letts.* **1995b**, *6*, 221-224.
Shao, Y.; Zhou, B.-N.; Lin, L.-Z.; Cordell, G.A. Triterpene saponins from *Aster yunnanensis*. *Phytochemistry* **1995c**, *38*, 1487-1492.
Sherman, M.; Borris, R.P.; Cordell, G.A.; Ogura, M.; Farnsworth, N.R. 3*S*,24*S*,25-Tri-hydroxytiruca11-7-ene from *Ailanthus excelsa*. *Phytochemistry* **1980**, *19*, 1499-1501.
Shoolery, J.N.; Rogers, M.J. Nuclear magnetic resonance spectra of steroids. *J. Am. Chem. Soc.* **1958**, *80*, 5121-5135.
Smith, L.L. Recognition of structure in hydroxy steroids. II. Nuclear magnetic resonance spectra. *Steroids* **1964** *4*, 395-414.
Summers, M.F.; Marzilli, L.G.; Bax, A. Complete ^{1}H and ^{13}C assignments of coenzyme B_{12} through the use of new two-dimensional nmr experiments. *J. Am. Chem. Soc.* **1986**, *108*, 4285-4294.
Yuan, J.-L.; Ding, W.-P.; Shi, J.-P.; Lu, Z.-Z.; Zhou, B.-N.; Erdelmeier, C.A.J.; Cordell, G.A.; Fong, H.H.S.; Farnsworth, N.R. Studies on the antifertility components of *Marsdenia koi*. *J. Tongji Med. Univ.* **1991**, *11*, 165-168.
Yuan, J.-L.; Lu, Z.-Z.; Chen, G.-X.; Ding, W.-P.; Zhou, B.-N.; Erdelmeier, C.A.J.; Hamburger, M.O.; Fong, H.H.S.; Cordell, G.A. The pregnane glycoside marsdekoiside A from *Marsdenia koi*. *Phytochemistry* **1992**, *31*, 1058-1060.
Zurcher, R.F. Protonresonanzspektroskopie und steroidsstruktur I. Das C-19-methylsignal in funktion der substituion. *Helv. Chim. Acta* **1961**, *44*, 1380-1395.

A SYSTEMATIC NMR APPROACH FOR THE DETERMINATION OF THE MOLECULAR STRUCTURE OF STEROIDAL SAPONINS

Pawan K. Agrawal
Central Institute of Medicinal and Aromatic Plants, Lucknow - 226 015, INDIA

INTRODUCTION

Steroidal saponins and their aglycones (steroidal sapogenins), which are widely distributed in various plant families, are attracting the attention of research workers not only as economically important raw material convertible into various steroid hormonal drugs, but recently also as biologically active materials having independent value.[1-6] All steroidal sapogenins possess a parent cholestane carbon skeleton (C_{27}), having perhydrocyclopentenophenanthrene nucleus (rings A, B, C, and D), the side chain of which undergoes cyclization resulting in either a hexacyclic system (four carbocyclic and two heterocyclic rings) or a pentacyclic system (four carbocyclic and one heterocyclic ring). Most of these contain, in addition, one 5-membered ring (E) and one 6- or 5-membered ring (F), both of which are heterocyclic and fixed in a spiran fashion at C-22. Those cholestane derivatives which are formed by the ring opening of both heterocyclic rings are also included in this class because these have been regarded as intermediate in the biosynthesis of steroidal sapogenins.[7-9]

In all of the natural steroidal sapogenins, the substituents at positions 8-9, 9-10, 8-14, and 13-14 are *trans*. By Fisher projection, the methyl groups attached to C-10 always have a β_f configuration, while the 20-methyl group possesses the 20α_f configuration. The natural spirostanes and their derivatives also, in most cases, have the same α_f configuration at C-22 but, occasionally, the β_f configuration. In general, steroidal sapogenins have been divided into four groups: spirostanes **1.1** (16,22; 22,26-bisepoxycholestane), furostanes **1.2** (16,22-epoxycholestane), furospirostanes **1.3** (16,22; 22,25-bisepoxycholestane), and 16,22-dihydroxycholestanes **1.4** (Fig. 1). The variation in steroidal structure generally arises from the configurations at C-5 and C-25, the number, position and configuration of nuclear hydroxyl groups, and also the occasional presence of an carbonyl group.

The problem of structure determination of these biomolecules has been attacked by many techniques in the past years. The conventional methods include hydrolytic studies followed by the characterization of the aglycone (sapogenin) and oligosaccharide moieties. The drawback of this procedure is the loss of information about the glycosylation site and sometimes about the anomeric configuration owing to anomerization of the reducing monosaccharide. The hydrolytic studies are usually performed under acidic conditions, the major disadvantage of which is the cyclization of furostanol to spirostanol aglycone. After the identification of the sapogenin and oligosaccharide residues, the final step requires deduction of the molecular structure of the saponin, that is, the determination of sapogenin - oligosaccharide linkage. Of all the modern structural methods for saponins, NMR yields the most complete picture of saponin structure and behavior in solution with or without prior structure knowledge.

The NMR chemical shifts and coupling constants are the easiest parameters to be measured for the saponins and they have been recognized to carry a wealth of structural information. However, the ^1H NMR spectra, even of steroidal sapogenins, are complicated by extensive interproton coupling; as a result only a few signals, such as the singlets for the angular methyl groups (H_3-18, H_3-19) and sometimes the doublets for secondary methyl groups (H_3-21, H_3-27), can be assigned. The rest of the methylene and methine resonances (typically > 20 protons) appear as a broad and almost featureless resonance from 1 - 2.5 ppm. This steroid hump, representing of nuclear and side chain methylene and methines, was universally ignored as hopelessly

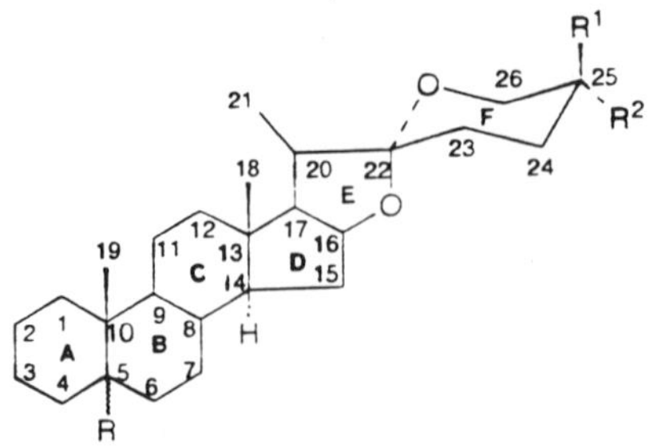

1.1 R = H or $\Delta^{5(6)}$, R^1, R^2 = H, Me; Spirostanoids
1.1.1 R = α-H, R^1, R^2 = H, Me; 5α-spirostane
1.1.2 R = β-H, R^1, R^2 = H, Me; 5β-spirostane
1.1.3 R = $\Delta^{5(6)}$, R^1, R^2 = H, Me; Δ^5-spirostane
25*R* 1.1 R = H or $\Delta^{5(6)}$; R^1 = H, R^2 = Me
25*S* 1.1 R = H or $\Delta^{5(6)}$, R^1 = Me, R^2 = H

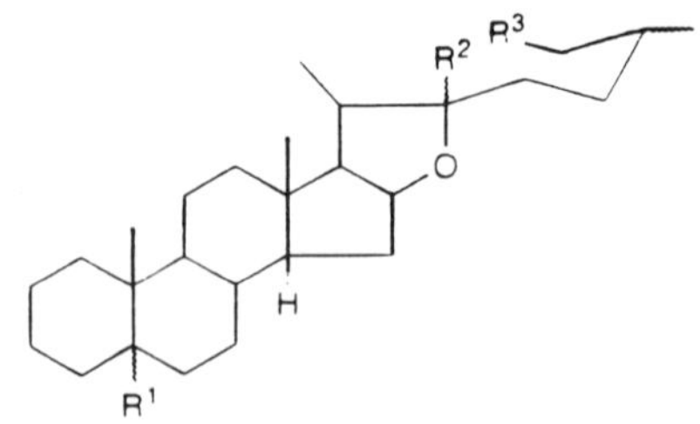

1.2 R^1 = H or $\Delta^{5(6)}$, R^2 = H, OH or OMe, R^3 = H or OH; Furostanoids
1.2.1 R^1 = H or $\Delta^{5(6)}$, R^2 = H, R^3 = OH; 22-deoxyfurostanol
1.2.2 R^1 = H or $\Delta^{5(6)}$, R^2 = R^3 = OH; 22-hydroxyfurostanol
1.2.3 R^1 = H or $\Delta^{5(6)}$, R^2 = OMe, R^3 = OH; 22-methoxyfurostanol
1.2.4 R^1 = H or $\Delta^{5(6)}$, R^2 = R^3 = H; 22-deoxyfurostane
1.2.5 R^1 = H or $\Delta^{5(6)}$, R^2 = OH, R^3 = H; 22-hydroxyfurostane
1.2.6 R^1 = H or $\Delta^{5(6)}$, R^2 = OMe, R^3 = H; 22-methoxyfurostane

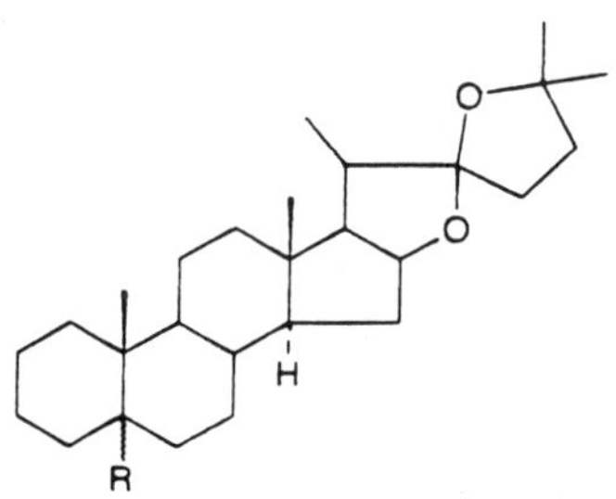

1.3 R = H or $\Delta^{5(6)}$; Furospirostanoids
1.3.1 R = α-H; 5α-furospirostane
1.3.2 R = β-H; 5β-furospirostane
1.3.3 R = $\Delta^{5(6)}$; Δ^5-furospirostane

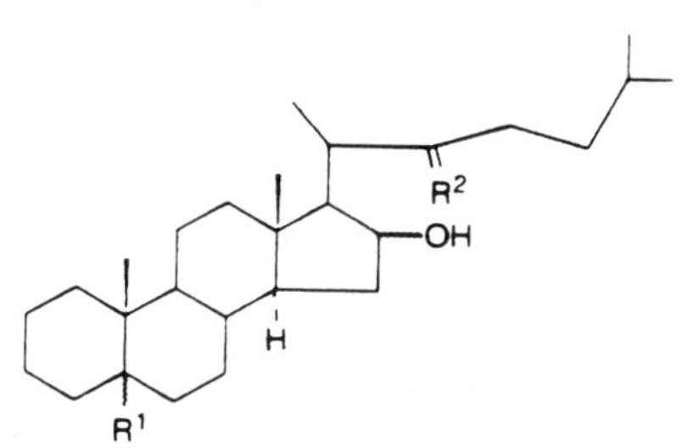

1.4 R^1 = H or $\Delta^{5(6)}$, R^2 = H, OH or H_2 or O; Cholestanoids
1.4.1 R^1 = H or $\Delta^{5(6)}$, R^2 = H, OH; 16, 22-dihydroxycholestane
1.4.2 R^1 = H or $\Delta^{5(6)}$, R^2 = H_2; 16-hydroxycholestane
1.4.3 R^1 = H or $\Delta^{5(6)}$, R^2 = O; 16-hydroxycholest-22-one

Figure 1. Parent skeletal types of steroidal sapogenins.

crowded with overlapping resonances, providing no structural information. This was the most probable reason why most of the early publications[10] incorporated only the ^{1}H NMR chemical shifts for the methyl signals, the substituent-induced shift resonances, such as olefinic, oxysubstituted methylene, and methine resonances appearing outside the 1.0 to 2.5 ppm region. In general, the magnitude of the geminal ($^2J_{HH}$) coupling ranges between -12 to -14 Hz whereas depending upon stereochemical relationships, vicinal coupling ($^3J_{HH}$) varies between 2 and 14 Hz ($^3J_{aa}$ = 10.5 - 14 Hz, $^3J_{ae}$ = 3.5 - 5.0 Hz, $^3J_{ee}$ = 2.5 - 4 Hz). These values are characteristic for six-membered rings (A, B, C, and F). Generally, axial protons are known to resonate at higher field than equatorial protons except those which occur in the vicinity of substituents. Signal line widths and spectral integration identify the kinds of protons and their relative abundance present in various spectral regions.

The proton-decoupled ^{13}C NMR spectral data, with a single line for each carbon atom, is usually resolved, and thus has become an integral part of the procedure for establishing the structure of new steroidal sapogenins and saponins. Alternatively, 1D ^{13}C NMR spectral editing methods (APT, DEPT, INEPT, etc.) readily reveal steroid methyl, methylene, methine, and quaternary carbons. It is worthwhile to mention that the ^{13}C NMR spectrum of a given steroidal sapogenin, as well as that of the saponin, is unique; mere comparison of the ^{13}C NMR spectra of compounds, even without detailed interpretation, can demonstrate the identity of compounds. Thus, a ^{13}C NMR study may suffice for primary structure determination or recognition of the characteristic fragments, if the sapogenin or saponin has been characterized previously by its ^{13}C NMR data. Moreover, the structural and substituent effects that control ^{13}C chemical shifts are well documented for steroids,[11-13] allowing good estimates to be made for individual carbon atoms. However, simple additivity of atomic increment can not always predict the true molecular constitution owing to steric factors and electronic influences of the substituents, etc., in polysubstituted steroidal sapogenins and therefore one often needs a more detailed analysis of chemical shift data. Few ^{13}C NMR data bases are available for this category of natural products[14,15] and when the spectrum does not match any of the spectra in the existing data bases, the attempts can be made to interpret the ^{13}C NMR spectrum in terms of partial elements of the primary structure.

Table 1. Selected NMR approaches used for the structure establishment of steroidal saponins.

NMR experiment (Acronyms)	Comments
Attached proton test (APT),[16] Distortionless enhancement by polarization transfer (DEPT),[17] Insensitive nuclei enhancement by polarization transfer (INEPT),[18]	Discriminates among carbon types; Spectral editing
Incredible natural abundance double-quantum transfer experiment (INADEQUATE)[19]	^{13}C - ^{13}C connectivity, Establishment of molecular skeleton
^1H, ^1H - COSY	Homonuclear shift correlation
a) normal[20,21]	Elucidation of direct couplings
b) with delays[22]	Detection of small couplings
c) double-quantum filtered-(DQF) - COSY[23]	Determination of vicinal and geminal coupling constants
d) Exclusive COSY (E. COSY)[24]	Accurate determination of *J*
e) Geminal COSY (Gem - COSY)[25]	Identification of geminal spin systems
f) Triple-quantum filtered (TQF)-COSY[26]	Detection of three or more mutually coupled spin systems
Relayed coherence transfer (RCT)[27], Total correlation (TOCSY)[28], and Hartmann-Hahn experiment (HOHAHA)[29]	Identification of all protons belonging to a single spin system; Coherence transfer across scalar connectivity (particularly useful in identifying monosaccharide residues)
Homonuclear nuclear Overhauser and exchange spectroscopy (NOESY[30] and ROESY[31])	Identification of protons that are within 5 A of one another (^1H, ^1H correlation through space); Stereochemical analysis (orientation of substituents); Intra- and inter-residual connectivities (sequence analysis in sugar chain including sugar - aglycone linkage)
$^1H\{^{13c}C\}$SBC (HETCOR[32] and HMQC[33])	Heteronuclear shift correlation; Assignments of directly bonded ^1H and ^{13}C shifts
HMQC-TOCSY and HMQC-RELAY[34]	Cross assignments of ^1H and ^{13}C shifts
$^1H\{^{13}C\}$MBC (Long-range HETCOR[35] and HMBC[36])	Assignment of quaternary C; Correlation of a proton resonance with a carbon resonance 2-4 bonds distant; Intra- and inter-residual assignments (inter-glycosidic and sugar-aglycone linkage); Confirmation of molecular structure

In practice, certain ^1H and ^{13}C NMR resonances such as those belonging to oxy-substituted and olefinic types can be identified and assigned on the basis of shift arguments, but for interpreting the results of NMR experiments in a rigorous manner, an NMR spectrum must be assigned unambiguously, which means establishing which peaks are associated with which carbon and/or hydrogen in the structure. This information, in most cases, can not be obtained from one-dimensional ^1H and ^{13}C NMR spctral data, but can better be determined with the aid of two-dimensional experiments. These experiments simplify spectral analysis by spreading out information into two frequency domains and by revealing interactions between nuclei. Despite the fact that the mechanisms on which the various pulse sequences are established may be intricate, the interpretation of two-dimensional NMR spectra is usually straightforward. A large number of different two-dimensional NMR experiments have been devised to solve chemical structures and it is not be possible to explain all of these techniques here. However, there are some widely used experiments,[16-36]; and these along with their acronyms, are presented in Table 1. Comprehensive and critical reviews covering the field have appeared.[37-48] It is important to mention that there is no definite sequence for the NMR experiments to be followed to solve a chemical structure and their choice depends, of course, upon the amount of the material available and the complexity of the structural problem to be solved.

DETERMINATION OF THE MOLECULAR STRUCTURE OF THE STEROIDAL SAPOGENIN MOIETY

To accomplish this, it is necessary to establish: a) the type of parent skeleton, b) the substitution pattern, and c) the configurations at C-5, C-22 and C-25.

DETERMINATION OF THE PARENT SKELETON

The common feature of all steroidal sapogenins is the presence of mostly four, but occasionally five, methyl resonances. These are distinguishable into two *tert.* and two *sec.* methyl groups in **1.1**, **1.2.2** and **1.2.3**, but into two *tert.* and three *sec.* methyl groups in **1.2.4** - **1.2.6** and **1.4**. The presence of four *tert.* methyl groups in addition to one *sec.* methyl groups identifies the furospirostane skeleton (**1.3**). The multiplicity and chemical shift of the C-22 acts as a structural marker for the determination of the parent skeletal type of the aglycone. A literature survey reveals that the chemical shift of C-22 is not only sensitive to the size and ring opening of ring F, but also exhibits a remarkable dependence on the presence of the hydroxyl groups in rings D and F (Table 2).

Table 2: ^{13}C NMR chemical shift ranges of C-22 for various steroidal sapogenins.

skeletal type		Chemical shift in ppm (carbon type)
Spirostane (**1.1**)		105.5 - 117.7
a.	General	108.9 - 110.0 (C)
b.	12-OH (and 14-OH/OAc)	105.5 - 105.7 (C)
c.	16α-OH/OMe	110.1 - 111.1 (C)
d.	17α-OH	112.0 (C)
e	23α-OH	117.7 (C)
f.	23β-OH	112.6 - 113.5 (C)
g.	24-OH	110.9 - 111.6 (C)
h.	25-OH	108.3 (C)
Furostane (**1.2**)		90.2 - 113.5
a.	22-H	90.2 - 90.6 (CH)
b.	22-OH/OMe	110.0 - 113.5 (C)
Furospirostane (**1.3**)		119.6 - 121.7 (C)
16-Hydroxycholestane (**1.4**)		34.8 - 219.5
a.	22-Deoxy	34.8 - 35.4 (CH_2)
b.	22-Deoxy, 23-oxo	50.4 (CH_2)
c.	22-Hydroxy	71.5 - 73.4 (CH)
d.	22-oxo	214.6 - 219.5 (C)

The naturally occurring furostane type of steroidal saponins occur mostly in the 26-*O*-glycosidic form and possess an OH/OMe group at C-22. The 22-OH and 22-OMe forms of **1.2** are distinguishable on the basis of the chemical shift of C-22 as it resonates at about 1-2 ppm higher field in the case of the former (**1.2.2**) relative to the latter (**1.2.3**), which may be further complemented by the absence of OMe resonance (^{13}C: δ 47.0-47.5; ^{1}H: δ 3.25±0.2) in the case of the former. A comparison of the ^{13}C NMR chemical shifts for spirostane and 22-hydroxylated, 26-*O*-glucosylated furostanes with identical substitution in the carbocyclic rings, reveals pronounced effects, particularly for the ring F resonances: C-20, -(1.3-1.6); C-21, +(0.9-1.6); C-22, +(1.3-1.5); C-23, +(5.1-5.4); C-24, -(0.8-1.0); C-25, +(3.5-3.7) and C-26, +(8.2-8.4) ppm, respectively.[49-52] These spectral features can be utilized in an empirical manner, for predicting the chemical shifts of furostanols derived from spirostanols. The C-22 absorbs at 90.5±0.1 ppm in furostane sapogenins lacking a 26-hydroxyl group and having an unsubstituted

22-position, which also led to the appearance of C-26 and C-27 methyl resonances between 22.7±0.2 ppm.[53,54] The resonances for C-26 and C-27 appear at 67.9±0.2 and 16.7±0.1 ppm, respectively in 26-hydroxylated furostanes.[8,54] The appearance of C-26 at about 2 ppm lower field (*ca.* 69.3 ppm) in the case of 26-*O*-esterified (acetylated) furostanes as compared to the 26-hydroxylated equivalent identifies the former.[55]

The appearance of C-25 at 84.0-85.5 ppm and its non-protonated behavior is quite peculiar for the identification of the **1.3** type skeleton. The hydroxyl substitution at both the C-26 and C-27 positions leads to the appearance at 65.4±0.5 ppm. The 16-hydroxylated cholestane subgroup of steroidal sapogenins lack ring E, but may possess ring F as in the 22,26-epoxycholestane skeleton. Polypodogenin is an example of this type in which C-22 and C-26 are methine carbons, appearing at δ78.1 and 101.3, respectively, due to hydroxyl substitution at the 26-position.[56,57]

DETERMINATION OF THE SUBSTITUTION PATTERN

Most of the naturally occurring steroidal sapogenins occur in hydroxylated form, having OH- substitution on any ring (A - D and F), and may possess additional oxo group(s). The OH substitution causes pronounced downfield shifts ranging between 1.8 and 2.9 ppm for the carbinol proton, which thus cover a wide chemical shift range (3.2 - 5.0 ppm). The ^{1}H NMR spectrum is therefore informative in determining the number and type of hydroxyl groups (*sec.* and/or primary), but requires the initial subtraction of resonances belonging to H-16 and H_2-26 from the above region. However, if the skeletal types **1.2.1**, **1.2.4**, and **1.4** are under consideration, then subtraction of H-22 is also required. The H-16 resonates between δ4.10 and 4.50 as a quartet, doublet of doublets, or a multiplet having $^3J_{16,15\alpha}$ = $^3J_{16,15\beta}$ = $^3J_{16,17}$ = 6.2-8.2 Hz in the case of both spirostanes and furostanes.[10,54] In 16,22-dihydroxycholestanes, H-16 and H-22 appear at *ca.* δ 4.80 as a doublet of triplets ($^3J_{16,17}$ = 8.0) and at δ 4.14 as a multiplet ($^3J_{22,23}$ = 8.5, $^3J_{22,23}$ = 4.5 Hz), respectively.[53] The geminal protons of H_2-26 appear between δ 3.20 - 3.90 as a triplet ($^2J_{26a,26e}$ = 9.0-12.0 Hz) and a doublet of doublets ($^3J_{26a,25e}$ = 3.0-4.9 Hz, $^3J_{26a,25a}$ = 9.0-12.0 Hz).[10] Despite the characteristic multiplicity pattern, H-16 and H_2-26 may not always be easily identifiable in polyhydroxylated steroidal sapogenins owing to the overlapping of these resonances with other hydroxy-bearing CH and CH_2 resonances. The situation becomes more complex in steroidal saponins, as all of the hydroxymethine and methylene resonances, excluding the anomeric-H of the sugar residues, also appear between 3.2 -4.5 ppm.

For allocating the number and type of hydroxyl groups, ^{13}C NMR data, combined with 13C NMR spectral editing techniques (e.g. DEPT), are of prime importance, as hydroxy-bearing carbon resonances absorb at relatively lower field position (δ 60-85). The fact that hydroxyl substitution at the nuclear CH_2, at ring junctions, and at methyl groups leads to an increase in CH, C, and CH_2 resonances in the ^{13}C NMR spectrum while reducing the number of CH_2, CH and CH_3 signals respectively, as primarily required for the parent skeleton (Table 3), facilitates the determination of the types of hydroxyl group. The hydroxylated methine resonances generally absorb between δ 67.4 and 79.5 and may appear at lower field position depending on the nature of the

Table 3. Type of carbon resonances constituting the parent skeleton of the steroidal sapogenins.[a]

Skeletal type	CH_3	CH_2		CH		C	
		Nonoxy	Oxy	Nonoxy	Oxy	Nonoxy	Oxy
Spirostane	4	11	1	7	1	2	1
Furostane	4	11	1	7	1	2	1
Furospirostane	5	11		7	1	1	2
16,22-Dihydroxycholestane	5	11		7	2	2	

[a] Nonoxy = nonoxygenated, Oxy = oxygenated

the substituents at the neighboring positions. Since C-16 and C-26 are also oxy-substituted, appearing between δ 80.7 and 81.4 and δ 65.1 and 67.1 respectively, these signals should be eliminated for ascertaining the number of hydroxyl groups. Hydroxyl substitution at the ring junctions A/B, C/D and D/E, i.e. at C-5, C-14, C-16 and C-17 and also at C-25, leads to the nonprotonated nature of these carbon atoms. The chemical shift ranges are δ 65.1-65.9 (C-25), 73.2-78.3 (C-5), 83.4-86.2 (C-5), 85.5-87.8 (C-14), 90.0-90.5 (C-17) and δ114.3-116.1 (C-16) for 25-OH, 5-OH, 5-OH and 6-oxo, 14-OH, 17-OH and 16-OH, respectively.[57-60] Concerning the hydroxyl substitution of the CH_3 group, 21- and 27-hydroxylated compounds have been characterized in which C-21 and C-27 resonate at δ 61.5 and δ 64.4-64.9, respectively.

One reliable approach for the assignment of ^{1}H NMR resonances utilizes the analysis of ^{1}H-^{1}H COSY spectral data involving oxy-substituted methine and methylene proton resonances as entry points for the assignment of vicinal and consequently other adjacent neighboring protons. The protons on ring F, including H_3-27, can be assigned by employing the H_2-26 geminal protons as the starting point, and, in analogous manner, H-16 acts as an entry point for the assignments of the ring D resonances. In Δ^5-spirostane, the connectivity to H-4 which in turn exhibits cross-peaks H-8 and H-3, and so on, resulting in gradual assignments of the protons belonging to rings B and A, etc.[37,61-63] Delayed COSY, which detects small long-range couplings, helps to penetrate into the heart of convoluted region (e.g. protons of an angular methyl group at the 18-position show characteristic coupling with H-12[61-65])and hence the long-range connectivities lead to verification of the ^{1}H NMR assignments. The E.COSY has also been proved to be useful in identifying $^3J_{HH}$ coupling constants and proton-proton coupling networks in steroidal compounds.[61] In the case of severe cross-peak overlapping, use of ω_1-decoupled COSY (COSYDEC) method can sometimes facilitate ^{1}H NMR analysis of steroids[66,67], but it is inherently dangerous, since if the constant time $\Delta = \sim 1/2J_{HH}$, then the cross-peaks will disappear. Additional information about the remote ^{1}H NMR connectivities is accessible by multi-step RCT or HOHAHA spectroscopy in which an individual proton resonance displays cross-peaks not only with vicinal protons, but also with protons that are not directly spin-coupled. A comparative study of COSY and RCT or HOHAHA spectra discriminates between the normal COSY cross-peaks and relayed peaks; hence assignments of the ^{1}H NMR resonances can be achieved. These techniques help in locating the site of OH substitution, but ambiguities may arise from coincident chemical shifts. Consequently accurate location of signals from middle ring protons without recourse to ^{13}C and heteronuclear correlated spectra is not normally trivial.

The ^{1}H{^{13}C}SBC correlations are useful for resolving overlap in the ^{1}H dimension, and often yield further information about the origin of proton assignments (based on the chemical shift of the attached ^{13}C), and hence identifying OH-bearing-carbon resonances. It means that quaternary carbons do not appear in such a spectrum owing to lack of directly attached protons. Hence tertiary hydroxyl-group-bearing carbons will not show up and thus, their presence or absence can be established by comparison of ^{13}C NMR and ^{1}H{^{13}C}SBC spectral data. The ^{1}H-detected HMQC experiments benefit from the fact that owing to the higher resonance frequency of protons than carbon-13, a given number of protons give rise to a larger signal than the same for carbon-13 atoms. Thus, the sensitivity of the HMQC experiment is much larger, an advantage which enables such experiment to be carried out with a rather small quantity of the material for which measurement of ^{13}C detected HETCOR is usually tedious or even not possible. Although ^{1}H-detected experiments have distinct sensitivity advantages, they suffer from limited ^{13}C resolution. By contrast, the ^{13}C-detected sequences can have excellent ^{13}C resolution and are therefore useful to distinguish resonances with closely similar chemical shifts. For example in the case of smilagenin (25*R*-3β-hydroxy-5β-spirostane), C-6 and C-7 exhibit coincidence at δ 26.5 (δ 26.56 and 26.54) and thus ^{1}H NMR chemical shifts for these methylene resonances could not be ascertained from the HMQC spectrum .[68] Once the chemical shifts of the hydroxy-bearing methine and methylene resonances and their vicinal partners have been identified, then ^{13}C NMR shifts for C-β (with respect to OH-bearing carbon) can be employed in establishment of the site of hydroxyl substitution as these exhibit appreciable deshielding. Iterative comparison between ^{1}H{^{13}C}SBC correlation and ^{1}H-^{1}H COSY spectra provides valuable information for differentiation between geminal and vicinal protons and the unambiguous assignment of the ^{13}C resonances.

The ^{1}H{^{13}C}MBC spectral data reveal connectivities between protons and carbons separated by two and four bonds and facilitate the assignment of quaternary carbons and carbon resonances across the heteronucleus. Such spectral data exhibit cross-peaks for the protons that couple to that carbon at the resonance position of the carbon signal. Many protons couple not only to one carbon, but also to several others as well, hence the ^{1}H{^{13}C}MBC spectrum contains connectivity information. The angular methyl protons exhibit correlation to quaternary carbon to which they are bonded; thus H_3-18 and H_3-19 exhibit correlation to C-13 and C-10, in addition to three-bond correlations with C-17 and C-14, and with C-1, C-5 and C-9, respectively.[61-63] The H_3-21 and H-20 exhibit their three- and two-bond correlations with C-22,[61,63,69-73] and in analogous manner, the assignment of the ^{13}C chemical shifts, postulated via heteronuclear methods such as ^{1}H{^{13}C}SBC etc., could be confirmed.

The LR-HETCOR experiments eventually suffer from sensitivity limitations, and when sufficient quantities of material are available and there exists no solubility problem, it is preferable to record heteronuclear detected ^{1}H-^{13}C correlation data. In the case of the ambiguity of the ^{13}C chemical shifts due to the overlapping of corresponding protons, a short heteronuclear correlation sequence with relayed transfer, as in the HMQC-TOCSY experiment is helpful in which individual ^{13}C resonance not only correlates with its directly bonded proton(s), but also with neighboring protons two and three bonds apart for a range of mixing times.[68] In the case of H,H COSY overlap, *Ge*-SELINCOR-TOCSY is helpful, as H,C COSY allows correlation between the proton and carbon chemical shifts, while H,H-TOCSY determines the long-range connectivity.[74]

To determine the orientation of the hydroxyl groups, ^3J coupling constants, ^{13}C NMR chemical shifts, and NOE connectivity experiments, such as one-dimensional NOE difference spectroscopy (1D NOEDS) and NOESY or ROESY are of significance. The 2D methods, such as phase-sensitive COSY or E.COSY, and *J*-resolved spectra, as well as 1D methods, such as simple spin decoupling, are usually employed for the establishment of coupling constants. All of the methods, however, have the common feature that they produce multiplets which connect different spins, and which still show additional spin couplings to other passive spins. Recently SERF (selective refocusing) has been reported to determine desired spin-coupling constants in steroid compounds.[75]

The NOE provides valuable evidence about the spatial proximity of protons which complements the information on through-bond connectivities and which is unobtainable from other experiments. For example, β-angular methyl (H_3-19) protons can produce NOE's at nearby β-axial protons (2β,4β,6β,8β,11β) and nearest-neighbor equatorial protons. Irradiation of axial methylene protons can yield NOEs at geminal partners, equatorial protons on adjacent carbons, and even nearby angular methyl protons. Saturation of equatorial methylenes can produce NOEs at all of the geminal and vicinal protons as well as structurally more distant ones. Across-ring proximities can also be revealed in difference NOE spectra. Other NOEs will appear when the ring conformation is perturbed by fuctionalization or by unsaturation. NOESY, while less demanding on spectrometer hardware, has problem that the NOE cross-peaks change from positive to negative as molecular tumbling slows with increasing molecular weight and/or solvent viscosity. Steroidal saponins having molecular masses between 600 and 1200, in particular, run the risk of tumbling at rates which give zero NOE effects in either NOESY or NOE-difference spectra,[70,76] thereby limiting the utility of the method. In contrast, nuclear Overhauser effects that occur under spin-lock conditions are known as the transverse or rotating-frame NOE which increase monotonically in a positive direction with increasing correlation time and provide a potential solution to the problem. The contributions from scalar coupling (TOCSY artifacts) near diagonal however, a disadvantage in ROESY experiments.[61]

The oxo group has been recognized at carbons 3, 6, 11, 12, 22 and 23, which leads to the appearance of the carbonyl carbon resonance between δ 200 and 218.1. The C-26 in 26-keto compounds, as being of ester type, appears at comparatively higher field, *ca.* 180.5 ppm.[77] The α,β-unsaturated carbonyl carbon appears at 5-10 ppm higher field relative to nonconjugated carbonyl.[78,79] The best way to locate the site of an oxo group is to identify the ^1H and ^{13}C NMR chemical shifts for the resonances representing the β-position, as these absorb at relatively deshielded position owing to the β-effect of the CO function. The $^1H\{^{13}C\}$MBC spectral data provide further evidence as resonances of neighboring protons, particularly those occupying the β-position, exhibit three-bond connectivity to the carbonyl resonance.[69,80]

Olefinic carbon resonances lie between δ 108.1 - 172.1, and their nature (C, CH, CH_2) identifies the type of olefinic bond (di-, tri- or tetra-substituted). So far, steroidal sapogenins having olefinic bonds at the 4, 5, 7, 9(11), 11, 17(20), and 25(27) positions have been reported. The Δ^5 olefinic bond, one of the most frequently encountered forms of unsaturation, causes the appearance of olefinic resonances at δ 139.7-142.1 (C) and δ 120.9-124.3 (CH) corresponding to C-5 and C-6 respectively, and can be supplemented by the doublet (J = 4-5 Hz) or multiplet at δ 5.21-5.62 for H-6. The chemical shift of the C-5 is almost unaffected by the hydroxyl substitution in ring A, whereas a 7-oxo group causes appreciable deshielding and the appearance of C-5 at δ166.5.[81] The olefinic resonances at *ca.* δ 116.0 (CH) and 139.3 (C), 147.5 (C), and 116.2 (CH), and 124.9 (CH) and 137.6 (CH) identify olefinic bonds at 7 and 9(11) position.[82,83] The olefinic bonds at the 17(20) and 20(22) positions introduce two quaternary carbons absorbing at δ 142.5 and 145.6, and 103.7 respectively.[84] The exocyclic methylene absorbing at 108.6±0.5 ppm (C-27), accompanied by a quaternary carbon at 144.4±0.5 ppm (C-25), identifies 25(27) olefinic unsaturation. Both C-20 and C-22 are of quaternary type, absorbing at δ 103.7-103.9 and δ 151.3-151.9, respectively, in $\Delta^{20(22)}$ furostane-type steroidal sapogenins,[86] whereas C-13 and C-14 absorb at δ 176.0 and 138.7 in 18-norspirostanol.[87]

Two-dimensional-INADEQUATE excperiments are of great importance, as these experiments provides independent proof for the entire carbon skeleton by exhibiting double quantum peaks indicating the presence of a covalent bond. However, such C-C connectivity information may be absent for carbon atoms with long relaxation times, strongly coupled carbons, and if the chemical shif difference is large. Nevertheless, because of the low sensitivity of this experiment, it has not been routinely employed.

To the best of our knowledge, the complete ^1H NMR assignments for three 5α- and 5β-steroidal sapogenins: hecogenin acetate **1** (25*R*, 3β-acetoxy-12-oxo-5α-spirostane)[67], smilagenin **2** (25*R*, 3β-hydroxy-5β-spirostane and sarsasapogenin **3** (25*R*, 3β-hydroxy-5β-spirostane)[68] have been reported so far. For comparison purposes, the NMR data for the steroidal sapogenin part of **4** of (22*S*,25*S*)-5α-spirostan-3β-ol 3-{-*O*-β-D-galactopyranosyl-(1→2)-*O*-β-D-xylopyranosyl-(1→3)]-*O*-β-D-glucopyranosyl-(1→4)-β-D-galactopyranoside[73] has also been incorporated. Concerning the Δ^5 spirostanoids, detailed NMR spectral analysis has been reported only for diosgenin **5** (25*R*, 3β-hydroxy-spirost-5-ene)[63] and for three saponins: (22*R*,25*S*), 1β-*O*-fucopyranosyl-3β-*O*-

α-L-rhamnopyranosyl-spirost-5-ene **6**, recurvoside A **7** (23*S*)-spirosta-5,25(27)-diene-1β,3β,23-triol 1-*O*-{α-L-rhamnopyranosyl-(1→2)-*O*-[β-D-xylopyranosyl-(1→3)-α-L -arabinopyranoside,[71] and aculeoside A **8** (23*S*,24*S*)-spirosta-5,25(27)-diene-1β,3β,23,24-tetraol 1-*O*-{*O*-2,3,4-triacetyl-α-L-rhamnopyranosyl-(1→2)-α-L-arabinopyranoside -24-*O*-6-deoxy-D-glycero-L-threo-4-hexosuloside.[72] The NMR data for the sapogenin part of the above mentioned saponins are also incorporated (Table 4.1). Concerning furostanoids, NMR data for the aglycone part of two furostane saponins, macrostemonoside G **9** (26-*O*-β-D-glucopyranosyl-22-hydroxy-5β-furost-25(27)-ene-3β,12β,26-triol 3-*O*-β-D-glucopyranosyl-(1→2)-*O*-β-D-glucopyrnoside and for macrostemonoside H **10**, which corresponds to 22-methoxy and 12-oxo product of **9**[69], have been incorporated in Table 4.2.

Table 4.1. Carbon and Proton Assignments of Selected Spirostanes.

Posi-tion	(22*R*, 25*R*), 3β-OAc,12-oxo, 5α **1**		(22*R*,25*S*), 3β-OH,5β **2**		(22*R*,25*R*) 3β-OH,5β **3**		(22*S*,25*S*) 3β-*O*-gly[a],5α **4**	
	^{13}C	^{1}H	^{13}C	^{1}H	^{13}C	^{1}H	^{13}C	^{1}H
1	36.2	1.03,1.59	29.95	1.40,1.53	29.95	1.39,1.50	37.5	0.82,1.52
2	27.2	1.83,1.50	27.82	1.40,1.58	27.79	1.40,1.50	30.2	1.62,2.02
3	72.8	4.68	67.12	4.11	67.12	4.11	78.0	3.89
4	33.8	1.67,1.37	33.54	1.33,1.98	33.51	1.32,1.97	35.1	1.35,1.79
5	44.6	1.19	36.58	1.72	36.51	1.72	35.0	0.91
6	28.2	1.33 x 2	26.56	1.16,1.91	26.56	1.15,1.90	29.2	1.13 x 2
7	31.4	0.95,1.78	26.54	1.07,1.58	26.54	1.04,1.39	32.8	0.80,1.55
8	34.4	1.91	35.28	1.58	35.27	1.58	35.3	1.40
9	55.3	1.13	39.85	1.33	39.85	1.32	55.9	0.54
10	36.0	-	35.28	-	35.28	-	36.1	-
11	37.6	2.23,2.40	20.90	1.25,1.40	20.90	1.25,1.39	21.6	1.38,1.99
12	211.7	-	40.30	1.16,1.72	40.31	1.15,1.72	40.8	1.05,1.64
13	54.9	-	40.70	-	40.70	-	41.6	-
14	55.6	1.44	56.48	1.16	56.47	1.15	56.5	0.94
15	31.4	1.45,2.11	31.80	1.25,1.98	31.74	1.25,1.97	33.4	1.43,2.01
16	78.9	4.32	80.93	4.40	81.02	4.40	81.1	4.31
17	53.6	2.50	62.27	1.77	62.09	1.75	62.9	1.59
18	15.9	1.00	16.49	0.76	16.50	0.76	17.2	0.92
19	11.8	0.88	23.92	0.98	23.92	0.97	12.5	0.64
20	42.2	1.74	41.61	1.86	42.12	1.81	42.4	2.29
21	13.2	1.06	14.51	0.97	14.34	0.99	16.8	2.29
22	108.7	-	109.26	-	109.74	-	110.8	-
23	31.1	1.58,1.68	31.39	1.58,1.67	25.94	1.39,1.87	28.4	1.46,1.68
24	28.9	1.56,1.42	28.80	1.46,1.64	25.77	2.02,1.39	28.3	1.52,1.58
25	30.2	1.62	30.31	1.64	27.08	1.69	31.0	1.63
26	66.6	3.35,3.49	66.86	3.48,3.38	66.13	3.95,3.28	69.9	3.67 x 2
27	17.1	0.80	17.14	0.79	16.05	1.08	17.5	0.70

[a]gly = *O*-β-D-galactopyranosyl-(1→2)-*O*-[β-D-xylopyranosyl]-(1→3)-*O*-β-D-glucopyranosyl-(1→4)-β-D-galactopyranoside.

DETERMINATION OF THE STEREOCHEMISTRY AT THE A/B RING JUNCTURE AND AT C-22 AND ORIENTATION OF THE 27-METHYL GROUP

The steroidal sapogenins possess either a *cis* or a *trans* fusion between ring A and B, or an olefinic bond between C-5 and C-6, leading to the 5α, 5β and Δ^5 series. The ^{13}C NMR chemical shifts for the resonances belonging to ring A and B are valuable for the purpose of discrimination between these three types. A comparison of ^{13}C NMR chemical shifts for the 5α and 5β series with analogous substitution pattern in ring A and B lead to the suggestion that C-5 (CH), C-7 (CH_2), C-9 (CH) and C-19 (CH_3) exhibit pronounced dependence ($\Delta\delta = \delta_{trans} - \delta_{cis}$) are +(9-13 ppm) for the C-9 and C-19 and +(3-6 ppm) for C-7 and C-5 respectively. The chemical shift values for C-5 and C-6 are highly characteristic for Δ^5-steroidal sapogenins and a comparison with 5α-spirostane reflects appreciable similarities for the ring A and B resonances, except for C-4, C-10 and C-19 which appear at 2-7 ppm deshielded. The C-8 and C-9 resonances resonate at 3-5 ppm shielded in the case of the latter. Hence on the basis of the ^{13}C NMR shielding behavior, the stereochemistry at the A/B ring junction can be deduced. The 5β can be identified by the characteristic cross-peak between H_3-19 and H-5, as such NOE connectivity is not typical of the 5α-series.

The NOE correlations between H-14/H-17 and H-16 are characteristic of a D/E *cis* ring junction, whereas the NOE correlations between H_3-18/H-20 establishes the orientation of the 21-Me group. Such a NOE correlation is characteristic of the 20*S* ($20\alpha_f$) configuration. Thus, the absence of a H_3-18/H-20 NOE identifies the 20*R* ($20\beta_f$) configuration. Intense NOE correlation between H-17 and H-23*e* and/or H-16 and H-23*e* is characteristic of the 22*S* (22β) configuration and hence the 22*R* (22α) configuration could be ascertained by the lack of such NOE correlations. Moreover, H-16 showed a NOE to H-26*a* in the case of steroidal sapogenins with 22*R* configuration.[71,73]

Table 4.1. Continued.

Position	25*R*, 3β-OH, Δ^5 **5** ^{13}C	^1H	22R,25*S*, 1β-*O*-fuc, 3β-*O*-rha, Δ^5 **6** ^{13}C	^1H	1β-*O*-gly[a],3β,23*S*(OH)$_2$, $\Delta^5,\Delta^{25(27)}$ **7** ^{13}C	^1H	1β-*O*-gly[b],3β,23*S*(OH)$_2$, 24*S*-*O*-hex[c], $\Delta^5,\Delta^{25(27)}$ **8** ^{13}C	^1H
1	37.43	1.05,1.82	84.3	3.46	83.7	3.74	83.6	3.79
2	31.79	1.81,1.48	35.8	2.23,1.70	37.4	2.63,2.27	37.6	2.68,2.22
3	71.89	3.51	74.5	3.47	68.3	3.79	68.0	3.82
4	42.46	2.22,2.27	39.8	2.37,2.25	43.9	2.61,2.52	43.8	2.68,2.60
5	141.02	-	139.1	-	139.6	-	139.4	-
6	121.60	5.30	126.4	5.62	124.7	5.56	125.1	5.63
7	32.25	1.52,1.95	32.8	2.02,1.58	32.1	1.86,1.50	32.1	1.83,1.53
8	31.64	1.63	34.0	1.61	33.1	1.52	33.1	1.52
9	50.25	0.94	51.7	1.30	50.4	1.47	50.4	1.49
10	36.84	-	43.8	-	42.9	-	43.0	-
11	21.07	1.51	24.6	2.58,1.47	24.1	2.87,1.57	24.1	2.86,1.57
12	39.99	1.17,1.72	41.4	1.71,1.24	40.6	1.57,1.29	40.6	1.59
13	40.46	-	41.1	-	40.8	-	40.9	-
14	56.72	-	58.1	1.17	56.9	1.14	56.8	1.09
15	32.04	1.96,1.26	32.9	2.01,1.58	32.4	2.01,1.50	32.6	1.85,1.45
16	81.03	4.38	82.4	4.43	82.0	4.58	83.1	4.61
17	62.28	1.78	63.9	1.75	62.5	1.80	61.7	1.72
18	16.49	0.76	17.1	0.85	16.9	1.00	16.9	1.00
19	19.63	1.04	14.9	1.12	15.0	1.34	15.0	1.36
20	41.80	1.85	43.5	1.91	35.8	2.94	37.5	2.88
21	14.73	0.92	14.7	1.03	14.6	1.08	14.8	1.08
22	109.50	-	111.2	-	111.8	-	111.8	-
23	31.59	1.58,1.68	27.0	1.96,1.39	68.6	3.83	70.3	3.86
24	28.99	1.64,1.43	26.8	2.07,1.46	38.9	2.86,2.76	82.8	4.81
25	30.49	1.61	28.5	1.72	144.8	-	143.7	-
26	67.04	3.38,3.36	66.1	3.97,3.31	64.3	4.38,3.96	61.5	4.82,4.00
27	17.35	0.76	16.4	1.13	109.3	4.84,4.81	114.3	5.27,5.12

[a]gly = *O*-α-L-rhamnopyranosyl-(1→2)-*O*-[β-D-xylopyranosyl-(1→3)]-α-L-arabinopyranoside.
[b]gly = *O*-2,3,4-triacetyl-α-L-rhamnopyranosyl-(1→2)-α-L-arabinopyranoside.
[c]hex = 6-deoxy-D-glycero-L-*threo*-hexosuloside.

The methyl group at C-25 in spirostanes (**1.1**) can have either an equatorial or an axial orientation, resulting in the (25*R*) or (25*S*) forms. The orientation of the methyl group at C-25 is reflected in the chemical shifts of H_3-27 ($\Delta H_3\text{-}27_e - H_3\text{-}27_a$ = *ca.* -0.3 ppm)[10,90], as well as in the ^{13}C NMR chemical shifts of the ring-F carbon resonances, since C-23, C-24, C-25 and C-26 absorb at 2.5-4.0 ppm and C-27 at 1.5-2.0 ppm lower field in the (25*R*)-compounds having an unsubstituted ring F in comparison to (25*S*)-spirostanes.[14,91,92] The diagnostic shifts for C-23, C-24, C-25, C-26 and C-27 are δ 31.3, 28.8, 30.3, 66.9 and 17.1, respectively for a (25*R*)-spirostane, and δ 26.6, 26.2, 27.5, 65.0 and 16.3 respectively for a (25*S*) spirostane with a standard deviation of ±0.4 ppm. These generalized features are, however, not applicable for achieving a differentiation between the (25*R*) and (25*S*) forms of 25-hydroxyspirostanes and ring F-substituted spirostanes.[77] A comparison of ^1H NMR chemical shifts for the 25-epimers 25*S*- and 25*R*-, 3β-hydroxy-5β-spirostane reveals that the chemical shift difference between the geminal protons of the 24- and 26 positions is about 0.15±0.05 ppm for the former but 0.65±0.05 for the latter.[68]

Most of the furospirostane-type steroidal sapogenins are 26-hydroxylated. Consequently the 27-methyl group, which may have either an *R*- or *S*-configuration, appears as a singlet at *ca.* 1.15 ppm.[93] The two 25-epimers exhibit marginal ^{13}C shielding variations for the ring F resonances; they are still significant as C-25 shows an upfield shift, whereas C-27 shows a downfield shift of about 1.5 ppm.[94]

Table 4.2. Carbon and Proton Assignments of Selected Furostanes.

Position	9: 3β-*O*-gly[a], 12β,22(OH)$_2$, 26-*O*-glc, $\Delta^{25(27)}$, 5β ^{13}C	9 ^1H	10: 3β-*O*-gly[a], 12-oxo, 22-OMe, 26-*O*-glc, $\Delta^{25(27)}$, 5β ^{13}C	10 ^1H	Position	9 ^{13}C	9 ^1H	10 ^{13}C	10 ^1H
1	31.0	1.44,1.79	30.9	1.43,1.80	15	32.1	1.55,2.05	31.7	1.50,2.00
2	26.6	1.44,1.90	26.7	1.43,1.90	16	81.4	5.04	81.6	4.58
3	75.4	4.28	75.4	4.25	17	63.7	2.33	64.2	2.13
4	31.0	1.44,1.82	30.9	1.43	18	11.3	1.10	11.2	1.04
5	36.7	2.16	36.7	2.15	19	23.9	0.97	23.9	0.96
6	27.0	1.21,1.76	27.0	1.20,1.77	20	41.6	2.49	41.4	2.49
7	26.6	0.91,1.24	26.7	0.90,1.25	21	15.6	1.59	15.5	1.44
8	34.5	1.49	34.6	1.47	22	110.5	-	112.6	-
9	39.4	1.41	39.3	1.41	23	38.0	2.27	31.7	2.06,2.16
10	35.2	-	35.2	-	24	28.4	2.70	28.2	2.39
11	31.4	1.51,1.72	31.4	1.45,1.72	25	147.2	-	146.8	-
12	79.5	3.53	79.4	3.49	26	72.0	4.29,4.59	72.0	4.34,4.60
13	46.5	-	46.9	-	27	110.6	5.02,5.32	110.8	5.03,5.34
14	55.1	1.08	55.1	1.05					

[a]gly = *O*-β-D-glucopyranosyl-(1→2)-*O*-β-D-galactopyranoside

DETERMINATION OF THE MOLECULAR STRUCTURE OF THE OLIGOSACCHARIDE MOIETY

A serious difficulty in assigning the structure of the saponins is in the identification of the oligosaccharide moiety, the structure of which can be highly variable owing to the diversity of the monosaccharide residues and the type of substitution. The characterization of the oligosaccharide moiety involves the establishment of : a) the monosaccharide composition, b) the anomeric configurations of each glycosidically linked monosaccharide unit and c) the interglycosidic linkage and sequence, and finally the site of attachment of the oligosaccharide moiety to the sapogenin residue.

DETERMINATION OF THE NUMBER AND TYPE OF MONOSACCHARIDE RESIDUES

A well known difficulty with the assignment of proton resonances of oligosaccharides is the overlap of constituent proton resonances in a narrow spectral width (δ 3.2-4.5). An exception is the anomeric (C-1) proton, which is found at lower field and is almost invariably well resolved, and the methyl groups (δ 1.0-1.5) of 6-deoxyhexoses. The subsequent overlap of these with some of the resonances of the sapogenin residue makes it difficult to assign these resonances specifically to the monosaccharide and sapogenin residues. Even for relatively small oligosaccharides having only two or three monosaccharide units, spectral overlap can severely limit the information available. Therefore, reduction of spectral overlap is of paramount importance in NMR experiments on saponins. In order to minimize signal overlap in the ^1H NMR spectrum, it is desirable to exchange the hydroxyl protons with deuterium prior to recording the ^1H and homonuclear spectral data in any deuterated solvent.

The anomeric resonances of the sugar moiety occur largely in a well defined region (^1H: δ 4.4-6.4; δ 92-108).[95,96] The ^{13}C NMR anomeric region is straightforward and hence the number of ^{13}C signals present in this region usually reflects the number of the monosaccharide residues. As mentioned already, C-22 resonances of the spirostanes and furostanes, in most cases, absorb in a well defined chemical shift range (δ 108-113.5). Therefore these do not complicate the anomeric region and can be differentiated by their quaternary nature (i.e.

no cross-peak in HETCOR/HMQC spectrum). The C-26 of 26-methoxy-spirost-5-en-3β-ol is also of methine type and resonates at δ 103.1. Hence the distinction of methine resonances of the sapogenin residue[78] is also required. The best way to achieve differentiation between the hydroxylated methine belonging to sapogenin and sugar residues is from the analysis of COSY, HOHAHA, and 1H{^{13}C}SBC spectral data.[96-99] Since anomeric carbon resonances of the aldopyranoses are of methine type and they show heteronuclear correlation with the anomeric-H, determination of cross-peaks in the anomeric region establishes the number of monosaccharide residues. This procedure becomes less significant owing to the quaternary nature of the anomeric-C in the case of ketosidic sugar, ^{1}H{^{13}C}MBC provides a solution.[48]

The prerequisite for the identification of individual monosaccharide residues and the interglycosidic linkages between them is the unambiguous assignment of the ^{1}H- and ^{13}C-resonances. The first task to accomplish this is to perform a through-bond connectivity analysis in order to determine the number of different spin systems corresponding to individual sugar residues by COSY experiments in which the magnetization is transferred via ^{1}H-^{1}H *J*-coupling. However, as the number of monosaccharide residues increases, the ^{1}H NMR spectrum becomes severely overlapped. It therefore becomes difficult to assign all of the protons of each individual sugar residue. Spectral editing techniques such as triple-quantum filtered (TQF) and geminal (GEM) COSY can be useful, as these delineate CH_n-CH ($n = 2$ or 3) and CH_2 spin systems, respectively. Phase sensitive (PS) and DQF-COSY spectra can unravel more easily the accurate measurement of chemical shifts and coupling constants from the appearance of the positive and negative signals by analyzing the fine structure of the cross-peaks, thereby delineating the spin systems and the relative configuration of the hydroxyl group at individual carbon atoms for each sugar residue. The anomeric protons, because of their appearance at lower field, act as entry points for the tracing out the spin systems belonging to each of the monosaccharide residues by the HOHAHA and multi-step RCT experiments.[96-100] In these experiments, proton resonances belonging to same monosaccharide residue are observed clearly and undisturbed and no signal of the other sugar/sapogenin residue usually appears. However, it is important to mention that propagation of the magnetization through the coupling network depends on the duration of the mixing time and one must use a long mixing time to sure that all of the correlation signals of the corresponding residue are detected.[101] To identify the spin system of an individual monosaccharide, DOUBLE TOCSY[102] selective TOCSY and 1D HOHAHA experiments[103] with increasing mixing times are worthwhile, since subspectra of an individual carbohydrate moiety can then be extracted from the crowded overlapping regions. The only prerequisite for the success of this experiment is that at least one of the protons should be sufficiently separated from all of the others. Once the proton resonances of each of the sugar unit have been assigned, assignment of the carbon resonances can be accomplished by tracing the ^{1}H-^{13}C connectivities observed in ^{1}H{^{13}C}SBC spectrum. Hence a combined application of these experiments leads to the assignment of the ^{1}H and ^{13}C NMR chemical shifts, coupling constants, and configurations and thus establishes the nature of the sugar moieties.

The anomeric configuration, particularly for pyranose sugars, can be inferred from the $^3J_{1,2}$ coupling constants, NOE correlation, one-bond ^{13}C-^{1}H coupling constants ($^1J_{CH}$), and long-range ^{1}H{^{13}C}MBC correlations. Broad singlet and doublets with small coupling constant ($^3J_{1,2}$ = 1-3 Hz) are usually due to an α-anomer, whereas doublets with large coupling constants ($^3J_{1,2}$ = 7-8 Hz) are due to the β-anomers of the pyranose sugars with a *gluco* or *galacto* configuration. However, both the α and β-anomeric forms of the monosaccharides with *manno* configuration possess $^3J_{1,2}$ = 1-3 Hz. The ^{13}C NMR chemical shifts of the anomeric carbons are also sensitive to the anomeric configuration as C-1 absorbs at *ca.* 5 - 7 ppm lower field in the β-anomer as compared with an α-anomer.[95,100] The one-bond ^{13}C-^{1}H coupling constants ($^1J_{CH}$) depend strictly on the orientation of the anomeric hydrogen. The ($^1J_{CH}$) for the anomeric carbon ranges between 160-175 Hz. For a pyranose with an axial H-1, the value is *ca.* 10 Hz lower than the corresponding value for and equatorial H-1. The standard ($^1J_{CH}$) values for monosaccharides with a *gluco* or a *galacto* configuration are *ca.* 160 Hz for the β-anomer and *ca.* 170 Hz for the α-anomer.[95,104] The intraresidual NOE also discriminates between the anomeric pair. In general, 1,3-diaxial and equatorial-axial proton pairs in pyranosyl rings produce intraresidual NOE cross peaks between H-1 and H-3 (H-5), whereas a strong cross peak is observed between H-1 and H-2 in an α-glucopyranosyl configuration in the 2D NOESY spectrum.[96,100]

INTERGLYCOSIDIC LINKAGE AND SEQUENCE

The determination of interglycosidic linkages and the sequential analysis of a given oligosaccharide via NMR spectroscopy requires the identification of those parameters that are affected simultaneously by two neighboring monosaccharide residues. The NMR parameters available to obtain this information will be presented in this section.

The carbon assignments ascertained by the combined application of homo- and heteronuclear NMR techniques can further verify the identification of monosaccharides by comparison with the reported values for reference methyl glycosides. This subsequently allows identification of the attachment positions in the sugar chains through the consideration of the glycosidation-induced shifts. This approach relies on the fact that glycosidation modifies the ^{13}C NMR chemical shifts for the internal sugar moieties. The glycosylated carbon shifts to lower field by 4-10 ppm (α-effect) and the resonances of carbon occupying adjacent positions shift upfield by a small amount (+0.9 to -4.6 ppm) whereas the other carbon resonances remain virtually unaffected. Despite the fact that these glycosylation effects depend on the configuration at the anomeric center of the glycosylating pyranose and on the absolute configuration of both pyranose residues, the reported general trend is normally applicable in identifying the glycosylating site. The regularities in the effects also led to the development of a computer-assisted approach to the structure analysis of linear and regular oligo- and poly-saccharides on the basis of ^{13}C NMR data.[105] Since the terminal sugar residues exhibit a remarkable resemblance to the respective methyl glycoside, the sequence of monosaccharide residues can also be inferred by following a systematic interpretation. In several cases, the ^{13}C resonance assignments could be aided considerably by using simpler saponins or prosapogenins having fewer monosaccharide residues for spectral comparison.

Comparison of the T_1 relaxation times of the terminal and interior sugars also provides information concerning monosaccharide sequence in an oligosaccharide moiety, since the correlation times decreases as their freedom of movement increases, i.e. as the sugars are increasingly distant from the aglycone, T_1s increases. Sometimes it is desirable to measure T_1 relaxation times at different temperatures for saponins differing in the number of sugar residues simply because saponin solutions at the same concentration becomes more viscous as the number of sugars increases, and the ^{13}C T_1 times of all sugar units are similar. Consequently information from T_1 measurements are limited, if not completely useless.[106]

Further confirmation regarding the sequence of the oligosaccharide chain can be obtained by measuring the dipolar coupling of neighboring anomeric protons and the linkage site protons of different residues by detecting inter-residual NOE or ROE cross peaks, since nuclear Overhauser enhancements traversing the glycosidic linkages are invariably observed. The advantage of the ROESY experiment over the NOESY have already been mentioned. But if the NOEs between each anomeric proton signal of an other substituted saccharide (the aglycone residue) are severely overlapped; then such studies become ambiguous. The $^1H\{^{13}C\}$MBC correlation for each anomeric proton to the carbon of another substituted monosaccharide traversing the glycosidic bond either observed in LR-HETCOR or HMBC is a very convenient method to resolve an ambiguity of this type and for making assignments.[107-110]

Three-dimensional NMR spectroscopy[111], in spite of its capability of giving rise to better resolution, has not been widely employed for the structural analysis of oligosaccharides.[112,113] In such a spectrum, there are three combinations possible to take 2D slices. In a 3D TOCSY-ROESY experiment, each of these slices taken at the constant frequency of an anomeric or linkage site proton contains information about the sequence and position of glycosidation.[114] Since 3D spectra make enormous demands on overall experimental time and computer data storage space, they are not a routine application. However, in the case of signal overlap, heteronuclear 3D NMR may be useful.

An interesting alternative is to use shaped pulses to improve selectivity. These have allowed the development of 1D equivalents of the 2D experiments[115]; the 2D equivalents of the 3D experiments offer further spectral simplification. The full sequence information can be gained in a few minutes by two semisoft 2D TOCSY-ROESY experiments.[116] Recently, the implementation of the pulsed field gradients has brought about a dramatic improvement in the execution of many two-dimensional experiments that have been in routine use for resonance assignment purpose.[117]

LOCATION OF THE SAPOGENIN-OLIGOSACCHARIDE LINKAGE

After the identification of the molecular structure of sapogenin and oligosaccharide residues, the next question is to determine the aglycone-sugar linkage, i.e., at which position(s) of the sapogenin residue are the sugar residue(s) linked? The effect of glycosylation is the difference between the chemical shift of a given carbon atom in a steroidal saponin and that in a free sapogenin, i.e. ($\delta_{saponin}$ minus $\delta_{sapogenin}$). These effects on the ^1H NMR spectra of saponins have been studied much less than those that on ^{13}C spectra, probably because the fully resolved ^1H spectra of the sugars have only recently become available. Characteristic downfield shifts (3-10 ppm) for the α-carbon resonances and upfield shifts (1 - 5ppm) for the resonances of the carbon atoms in the adjacent positions are usually associated with glycosidic bond formation (Table 5). Since the spectra of the glycosides are usually measured in pyridine, these values need to be treated with caution because of the influence of the solvent on the chemical shift, if the ^{13}C NMR chemical shifts for the aglycone were obtained in another solvent such as chloroform.

The inter-residual NOE between the anomeric-H of the sugar and the sapogenin signal observed in NOESY, ROESY, or one-dimensional NOE-difference experiments also identifies the monosaccharide residue directly involved in glycosylation with the aglycone. The anomeric H of the sugar moiety directly bonded to the aglycone moiety exhibits an $^{1}H\{^{13}C\}$MBC cross-peak with the aglycone carbon or *vice versa*; i.e., anomeric-H of that particular monosaccharide which is directly bonded to the sapogenin residue exhibits its long-range connectivity to aglycone C, thereby identifying the sugar-aglycone linkage.[61,71-73] Thus a concerted and systematic application of homo- and heteronuclear spectroscopic techniques can lead to unambiguous ^{1}H and ^{13}C assignments, as well as the establishment of molecular structure of the saponin.

Table 5. Glycosidation-induced shifts in steroidal saponins

Site	C-α		C-β				C-γ
1β-OH[a]	+(5.7-6.4)	C-1	−(5.1-7.8) C-2		−(0.6-0.7) C-10		−(0.9-1.1) C-19
1β-OH[b]	+12.3	C-1	+0.5 C-2		+0.5 C-10		−0.2 C-19
2α-OH	+(11.1-11.4)	C-2	−(1.8-2.7) C-1		−(1.4-2.3) C-3		
3β-OH	+(6.3-7.9)	C-3	−(0.8-2.4) C-2		−(2.9-4.8) C-4		
5β-OH[c]	+(9.4-9.7)	C-5	−0.7 C-4		−5.4 C-6		
					−1.6 C-10		
6α-OH	+(11.1-12.1)	C-6	−(0.5-1.6) C-5		−(1.4-2.3) C-7		
12β-OH	+11.1	C-12	−(1.7-1.8) C-11		−(0.3-0.4) C-13		
24-OH	+(8.8-10.9)	C-24	−(0.8-1.7) C-23		−(0.9-1.7) C-25		
27-OH	+7.9	C-26	−2.6 C-25		−(0.2-0.7) C-24, C-27		

[a] In the 3β-hydroxylated Δ^5 series.
[b] In the 2β,3α-dihydroxylated 5β-series.
[c] In the 1β,2β,3β,4β-tetrahydroxylated 5β-series.

CONCLUDING REMARKS

An essential prerequisite for deducing the structures of saponins by NMR spectroscopic studies is the unambiguous assignment of ^{1}H and ^{13}C NMR resonances which, in principle, involves three main steps. In the first step, networks of protons which interact with one another through chemical bonds are identified. Each such proton spin network corresponds to either the structural fragment of the sapogenin or the monosaccharide residue. Determination of the further connectivity of these structural fragments by establishing their respective positions in the sapogenin and/or oligosaccharide system is the next step. In the final step, the connectivity between the monosaccharide residues and to the sapogenin residue is determined.

For the identification of the skeletal type, the chemical shift of C-22 and its nature (C, CH and CH_2) (Table 2) are uniquely characteristic and a comparison of the observed multiplicities of the ^{13}C NMR resonances with those for the parent skeleton (Table 3) can initially be utilized in determining the number and type of the functional groups, including the type of olefinic bond present, which can later be verified by the ^{1}H-^{1}H COSY spectrum. For example in the case of a *sec.*-hydroxyl group, the hydroxymethine resonance exhibits its cross-peak connectivities with the adjacent vicinal proton partners. Such a procedure, however becomes less significant for locating the site of a *tert.*-hydroxyl group owing to its substitution on a quaternary carbon.

The oxy-substituted methine and methylene resonances and H-6 in the case of Δ^5 steroidal compounds provide entry points for the assignment of the proton resonances of various ring systems. If there are uncertainties in the ^{1}H NMR assignments, for example, to differentiate between geminal and vicinal protons, an iterative comparison of the ^{1}H-^{1}H COSY and $^{1}H\{^{13}C\}$SBC spectra may be of significance, since in the latter spectrum, cross-peak connectivity between the ^{13}C and directly bonded ^{1}H resonances are observed. For the assignment of quaternary carbons, the best way is by the observation of long-range ^{1}H-^{13}C connectivities in the $^{1}H\{^{13}C\}$MBC experiments (LR-HETCOR or HMBC). The site of substitution and the orientation of the hydroxyl group can then be established by consideration of the ^{13}C NMR chemical shifts involving OH-induced substituent effects, analysis of the ^{1}H NMR coupling constants, or by intra-residual NOE proximity.

For the structural analysis of the oligosaccharide moiety, one needs to identify the anomerically defined individual monosaccharide residue(s), the interglycosidic linkages, and the sequence between them. Cross-peak connectivity within COSY (DQF/PS) and HOHAHA spectra identify the intraresidual spin system of each monosaccharide unit, whereas the NOESY and/or ROESY reflects sequential inter-residue effects including sugar-sapogenin connectivity information. Another approach involves the assignment of the ^{13}C resonances based on the $^{1}H\{^{13}C\}$SBC spectral data and by following the $^{1}H\{^{13}C\}$MBC connectivity. Thus, a concerted application of the homo- and heteronuclear NMR approaches not only lead to the definitive assignments of ^{1}H and ^{13}C NMR resonances, but also to the structural analysis of saponins. The rapid developments in new NMR experimental pulse schemes and equipment continue to increase the power and applicability of high-field NMR to the structural characterization of saponins, as well as other complex natural products.

Acknowledgement. I am grateful to Prof. Sushil Kumar, Director, Dr. R. P. Sharma, Head, Phytochemical Technology Division of this institute and Prof. N. Rama krishna, University of Alabama at Birmingham, BIrmingham, U.S.A. for constant encouragement, and to Dr. G. A. Morris, University of Manchester, Manchester, U.K. for his continued cooperation. This work was partially supported by NCI grant CA-13148 and the Athritis Foundation.

REFERENCES

1. R. Hardman, Recent developments in our knowledge of steroids, *Planta Med.*, 53:233 (1987).
2. S.B. Mahato, A.N. Ganguli and N.P. Sahu, Steroidal saponins, *Phytochemistry*, 21:639 (1987).
3. G. Voigt and K. Hiller, Neuere Ergebnisse zur Chemie und Biologie der Steroid Saponine, *Sci. Pharm.* 55:201 (1987).
4. A. Martson and K. Hostettmann, Plant molluscicides, *Phytochemistry*, 24:639 (1985).
5. R. Tschesche and G. Wulff, Chemie und Biologie der Saponins, in *Fortschritte der Chemie Organische Naturstoffe*, W. Herz, H. Grisebach and G.W. Kirby, Eds, Springer Verlag, Berlin, 30:461 (1973).
6. S.D. Kravets, Yu.S. Vollemer, M.B. Gorovits and N.K. Abubikarov, Steroids of spirostan and furostan series from plants of the genus *Allium*, *Khim. Prir. Soed.*, 429 (1990).
7. E. Heftmann, Biogenesis of steroids in Solanaceae, *Phytochemistry*, 22:1843 (1983).
8. S. Seo, A. Uomori, Y. Yoshimura and K. Tori, Biosynthesis of (25*S*)- and (25*R*)-furostanol glycosides from [1,2-$^{13}C_2$] acetate in *Dioscorea tokoro* tissue culture, *J. Chem. Soc., Perkin Trans I*, 869 (1984).
9. A. Uomori, S. Seo, T. Sato, Y. Yoshimura and K. Takeda, Synthesis of (25*R*)-[26-$^{2}H_1$] cholesterol and ^{1}H NMR and HPLC resolution of (25*R*)- and (25*S*)-26-hydroxycholesterol, *J. Chem. Soc., Perkin Trans I*, 1713 (1987).
10. A.V. Patel, G. Blunden, T. Crabb, Y. Sauvaire and Y. Baccou, A. review of naturally occurring steroidal sapogenins, *Fitoterapia*, 58:56 (1987).
11. J. W. Blunt and J. B. Stothers, ^{13}C NMR of steroids, *Org. Magn. Reson.*, 9:439 (1977).
12. D.H. William and I. Flemming, *Spectroscopic Methods in Organic Chemistry*, 4th edn., McGraw Hill, London (1987).
13. P.K. Agrawal, S.K. Srivastava and W. Gaffield, Carbon-13 NMR spectroscopy of steroidal alkaloids, in *Alkaloids: Chemical and Biological Perspectives,* S.W. Pelletier Ed., Springer Verlag, New York, Vol 7, pp.49 (1991).
14. P.K. Agrawal, D.C. Jain, R.K. Gupta and R.S. Thakur, Carbon-13 NMR steroidal sapogenins and steroidal saponins, *Phytochemistry*, 24:2479 (1985).
15. P.K. Agrawal, D.C. Jain and A.K. Pathak, NMR spectroscopy of steroidal sapogenins and steroidal saponins: an update, *Magn. Reson. Chem.*, in press .
16. S.L. Patt and J.N. Shoolery, Attached proton test for carbon-13 NMR. *J. Magn. Reson.*, 46:435 (1982).
17. D.M. Doddrell, D.T. Pegg and M.R. Bendall, Distortionless enhancement of NMR signals by polarization transfer, *J. Magn. Reson.*, 48:323 (1982).
18. G.A. Morris, Sensitivity enhancement in ^{15}N NMR: polarization transfer using the INEPT pulse sequence, *J. Am. Chem. Soc.*, 102:428 (1980).
19. A. Bax, R. Freeman and S.P. Kempsell, Natural abundance ^{13}C-^{13}C coupling observed via double-quantum coherence, *J. Am. Chem. Soc.*, 102:4849 (1980).
20. This experiment was first proposed by Jeener (J. Jeener, Ampere International Summer School II, Basko Polje), Yugoslavia, (1971).
21. W.P. Aue, J. Karhan and R.R. Ernst, Homonuclear broad band decoupling and two-dimensional *J*-resolved NMR spectroscopy, *J. Chem. Phys.*, 64:4226 (1976).
22. A. Bax and R. Freeman, Investigation of complex networks of spin-spin coupling by two-dimensional NMR. *J. Magn. Reson.*, 44:542 (1981).
23. M. Rance, O.W. Sorensen, G. Bodenhausen, G. Wagner, R.R. Ernst and K. Wuthrich, Improved spectral resolution COSY ^{1}H NMR spectra of proteins via double-quantum filtering. *Biochem. Biophys. Res. Commun.*, 17:479 (1983).
24. C. Griesinger, O.W. Sorensen and R.R. Ernst, Two-dimensional correlation of connected NMR transitions. *J. Am. Chem. Soc.*, 107:6394 (1985).
25. T. Domke, P. Xu and R. Freeman, Geminal filtered correlation spectroscopy, *J. Magn. Reson.*, 92:218 (1991).
26. U. Piantini, O.W. Sorensen and R.R. Ernst, Simplification of ^{1}H NMR spectra by selective excitation of experimental subspectra, *J. Am. Chem. Soc.*, 104:6800 (1982).
27. G. Wagner, Two-dimensional relayed coherence transfer spectroscopy, *J. Magn. Reson.*, 55:151 (1983).
28. S.W. Homans, R.A. Dwek, D.L. Fernandes and T.W. Rademacher, Multi-step relayed correlation spectroscopy: sequential resonance assignments in oligosaccharides, *Proc. Natl. Acad. Sci. U.S.A.*, 81:6286 (1984).
29. D.G. Davis and A. Bax, Assignment of complex ^{1}H NMR spectra via two-dimensional homonuclear Hartmann-Hahn spectroscopy, *J. Am. Chem. Soc.*, 107:2820 (1985).
30. A. Kumar, R.R. Ernst and K. Wuthrich, A two-dimensional nuclear Overhauser enhancement (2D NOE) experiment for the evolution of complete proton-proton cross relaxation networks in biological molecules, *Biochem. Biophys. Res. Commun.*, 95:1 (1980).

31. A.A. Bothner-By, R.L. Stephens, J.T. Lee, C.D. Warren and R.W. Jeanloz, Structure determination of a tetrasaccharide by transient nuclear Overhauser effects in the rotating frame, *J. Am. Chem. Soc.*, 106:811 (1984).
32. G. Bodenhausen and R. Freeman, Correlation of protonated carbon-13 NMR spectra by heteronuclear two-dimensional spectroscopy, *J. Magn. Reson.*, 28:471 (1977).
33. A. Bax, W. Egan and P. Kovac, New NMR techniques for structure determination and resonance assignments of complex carbohydrates, *J. Carbohydr. Chem.*, 3:593 (1984).
34. A. Bax and S. Subramanian, Sensitivity-enhanced two-dimensional heteronuclear shift correlation NMR spectroscopy, *J. Magn. Reson.*, 67:565 (1986).
35. G.E. Martin and A.S. Zektzer, Long-range two-dimensional heteronuclear shift correlation, *Magn. Reson. Chem.*, 26:631 (1988).
36. A. Bax and M.F. Summers, ^{1}H and ^{13}C assignments from sensitivity enhanced detection of heteronuclear multiple-bond connectivity by 2D multiple-quantum NMR, *J. Am. Chem. Soc.*, 108:2093 (1986).
37. W.R. Croasmum and R.M.K. Carlson, Steroid structural analysis by two-dimensional NMR, In *Two-Dimensional NMR spectroscopy, Application for Chemists and Biochemists,* Verlag Chemie, New York, pp 387 (1987).
38. G.A. Morris, Modern NMR techniques for structure elucidation, *Magn. Reson. Chem.*, 24:371 (1986).
39. A.E. Derome, *Modern NMR Techniques for Chemical Research*, Pergamon Press, New York (1987).
40. J.K.M. Sanders and B.K. Hunter, *Modern NMR Spectroscopy: A Guide for Chemists*, Oxford University Press, Oxford (1987).
41. J. Keeler, Two-dimensional nuclear magnetic resonance spectroscopy, *Chem. Soc. Rev.*, 19:381 (1990).
42. P.K. Agrawal, Two-dimensional NMR spectroscopy in the structure determination of natural products. *J. Sci. Indust. Res.*, 53:329 (1994).
43. A.E. Derome, The use of NMR spectroscopy in the structure determination of natural products: two-dimensional methods, *Nat. Prod. Rep.*, 6:111 (1989).
44. J.N. Shoolery, Recent developments in ^{13}C and proton NMR, *J. Nat. Prod.*, 47:226 (1984).
45. I.H. Sadler, The use of NMR spectroscopy in the structure determination of natural products: one-dimensional methods, *Nat. Prod. Rep.*, 5:101 (1988).
46. W.S. Brey, Ed., *Pulse Methods in 1D and 2D Liquid-phase NMR*, Academic Press, San Diego (1988).
47. R.R. Ernst, G. Bodenhausen and A. Wokun, *Principles of Nuclear Magnetic Resonance in One and Two Dimensions*, Oxford University Press, Oxford (1987).
48. P.K. Agrawal, C.A. Bush, N. Qureshi and K. Takayama, Structural analysis of lipid A and re-lipopolysaccharides by NMR spectroscopic methods, *Adv. Biophys. Chem.*, 4:179 (1994).
49. S. Kubo, Y. Mimaki, Y. Sashida, T. Nikiado and T. Ohmoto, Steroidal saponins from the rhizomes of *Smilax sieboldii*, *Phytochemistry*, 31:2445 (1992).
50. T. Morita, T. Ushiroguchi, N. Hayashi, H. Matsuura, Y. Itakura and T. Fuwa, Steroidal saponins from elephant garlic, bulbs of *Allium ampeloprasum* L., *Chem. Pharm. Bull.*, 36:675 (1993).
51. Y. Mimaki, Y. Sashida, O. Nakamura, T. Nikiado and T. Ohmoto, Steroidal saponin from the bulbs of *Lilium regal* and *L. henryi*. *Phytochemistry*, 33:675 (1993).
52. X.C. Li, C.R. Yang, M. Ichikawa, H. Matsuura, R. Kasai and K. Yamasaki, Steroidal saponins from *Polygonatum kingianum*, *Phytochemistry*, 31:3559 (1992).
53. Yu. S. Vollerner, S.D. Kravets, A.S. Sashkov, B. Tashkhodzhaev, M.B. Gorovits, M.R. Yagudaev and N.K. Abubikarov, Alliosterol and alliosides A and B from *Allium suvorovii* and *Allium stipitatum* - Structural analog of furostanols. *Khim. Prir. Soed.*, 231 (1991).
54. M.H. Yang, G. Blunden, A. Patel, T.A. Crabb, K. Brain and W.J. Griffin, Two furostane saponins from *Cordyline rubra*. *Phytochemistry*, 29:1332 (1990).
55. H. Shimomura, Y. Sashida and Y. Mimaki, 26-*O*-acylated furostanol saponins pardarinoside A and B from bulbs of *Lilium pardarinum*, *Chem. Pharm. Bull.*, 36:3226 (1988).
56. J. Kim, J.M. Pezzuto, D.D. Soejarto, F.A. Lang and A.D. Kinghorn, Polypodoside A, an intensely sweet constituent of the rhizomes of *Polypodium glycyrrhiza*, *J. Nat. Prod.*, 51:1166 (1988).
57. M. Nishizawa, H. Yamada, Y. Yamaguchi and S. Hatakeyama, Structure revision of polypodoside A : major sweet principle of *Polypodium glycyrrhiza*, *Chem. Lett.*, 1555 (1994).
58. H. Shimomura, Y. Sashida, Y. Mimaki and Y. Iitaka, Studies on the constituents of *Lilium henryi* B, *Chem. Pharm. Bull.*, 36:2430 (1988).
59. S. Okamura, K. Shingu, S. Yahara, H. Kohda and T. Nohara, Two steroidal glycosides from *Scopolia japonica*, *Chem. Pharm. Bull.*, 40:2981 (1992).
60. Y. Ju and Z. Jia, Steroidal saponins from the rhizomes of *Smilax menispermoidea*, *Phytochemistry*, 31: 1349 (1992).
61. W. Willker and D. Liebfritz, Complete assignment and conformational analysis of tomatine and tomatidine, *Magn. Reson. Chem.*, 30:645 (1992).
62. A.K. Pathak, P.K. Agrawal, D.C. Jain, R.P. Sharma and O.W. Howarh, NMR studies of 20β-hydroxyecdysone, a steroid isolated from *Tinospora cordifolia*. *Indian J. Chem.*, 34B:674 (1995).
63. R. Puri, T.C. Wong and R.K. Puri, Solasodine and diosgenin: ^{1}H and ^{13}C assignments by two-dimensional NMR spectroscopy, *Magn. Reson. Chem.*, 31:278 (1993).
64. N. Platzer, N. Goasdoue and D. Dovoust, Long-range ^{1}H coupling interactions: Identification of different pathways by 2D NMR δ-δ correlated spectroscopy. Application in structural analysis, *Magn. Reson. Chem.*, 25:311 (1987).
65. H. Schroder and E. Haslinger, Long-range proton spin-spin coupling in rigid cyclic structures by 2D NMR through space coupling. *Magn. Reson. Chem.*, 32:12 (1994).
66. D.N. Krik and H.C. Toms, ω_1-decoupled ^{1}H homonuclear shift correlated nuclear magnetic resonance spectroscopy (COSYDEC) applied to steroids. *Steroids*, 56:195 (1991).
67. H.C. Toms, C. Douglas, K.A. White, K.E. Smith, S. Latif and R.W.P. Hubbard, A survey of the high-field ^{1}H NMR spectra of the steroid hormones, their hydroxylated derivatives, and related compounds, *J. Chem. Soc., Perkin Trans. II*, 1567 (1990).

68. P.K. Agrawal, G.A. Morris and D. Bunsawansong, Unambiguous ^{1}H and ^{13}C NMR assignments for a stereoisomeric pair of steroidal sapogenins: smilagenin and sarsasapogenin, Manuscript in preparation.
69. J. Peng, X. Yao, H. Kobayashi and C. Ma, Macrostemonosides, novel furostanol glycosides from *Allium macrostemon*, *Planta Med.*, 61:58 (1995).
70. J.T. Branke and E. Haslinger, Spirostanol glycoside from the tuber *Ophiopogon japonicus*, *Liebigs Ann. Chem.*, 587 (1995).
71. Y. Takashi, Y. Mimaki, A. Kameyama, Y. Sashida, T. Nikaido and T. Ohmoto, Recurvosides A-E, new polyhydroxylated steroidal saponins from *Nolina recurvata* stems, *Tetrahedron*, 51:2281 91995)
72. T. Horikawa, Y. Mimaki, Y. Sashida, T. Nikaido and T. Ohmoto, Aculeoside A, a novel steroidal saponin containing a deoxyaldoketose from *Ruscus aculeatus*, *Chem. Lett.*, 2303 (1994)
73. T. Inoue, Y. Mimaki, Y. Sashida, T. Nikaido and T. Ohmoto, Steroidal saponins from the tubers of *Dechelostemma multiflorum* and their inhibitory activity on cyclic -AMP phosphodiesterase. *Phytochemistry*, 39:1103 (1995).
74. T. Faecke and S. Berger, Complete proton assignment in acetylcholesterol using ge-SELINCOR-TOCSY, *Tetrahedron*, 51:3521 (1995).
75. T. Faecke and S. Berger, SERF, a new method for H, H spin coupling measurement in organic chemistry, *J. Mag. Reson., Ser. A.* 113:114 (1995).
76. G. Pant, M.S. Panwar, D.S. Negi, M.S.M. Rawat, G.A. Morris and R.I.G. Thompsen, Structure elucidation of a spirostanol glycoside from *Aspergus officinalis* fruits by concerted use of two-dimensional NMR techniques, *Magn. Reson. Chem.*, 26:911 (1988).
77. T. Yamashita, T. Matsumoto, S. Yahara, N. Yoshida and T. Nohara, Structures of two new steroidal glycosides, soladulcosides A and B from *Solanum dulcamara*, *Chem. Pharm. Bull.*, 39:1626 (1991).
78. Y. Mimaki, N. Ishibashi, K. Ori and Y. Sashida, Steroidal glycosides from the bulbs of *Lilium dauricum*, *Phytochemistry*, 31:1753 (1992).
79. P.K. Agrawal and D.C. Jain, ^{13}C NMR spectroscopy of oleanane triterpenoids, *Progr. NMR Spectrosc.*, 24:1 (1992)
80. V. Dirsch, M. Lacaille-Dubois and H. Wagner, Dracoside, a new steroidal saponin from *Helleboru purpurascens*, *Nat. Prod. Lett.*, 4:29 (1994).
81. G. Blunden, A.V. Patel and T.A. Crabb, Microbiological transformations of hecogenin and diosgenin by *Cunninghamella elegans*, *Phytochemistry*, 29:1771 (1990).
82. J.A.R. Garcia and H.T.V. Castro, ^{13}C NMR spectroscopy of (25*R*)-5α-spirostanes, *Magn. Reson. Chem.*, 25:831 (1987).
83. O. Nakamura, Y. Mimaki, Y. Sashida, T. Nikaido and T. Ohmoto, Agapanthussaponins A-D, new potent cAMP phosphodiesterase inhibitors from the underground parts of *Agapenthus inapertus*, *Chem. Pharm. Bull.*, 41:1784 (1993).
84. K. Nakano, K. Murakami, Y. Takaishi, T. Tomimatsu and T. Nohara, Studies on the constituents of *Haloniopsis orientalis* (Thunb) c. *Tanaka*, *Chem. Pharm. Bull.*, 37:116 (1989).
85. Y. Renzhou, W. Dezu and F. Jian, Carbon-13 NMR spectra of ten steroidal sapogenins. *Acta. Bot. Yunnan.*, 9:374 (1987).
86. T. Konishi, S. Kiyosawa and J. Shoji, Studies on the coloration mechanism of furostanol derivatives with Ehrlich reagent. I. On the reaction of 3,26-dimethoxyfurosta-5,20-diene with Ehrlich reagent. *Chem. Pharm. Bull.*, 32:2111 (1987).
87. K. Nakano, T. Nohara, T. Tomimatsu and T. Kawasaki, 18-Norspirostanol derivatives from *Trillium tschonoskii*, *Phytochemistry*, 22:1047 (1983).
88. Y. Sashida, K. Kawashima and Y. Mimaki, Novel polyhydroxylated steroidal saponins from *Allium giganteum*, *Chem. Pharm. Bull.*, 39:698 (1991).
89. J. Buddrus and H. Bauer, Direct identification of the carbon skeleton of organic compounds using double quantum coherence ^{13}C NMR spectroscopy. The INADEQUATE pulse sequence, *Angew. Chem. Int. Ed. Engl.*, 26:625 (1987).
90. K. Kudo, K. Miyahara, N. Marubayashi and T. Kawasaki, Characterization of a minor compound which accompanies the usual 22α(*R*)-*O*,25β(*S*)-spirostanol glycoside, as a novel type of 22β(*S*)-*O*-25α(*S*)analog. *Chem. Pharm. Bull.*, 32:4229 (1984).
91. J.M.B. Filho, D.F. Medeiros, M.D.F. Agra and J. Bhattacharya, Spirostanes of *Kallistromea tribulodes*: Identification of C-25 epimers in mixtures by ^{13}C NMR spectroscopy, *Phytochemisrty*, 28:1985 (1989).
92. M. Ono, K. Shoyama and T. Nohara, The constituents of chinese *Polygonati officinalis* rhizome and *Poligonati rhizoma*, *Shoyakugaku Zasshi*, 42:135 (1988).
93. A.M.E. Abdel-Aziz, K. R. Brain, G. Blunden, T. Crabb and A.K. Bashir, Steroidal saponins from *Tacca leontopetaloides*, *Planta Med.*, 56:218 (1990).
94. A. Abdel-Aziz, K. Brain, G. Blunden, T. Crabb and A.K. Bashir, Isomeric F-ring dihydroxylated 22,25-epoxyfurostane an spirostane sapogenins from *Tacca leontopetaloides*, *Phytochemistry*, 29:1643 (1990).
95. K. Bock, C. Pedersen and H. Pedersen, Carbon-13 NMR data for oligosaccharides, *Adv. Carbohydr. Chem. Biochem.*, 42:193 (1984).
96. P.K. Agrawal, NMR spectroscopy in the structural elucidation of oligosaccharides and glycosides, *Phytochemistry*, 31:3307 (1992).
97. J. Dabrowski, Two-dimensional proton magnetic resonances spectroscopy, *Meth. Enzymol.*, 179:122 (1989).
98. C.A. Bush, High resolution NMR in the determination of the structure of complex carbohydrates. *Bull. Magn. Reson.*, 10:73 (1989).
99. S.W. Homans, Oligosaccharide conformations: application of NMR and energy calculations. *Progr. NMR Spectrosc.*,22:55(1990).
100. P.K. Agrawal and C.A. Bush, ^{1}H and ^{13}C NMR assignments of 2,3-diacetylamido-2,3-dideoxy-D-glucopyranose. *J. Carbohydr. Chem.*, 11:945 (1992).
101. H. Schroeder, M. Schubert-Zsilaveez, G. Reznicek, J. Cart, J. Jurenitisch and E. Haslinger, A triterpene saponin from *Herniaria glabra*, *Phytochemistry*, 34:1609 (1993).
102. R. Bazzo, C.J. Edge, R.A. Dwek and T.W. Rademacher, Extracting subspectra from overlapping regions DOUBLE TOCSY, *J. Mag. Reson.*, 86:199 (1990).
103. F. Inagaki, I. Shimada, D. Kohda, A. Suzuki and A. Bax, Relayed HOHAHA, a useful method for extracting subspectra of individual components of sgar chains. *J. Magn. Reson.*, 81:186 (1989).
104. P.K. Agrawal and M.C. Bansal, in *Studies in Organic Chemistry*, Vol 39 *Carbon-13 NMR of Flavonoids*, P.K. Agrawal Ed Elsevier, New York, pp. 283 (1989).
105. N.K. Kochetkov, Microbial polysaccharides: new approaches, *Chem. Soc. Rev.*, 19:29 (1990).

106. S. Chen and J.K. Snyder, Diosgenin-bearing molluscidal saponins from *Allium vineale*: an NMR approach for the structural assignment of oligosaccharide units. *J. Org. Chem.*, 54:3679 (1988).
107. V.U. Ahmad, E.T. Baqai, I. Fatima and R. Ahmad, A spirostanol glycoside from *Cestrum nocturnum*, *Phytochemistry*, 30:3057 (1991).
108. Y. Mimaki, O. Nakamura, Y. Sashida, T. Nikaido and T. Ohmoto, Steroidal saponins from the bulbs of *Triteleva lacta* and their inhibitory activity on cyclic AMP phosphodiesterase. *Phytochemistry*, 30:1279 (1995).
109. P.K. Agrawal and A.K. Pathak, NMR spectroscopic approaches for the determination of interglycosidic linkage and sequence in oligosaccharides, *Phytochem. Anal.*, accepted
110. G. Pant, M.C. Purohit, G.A. Morris, A.G.W. Halstead and R.I.G. Thompson, Structure elucidation and proton and carbon-13 assignments of a pentasaccharide glycoside from *Agave americana*, *Magn. Reson. Chem.*, 32:213 (1994).
111. G.M. Clore and A.M. Gronenborn, Applications of three- and four-dimensional heteronuclear NMR spectroscopy to protein structure determination, *Progr. NMR Spectrosc.*, 23:108 (1991).
112. G.W. Vuister, P.D. Waard, R. Boelens, J.F.G. Vliegenthart and R. Kaptein, The use of 3D NMR in structural studies of oligosaccharides, *J. Am. Chem. Soc.*, 111:772 (1989).
113. P.D. Waard, R. Boelens, G.W. Vuister and J.F.G. Vliegenthart, Structural studies by $^{1}H/^{13}C$ two-dimensional and three-dimensional HMQC-NOE at natural abundance on complex carbohydrates, *J. Am. Chem. Soc.*, 112:3232 (1990).
114. D. Uhrin, J.R. Brisson and D.R. Buundle, Pseudo-3D NMR spectroscopy: application to oligo- and polysaccharides, *J. Biomol. NMR*, 3:367 (1993).
115. H. Kessler, S. Mronga and G. Gemmecker, Mutidimensional NMR experiments using selective pulses. *Magn. Reson. Chem.*, 29:527 (1991).
116. H. Schroeder and E. Haslinger, Sequential analysis of oligosaccharide structures in a few minutes, *Angew. Chem. Int. Ed. Engl.*, 32:1349 (1993).
117. T.A. Carpenter, L.D. Colebrook, L.D. Hall and G.K. Pierens, Application of gradient-selective COSY and double-quantum filtered gradient-selective COSY experiments to carbohydrates: 2-deoxy-D-*arabino*-hexose (2-deoxy-D-glucose). *Carbohydr. Res.*, 241:267 (1993).

APPLICATION OF TANDEM MASS SPECTRAL APPROACHES TO STRUCTURAL DETERMINATIONS OF SAPONINS

Catherine E. Costello

Mass Spectrometry Resource
Dept. of Biophysics
Boston University School of Medicine
Boston, MA 02118-2394

INTRODUCTION

Structural determinations of saponins present challenges to the analyst because of the variety of potential modification sites, the incorporation of both carbohydrate and non-carbohydrate moieties, and the occurrence of mixtures that may include isomeric species and highly-active minor components. These quite daunting features of the research problem represent an excellent opportunity for effective application of new analytical methodologies. Tandem mass spectrometry is a particularly logical choice: advances in sample preparation, ionization methods and analyzer techniques can now lessen the sample requirements, shorten the analysis time and provide detailed structural information for individual mixture components. We are utilizing liquid secondary ionization, matrix-assisted laser desorption/ionization and electrospray ionization in conjunction with high- and low-energy collision-induced decomposition and post-source decay tandem mass analysis to probe the structures of saponins with established biological activities. Methods developed with known compounds are then used for the structural elucidation of unknowns. In the work described here, structures of saponins from a cardivascular active Chinese medicine were determined first and the fragmentation information was used as a guide in the interpretation of the spectra from more complex saponins found in extracts of the bark of *Quillaja saponaria* Molina that may have use as adjuvants in vaccines.

EXPERIMENTAL

Materials

Saponins from Chinese medicines were supplied by S. Du as chromatographically isolated fractions that were known to have oleanolic acid as the aglycone and the sugars glurcuronic acid, rhamnose and hexose, in unknown ratios. About 200-500 μg of each saponin was provided for mass spectral analysis.

Fractions of the saponin extract from *Quillaja saponaria* Molina (Quil A) were prepared by J. Thurin and K. Vogele by reverse phase and silicic acid chromatographic steps, monitored by bioassay. These had quillaic acid as the aglycone, and contained glucuronic acid, rhamnose, hexose and a fatty acyl component in varying ratios. NMR studies indicated that the structures were similar to those proposed by Higuchi *et al.* Fifty to 200 μg of each fraction were provided for mass spectral analysis.

Water was purified with a Millipore Milli-Q system; other solvents were HPLC grade. Glycerol and *N,N,N*-triethanolamine matrices for LSIMS were obtained from Fluka and Aldrich, respectively. Sinapinic acid and 2,5-dihydroxybenzoic acid matrices for MALDI-TOFMS were purchased from Sigma.

Mass Spectrometry

Liquid secondary ionization mass spectra and tandem mass spectra were acquired with an HX110/HX110 (E_1B_1[MS1]-E_2B_2[MS2]) tandem mass spectrometer (JEOL USA, Peabody, MA) with JEOL Complement data system. The primary beam was 6 keV Xe^0 or 15 keV Cs^+. Samples were dissolved in 3:1 $CHCl_3$/MeOH (1-2 μg/μL) and 0.5 μL of this solution was mixed on the stainless steel probe tip with 0.5 μL of the glycerol matrix. MS1 was operated at or above unit resolution; MS2 was operated at 1:1000 resolution. The accelerating voltage was ±10 kV and ±18 kV postacceleration was used at the detector. The collision cell between MS1 and MS2 was floated at ±3 kV, so that collision energy was 7 keV. MS1 spectra were acquired during magnetic field scans. Collision-induced decomposition (CID) tandem mass spectra were recorded during linked scans of MS2 at constant B/E, using He as the collision gas at a pressure sufficient to reduce the abundance of the precursor ion to 20% of its initial value.

Matrix-assisted laser desorption/ionization (MALDI) mass spectra were acquired with a Vision 2000 (Thermo BioAnalysis Ltd. [formerly Finnigan MAT, Ltd.], Hemel Hempstead, UK) reflectron time-of-flight (RTOF) mass spectrometer. Samples were dissolved in 3:1 $CHCl_3$/MeOH (1-2 pmol/μL) and 1-2 μL of this solution was mixed with 1 μL of a solution (10 g/L) of 2,5-dihydroxybenzoic acid in the same solvents. The nitrogen laser (Laser Science, Inc., Newton, MA) irradiated the sample at 337 nm with 3-ns pulses at a manually controlled rate during observation with a videocamera; data were acquired at 500 MHz and spectra were added together as noted. Linear spectra were recorded with ±30 kV accelerating voltage. For RTOF and PSD spectra, the accelerating voltage was ±5 kV and ±5-18 kV postacceleration was applied at the detector. Calibration was external.

Electrospray ionization (ESI) mass spectra were acquired with a Quattro II (VG/Fisons, Danvers, MA) triple quadrupole mass spectrometer. The peracetylated saponin sample was dissolved in methanol/water containing 0.05% NaOAc and the solution was introduced with a syringe pump.

RESULTS AND DISCUSSION

The *Quillaja* saponins (Figure 1) have potential use in vaccines, largely because of the adjuvant properties of some components (Kensil *et al.*, 1991, 1995). An additional property, that of forming small (35 nM) particles together with cholesterol and antigen, is also pertinent (Thurin *et al.*, 1990). The *Quil A* extracts are complex mixtures of compounds that vary in their therapeutic effects and include toxic components; it is thus necessary to carry out structure-activity studies to thoroughly characterize both the variants that show promise for drug development and structural elements that may be deleterious.

As part of our investigations of the tandem mass spectra of saponins, five fractions from a Chinese medicine and sixteen fractions from *Quillaja* were analyzed by mass spectrometry. We have employed LSI (introduced as fast atom bombardment, FAB, by Barber *et al.,* 1981), MALDI (Karas *et al.*, 1987), and ESI (Fenn *et al.*, 1989) as ionization methods. Numerous mass spectral methods have been reported useful for structural studies of saponins: for example, Budziekiewicz *et al.* used electron ionization (1963), as did Haddon (1980), whereas Schulten and coworkers (1977, 1978) recorded spectra on photoplates after field desorption (FD). FAB was used by Adinolfi *et al.* (1984), Fraisse *et al.* (1986), Higuchi and his colleagues (1987a, 1987b, 1988), and Warashina *et al.* (1991). Tandem mass spectrometry with FAB ionization as has been utilized by Crow *et al.* (1986) and Chen *et al.* (1987), and by the group of van der Werken (van der Werken, 1991; van Setten, 1995). Prome and his colleagues have discussed negative-ion FABMS fragmentation pathways (1987). Coates and Wilkins (1985) reported laser desorption mass spectra of steroidal oligoglycosides, recorded at high resolution on a Fourier transform ion cyclotron resonance instrument, without a matrix. A capsule of the extensive studies of Hostettmann *et al.* is included in a new book edited by this investigator (1995). A summary of our LSIMS/MS structural studies for the cardiac active saponins has been reported elsewhere (Costello *et al.*, 1987). Accounts of the use of ESI or MALDI as ionization modes for saponin analysis have not previously appeared in the literature, although these techniques are beginning to be employed for this purpose, by other groups as well as our own.

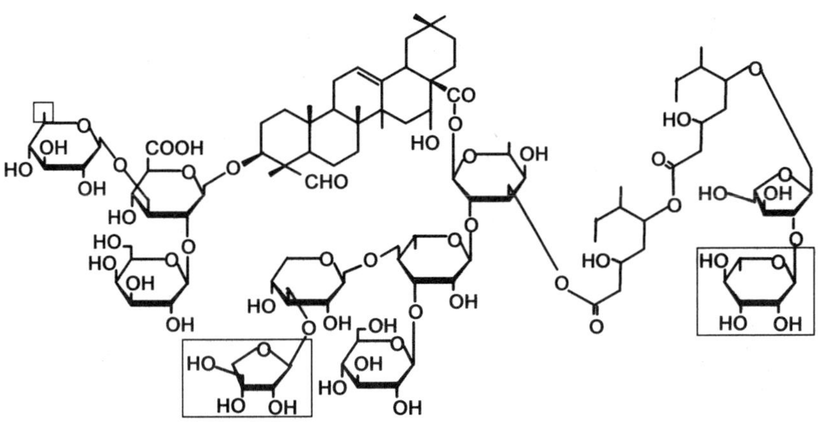

Figure 1. Structure of the *Quillaja* saponin proposed by Higuchi *et al.* (1987a, 1987b, 1988). Boxes are discussed in text.

Our preliminary MS/MS results for the *Quillaja* saponins have been presented (Costello *et al.*, 1993); representative LSIMS and MS/MS data and MALDI-TOF mass spectra for some of the fractions from *Quillaja* are discussed here in order to illustrate the methodology. In general, our results support the structure (Figure 1) suggested by Higuchi *et al.* (1987a, 1987b, 1988), but we have identified numerous variations on the basic structure, as indicated by the boxed areas shown on the structure in Figure 1. None of the bioactive fractions we examined contain the complete Higuchi structure, although every element appeared in one or more fractions. Further, none of the active fractions contain the unknown component ($C_8H_{12}O_5$) reported by van der Werken and colleagues (1991,1995).

MALDI-TOF Mass Spectrometry

Molecular weights of the fractions were determined by positive- and negative-ion MALDI-TOF mass spectrometry. The results for an active fraction of the native *Quil A* saponins are shown in Figures 2 and 3. Figure 2A contains the positive-ion linear TOF mass spectrum; note that the dominant species for each component is the $[M+Na]^+$ ion. In the positive-ion reflectron TOF mass spectrum, Figure 2B, the $[M+2Na-H]^+$ is the more abundant species. This behavior reflects the relative instability of the $[M+H]^+$ and $[M+Na]^+$ species. The protonated species (if formed at all) undergoes prompt fragmentation in the ion source, prior to acceleration, and therefore does not appear even in the linear TOF mass spectrum. The monosodiated species undergoes substantial metastable ion decomposition during flight, but, since intact and fragmented species continue to travel at their initial speed, both reach the linear detector at the time appropriate for the intact ion. However, the decomposition products are not focused by the reflector at its normal voltage, so the unstable species are depleted in the RTOF spectrum, relative to the more stable disodiated species. The negative-ion linear TOF mass spectrum (Figure 3) contains abundant peaks corresponding to the $[M-H]^-$ for the two species that can easily bear a negative charge. The spectrum is simple, because there is no distribution of species adducted by various numbers of cations. It shows only a very low-abundance signal for the higher molecular weight species which we have assigned to the component having a pentosyl ester instead of a free carboxyl group on the glucuronic acid residue at C-3. For profiling of molecular weights,

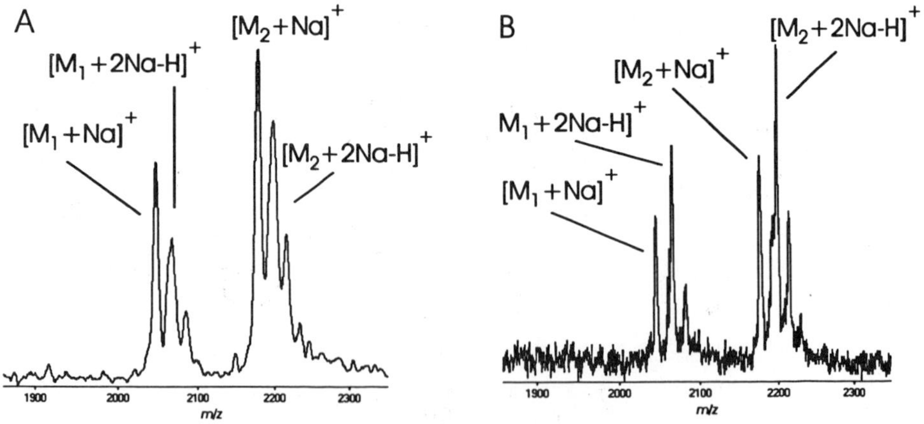

Figure 2. Molecular ion regions of positive-ion (A) linear and (B) reflectron MALDI-TOF mass spectra of a bioactive *Quil A* fraction in 2,5-DHB matrix, 337 nm irradiation.

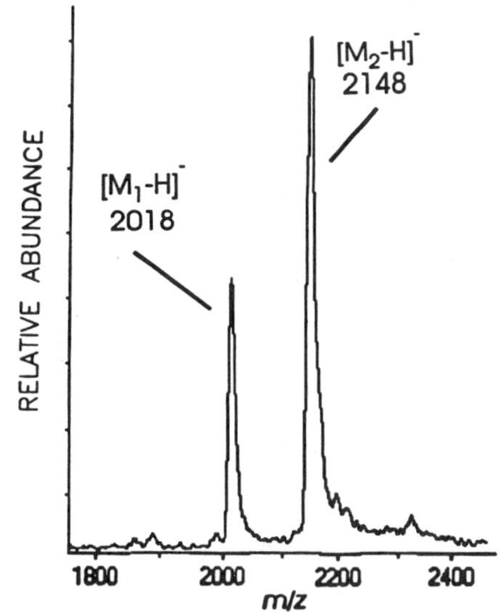

Figure 3. Negative-ion linear MALDI-TOF mass spectrum of a *Quil A* fraction.

negative-ion MALDI-TOF MS offers a straightforward and sensitive approach.

LSI mass spectrometry

For the native compounds the sample requirement for the LSIMS experiments is 2-3 orders or magnitude greater than for MALDI-TOFMS, that is, samples in the nmol range are required instead of the low pmol range. ESI sensitivity for the native compounds is relatively poor. For peracetylated derivatives, the sensitivity for LSIMS and MALDI each improves about a hundred-fold, and the sensitivity for ESI-MS (data not shown) becomes equal to or greater than that for MALDI. MALDI and ESI are not so subject to matrix and competitive suppression effects as is LSIMS. These newer ionization methods are therefore favored for screening purposes, for analysis of scarce samples and the structural determinations of minor components in complex mixtures. When adequate

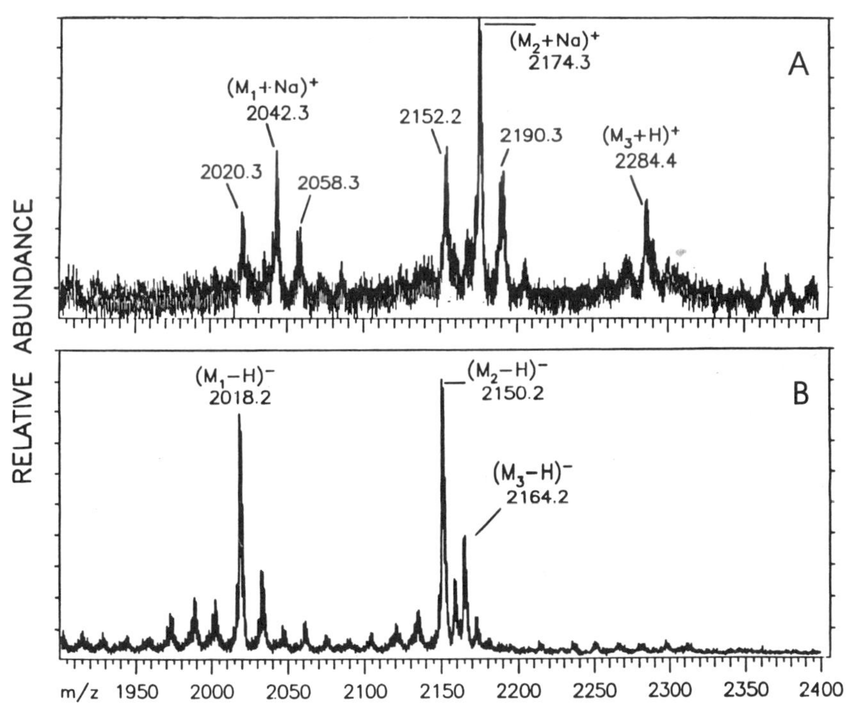

Figure 4. Molecular ion regions of the (A) positive- and (B) negative-ion LSIMS of the *Quil A* fraction analyzed above by MALDI-TOF MS.

samples are available, however, LSIMS gives excellent results that can be obtained with high resolution for accurate mass determinations (van Setten *et al.*, 1995) and with unit mass resolution for the selection of precursor ions and recording of product ions in MS/MS experiments. Figure 4 shows the molecular ion regions of the positive- and negative-ion LSI mass spectra of the same fraction for which MALDI-TOF mass spectra were illustrated in Figures 2 and 3.

The results of the LSIMS experiments and the molecular weights deduced for the *Quil A* fractions discussed in the text are listed in Table 1. In the positive-ion LSI mass spectra, the $[M+H]^+$ peaks are present at significant levels, although the monocationized species are once again the more abundant. The observation of dicationized species is less favored than in MALDI-TOF mass spectra. Discrimination against the species that lacks a free carboxyl group is apparent in the negative-ion LSI mass spectra, as it was in the case of MALDI.

The full-mass-range negative-ion mass spectra obtained for the sample shown in Figures 2-4 and another fraction of the native *Quil A* are presented in Figure 5. Evidence for the presence of mixtures is given in the fragment ion regions of the spectra as well as in the molecular ion regions. The heterogeneity of the intact molecules is reflected in the relative abundances of peaks in the ranges *m/z* 400-700 and *m/z* 900-950, among others. In order to ascertain which structural features give rise to the multiplicity of ions, MS/MS spectra of fragment ion peaks which showed significant abundance differences were obtained and compared to the MS/MS spectra of apparently related peaks that were common to both spectra. Two such pairs and their structural assignments are discussed below.

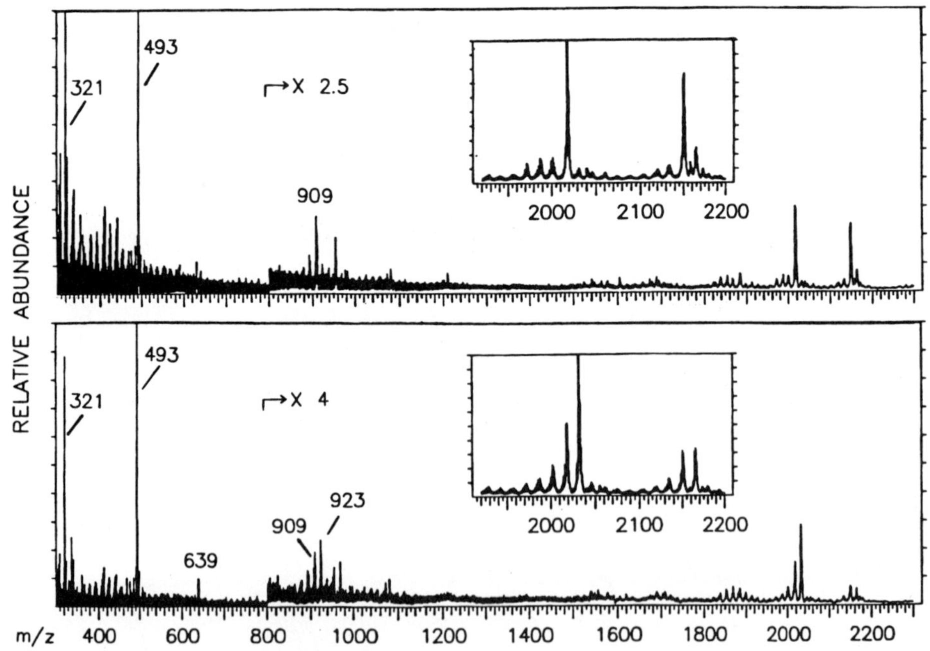

Figure 5. Negative-ion LSI mass spectra of two fractions of Quil A. Insets show expansions of the molecular ion regions.

Table 1. LSIMS Molecular Weight Determinations for *Quillaja saponaria* Molina Fractions

Fraction	Pos. LSIMS *m/z*	Assign.	Neg. LSIMS *m/z*	Assign.	Sugar Comp.	MW calc. (^{12}C)
native (acylated) fraction B3ba	2020.3 2042.3 2058.3	$[M_1+H]^+$ $[M_1+Na]^+$ $[M_1+K]^+$	2018.2	$[M_1-H]^-$	Pen(3) dHex(2) Pen(3) GlcA(1)	2018.9
	2152.2 2174.3 2190.3	$[M_2+H]^+$ $[M_2+Na]^+$ $[M_2+K]^+$	2150.2	$[M_2-H]^-$	Pen(4) dHex(2) Hex(2) GlcA(1)	2151.0
	2188.3	$[M_3+Na]^+$	2164.2	$[M_3-H]^-$	Pen(3) dHex(3) Hex(2) GlcA(1)	2165.0
	2284.4	$[M_4+H]^+$			Pen(5) dHex(2) Hex(2) GlcA(1)	2283.0
desacyl fraction F-11-3	1543.6 1565.6 1581.6	$[M_1+H]^+$ $[M_1+Na]^+$ $[M_1+K]^+$	1541.8	$[M_1-H]^-$	Pen(2) dHex(2) Hex(2) GlcA(1)	1542.7
	1675.7 1697.7 1713.7	$[M_2+H]^+$ $[M_2+Na]^+$ $[M_2+K]^+$	1673.8	$[M_2-H]^-$	Pen(3) dHex(2) Hex(2) GlcA(1)	1674.7
	1807.8	$[M_3+H]^+$			Pen(4) dHex(2) Hex(2) GlcA(1)	1806.8

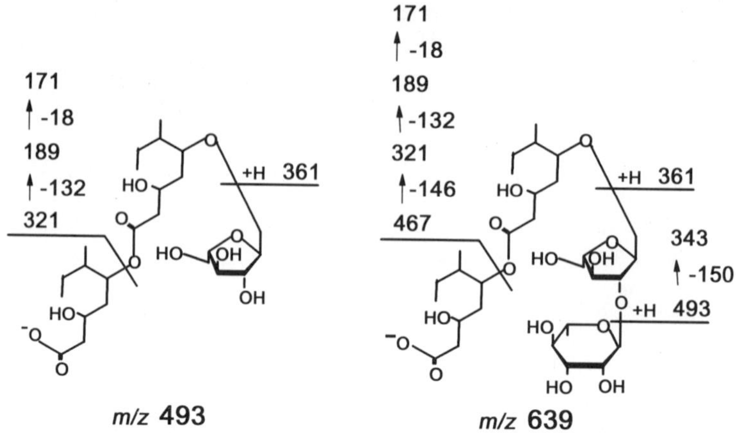

Scheme 1. Product ion assignments for CID of *Quil A* negative-ion acyl fragments.

On the basis of the CID MS/MS spectra shown in Figure 6, the fragment ions at *m/z* 493 and 639 can be assigned to the acyl group, as indicated in Scheme 1. The higher-mass species contains an additional deoxyhexose unit in the terminal position. The observed fragments are consistent with the structure of the pendant acyl group proposed by Higuchi *et al.* (1987b, 1988). Evidence for the branching within the hydrocarbon portion is not provided by these spectra, because the availability of low-energy fragmentation pathways through cleavage at glycosidic and ester bonds favors these routes over high-energy scission of C-C bonds. The unusual loss of 172 u from either *m/z* 493 or 639 makes it clear that the fragments selected are not simple oligosaccharide ions, because this value does not correspond to a sugar residue. It is appropriate for stepwise loss of the two 3,5-dihydroxy-6-methyloctanoic acid moieties proposed by the Japanese group.

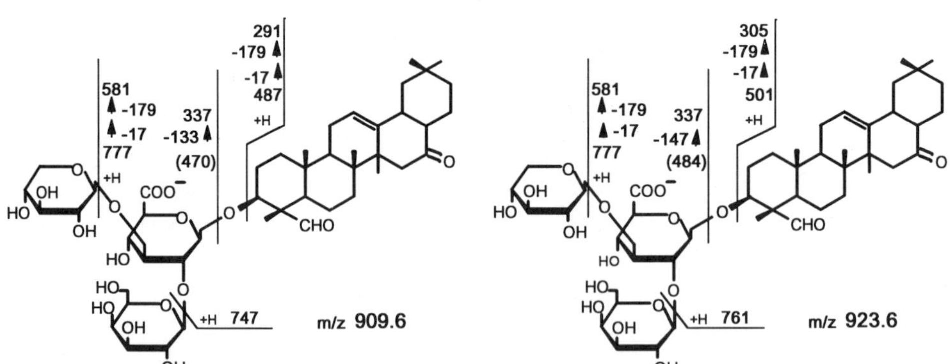

Scheme 2. Product ion assignments for CID of *Quil A* negative-ion fragments formed upon elimination of the prosapogenin C-28 carboxyl moiety.

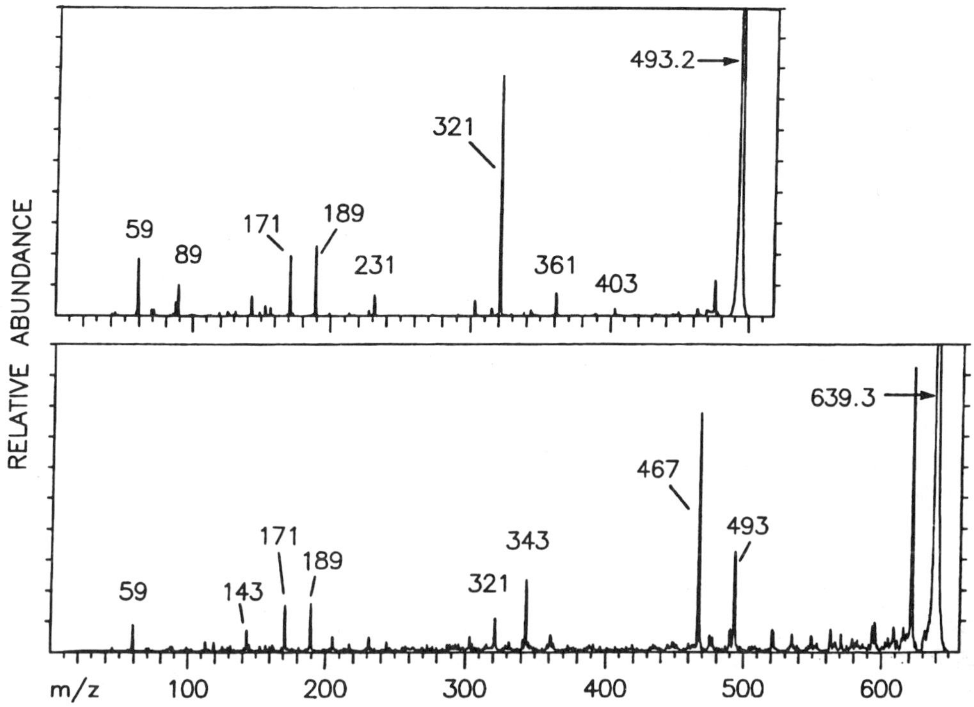

Figure 6. CID MS/MS spectra of two acyloxy fragment ions in the LSIMS of a *Quil A* fraction.

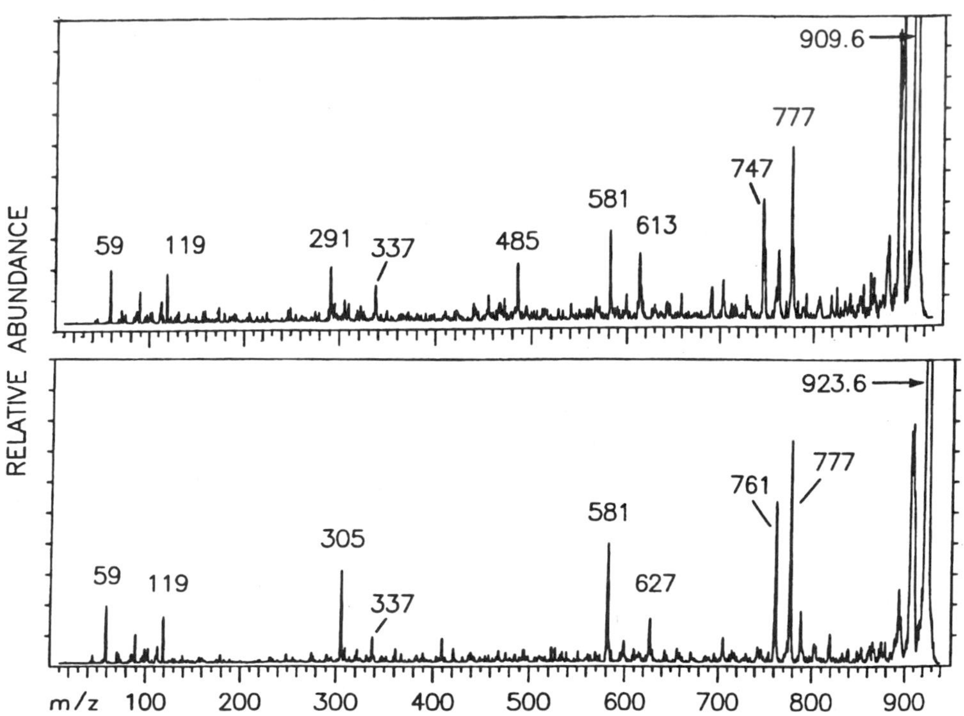

Figure 7. CID MS/MS spectra of selected fragment ions in the LSIMS of a Quil A fraction.

The fragment ions at *m/z* 909 and 923 contain the decarboxylated aglycone and the 3-*O*-linked saccharide. In this case, the variation consists of an exchange of a the pentose 3-linked to glucuronic acid in the lower-mass species for a deoxyhexose in the higher mass analog. Evidence for this assignment comes from observation that the product ions at *m/z* 291, 487 and 747 undergo 14 u mass shifts, whereas those at *m/z* 337, 581 and 777 retain their mass values. The fragmentation pathways are indicated in Scheme 2.

The heterogeneity in the saccharide substituent at C-28 proved easier to examine after removal of the fatty acyloxy group. This substituent can easily be removed by hydrolysis; its ester bond is more labile than that of the hindered ester at C-28 (Kensil, 1995). The molecular ion regions of the positive- and negative-ion LSIMS spectra of the hydrolysis products from the fraction under discussion are shown in Figure 8 and the peak assignments are given in Table 1. In the positive-ion spectrum, the most abundant ions are the $[M+Na]^+$, *m/z* 1565 and 1697. The corresponding $[M+K]^+$ ions appear at *m/z* 1581 and 1713. Protonated ions appear at *m/z* 1543, 1675 and 1807. The last is assigned to the species with a pentosyl ester on the glucuronic acid, hence lacking a free carboxyl group. As demonstrated by the CID MS/MS spectra of the $[M+Na]^+$ ions (Figure 9), the spacing of 132 u between the major species is due to the variable presence of the terminal pentose on the C-28 oligosaccharide. The hydrolysis products include minor components with an additional 14 u that can be attributed to the heterogeneity of the 3-*O*-linked saccharide described above, on the basis of positive- and negative-ion CID MS/MS spectra (not shown). Comparison of the CID MS/MS spectra of the $[M+Na]^+$ ions at *m/z* 1565 and

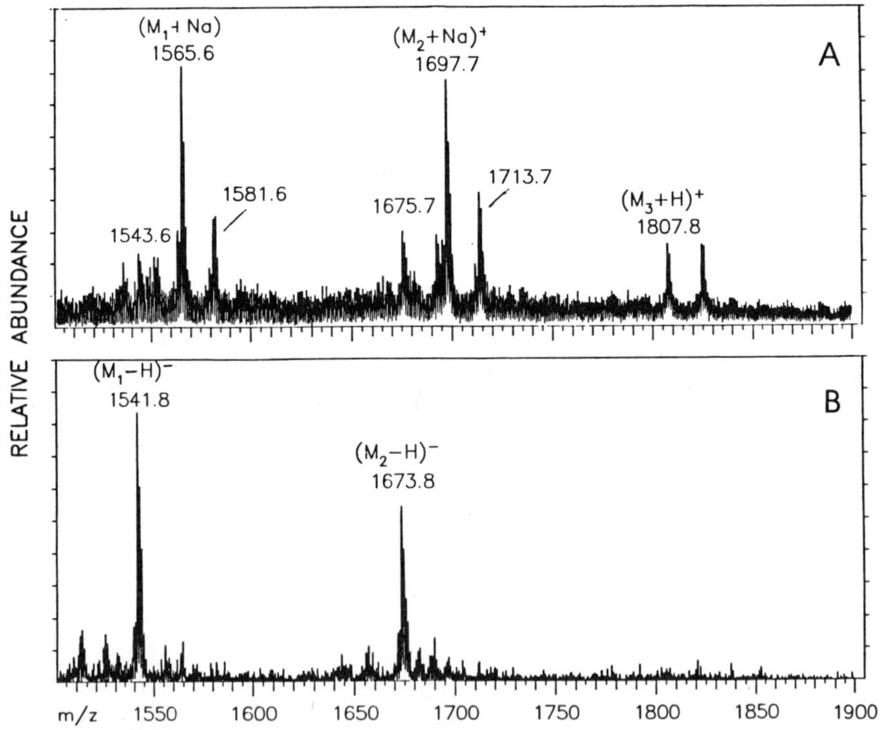

Figure 8. (A) Positive- and (B) negative-ion LSI mass spectra of a desacyl saponin mixture obtained by hydrolysis of a *Quil A* fraction.

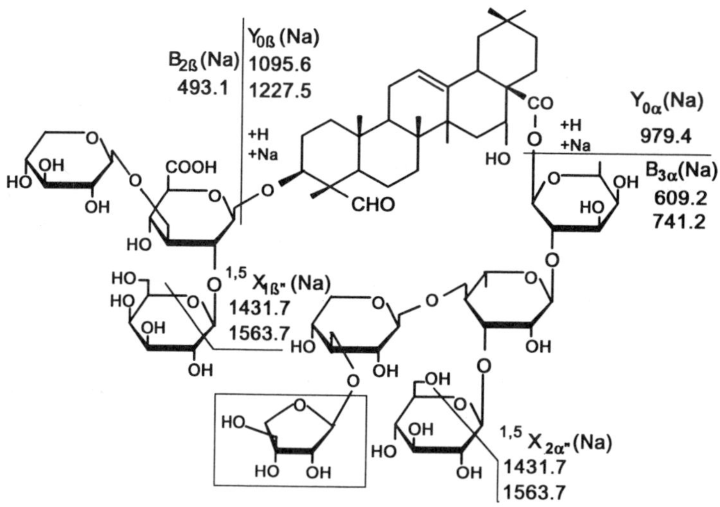

Scheme 3. Product ion assignments for CID of the $[M+Na]^+$ ions of two components from a desacylated *Quil A* fraction.

1697 allows the assignment of the location of the variable pentose residue, as indicated in the Scheme 3, in which the residue is boxed and the fragments that undergo mass shifts (*m/z* 609→741, *m/z* 1095→1227, *m/z* 1431→1563) are designated with two mass assignments, depending on the presence or absence of the residue. Fragments that have a constant value (*m/z* 493, 979) do not include the variable residue. The nomenclature system for the fragment ions follows that proposed by Domon and Costello (1988). In this instance, the

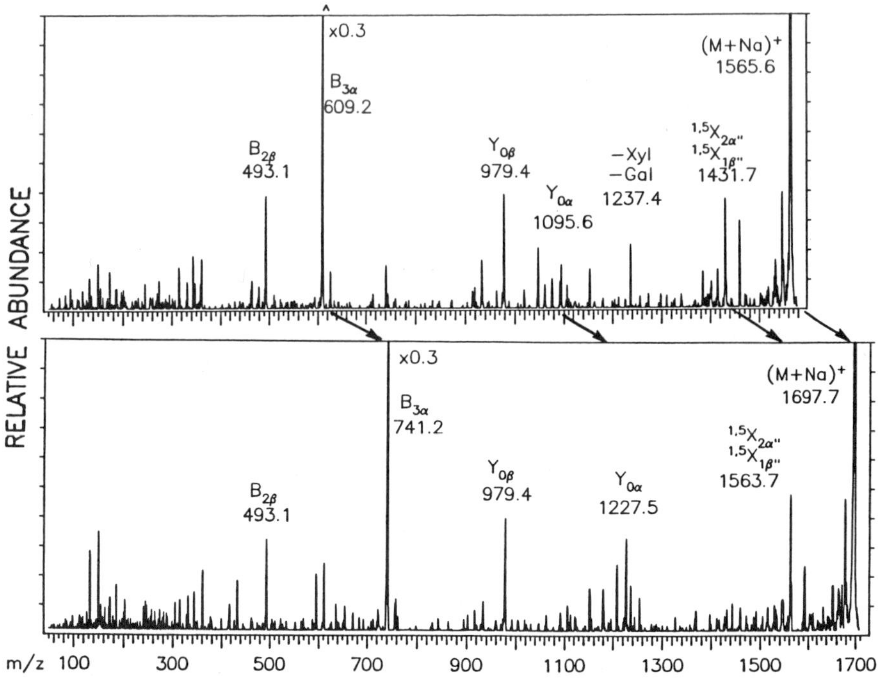

Figure 9. CID MS/MS spectra of the two abundant $(M+Na)^+$ ions in the LSI mass spectrum of the hydrolysis products of a *Quil A* fraction.

α-chain designation has been used for the longer C-28 oligosaccharide, and the β-designation for the shorter C-3 oligosaccharide. Negative-ion CID MS/MS spectra showed prominent ions with the free carboxyl group at *m/z* 955 resulting from cleavage of the ester and subsequent losses of 46 u (-HCOOH) to *m/z* 909 and 62 u (-CO_2, H_2O) to *m/z* 893. The negative-ion spectra showed retro-Diels-Alder cleavage of the C-ring of the triterpene, to yield a fragment pair at *m/z* 689/690. We had earlier observed (Costello *et al.*, 1987) that this cleavage was useful for determining the total weights of substituents on either side of the aglycone. Such ring-opening was reported by Haddon (1980) for positive-ion electron ionization mass spectra of oleanolic acid.

The sets of CID MS/MS spectra shown here and parallel experiments with the remaining fractions from the *Quil A* extracts and the cardiac glycoside have enabled us to assign the structures of the major and some minor components in the many fractions. In making these assignments, we have been greatly aided by the information available from carbohydrate and NMR analyses that defined the specific sugars present, and provided details about linkages. The complexity of the mixtures made interpretation of the NMR spectra difficult, and in this instance the molecular weight and partial structural information from the MS data facilitated interpretation of the NMR spectra. Initial experiments utilizing ESI with collision-induced decomposition on a triple quadrupole instrument and MALDI with postsource decay (Spengler *et al.*, 1995) on the RTOF mass spectrometer have indicated that substantial fragmentation is present that will provide high sensitivity structural information as the rules for the interpretation of these spectra are worked out to a level comparable to that now available for high-energy collision studies.

CONCLUSIONS

Mass spectrometry, especially tandem mass spectrometry, is particularly well suited for structural studies of saponins. The investigator has the choice of several ionization methods. The amount of material required for the LSIMS and MS/MS experiments (*ca.* 1 nmol of native saponin) is substantially less than that needed for recording NMR spectra. Both molecular weight and sugar sequence information are available from these spectra, and the presence and location of unusual structural components can be determined. For MALDI and ESI experiments, the sample requirement for molecular weight profiling is reduced by another two or more orders of magnitude. With the recording of post-source decay (PSD) spectra in the MALDI experiment or CID MS/MS spectra with ESI, detailed structural information is also available at high sensitivity. For the latter methods, the reference data base is yet quite small, but initial experiments in our laboratory and elsewhere indicate that these types of tandem MS spectra will also prove highly useful for the characterization of minute quantities of saponins, even when they are present as components of mixtures.

ACKNOWLEDGEMENTS

The author wishes to acknowledge the contributions of her colleagues B. Domon, C.-H. Zeng, J. Helin and B. B. Reinhold and her collaborators S. Du (Shanghai Inst. of Materia Medica), and J. Thurin and K. Vogele-Prammer (Wistar Institute) to the projects described. The research has been supported by the National Institutes of Health, Grant Nos. RR00317 (to K. Biemann), RR10493 (to C. E. Costello) and CA09171 (to H. Koprowski) and by Thermo BioAnalysis, Ltd.

REFERENCES

Adinolfi, M., Mangoni, L., Parrilli, M. and Self, R., 1984, Fast atom bombardment mass spectrometry of the muscarosides, *Biomed. Mass Spectrom.* 11:310.

Barber, M., Bordoli, R. S., Elliott, G. J., Sedgwick, R. D. and Tyler, A. N., 1981, Fast atom bombardment of solids as an ion source in mass spectrometry, *Nature* 293:270.

Budziekiewicz, H., Wilson, J. M. and Djerassi, C., 1963, Mass spectrometry in structural and stereochemical problems. XXXII. Pentacyclic triterpenes, *J. Am. Chem. Soc.* 85:3088.

Chen, Y., Chen, N., Li, H., Zhao, F. and Chen, N., 1987, Fast atom bombardment and collisional activation mass spectrometry in the structural analysis of steroidal oligoglycosides, *Biomed. Env. Mass Spectrom.* 14:9.

Coates, M. L. and Wilkins, C. L., 1985, Laser desorption/Fourier transform mass spectra of glycoalkaloids and steroid glycosides, *Biomed. Env. Mass Spectrom.* 13:199.

Costello, C. E., Domon, B., Zeng, C.-H., Du, S., Gian, F. and Li, H., 1987, FABMS and MS/MS analyses of glycosidic triterpenes from Chinese medicines, Proc. 35th ASMS Conf. on Mass Spectrom. and Allied Topics, Denver, CO , p. 1-2.

Costello, C. E., Vogele, K. E., Thurin, J. and Morein, B., 1993, Structural characterization of the adjuvant active compounds in *Quillaja saponaria* Molina Extracts, Proc. 41st ASMS Conf. on Mass Spectrom. and Allied Topics, San Francisco, CA, p. 1080.

Domon, B. and Costello, C. E., 1988, A systematic nomenclature for carbohydrate fragmentations in FABMS/MS of glycoconjugates, *Glycoconj. J.* 5:397.

Crow, F. W., Tomer, K. B., Looker, J. H. and Gross, M. L., 1986, Fast atom bombardment and tandem mass spectrometry for structure determination of steroid and flavonoid glycosides, *Anal. Biochem.* 155:286.

Fenn, J. B., Mann, M., Meng, C. K., Wong, S. F. and Whitehouse, C. M., 1989, Electrospray ionization for mass spectrometry of large biomolecules, *Science* 246: 64.

Fraisse, D., Tabet, J. C., Becchi, M. and Raynaud, J., 1986, Fast atom bombardment mass spectrometry of triterpenic saponins, *Biomed. Env. Mass Spectrom.* 13:1.

Haddon, W. F., 1980, The constant neutral linked magnetic field-electric sector scan, *Org. Mass Spectrom.* 15:539.

Hostettmann, K. and Marston, A., 1995, *Saponins*, Cambridge University Press, Cambridge, UK.

Higuchi, R., 1987a, Structure of desacylsaponins obtained from the bark of *Quillaja saponaria*, *Phytochemistry* 26:229.

Higuchi, R. and Komoni, T., 1987b, Structures of compounds derived from the acyl moieties of Quillajasaponin, *Phytochemistry* 26:2357.

Higuchi, R., Tokimitsu, Y. and Komoni, T., 1988, An acylated triterpenoid saponin from *Quillaja saponaria*, *Phytochemistry*, 27:1166.

Karas, M., Bachmann, D., Bahr, U. and Hillenkamp, F., 1987, Matrix-assisted ultraviolet laser desorption of non-volatile compounds, *Int. J. Mass Spectrom. Ion Processes* 78: 53.

Kensil, C. R., Patel, U., Lennick, M. and Marciani, D., 1991, Separation and characterization of saponins with adjuvant activity from *Quillaja saponaria* Molina cortex, *J. Immunol.* 146:431.

Kensil, C. R., Wu, J.-Y. and Soltysik, S., 1995, Structural and immunological characterization of the vaccine adjuvant QS-21, in: *Vaccine Design: The Subunit and Adjuvant Approach*, M. F. Powell and M. J. Newman, eds., Plenum Press, New York, p. 525.

Prome, J.-C., Aurelle, H., Prome, D. and Savagnac, A., 1987, Gas phase glycosidic cleavage of oxyanions from alkyl glycosides, *Org. Mass Spectrom.* 22:6.

Schulten, H.-R., Komori, T. and Kawasaki, T., 1977, Field desorption mass spectrometry of natural products - I. Steroid and triterpene saponins, *Tetrahedron* 33:2595.

Schulten, H.-R., Komori, T., Nohara, T., Higuchi, R. and Kawasaki, T., 1978, Field desorption mass spectrometry of natural products - II. Physiologically active pennogenin and hederagenin glycosides, *Tetrahedron* 34:7.

Spengler, B., Kirsch, D., Kaufmann, R. and Lemoine, J., 1995, Structure analysis of branched oligosaccharides using post-source decay in matrix-assisted laser desorption ionization mass spectrometry, *J. Mass Spectrom.* 30:782.

Thurin, J., Wroblewski, K., Morein, B. and Lövgren, K., 1990, The ISCOM glycoside adjuvant and HIV vaccine, *AIDS Hum. Res. Retrovir.* 5:45.

van der Werken, G., Kersten, G. F. A., Evers, E. A. I. M., ten Hove, G. J., Beuvery, E. C. and de Jong A. P., 1991, Structure analysis of components of Quil-A. Abstr., Int. Mass Spectrom. Meeting, Amsterdam, 1991.

van Setten, D. C., van der Werken, G., Zomer, G. and Kersten, G. F. A., 1995, Glycosyl compositions and structural characteristics of the potential immuno-adjuvant active saponins in the *Quillaja saponaria* Molina extract Quil A, *Rapid Commun. Mass Spectrom.*, 9:660.

STRUCTURAL DETERMINATION OF SAPONINS FROM MUNGBEAN SPROUTS BY TANDEM MASS SPECTROMETRY

M.R. Lee, J.S. Lee, J.C. Wang, and G.R. Waller

[1]Department of Chemistry
National Chung Hsing University
Taichung, Taiwan, R.O.C.
[2]Department of Biochemistry and Molecular Biology
Oklahoma Agricultural Experiment Station
Oklahoma State University
Stillwater, Oklahoma 74078, USA

INTRODUCTION

Saponins are glycosides that commonly occur in higher plants. Saponins are classified into triterpene and steroid groups; the aglycone usually has an oleanane, ursane, or dammarane skeleton. The sugars encountered in saponins are hexoses (including glucose, galactose, mannose, and 6-deoxyhexoses), pentoses, uronic acids, or amino sugars. Sugars may be linked to the sapogenin at one or two glycosylation sites, giving corresponding monodesmosidic or bidesmosidic saponins, respectively.

Mass spectrometry has been used extensively for saponins characterization and structural confirmation. Sapogenol composition of the seeds of 13 varieties of legumes has been studied by Price and coworkers using electron impact (EI) ionization gas chromatography/mass spectrometry[1] and fast atom bombardment mass spectrometry (FAB)[2]. Maillard and Hostettmann utilized liquid chromatography/thermospray mass spectrometry for identification of saponins in crude plant extracts[3]. Kitagawa *et al.*[4] used secondary ion mass spectrometry for structure elucidation of acetylsoyasaponin A_3 from American soybeans, the seed of *Glycine max* Merrill. Waller *et al.*[5,6], West *et al.*[7], and Lee *et al.*[8] also used secondary ion mass spectrometry to study allelopathic activity of root saponins from alfalfa.

Tandem mass spectrometry techniques have also been used for detailed structure determination of saponins and sapogenins. Fraisse *et al.*[9] utilized collisionally activated decomposition mass-analyzed ion kinetic energy combined with FAB for triterpenic saponins with ester glycosides or glycosylated tertiary alcohols. Facino *et al.*[10] studied saponins in raw plant extracts and Chen *et al.*[11] examined steroidal oligoglycosides also using the MS/MS technique. Linked scanning at constant B/E has been used by Waller *et al.*[12] for determination of soyasaponin I in *Vigna radiata* L. sprouts. This technique was used by Mil'grom *et al.*[13] for the characterizations of metastable ions resulting from the fragmentation of steroid sapogenins and by Takayama *et al.*[14] in a study of the formation and fragmentation of $(M+Na)^+$ ions of flavonol and steroid glycosides. In this book there are many examples of mass spectrometry of the structural determination of saponins; the tandem method by Costello[15] and the use in search for molluscocidal and antifungal saponins by Hostettmann[16] are two examples.

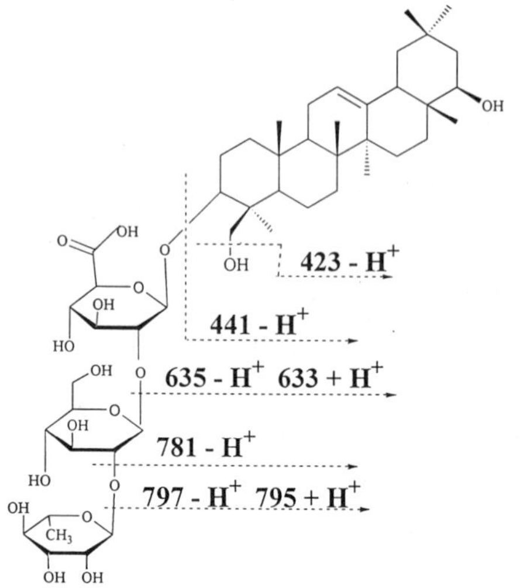

Figure 1. The structure of soyasaponin I. Charge retention is indicated by the direction of the arrow and *m/z* values for fragments (F) are shown with superscripts indicating adduct and charge as follows: $[F+H]^+ = m/z^{+H+}$; $[F–H] = m/z^{-H-}$

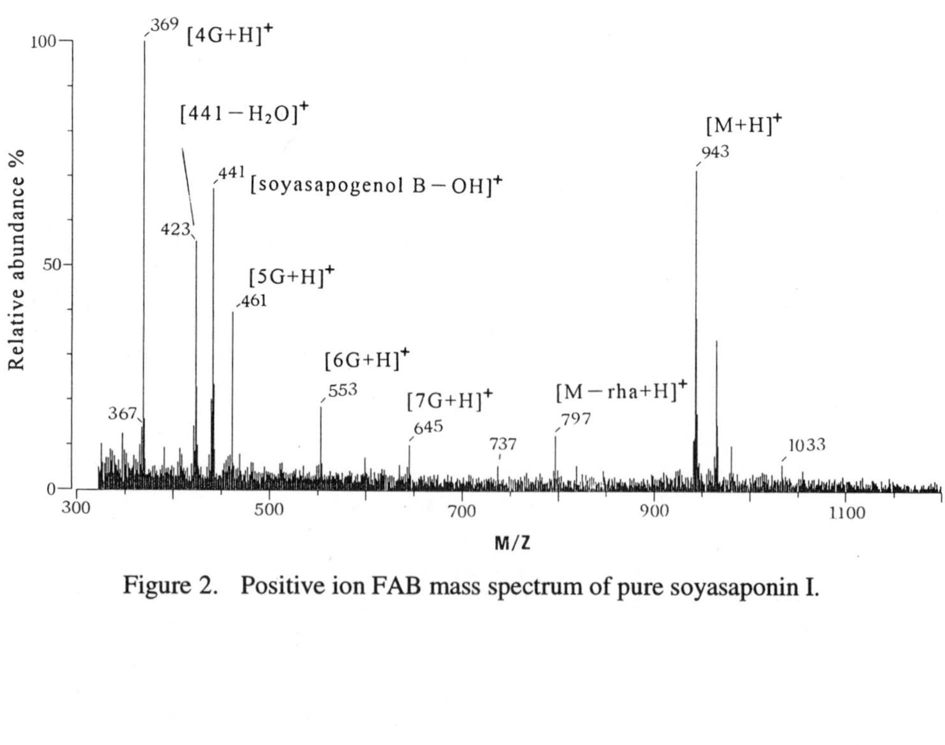

Figure 2. Positive ion FAB mass spectrum of pure soyasaponin I.

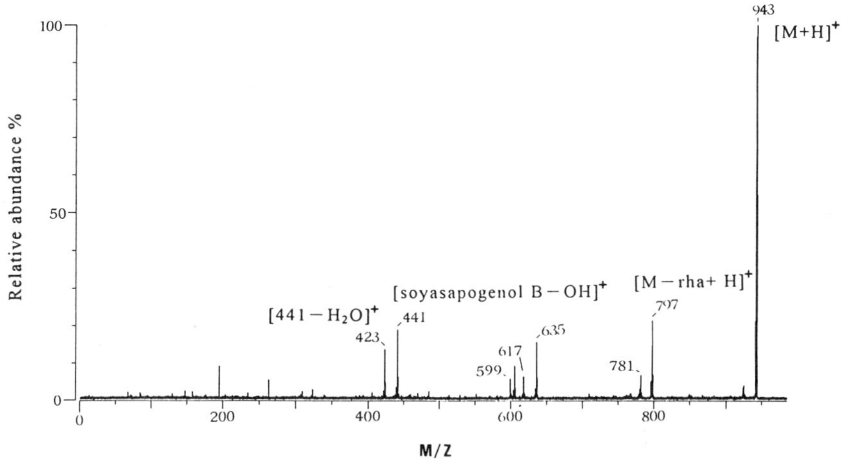

Figure 3. Daughter spectrum of $(M+H)^+$ ion of *m/z* 943 of pure soyasaponin I by FAB.

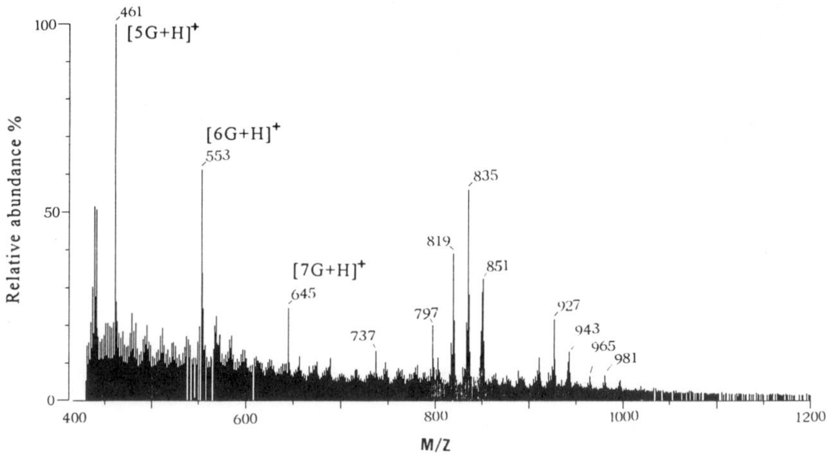

Figure 4. Positive ion FAB mass spectrum of mungbean sprout saponins after 5-h dialysis.

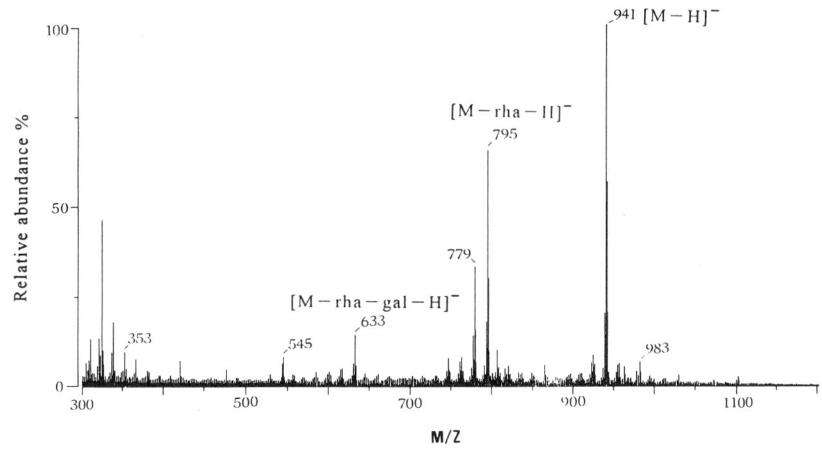

Figure 5. Negative ion FAB mass spectrum of mungbean sprout saponins after 5-h dialysis.

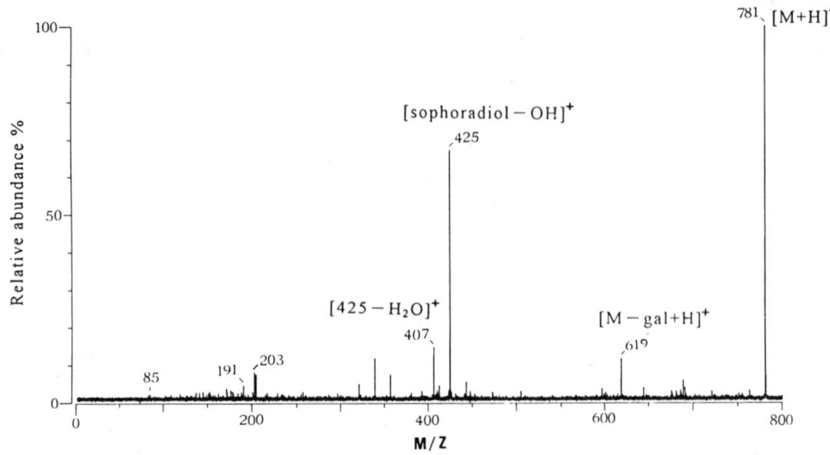

Figure 6. Daughter spectrum of ion of *m/z* 781 of mungbean sprout by FAB saponins after 5-h dialysis.

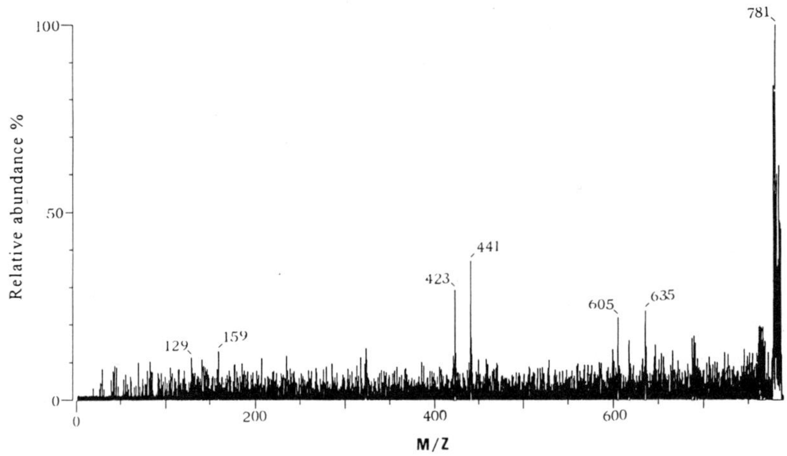

Figure 7. Daughter spectrum of ion of *m/z* 781 of pure soyasaponin I by FAB.

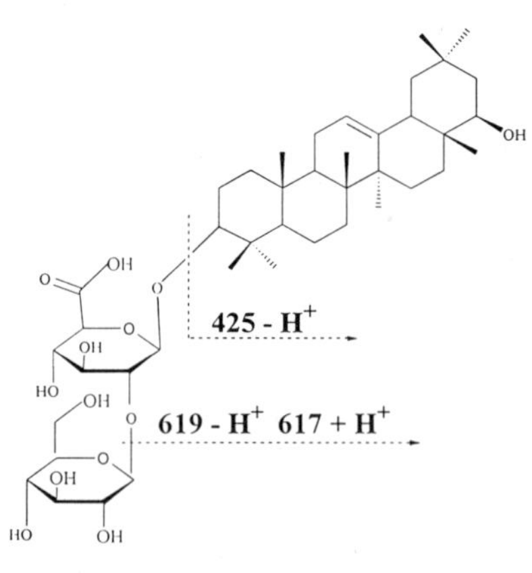

Figure 8. The structure of 3–*O*–[β–*D*–galactopyranosyl–(1→2)–β–*D*–glucuronopyranosyl]sophoradiol from mungbean sprout saponins after 5-h dialysis. Charge retention is indicated by the direction of the arrow and *m/z* values for fragments (F) are shown with superscripts indicating adduct and charge as follows: $[F+H]^+ = m/z^{+H+}$; $[F–H] = m/z^{-H-}$

Mungbean (*Vigna radiata* L.) saponins are a mixture of several triterpenoid (C_{30} pentacyclic) glycosides, which upon hydrolysis yield pentose(s), hexose(s), uronic acid(s), and the aglycone that is the nonsugar portion of the saponin moiety. Most of the saponins in mungbean are not easy to identify because of the trace amounts present and the lack of adequate analytical procedures. In this study the saponins from mungbean sprouts were analyzed using positive and negative fast atom bombardment mass spectrometry. Mixtures from mungbean sprouts were shown by FAB/MS/MS to contain soyasaponin I as the predominant saponin. One of the unknowns has been identified as 3-*O*-[β-*D*-galactopyranosyl-(1→2)-β-*D*-glucuronopyranosyl]sophoradiol (I), molecular weight 780.

EXPERIMENTAL

All solvents used were of liquid chromatographic grade. Solvent for the FAB matrix (glycerol) was obtained from Aldrich Chemical Co. and used without further purification. Sodium chloride or silver nitrate was added to help confirm the structure of saponins.

Methanol was used to extract the saponins from defatted mungbean sprouts. The mixtures were extracted for 4 h and filtered; the filtrate was reduced in volume with a rotary evaporator, extracted three times with 1-butanol saturated with water to remove the saponins, and the 1-butanol fraction rotary-evaporated to dryness. This fraction containing crude mungbean saponins was dialyzed with 3500-MW-cut-off tubing[16]. The dialysate was analyzed using TLC, HPLC, and mass spectrometry[16].

Mass spectra were obtained with a JEOL JMS SX/SX 102A high-resolution double-focusing four-sector tandem mass spectrometer. The mass spectrometer was operated at a full accelerating voltage of 10 KeV with a standard JEOL FAB source. Xenon was used as the primary atom beam accelerated to 5 KeV with an ion current of 20 mA. Spectra were obtained with a magnet scan rate of 10 s decade^{-1}. The source pressure was typically 1.5 x 10^{-3} Pa (10^{-4} torr). Helium was used as the collision gas.

RESULTS AND DISCUSSION

Mass spectrometry has been used in many laboratories to elucidate the structure of saponins. The chemical structure of soyasaponin I is 3-*O*-[α-*L*-rhamnopyranosyl-(1→2)-β-*D*-galactopyranosyl-(1→2)-β-D-glucoronopyranosyl]-soyasapogenol B (Fig. 1). It is not easy to determine individual sugar identities, linkages, or α or β isomers in this way; however, the mass spectra are very useful for structural confirmation. The normal positive FAB mass spectrum of standard soyasaponin I is characterized by the protonated molecular ion $(M+H)^+$, *m/z* 943, and numerous fragment ions (Fig. 2). The most abundant ions observed under positive ion conditions are at *m/z* 441 and 423, corresponding to the soyasapogenol B moiety loss of one and two molecules of water, respectively. It is also possible to identify some peaks consistent with the loss of sugar moieties from soyasaponin I. The peak at *m/z* 797, $(M+H\text{–rhamnose})^+$, indicates this loss. The daughter spectrum of the protonated molecular ion $(M+H)^+$, *m/z* 943, of soyasaponin I is shown in Fig. 3. From the daughter spectrum, the fragmentation is similar to that in the normal FAB spectrum. The major ion at *m/z* 441 corresponds to the loss of OH from the aglycone (soyasapogenol B). With successive or simultaneous loss of H_2O, an ion at *m/z* 423 is produced. It must be emphasized that the presence of a peak at *m/z* 797 might arise through expulsion of a rhamnose fragment from the protonated molecular ion. Direct loss of *O*-rhamnose from the protonated molecular ion at *m/z* 943 is observed to produce the specific ion at *m/z* 781.

A number of the characteristic ions of soyasaponin I, such as those at *m/z* 943, $(M+H)^+$; *m/z* 797, $(M\text{–rha}+H)^+$; and *m/z* 981, $(M+K)^+$, were found in the positive ion FAB mass spectrum of mungbean sprouts extract (Fig. 4). In the negative ion FAB mass spectrum of mungbean sprouts extract (Fig. 5), a series of ions consistent with the presence of soyasaponin I are shown at *m/z* 941, $(M–H)^-$; *m/z* 795, $(M–rha–H)^-$; and *m/z* 663, $(M–rha–gal–H)^-$. Thus soyasaponin I is shown as the predominant saponin of mungbean sprouts. A characteristic ion at *m/z* 779 was found in the negative FAB spectrum of the mungbean sprouts extract (Fig. 5).

Under collisionally activated dissociation (CAD) conditions, fragmentation of the ion at *m/z* 781 of the mungbean sprouts in positive FAB spectrum yielded several abundant daughter ions (Fig. 6). The major abundant ions are *m/z* 619, 425, and 407. The daughter ions of the ion at *m/z* 781 of the standard soyasaponin I is shown (Fig. 7). The characteristic daughter ions of *m/z* 781 of the standard soyasaponin I are at *m/z* 635, 441, and 423. Therefore, the specific ion at *m/z* 781 may come from the other components in mungbean sprouts. The daughter ions at *m/z* 619 and 425 from the precursor ion at *m/z* 781 of the mungbean sprouts (Fig. 6) are proposed to correspond to $[M–gal+H]^+$ and $[sophoradiol–OH]^+$ respectively. The ion at *m/z* 407 is derived from ion at *m/z* 425 with loss of H_2O or with simultaneous loss of ions at *m/z* of 425 and H_2O from ion at *m/z* 781. From the daughter spectra of mungbean sprouts extract, the structure of this unknown saponin was determined as 3–*O*–[β–*D*–galactopyranosyl–(1→2)–β–*D*–glucuronopyranosyl] sophoradiol, *m/z* 780 (Fig. 8).

CONCLUSION

The saponins from mungbean (*Vigna radiata* L.) sprouts were analyzed by using positive and negative fast atom bombardment mass spectrometry. Mixtures from mungbean sprouts were shown by FAB/MS/MS to contain soyasaponin I as the predominant saponin. On the basis of tandem mass spectrometric analysis the structure of one of the unknowns from mungbean sprouts has been determined as 3–*O*–[β–*D*–galactopyranosyl–(1→2)–β–*D*–glucuronopyranosyl]sophoradiol (I), *m/z* 780. Tandem mass spectrometry provides unique structural information that was helpful in identifying the aglycones and in the determination of sugar sequences of mungbean saponins.

ACKNOWLEDGEMENT

Financial support from the National Science Council of the Republic of China, Grant No. NSC-83-0208M-005-017 is greatly appreciated.

REFERENCE

1. K.R. Price, C.L. Curl, and G.R. Fenwick, The saponin content and sapogenol composition of the seed of thirteen varieties of legume, *J. Sci. Food. Agric.* 37: 1185 (1986).

2. K.R Price, J. Eagles, and G.R. Fenwick, Saponin composition of 13 varieties of legume seed using fast atom bombardment mass spectrometry, *J. Sci. Food. Agric.* 42: 183 (1988).

3. M.R. Maillard and K. Hostettmann, Determination of saponins in crude extracts by liquid chromatography-thermospray mass spectrometry, *J. Chromatogr.* 647: 137 (1993).

4. I. Kitagawa, T. Taniyama, Y. Nagahama, K. Okubo, F. Yamauchi, and M. Yoshikawa, Saponin and sapogenol XLII. Structures of acetyl-soyasaponins A_1, A_2, and A_3, astringent partially acetylated bisdesmosides of soyasapogenol A, from American soybean, the seeds of *Glycine max* Merrill, *Chem. Pharm. Bull.* 36: 2819 (1988).

5. G.R. Waller, M. Jurzysta, and R.L. Thorne, Allelopathic activity of root saponins from alfalfa (*Medicago sativa* L.) on weeds and wheat, *Bot. Bull. Acad. Sin.* 34: 1 (1993).

6. G.R. Waller, M. Jurzysta, and R.L.Z. Thorne, Root saponins from alfalfa (*Medicago sativa* L.) and their allelopathic activity on weeds and wheat, *Allelopathy J.* **2**: 21 (1995).

7. P.R. West, P.W. Geno, G.R. Waller, W. Oleszek, and M. Jurzysta, Liquid secondary ion mass spectrometry and linked scanning at constant B/E LSIMS/MS for structure confirmation of saponins in *Medicago sativa* (alfalfa), *Saponins Used in Food and Agriculture*, G.R. Waller and K. Yamasaki, eds., Plenum Publishing Co., New York, NY. In press, (1996).

8. M.K. Lee, Y.C. Ling, M. Jurzysta, and G.R. Waller, Saponins from alfalfa, clover, and mungbeans analyzed by electrospray ionization–mass spectrometry compared with positive and negative FAB-mass spectrometry, G.R. Waller and K. Yamasaki, eds., Plenum Publishing Co., New York, NY. In press, (1996).

9. D. Fraisse, J.C. Tabet, M. Becch, and J. Raynaud, Fast atom bombardment mass spectrometry of triterpenoid saponins, *Biomed. Environ. Mass Spectrom.* 13, 1 (1986).

10. R.M. Facino, M. Carini, P. Trald, B. Belli, B. Gioia, and E. Arlandini, Confirmatory assay and quantitative determination of *Hedera helix* L. saponins (α-hederin, hederacosides B and C) in raw plant extracts and in cosmetic formulations by EI and CAD MIKES spectrometry, *Biomed. Environ. Mass Spectrom.* 14, 9 (1987).

11. Y. Chen, N. Chen, H.Li, F. Zhao, and N. Chen, Fast atom bombardment and collisional activation mass spectrometry in the structural analysis of steroidal oligoglycosides, *Biomed. Environ. Mass Spectrom.* 14, 9 (1987).

12. G.R. Waller, P.R. West, C.S. Cheng, Y.C. Ling, and C.H. Chou, The occurrence of soyasaponin I in *Vigna radiata* (mungbean) sprouts as determined by fast atom bombardment, liquid secondary ion mass spectrometry, and linked scanning at constant B/E MS/MS. *Bot. Bull. Acad. Sin.* 34: 323 (1993).

13. Y. Mil'grom, Y Rashkes, G. Fridlyanskii, and B. Voronin, Spectra of the metastable ions of steroid sapogenins and their dihydro derivatives, *Chem. Nat. Compds.* **26**, 412 (1990).

14. M. Takayama, T. Fukai, T. Nomura, and T. Yamaguchi, Formation and fragmentation of the $[M+Na]^+$ ion of glycosides in fast atom bombardment mass spectrometry, *Org. Mass Spectrom.* 26: 655 (1991).

15. C.E. Costello, Application of tandem mass spectrometry spectral approaches to structural determinations of saponins, *Saponins Used in Food and Agriculture*, G.R. Waller and K. Yamasaki, eds., Plenum Publishing Co., New York, NY. In press, (1996).

16. K. Hostettmann, Search for molluscocidal and antifungal saponins from tropical plants, in *Saponins Used in Traditional and Modern Medicine,* G.R. Waller and K. Yamasaki, eds., Plenum Publishing Co., New York, NY. In press, (1996).

LIQUID SECONDARY ION MASS SPECTROMETRY AND LINKED SCANNING AT CONSTANT B/E LSIMS/MS FOR STRUCTURE CONFIRMATION OF SAPONINS IN MEDICAGO SATIVA (ALFALFA)

P. R. West[1], G. R. Waller[2], P. W. Geno[3], W. Oleszek[4] and M. Jurzysta[4]

[1]Structural Chemistry, Department 418, AP31
Abbott Laboratories
Abbott Park, IL 60064
USA

[2]Department of Biochemistry and Molecular Biology
Oklahoma State University
Stillwater, OK 74078
USA

[3]Department of Environmental Chemistry
Southwest Research Institute
San Antonio, TX 78228
USA

[4]Department of Biochemistry
Institute of Soil Science and Plant Cultivation
Pulawy 24-100
Poland

INTRODUCTION

Mass spectrometry has been used extensively for saponin determination and structural confirmation. Sapogenins (saponin aglycones) were first studied by Budzikiewicz and coworkers using electron impact (EI) ionization.[1] Hostettmann utilized desorption/chemical ionization mass spectrometry for the structure determination of triterpenoid and spirostanol saponins.[2] Schulten and Soldati applied field desorption mass spectrometry for the confirmation structure of saponins from *Gleditsia japonica*.[3] Mostad and Doehl used liquid chromatography mass spectrometry (LC-MS) with chemical ionization (CI) for the characterization of sapogenins from *Gypsophila arrostii*.[4] Massiot et al. used californium plasma desorption mass spectrometry (^{252}Cf-PDMS) for structure elucidation of alfalfa root saponins.[5] Numerous investigators have used fast atom bombardment (FAB) or LSIMS for saponin determination.[6,7]

Tandem mass spectrometry techniques have also been used for detailed structure determination of saponins and sapogenins. Facino and coworkers studied saponins in raw plant extracts and Chen et al. examined steroidal oligoglycosides, both investigators using collisionally activated dissociation mass-analyzed ion kinetic energy spectroscopy (CADMIKES).[8,9] Fraisse et al. also utilized this technique combined with FAB for triterpenic saponins with ester-glycosides or glycosylated tertiary alcohols.[10] Crow et al. in addition to Tomer and Gross also used FAB combined with CAD MS/MS for structure determination of steroid glycosides on a triple analyzer mass spectrometer.[11,12] Linked scanning at constant B/E has been employed by Mil'grom and coworkers for the determination of metastable ions resulting from the fragmentation of steroid sapogenins.[13] This method was also used by Madhusudanan and Singh to study charge-remote fragmentation of lithiated, bariated, and deprotonated triterpenoid compounds oleanolic acid and hederagenin, which are both also sapogenins.[14] B/E linked scanning combined with FAB was also employed by Takayama et al. in a study of the formation and fragmentation of $[M+Na]^+$ ions of flavonol and steroid glycosides.[15] We have previously reported the use of CAD combined with B/E liquid secondary ion mass spectrometry/mass spectrometry (LSIMS/MS) for the structural determination of intact mungbean saponins in pure samples and crude mixtures.[16]

Recently, a new sample preparation procedure was developed by Oleszek that facilitates the separation and purification of individual alfalfa saponin compounds.[17] We have used this procedure to separate and extract saponins from alfalfa (*Medicago sativa* L.) roots.[18] In this paper, we report link-scanning at constant B/E liquid secondary ion mass spectrometry to obtain structural information on saponins. Nine saponins from alfalfa roots were analyzed by LSIMS and the protonated, deprotonated, and natriated precursor ions were subjected to CAD and then their product ion spectra were compared.

EXPERIMENTAL

Alfalfa Saponins

Pure saponin and sapogenin samples from alfalfa (*Medicago sativa*) roots were obtained by extraction using the previously described method.[18] The following sapogenin and saponin compounds (See Figure 1) were used for this study:

I. medicagenic acid.
II. 3-O-β-D-glucopyranosyl medicagenate.
III. 28-O-β-D-glucopyranosyl-3-O-β-D-glucopyranosylmedicagenate.
IV. 28-O-[β-D-xylopyranosyl-(1→4)-α-L-rhamnopyranosyl-(1→2)-α-L-arabinopyranosyl]-3-O-β-D-glucopyranosylmedicagenate.
V. 28-O-[β-D-xylopyranosyl-(1→4)-α-L-rhamnopyranosyl-(1→2)-α-L-arabinopyranosyl]-3-O-[β-D-glucopyranosyl-(1→2)-β-D-glucopyranosyl medicagenate.
VI. 3-O-β-D-glucuronopyranosylmedicagenate.
VII. 28-O-[β-D-xylopyranosyl-(1→4)-α-L-rhamnopyranosyl-(1→2)-α-L-arabinopyranosyl]-3-O-β-D-glucuronopyranosylmedicagenate.
VIII. soyasapogenol B
IX. 3-O-[α-L-rhamnopyranosyl-(1→2)-β-D-galactopyranosyl-(1→2)-β-D-glucuronopyranosyl]soyasapogenol B.

Chemical Reagents

Thioglycerol (Fluka), glycerol (Aldrich), sodium chloride (J. T. Baker).

All mass spectral figures were labeled and printed using "SpectraGraph" on an Apple Macintosh© LC III or Apple Macintosh© Classic II personal computer.[20]

Medicagenic Acid and Saponin Glycosides

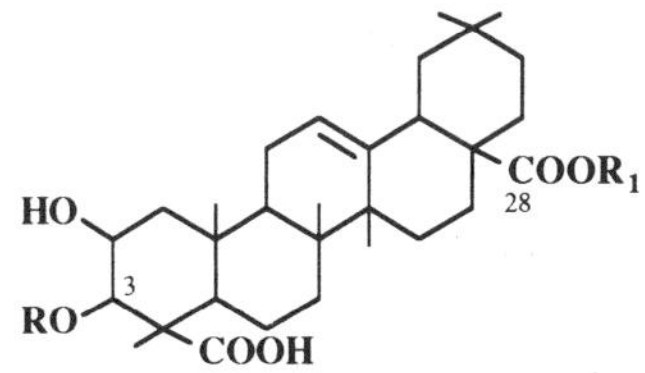

	Compound	Mol. Wt.	-R	-R1
I	Medicagenic Acid (MA)	502	-H	-H
II	3-Glc MA	664	-Glc	-H
III	3-Glc,28-Glc MA	826	-Glc	-Glc
IV	3-Glc,28-Ara Rha Xyl MA	1074	-Glc	-Ara-Rha-Xyl
V	3-GLc Glc,28-Ara Rha Xyl MA	1236	-Glc-Glc	-Ara-Rha-Xyl
VI	3-GlcA MA	678	-GlcA	-H
VII	3-GlcA, 28-Ara Rha Xyl MA	1088	-GlcA	-Ara-Rha-Xyl

Soyasapogenol B and Saponin Glycoside

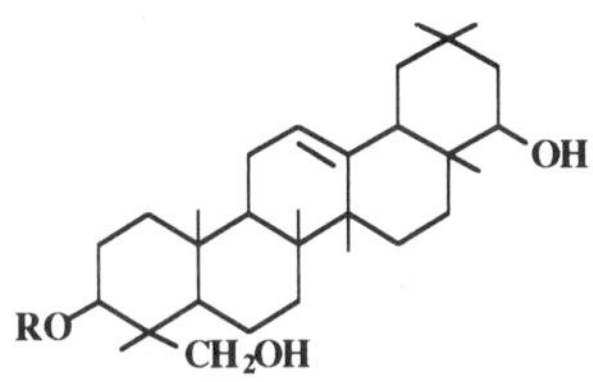

	Compound	Mol. Wt.	-R
VIII	Soyasapogenol B (SB)	458	-H
IX	3-GlcA Gal Rha SB (Soyasaponin I)	942	-GlcA-Gal-Rha

Figure 1. Structures of sapogenins and saponins, compounds I-IX.

Mass Spectrometry

Positive liquid secondary ion mass spectrometry ((+)LSIMS) and negative (–)LSIMS studies were performed on a VG ZAB2-SE mass spectrometer at 8-kV acceleration voltage, using a 35-kV Cs^+ primary ion beam. Typically, a 10- to 20- µg sample was dissolved in either a thioglycerol or 1:1 thioglycerol/glycerol matrix on the LSIMS probe tip. 1 µL of 0.1 M NaCl was added directly to the sample matrix on the probe tip when spectra of natriated ions were desired. For electron impact ionization (EI) of the sapogenins (aglycones), approximately 5 µg of the sample was introduced into the mass spectrometer EI source via direct insertion probe. The sample was then heated until volatilized at a temperature of between 250 and 300 °C. EI spectra for the sapogenins were obtained at 70-eV electron beam energy.

Linked scanning at constant B/E LSIMS/MS spectra were obtained in positive and negative mode for metastable product ions produced in the first field-free region of the instrument. This linked scanning technique was performed for both spontaneously formed product ions and those resulting from CAD of a selected precursor ion. Helium was introduced into the first field-free-region collision cell for CAD of the $[M+H]^+$, $[M-H]^+$ or $[M-H]^-$ precursor ions and argon was used for the $[M+Na]^+$ precursor ion, since it was believed that a heavier collision gas would be needed owing to the increased stability of the natriated species.[19] Sufficient collision gas was used in order to bring the pressure in the analyzer, measured near the collision cell, to about 1 x 10^{-6} mbar. This resulted in approximately a 50 to 75% reduction of the precursor ion peak intensity. All spectra resulted from the addition of 3 to 5 scans acquired at a rate of 15 s per decade in multi-channel averaging (MCA) mode, summed and mass measured by the VG 11-250 data system. Product ion spectral data were acquired by the computer-controlled simultaneous scanning of the magnetic field (B) and the electric sector field (E), while a constant B/E ratio was maintained. The data system was calibrated in positive ion mode using a 2:1 mixture of NaI and CsI, dried on the probe tip. Negative ion mode scanning was calibrated with a similar mixture of NaF and CsI.

RESULTS AND DISCUSSION

LSIMS and EI of Sapogenins

Initially, LSIM spectra were acquired for the two sapogenins, medicagenic acid and soyasapogenol B (compounds I and VIII). The sapogenins typically show a $[M-H]^+$ or $[M-H]^-$ but do not show nearly as high an affinity for sodium as the saponins. This is indicated by the lack of $[M+Na]^+$ in the sapogenin spectra obtained by using the non-natriated thioglycerol/glycerol matrix. The (+)LSIMS spectra did however show increased natriated pseudomolecular ions when sodium was added to the matrix.

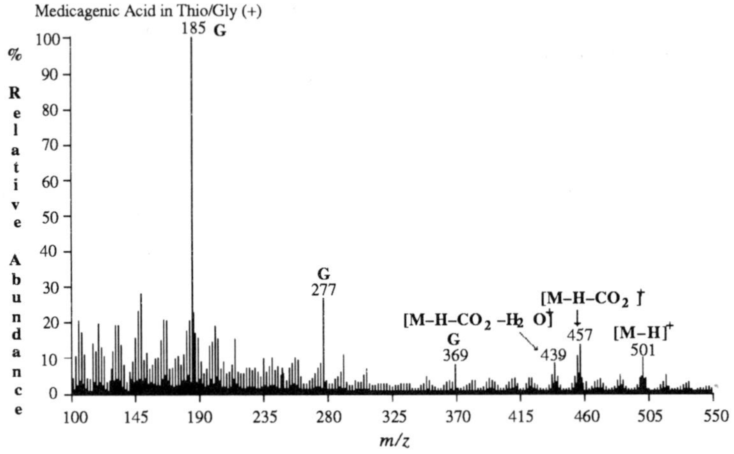

Figure 2. (+)LSIM spectrum of medicagenic acid (compound I). G denotes glycerol peaks.

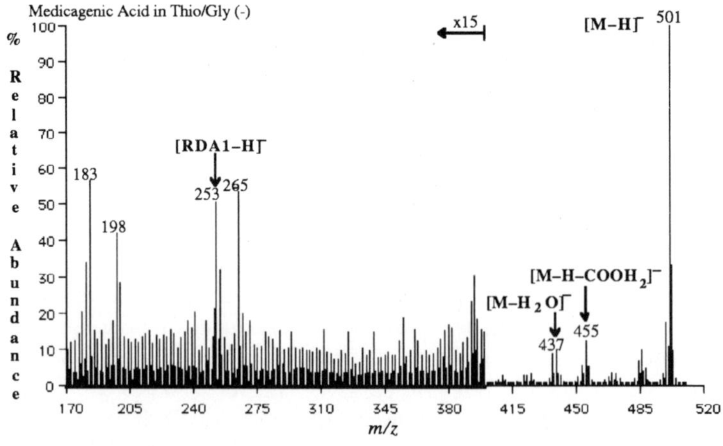

Figure 3. (-) LSIM spectrum of medicagenic acid. See Figure 4 for RDA1 Structure.

The (+)LSIM spectrum of medicagenic acid (Figure 2) showed a weak $[M–H]^+$ instead of the much more common $[M+H]^+$ generally found in most (+)LSIM spectra. Several unidentified peaks were also present at higher mass. Peaks at *m/z* 185, 277, and 369 are attributed to the glycerol/thioglycerol matrix. Other peaks of interest are due to neutral losses of CO_2 and H_2O. The (–)LSIM spectrum (Figure 3) of medicagenic acid shows an $[M–H]^-$ base peak, giving a much more abundant pseudomolecular ion than in the positive spectrum, owing to the negatively charged acid groups on the molecule. This spectrum also shows a peak at *m/z* 253 resulting from retro Diels-Alder (RDA) fragmentation. The RDA fragmentation scheme for medicagenic acid is illustrated in Figure 4.

The (+)LSIM spectrum of soyasapogenol B (Figure 5) also shows a predominant $[M–H]^+$ in addition to peaks arising from RDA fragments and neutral losses as labeled in the figure. The RDA fragmentation scheme for soyasapogenol B is shown in Figure 6. Other investigators have attributed the CAD RDA fragmentation of similar sapogenin compounds to be a charge-remote fragmentation process, with the presence of a carboxyl group as a key requirement.[13] This is a satisfactory explanation for RDA of medicagenic acid, but soyasapogenol B does not contain a carboxyl group. The possibility of charge-remote fragmentation cannot be excluded; however, we believe the RDA fragmentation of soyasapogenol B may be a result of energy deposition during ionization or localized charge as a result of hydride abstraction at another location. The (–)LSIM spectrum of soyasapogenol B showed a $[M–H]^-$ pseudomolecular ion at 10% abundance, which is much lower than for medicagenic acid.

(+)EI spectra of medicagenic acid and soyasapogenol B were also obtained for comparison to the LSIM spectra. These compounds required heating to between 250° to 300°C for volatilization. The EI spectra showed significant levels of background ions and generally less meaningful fragment information. However, some peaks corresponding to RDA fragment ions were once again present. The EI spectrum of medicagenic acid (Figure 7) did not yield a molecular ion, but intense peaks were present at *m/z* 248 and 203, corresponding to fragments $RDA2^+$ and $RDA2a^+$. Similarly, the EI spectrum of soyasapogenol B, shown in Figure 8, showed significant peaks at *m/z* 234, 224, and 219 corresponding to fragments RDA3, RDA4 and RDA3a respectively. A low abundance molecular ion at *m/z* 458 was also present in the spectrum.

LSIMS of Saponins

The (+)LSIM spectra of saponin compounds (II–VII and IX) showed $[M+H]^+$, $[M–H]^+$, and natriated pseudomolecular ions. We attribute this to the fact that the sapogenins showed primarily $[M–H]^+$, whereas for saccharides, $[M+H]^+$ and natriated species were dominant and $[M–H]^+$ was rarely present. These glycoconjugates also show abundant $[M+Na]^+$, $[M+2Na-H]^+$ and even $[M+3Na-2H]^+$ pseudomolecular ions. The abundance and number of sodium adducts generally increases relative to the number of sugar moieties present in the molecule. These results, in addition to the low abundance of natriated aglycones, agree with those of a previous investigation of flavonol and steroidal glycosides.[15] The sodium adduct abundance and number also increases proportionally with the presence of acidic sugars (glucuronic acid) on the sugar side chain(s), as shown for compounds VI, VII, and IX. For these compounds, the predominant pseudomolecular ion in the (–)LSIM spectra was $[M–H]^-$. However, the sodium affinity was so great that the (–)LSIM spectra of compounds V, VI, and IX in the non-natriated matrix also showed peaks corresponding to $[M+Na-2H]^-$, $[M+2Na-3H]^-$ and $[M+3Na-4H]^-$. It is probable that these samples initially contained varying amounts of sodium but the overall tendencies for the sodium adduct ion abundancies still held true. This was verified by obtaining (+)LSIM spectra resulting from the addition of NaCl to the matrix, where up to four sodium adducts were observed, depending upon the number and acidity of the sugars present.

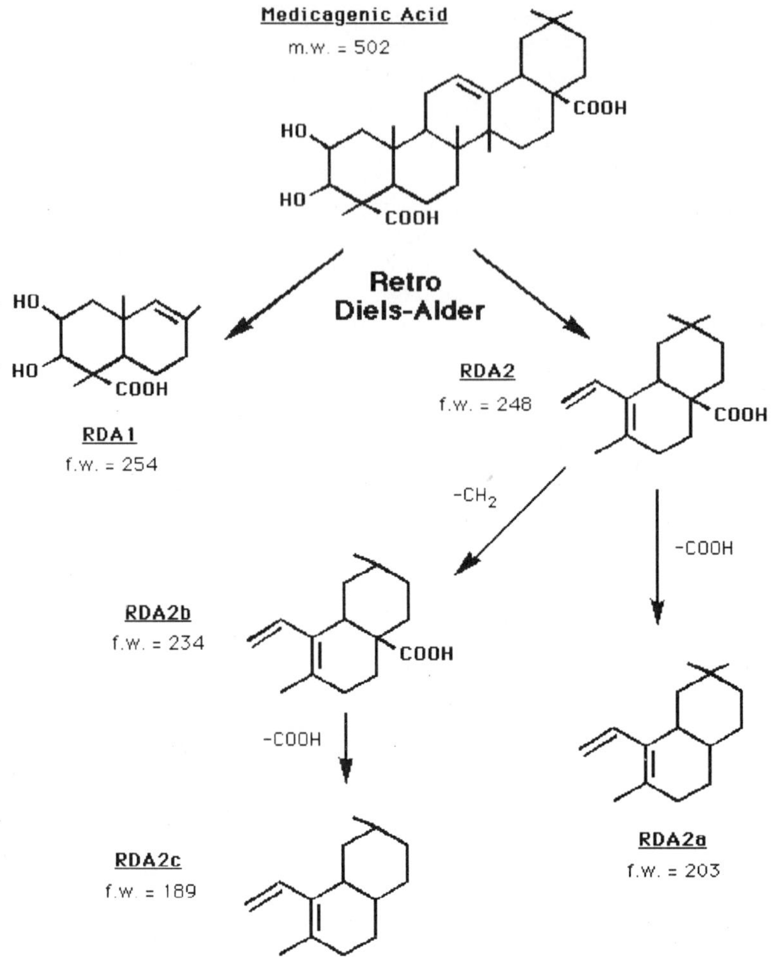

Figure 4. Retro-Diels-Alder fragmentation mechanism for medicagenic acid.

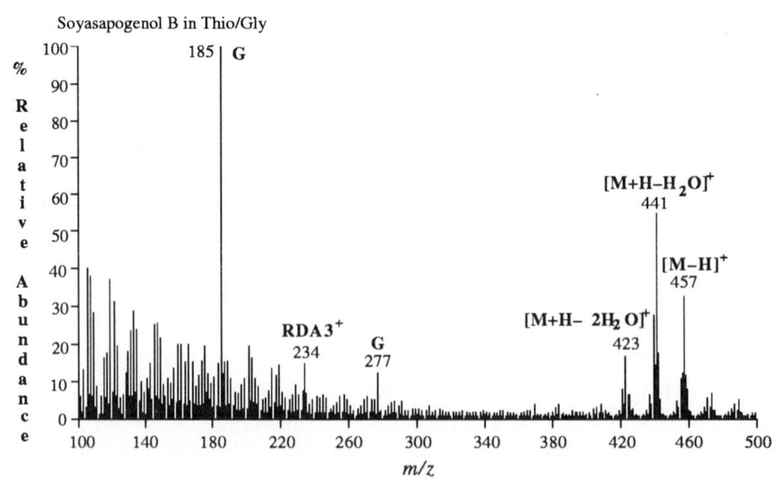

Figure 5. (+) LSIM spectrum of compound VIII. See Figure 6 for RDA structures.

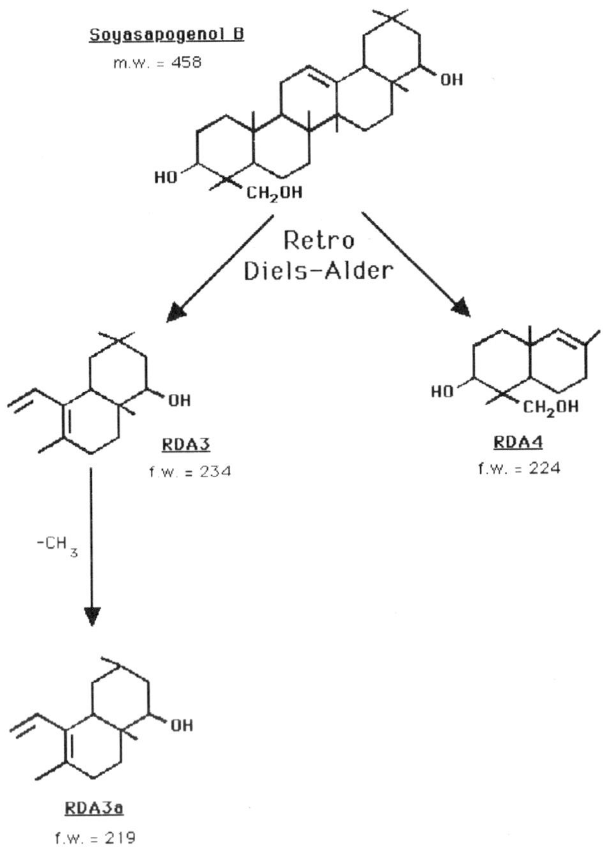

Figure 6. Retro Diels-Alder fragmentation mechanism for soyasapogenol B (compound VIII).

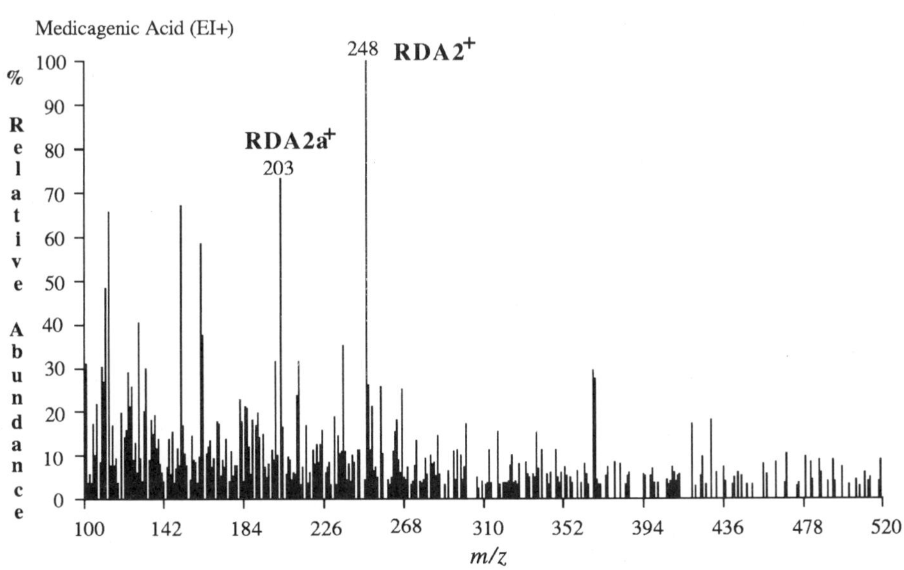

Figure 7. (+)EI spectrum of medicagenic acid (compound I). See Figure 3 for RDA structures.

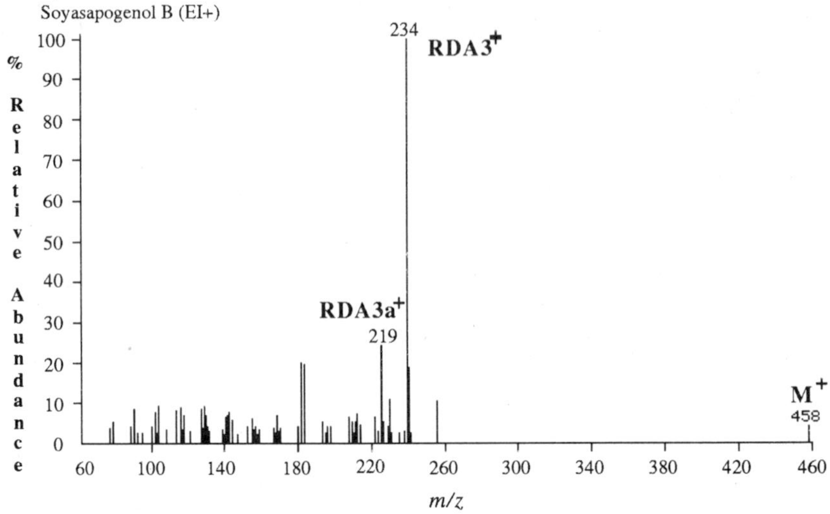

Figure 8. (+)EI spectrum of soyasapogenol B (compound VIII). See Figure 6 for RDA structures.

The LSIMS spectra also provided some information about the fragmentation of the molecular species. However intense peaks seemingly resulting from fragmentation of the molecular species (but in reality often resulting from the matrix or other independent species in the sample) often made interpretation difficult. In addition, the LSIMS spectra often lacked an intense peak for the aglycone moiety. These spectra instead usually showed peaks corresponding to the loss of CO_2, H_2O, CH_2O, or other neutral fragment from the sapogenin. This can lead to a misinterpretation of the data concerning the identity of the aglycone.

The (+)LSIMS spectra of saponin compounds II, III, IV, VI, and VII in thioglycerol/glycerol matrix are summarized in Table 1. The corresponding spectra for compounds V and IX are not included since in positive ion mode, these two compounds showed only natriated peaks and did not show significant protonated or deprotonated species. In Table 1, *m/z* values marked with an asterisk were observed only in the LSIMS/MS spectra and not the LSIMS spectra. Table 1 shows the typical losses of H_2O and CO_2 and successive losses of individual sugars resulting from cleavage at the glycosidic linkages with retention of the glycosidic oxygen atom by the species formed from the reducing terminus. Charge retention is almost always with the species containing the aglycone, therefore ions consisting of sugar fragments alone were not present. The protonated or deprotonated medicagenic acid aglycone ion was not observed with sufficient abundances. The most predominant aglycone ion peak resulted from the loss of CO_2 from the deprotonated or protonated medicagenic acid. As evident in Table 1, reliable fragmentation information cannot be obtained from the (+)LSIMS spectra of saponin compounds.

The (–)LSIMS spectra typically showed losses of H_2O and CO_2 for medicagenic acid saponins and a loss of CH_2OH for compound IX, which contains soyasapogenol B. Also evident are peaks resulting from successive losses of individual sugars resulting from cleavage at the glycosidic linkages. However, in this case, the glycosidic oxygen atom can be retained by the species formed from either the reducing or the non-reducing terminus. Unlike the situation in the positive ion spectra, charge retention is not always with the fragment containing the aglycone; therefore, ions resulting from sugar fragments were also present. The deprotonated aglycone ion is much more prevalent, especially for the medicagenic acid-containing species, owing to the presence of the negatively charged acid group. By far the most predominant medicagenic acid ion at *m/z* 439 results from the loss of both H_2O and CO_2 from the deprotonated aglycone.

Table 1: LSIMS/MS spectra for $[M+H]^+$ or $[M-H]^+$ precursor ions of compounds II, III, IV, IV, VI, and VII. Data for compounds V and IX are not shown owing to a predominance of natriated and lack of protonated or deprotonated ions. All listed *m/z* values were present in LSIMS/MS spectra; those marked with an asterisk (*) were not present in the (+)LSIMS spectra. Approximate abundances are indicated as: strong = bold text; moderate = normal text; weak = italicized text. Pen = pentose; Hex - hexose; HexA = hexuronic acid; Dehex = deoxyhexose. The aglycone (a) is medicagenic acid (MA) or soyasapogenol B (SB). p = parent ion or pseudomolecular ion.

Compound	II	III	IV	VI	VII
Precursor Ion (p) $[M+H]^+$		**827**			**1089**
Precursor Ion (p) $[M-H]^+$	**663**		**1073**	**677**	
Fragment or Product Ion					
p–H_2O	**645**	809*	**1055***	**659**	**1071**
P–$2H_2O$			1037		1053
p–$COOH_2$	619			**631***	
p–$COOH_2$–H_2O	*601**			*615**	
p–Pen			**941***		**957**
p–Hex		**665**	**913***		
p–HexA					913
p–Pen–Dehex			*795**		811
p–Pen–Hex			*779**		
p–Pen–HexA					*781*
p–Pen–Dehex–Pen			*663**		**679**
p–Pen–Dehex–Hex			*633**		
Aglycone (a)	(MA)	(MA)	(MA)	(MA)	(MA)
$[a+H]^+$		**503**			*503*
$[a-H]^+$	**501**		*501*	**501**	
$[a-H_2O+H]^+$		*485*			
$[a-H_2O-H]^+$	**483**			**483***	
$[a-COOH]^+$	**457**	457			
$[a-COOH_2-H]^+$			455	**455**	**455**
$[a-H_2O-COOH]^+$	439	*439*			*439*
$[a-2H_2O-COOH]^+$	*421*		*421*	*421*	*421*

(+)LSIMS spectra were also obtained for saponin compounds after the addition of NaCl to the matrix. Unlike the (+)LSIMS and (–)LSIMS spectra without the addition of sodium, the spectra from a natriated matrix do not show the typical losses of H_2O and CO_2. Successive losses of individual sugars were not always observed. The observed losses resulted from cleavage at the glycosidic linkages with the glycosidic oxygen atom retained by the ion formed from either the reducing or the non-reducing terminus. Charge retention is always with the species or fragment coordinated with the sodium ion. This resulted in the prevalent formation of fragment ions consisting of only sugars, or sugar fragments attached to the aglycone. No natriated aglycone ion(s) were observed without sugars. As a result, structural information about the aglycone could not be determined. The overall fragmentation information from the natriated spectra without the aid of CAD is inconsistent at best, with the spectra being dominated by matrix peaks at lower masses. LSIMS/MS of Sapogenins and Saponins

In order to obtain mass spectra with reduced matrix peaks and other peaks not resultant from the analyte of interest, and in order to obtain reproducible and structurally significant product ion spectra, linked scanning at constant B/E MS/MS was used. Tandem mass spectra (MS/MS) of the metastable ions produced by fragmentation of the $[M+H]^+$ or $[M-H]^+$, $[M+Na]^+$ and $[M-H]^-$ precursors for compounds I through IX were acquired, both with and without collisionally activated dissociation (CAD). The CAD MS/MS spectra generally showed a greater number of more abundant product ions than the metastable ion spectra without CAD. This difference in product ion abundance was especially pronounced in the product ion spectra of the natriated precursors of these compounds compared to those of the protonated precursors. Without the aid of CAD, the MS/MS spectra of the $[M+Na]^+$ metastable ions showed virtually no product ions. However, the corresponding $[M+H]^+$ or $[M-H]^+$ MS/MS spectra showed some meaningful fragmentation. This indicates that the natriated saponin precursor ions were more stable and less likely to fragment than their protonated counterparts.[19]

Comparison of (+) and (–) LSIMS and LSIMS for compound III is shown in Figure 9. A significant reduction in background peaks for the B/E CAD LSIMS/MS spectrum is observed when compared to LSIMS. Also, there is an increase in structurally meaningful fragmentation, particularly for the natriated species.

The B/E CAD (+)LSIMS/MS product ion spectrum of the sapogenin medicagenic acid $[M-H]^+$ (Figure 10) shows many structurally significant peaks. These include several product ions resulting from RDA fragmentation, similar to those in the EI spectrum. Other informative peaks include losses of CO_2 and H_2O, similar to those present in the (+)LSIM spectrum. The B/E CAD MS/MS spectrum of soyasapogenol B $[M-H]^+$ also showed several RDA product ion peaks. The B/E CAD LSIMS/MS data for medicagenic acid and soyasapogenol B show that the CAD spectra of the $[M-H]^+$ precursors generally provided more structural information than the $[M-H]^-$ product ion spectra. As stated earlier, the aglycones showed little affinity for sodium and as a result the $[M+Na]^+$ product ion spectra of these compounds yielded little useful information.

The B/E product ion spectra resulting from CAD of the $[M+H]^+$ or $[M-H]^+$ saponin precursor ions (not present from compounds V and IX owing to the predominance of natriated species) showed losses of H_2O and CO_2 similar to the LSIM spectra. Also evident were successive losses of individual sugars resulting from cleavage at the glycosidic linkages, with retention of the glycosidic oxygen atom by the ion formed from the reducing terminus. Charge is again retained on the fragment containing the aglycone, and ions consisting of sugar fragments alone were not observed. A protonated or deprotonated aglycone (medicagenic acid) ion was not always observed. The most predominant aglycone ion resulted from the loss of CO_2 from the deprotonated or protonated medicagenic acid. The number of structurally significant fragment ions is generally greater in the MS/MS spectra than in the corresponding LSIMS spectra.

B/E CAD (–)LSIMS/MS product ion spectra for the saponin $[M-H]^-$ precursor ions showed the typical losses of H_2O and CO_2 for medicagenic acid saponins and a loss of CH_2OH for compound IX containing soyasapogenol B. Interestingly, a neutral loss of CH_3 from the $[M-H]^-$ precursor is present in all B/E CAD (–) LSIMS/MS spectra. It would be expected that this would be the case in the positive spectrum, owing to the formation of a stable tertiary carbocation. However, this is not evident. Also shown were successive losses of individual sugars resulting from cleavage at the glycosidic linkages with retention of the glycosidic oxygen atom by the species formed from either the reducing or the non-reducing terminus. As in the (–)LSIMS spectra, charge retention was shown to be on both the species with and without the aglycone. This resulted in the observation of sugar fragment ion peaks. In many cases the deprotonated aglycone ion is predominant, especially for the medicagenic acid-containing species. The most abundant medicagenic acid ion is at *m/z* 439, resulting from the loss of both H_2O and CO_2 from the deprotonated aglycone. Product ions were also

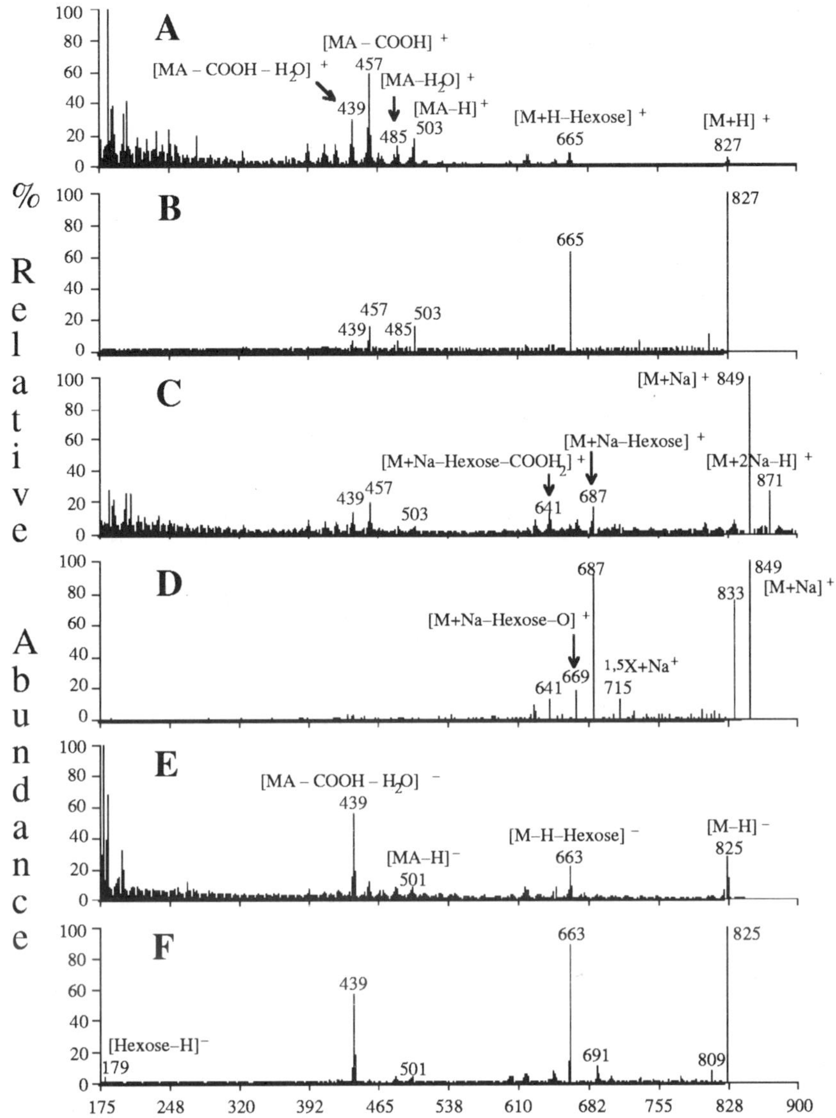

Figure 9. Mass spectra of compound III. **A**. (+)LSIM spectrum. **B**. B/E CAD LSIMS/MS of $[M+H]^+$ precursor. **C**. (+)LSIM spectrum with 0.1 M NaCl added to matrix. **D**. B/E CAD LSIMS/MS of $[M+Na]^+$ precursor. **E**. (-)LSIM spectrum. **F**. B/E CAD LSIMS/MS of $[M-H]^-$ precursor. (MA denotes medicagenic acid).

present corresponding to RDA fragments with an entire sugar side chain attached. Fragments resulting from an energetically unfavorable retro Diels-alder fragmentation followed by an additional loss of a sugar side chain were not present. Overall, the B/E CAD LSIMS/MS product ion spectra had a higher signal-to-background ratio for the $[M-H]^-$ precursors than for their $[M-H]^+$ or $[M+H]^+$ counterparts.

A minor problem was encountered due to the presence of both an $[M+H]^+$ and a $[M-H]^+$ ion for some of the saponins. The instrument was tuned to a resolution of 1000; however, as is often the case for B/E product ion spectra, product ion peaks were occasionally present in the $[M+H]^+$ spectra resulting from the fragmentation of the nearby $[M-H]^+$ and vice versa. Another problem encountered was the presence of peaks resulting from the fragmentation of a matrix ion at the same mass as the desired saponin precursor ion. This

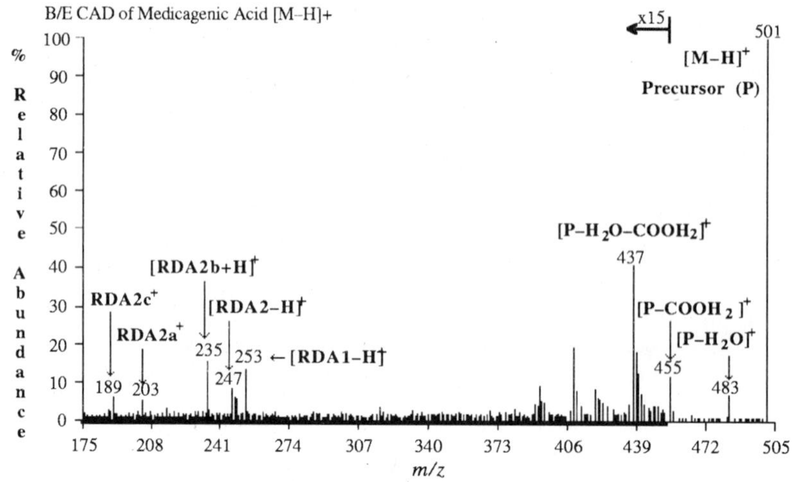

Figure 10. B/E CAD LSIMS/MS product ion spectrum of medicagenic acid [M-H] precursor ion.

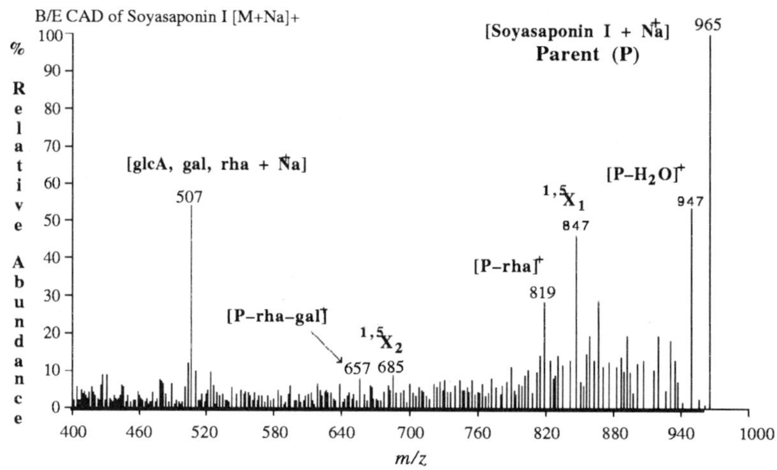

Figure 11. B/E CAD LSIMS/MS product ion spectrum of medicagenic acid $[M+Na]^+$ precursor ion.

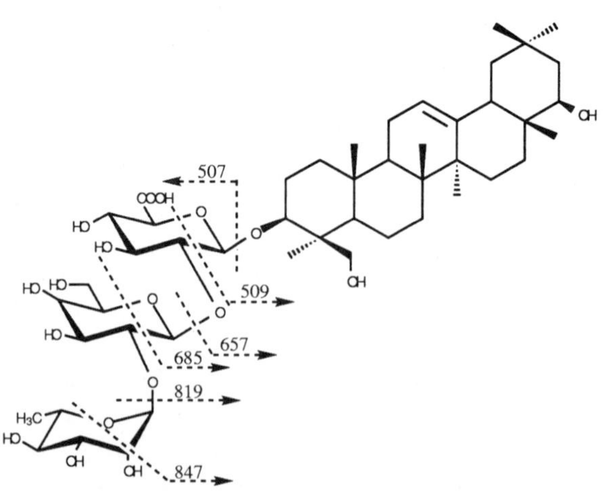

Figure 12. Fragmentation pattern for compound IX consistent with *m/z* values observed in the B/E CAD LSIMS/MS product ion spectrum of the $[M-Na]^+$ precusor. Charge retention is indicated by the direction of the arrows.

resulted in intense peaks arising from neutral losses corresponding to the mass of the matrix (usually glycerol) or peaks at masses corresponding to stable matrix cluster ions. This was particularly confusing for the neutral loss of glycerol (92 amu) in the spectra for medicagenic acid saponins since 92 amu also corresponds to $(COOH_2)_2$.

The B/E CAD LSIMS/MS product ion spectra of the natriated saponin precursor ions showed the loss of CH_3–H in addition to the losses of H_2O and CO_2, the latter two being similar to those in the $[M-H]^-$ product ion spectra. Successive losses of individual sugars resulting from cleavage at the glycosidic linkages were shown, as usual with the glycosidic oxygen atom retained by either the reducing or the non reducing terminus. In addition to glycosidic bond ruptures, $^{1,5}X$ fragment ions (in the nomenclature as suggested by Domon and Costello) were present.[21] These product ions, resulting from two-bond sugar ring cleavages, were not present in the MS/MS spectra of the protonated or deprotonated species. The formation of natriated fragment ions consisting of only sugars or sugar fragments attached to the aglycone and relatively few natriated aglycone ions, indicates that charge retention is usually on the fragment coordinated with the sodium ion, which predominantly contains one or more sugar moieties.

Peaks corresponding to the natriation of complete sugar side chains were often quite intense in the product ion spectra of the natriated saponin precursors. An excellent example of this is shown in the CAD spectrum for the $[M+Na]^+$ precursor ion of compound IX (soyasaponin I) (Figure 11). Figure 12 shows the corresponding fragmentation pattern for soyasaponin I $[M+Na]^+$. In addition to peaks corresponding to the loss of rhamnose at *m/z* 819 and the additional loss of galactose at *m/z* 657, the predominant fragment ion at *m/z* 507 results from the sodium adduct ion of the entire –GlcA–Gal–Rha side chain. This indicates that the sodium is coordinated to soyasaponin I at one or more glycosidic bonds along the sugar chain and not at the aglycone. B/E spectra typically show a reduction in intensity proportional to the reduction in energy. The *m/z* 507 peak is very intense, especially considering that the energy of this product ion is only a little more than half that of the precursor ion. Peaks at *m/z* 685 and *m/z* 847 show $^{1,5}X$ fragmentation of the individual sugar rings.

The predominance of sugar-containing natriated product ions and $^{1,5}X$ fragment ions and the lack of natriated aglycone ions indicates that the sodium ion is coordinated principally at a glycosidic linkage on a sugar side chain and not to the aglycone. Similar conclusions were drawn by Takayama et al. for steroidal and flavonol glycosides.[15] A mechanism for the sodium-directed fragmentation has previously been proposed by Orlando et al.[22]

CONCLUSIONS

In summary, these results show that LSIMS and linked scanning at constant B/E LSIMS/MS can be used for structure confirmation of saponins and sapogenins in alfalfa roots. This research demonstrates that when positive and negative LSIMS and LSIMS/MS are performed on the protonated, deprotonated, and natriated molecular ions, valuable structural information can be obtained by observing the differences in fragmentation patterns. The addition of sodium to the matrix intensifies the various natriated molecular ions and the LSIMS/MS spectra of these species show fragmentation patterns different from those of the protonated or deprotonated species, including two-bond sugar ring cleavages and abundant natriated sugar side chain product ions. These results are principally due to the significant differences in sodium affinities exhibited by the sapogenin and the saccharide side chains.

ACKNOWLEDGMENTS

The assistance of A. J. Mort and Otis Dermer is appreciated for critically reviewing this manuscript, and Colette Spriggs for its preparation. This research was supported by the Oklahoma Agricultural Experiment Station, Oklahoma State University, Stillwater, OK 74078. We also thank the Oklahoma State University Department of Chemistry in addition to the National Science Foundation for Grant BBS-8704089, which aided in the purchase of the mass spectrometer, and Grant DMB-9006052, which helped to fund this study.

REFERENCES

1. H. J. Budzikiewicz, J. M. Wilson and C. Djerassi, Mass spectrometry in structural and stereochemical problems. XXXII. Pentacyclic triterpenes, *J. Am. Chem. Soc.* 85:3688 (1963).
2. K. Hostettmann, J. Doumas and M. Hardy, Desorption/chemical ionization mass spectrometry of naturally occurring glycosides, *Helv. Chim. Acta* 64:297 (1981).
3. H.-R Schulten and F. Soldati, Identification of ginsenosides from *panax ginseng* in fractions obtained by high-performance liquid chromatography by field desorption mass spectrometry, multiple internal reflection infrared spectroscopy and thin-layer chromatography, *Planta Med.* 46:67 (1982).
4. H. Mostad and J. Doehl, Separation and characterization of oleanene-type-pentacyclic triterpenes from *gypsophila arrostii* by liquid chromatography mass spectrometry, *J. Chromatogr.* 396:157 (1987).
5. G. Massiot, C. Lavaud, L. Le Men-Olivier, G. Van Binst, S. Miller and H. Fales, Structural elucidation of alfalfa root saponins by mass spectrometry and nuclear magnetic resonance analysis, *J. Chem. Soc. Perkin Trans.* 12:3071 (1988).
6. K. R. Price, J. Eagles, and G. R. Fenwick, Saponin composition of 13 varieties of legume seed using fast atom bombardment mass spectrometry, *J. Sci. Food Agric.* 42:183 (1988).
7. G. R. Waller, M. Jurzysta and R. L. Z. Thorne, Allelopathic activity of root saponins from alfalfa (*Medicago sativa* L.) on weeds and wheat, *Bot. Bull. Acad. Sin.* 34:1 (1993).
8. R. M. Facino, M. Carini, P. Traldi, B. Belli, B. Gioia and E. Arlandini, Confirmative assay and quantitative determination of *hedera helix* L. saponins (α-hederin, hederacosides B and C) in raw plant extracts and in cosmetic formulations by EI and CAD MIKE spectrometry, *Biomed. Environ. Mass Spectrom.* 14:187 (1987).
9. Y. Chen, N. Chen, H. Li, F. Zhao and N. Chen, Fast atom bombardment and collisional activation mass spectrometry in the structural analysis of steroidal oligoglycosides, *Biomed. Environ. Mass.Spectrom.* 14:9 (1987).
10. D. Fraisse, J. C. Tabet, M. Becchi and J. Raynaud, Fast atom bombardment mass spectrometry of triterpenic saponins, *Biomed. Environ. Mass Spectrom.* 13:1 (1986).
11. F. Crow, K. B. Tomer, J. Looker and M. L. Gross, Fast atom bombardment and tandem mass spectrometry for structure determination of steroid and flavonoid glycosides, *Anal. Biochem.* 155:286 (1986).
12. K. B. Tomer and M. L. Gross., Fast atom bombardment and tandem mass spectrometry for structure determination: remote site fragmentation of steroid conjugates and bile salts, *Biomed. Environ. Mass Spectrom.* 15:89 (1988).
13. Y. Mil'grom, Y. Rashkes, G. Fridlyanskii and B. Voronin, Spectra of the metastable ions on steroid sapogenins and their dihydro derivatives, *Chem. Nat. Compds.* 26:412 (1990).
14. K. P. Madhusudanan and C. Singh, Collisional activation of metal cationized and deprotonated triterpenoids: charge remote fragmentations, *Org. Mass Spectrom.* 27:1329 (1992).
15. M. Takayama, T. Fukai, T. Nomura, and T. Yamaguchi, Formation and fragmentation of the $[M+Na]^+$ ion of glycosides in fast atom bombardment mass spectrometry, *Org. Mass Spectrom.* 26:655 (1991).
16. G. R. Waller, P.R. West, C. S. Cheng, Y. C. Ling and C. H. Chou, The occurrence of soyasaponin I in *vigna radiata* L. (mungbean) sprouts of as determined by fast atom bombardment, liquid secondary ion mass spectrometry, and linked scanning at constant B/E MS/MS, *Bot. Bull. Acad. Sin.* 34:323 (1993).
17. W. Oleszek, Solid-phase extraction-fractionation of alfalfa saponins, *J. Sci. Food Agric.* 44:43 (1988).
18. W. Oleszek, K. R. Price, I. J. Colquhoun, M. Jurzysta, M. Ploszynski and G R. Fenwick, Isolation and identification of alfalfa (*Medicago sativa* L.) root saponins: their activity in relation to a fungal bioassay, *J. Agric. Food Chem.* 38:1810 (1990).
19. J. Adams and M. L. Gross, Energy requirements for remote charge site ion decompositions and structural information from collisional activation of alkali metal cationized fatty alcohols, *J. Am. Chem. Soc.* 108:6915 (1986).
20. P. R. West and A. J. Mort, SpectraGraph and SpectraSort: mass spectral display and interpretation software for the Macintosh, *J. Chem. Inf. Comput. Sci.* 33:234 (1993).
21. B. Domon and C. Costello, A systematic nomenclature for carbohydrate fragmentations in FAB-MS/MS spectra of glycoconjugates, *Glycoconjugate* 5:397 (1988).
22. R. Orlando, C. Bush and C. Fenselau, Structural analysis of oligosaccharides by tandem mass spectrometry: collisional activation of sodium adduct ions, *Biomed. Environ. Mass Spectrom.* 19:747 (1990).

SAPONINS FROM ALFALFA, CLOVER, AND MUNGBEANS ANALYZED BY ELECTROSPRAY IONIZATION-MASS SPECTROMETRY AS COMPARED WITH POSITIVE AND NEGATIVE FAB-MASS SPECTROMETRY

M.K. Lee,[1] Y.C. Ling,[1] M. Jurzysta,[2] and G.R. Waller[3]

[1]Department of Chemistry
National Tsing Hua University
Hsinchu 30043, Taiwan, R.O.C.
[2]Institute of Soil Science and Plant Cultivation
Department of Biochemistry
Pulawy, Poland
[3]Department of Biochemistry and Molecular Biology
Oklahoma State University
Stillwater, OK 74078, USA

INTRODUCTION

Saponins are mixtures of several triterpenoid glycosides and can serve as allelochemicals. The variety of biological properties inherent in saponins has attracted interest in the development of new analytical methodologies. Mass spectrometry has been used for determination and structural characterization of saponins. They were ionized by ionization modes such as electron impact (EI)[1], chemical ionization (CI)[2], field desorption (FD)[3], Californium plasma desorption (^{252}Cf-PDMS)[4], fast atom bombardment (FAB)[5,6], and energetic ions (LSIMS)[7] coupled with liquid chromatography using thermospray (TSP) ionization[8] and reported in this book by ESI[9,10]. In this paper, we reported using electrospray ionization (ESI) to obtain structural information on saponins. Samples of soyasaponins from mungbeans, *Vigna radiata* L. (**I** and **II**), root saponins from *Medicago sativa* L. Cimarron cultivar grown in Oklahoma (**III**), and seed saponins from *Trifolium incarnatum* L. (**IV-VII**) were analyzed by ESI/MS and FAB/MS. The results were compared with the positive and negative FAB to illustrate the potentials of using ESI/MS to characterize saponins.

Table 1. Ion peaks and structures in (–)FAB/MS and ESI/MS spectra of sample III[1,2,3]

FAB(–)		ESI(+)		
m/z	structure	*m/z*	structure	M.W.
663	[***A***–H]⁻	687	[***A***+Na]⁺	664
		701	[***B***+Na]⁺	678
765	[***C***–H]⁻	789	[***C***+Na]⁺	766
825	[***D***–H]⁻	849	[***D***+Na]⁺	826
897	[***E***–H]⁻	921	[***E***+Na]⁺	898
941	[***SSI***–H]⁻	965	[***SSI***+Na]⁺	942
		981	[***SSI***+K]⁺	
1059	[***F***–H]⁻	1083	[***F***+Na]⁺	1060
1073	[***G***–H]⁻	1097	[***G***+Na]⁺	1074
1087	[***H***–H]⁻	1111	[***H***+Na]⁺	1088
		1127	[***H***+K]⁺	
1149	[***I***–H]⁻	1173	[***I***+Na]⁺	1150
1235	[***J***–H]⁻	1259	[***J***+Na]⁺	1236
		1275	[***J***+K]⁺	
		1421	[***K***+Na]⁺	1398

[1]Structures are shown in Fig. 3: A = 3-Glc MA, ***B*** = 3-GlcA MA, ***C***= 3 -Ara Glc(H), ***D*** = 3-Glc,28-Glc MA, ***E*** = 3-Ara Glc Ara(H), ***G*** = 3-Glc,28-Ara Rha Xyl MA, ***H*** = 3-Glc***A***,28Ara Rha Xyl MA, ***J*** = 3-Glc Glc,28-Ara Rha Xyl MA
[2]MA = Medicagenic Acid, H = Hederagenin, SSI = Soyasaponin I
[3]Structures ***F***, ***I***, ***K*** have not been identified

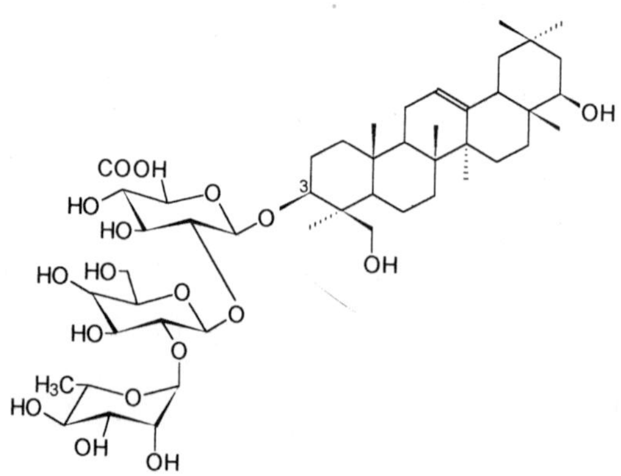

Figure 1. Structure of Sample I (Soyasaponin I).

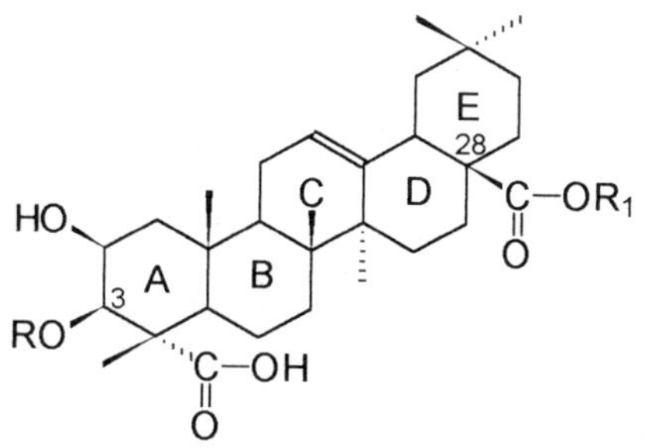

(a) Medicagenic Acid and Its Glycosides

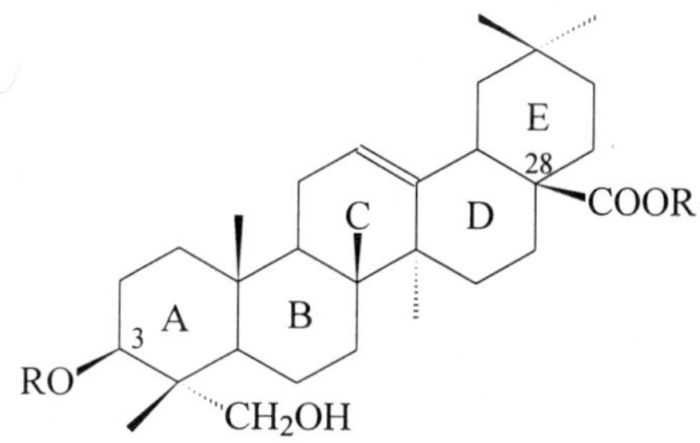

(b) Hederagenin and Its Glycosides

Comp. No.	Compound	Mol. Wt.	-R	-R1
–	Medicagenic Acid (MA)	502	-H	-H
III A –	3-Glc MA	664	-Glc	-H
III G –	3-Glc,28-Glc MA	826	-Glc	-Gla
III F –	3-Glc,28-Ara Rha Xyl MA	1074	-Glc	-Ara-Rha-Xyl
III J –	3-Glc Glc,28-Ara Rha Xyl MA	1236	-Glc-Glc	-Ara-Rha-Xyl
III B –	3-GlcA MA	678	-GlcA	-H
III H –	3-GlcA,28-Ara Rha Xyl MA	1088	-GlcA	-Ara-Rha-Xyl
–	Soyasapogenol B	458	-H	
III E –	3-GlcA Gal Rha SB (Soyasaponin I)	942	-GlcA-Gal-Rha	
–	Hederagenin (H)	488	-H	-H
III D –	3-Ara Glc Ara (H)	898	-Ara-Glc-Ara	-H
III C –	3-Ara Glc (H)	766	-Ara-Glc	-H

Figure 2a,b. Structure of major saponins isolated from alfalfa 'Cimarron' roots as determined by LSIMS (5).

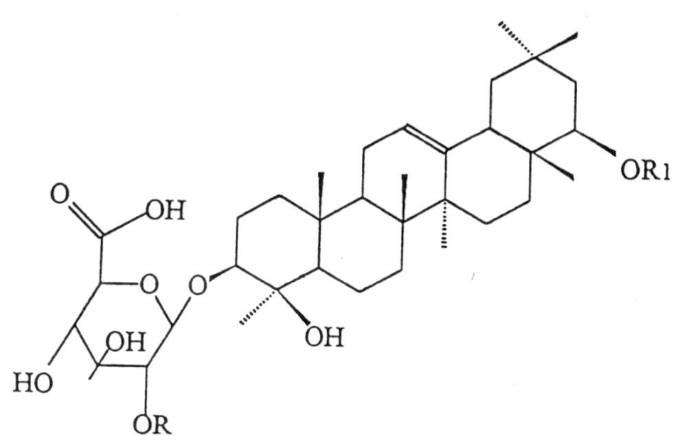

Figure 3. Structures of pure seed saponins from clover, *Trifolium incarnatum* L., samples IV-VII; IV. R = H, R_1 = H; V. R = β-D-Gal(1→2), R_1 = H; VI. R = α-L-Rha(1→2) β-D-Gal(1→2), R_1 = H; VII. R = α -L-Rha(1→2) β-D-Gal(1→2), R_1 = β-D-Gal(1→2) β-D-Glu(1→2).

EXPERIMENTAL

Source of Saponins

The saponins were obtained by extraction and purified by using the previously described methods for alfalfa roots [6] and clover[11]. The soyasaponin I in H^+ and Na^+ form were gifts from K.R. Price. The following saponin samples were investigated, and the chemical structure of sample **I** is shown in Figure 1; those of samples **III** are shown in Figure 2; those of samples **IV-VII** are shown in Figure 3:

I. Soyasaponin I (from mungbeans *Vigna radiata* L.) in H^+ form; 3-O-(α–L-rhamnopyranosyl(1→2)β-D-galactopyranosyl)(1→2)-β-D-glucuronopyranosyl) olean-12-ene-22, 24 diol;

II. Soyasaponin I (from mungbeans, *Vigna radiata* L.) in Na^+ form;

III. Root saponins from *Medicago sativa* L. Cimarron cultivar grown in Oklahoma;

IV. Saponin from clover, *Trifolium incarnatum* L., 3-O-(β-D-glucuronopyranosyl)-olean-12-ene-22, 24-diol;

V. Saponin from clover, *Trifolium incarnatum* L., 3-O-(β-D-galactopyranosyl)(1→2)-β-D-glucuronopyranosyl)olean-12-ene-22,24-diol (soyasaponin III);

VI. Saponin from clover, *Trifolium incarnatum* L., 3-O-(α-L-rhamnopyranosyl(1→2)-β-D-galactopyranosyl)(1→2)-β-D-glucuronopyranosyl)olean-12-ene-22,24-diol (soyasaponin I);

VII. Saponin from clover, *Trifolium incarnatum* L.., 3-O-[α-L-rhamnopyranosyl (1→2)-β-D-galactopyranosyl(1→2)-β-D-glucuronopyranosyl]-β-22-O-[β-D-glucopyranosyl-(1→2) -β-D-glucopyranosyl]olean-12-ene-24-ol.

Analysis by Mass Spectrometry

About 5 mg of sample was dissolved in 5 ml MeOH. Aliquots of the MeOH (~5 μl) solution were used for ESI/MS analyses, and mixed with glycerol before FAB/MS analyses.

ESI/MS experiments were conducted on a reversed-geometry double-focusing mass spectrometer (JMS SX-102A, JEOL, Tokyo, Japan) equipped with an ESI source (Analytica, Branford, CT, USA). Samples were introduced into the ESI source at a flow rate of 1 μl/min with a syringe pump (Harvard Apparatus model 22, South Natick, MA, USA). The sample solution was electrosprayed through a triple-layer needle. The electrospray needle potential (V_m) was held at 3.0 kV. A coaxial nitrogen flow through the middle layer along the inner needle sprayer was provided to assist nebulization of the solution. N_2 drying gas heated to 120 °C was introduced against the needle flow at a flow rate of 20 L/min. The potential difference between the capillary and skimmer 1 (V_{cs}) was held at ground. Mass spectra were recorded in the range 100 - 3500 *m/z* by scanning the magnetic field in 20 s with mass resolution 1000 and an accelerating voltage of 7 kV. The scans were averaged for 5 min to yield the final spectra. Detailed experimental conditions were described elsewhere[12].

FAB/MS experiments were conducted on the same instrument using 6.0-kV Xe atoms and glycerol matrix. Samples were homogeneously mixed with the glycerol on the FAB probe before mass analyses. Mass spectra were recorded in the range 100 - 1500 *m/z* by scanning the magnetic field in 10 s with mass resolution 1000 and an accelerating voltage of 10 kV.

RESULTS

Soyasaponin in H^+ form (Sample I) and Soyasaponin in Na^+ form (Sample II)

The ESI/MS spectrum of **I** (Fig. 4a) typically shows a strong protonated ion $[M+H]^+$ and adduct molecular ions $[M+Na]^+$ and $[M+K]^+$. Weak doubly charged ions of $[3M+2Na]^{2+}$ at *m/z* 1436, $[4M+2Na]^{2+}$ at *m/z* 1908, and $[5M+2Na]^{2+}$ at *m/z* 2378 also appear. By combining with many clusters of molecular ions of $[2M+Na]^+$ at *m/z* 1908, and $[3M+Na]^+$ at *m/z* 2852, the molecular weight of **I** could be readily determined to be 942. The (+)FAB/MS spectrum (Fig. 5a) shows $[M+H]^+$ at *m/z* 943, $[M–Rha+H]^+$ at *m/z* 797, $[M–Rha-Gal+H]^+$ at *m/z* 635, [soyasapogenol $B–OH]^+$ at *m/z* 441, and [soyasapogenol $B–OH–H_2O]^+$ at *m/z* 423. The other peaks are from matrices. The (–)FAB/MS spectrum (Fig. 5c) shows $[M–H]^-$ at *m/z* 941, $[M–Rha–H]^-$ at *m/z* 795, and $[M–Rha–Gal–H]^-$ at *m/z* 633. The other peaks are from matrices.

Sample **II** was obtained by replacing one H in **I** by Na. Its ESI/MS spectrum (Fig. 4b) is dominated by the $[M+Na]^+$ at *m/z* 987, and no $[M+H]^+$ is observed, indicating that the H in **I** is indeed replaced by the Na. The presence of $[2M+Na]^+$ at *m/z* 1953 indicates that **II** has higher affinity towards Na. The (+)-FAB/MS spectrum of **II** (Fig. 5b) appears similar to that of **I** (Fig. 5a), except the appearance of $[M+Na]^+$ at *m/z* 965. The reduced abundance of $[M+H]^+$ at *m/z* 943 is presumably due to the exchange of H in soyasaponin for Na. The (–)FAB/MS spectrum (Fig. 5d) is similar to that of **I** (5c) except that $[M–H]^-$ at *m/z* 941 becomes the base peak.

Alfalfa Root Saponins (Sample III)

The (+)FAB/MS and (–)FAB/MS spectra of **III** shown in Figs. 6a and 6b are quite different, indicating that the (–)FAB/MS spectra provide more structural information. This indicates that the composition of **III** is rather complicated. This is shown in the corresponding ESI/MS spectrum (Fig. 6c) as evidenced by the many ions. To minimize the interference from impurities formed by the attachment of small molecules or ions to the molecular ions, we increased V_{cs} from ground to 200 V to decompose the impurities. The resultant ESI/MS spectrum is shown in Fig. 7. Peaks representing the major components in **III** are retained. The reduced abundance at *m/z* 1097 ion (**III** G+Na)$^+$ indicates that this component is structurally different from other components. It is less stable under high capillary/skimmer voltage. This verifies that **III** indeed contains different kinds of saponins. The combined (–)FAB/MS and ESI/MS results are listed in Table 1.

Clover Seed Saponins (Samples IV-VII)

The molecular weights of **IV-VII** have been reported to be 634, 796, 942, and 1266. The FAB/MS spectra consist mainly of protonated and deprotonated molecular ions[11]. The ESI/MS spectra (Fig. 8) consist mainly of natriated molecular ions. All ESI/MS spectra were of good signal-to-noise ratio, except for **VII**.

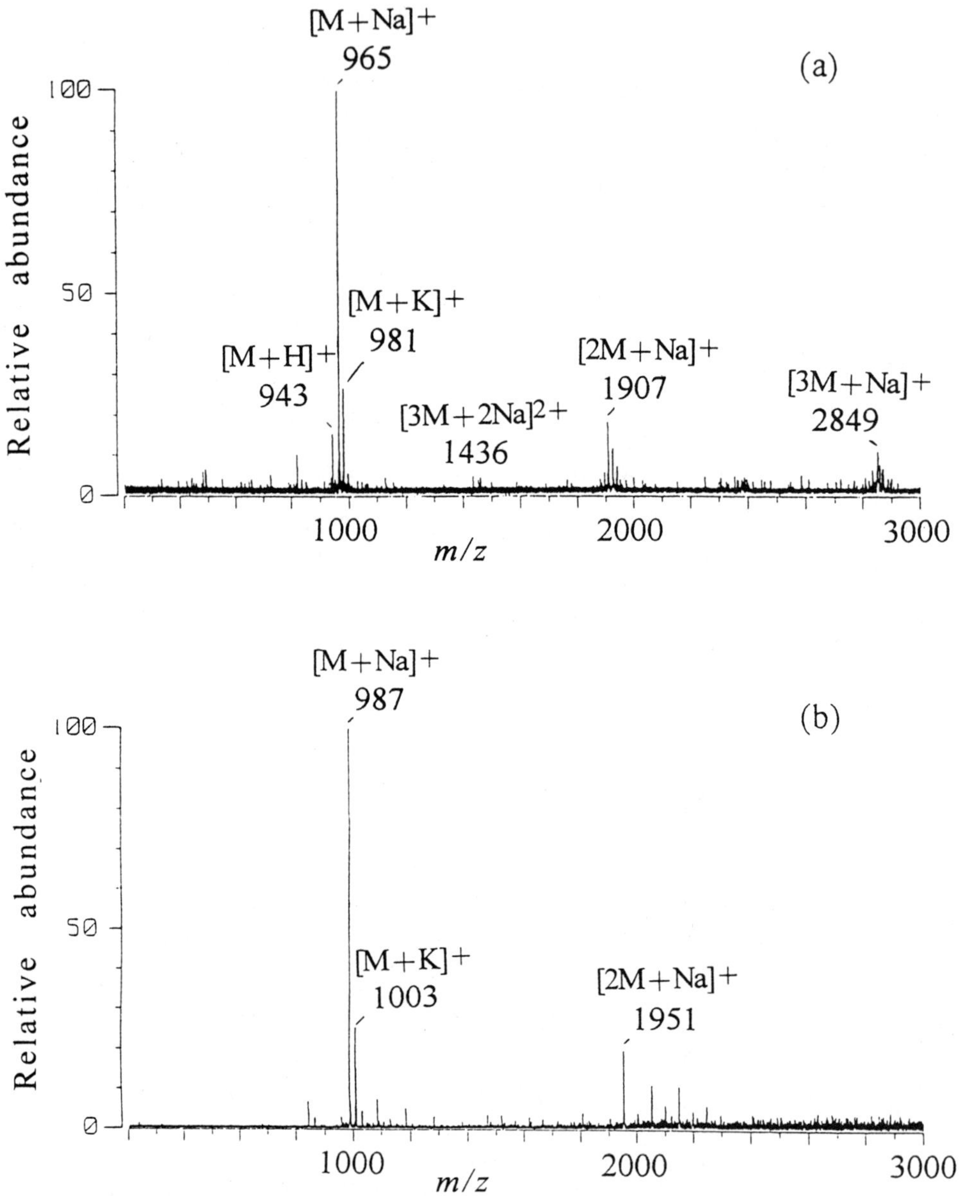

Figure 4. ESI/MS Spectra of (a) Sample **I**, and (b) Sample **II**.

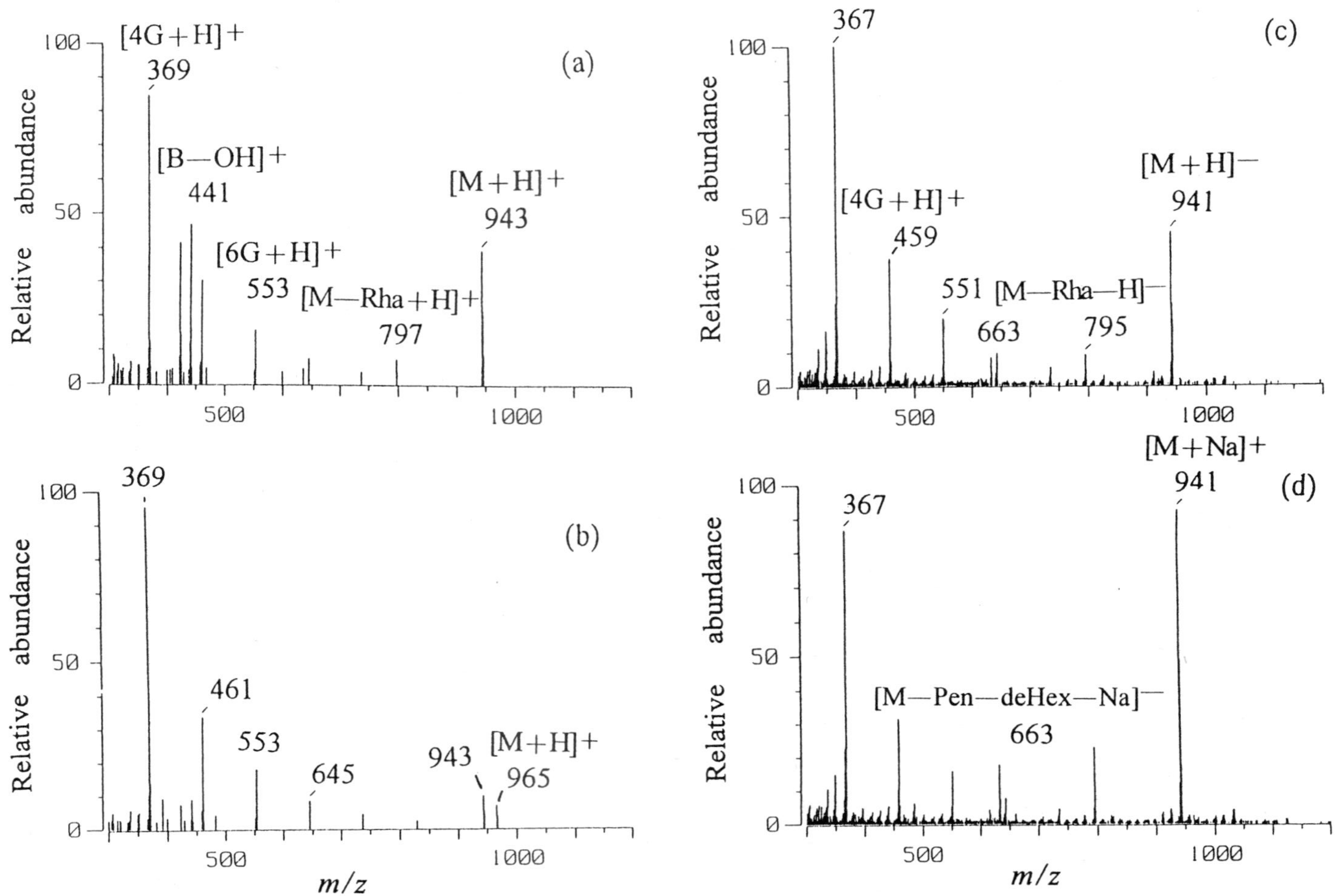

Figure 5. (+)FAB/MS Spectra of (a) Sample **I**, (b) Sample **II**, and (–)FAB/MS Spectra of (c) Sample **I**, (d) Sample **II**.

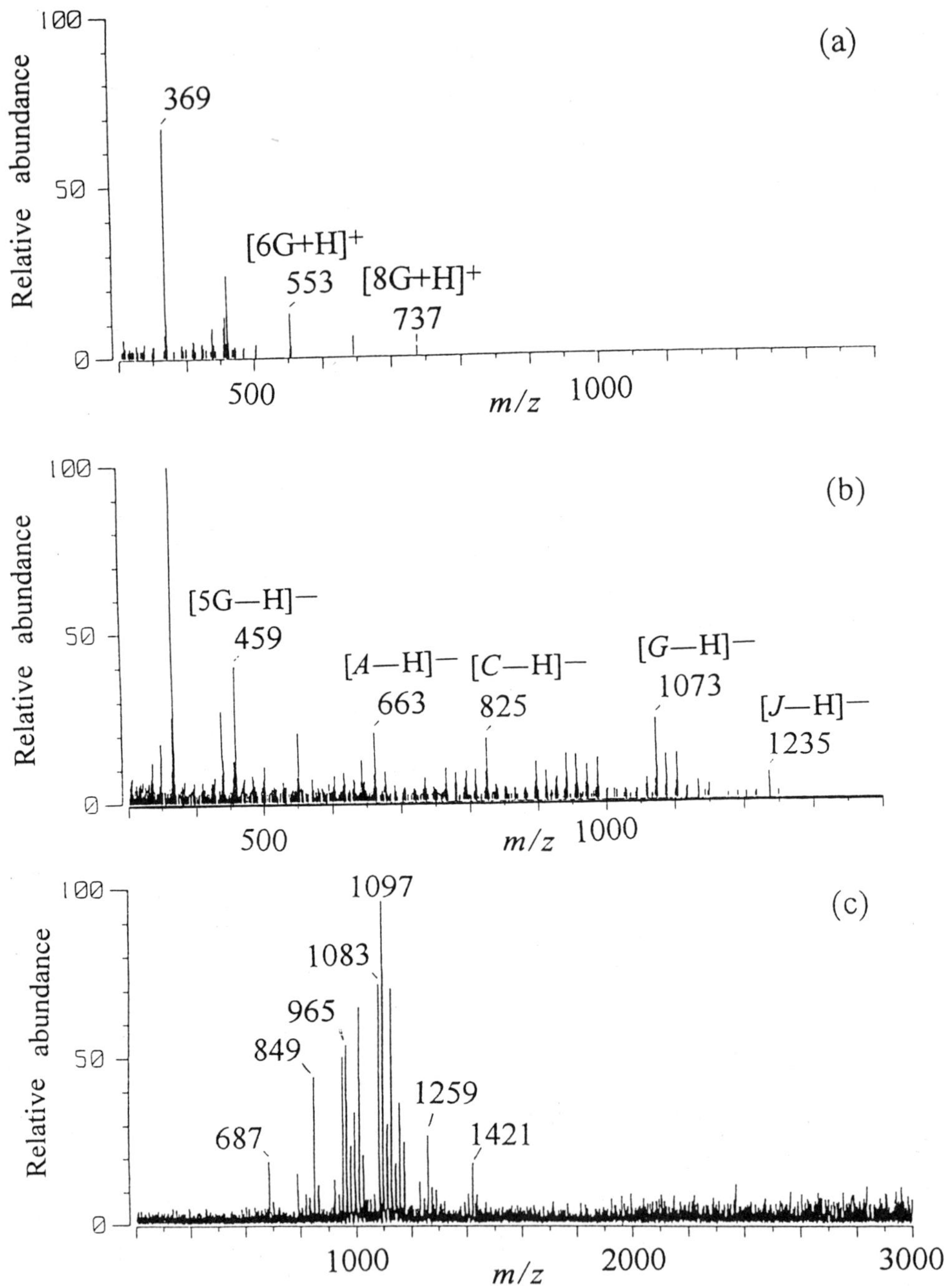

Figure 6. Spectra of Sample **III**: (a) (+)FAB/MS, (b) (–)FAB/MS, and (c) ESI/MS.

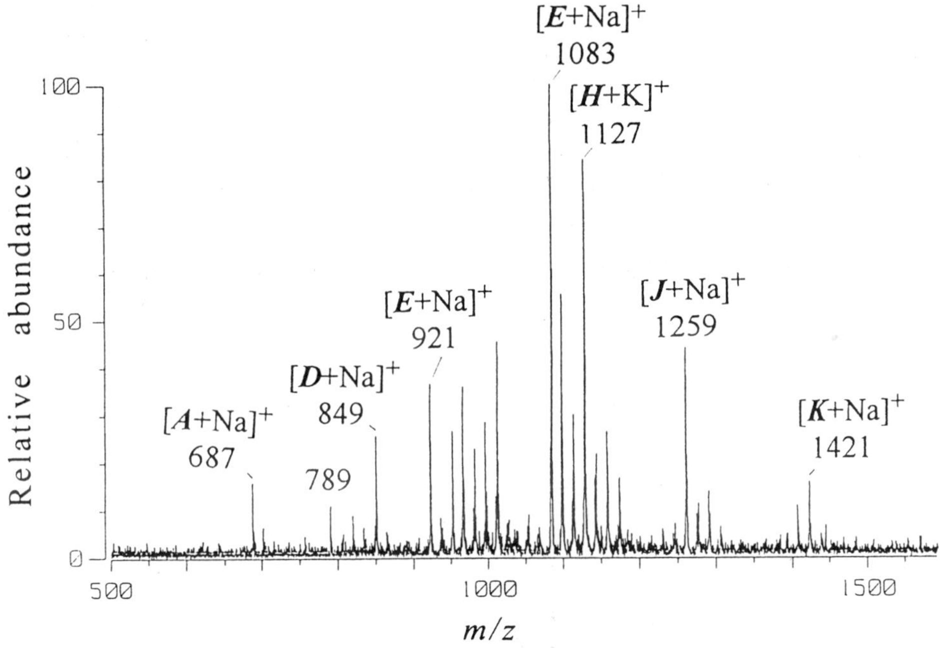

Figure 7. ESI/MS Spectrum of Sample **III**, $V_{cs} = 200$ V.

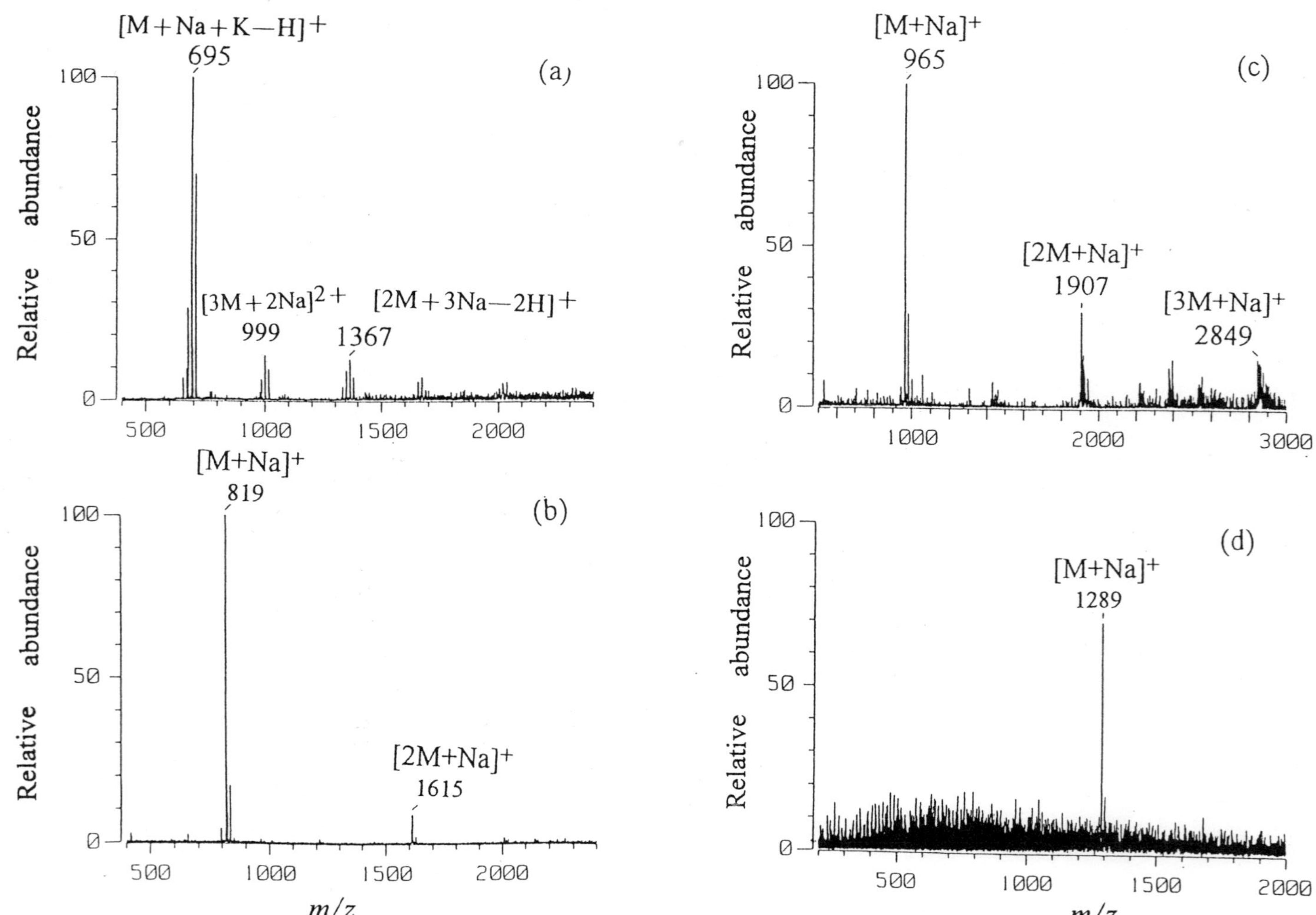

Figure 8. ESI/MS Spectra of Samples **IV-VII** (a-d)

DISCUSSION

Soyasaponin in H^+ form (Sample I) and Soyasaponin in Na^+ form (Sample II)

Compared to the ESI/MS spectra, the FAB/MS spectra provide more structural information. However, correct assignment of molecular weight is more difficult with these FAB/MS spectra owing to interference from the matrices. Clear distinction between **I** and **II** could be readily achieved from the corresponding ESI/MS spectra. It is difficult to draw the same conclusion from the FAB/MS spectra. In short, the (–)FAB/MS spectra as well as the (+)FAB/MS spectra could not provide a clear distinction between **I** and **II**.

Alfalfa Root Saponins (Sample III)

The (–)FAB/MS spectrum consists mainly of deprotonated molecular ions, whereas the ESI/MS spectrum consists mainly of natriated ions. It indicates that there are about ten saponins in **III**. Eight of them consistent with the previous report were verified by FAB/MS and MS/MS analysis (Fig. 2)[6]. Three saponins are of molecular weight 1060, 1150, and 1398 but it is not known if they are bidesmosidic saponins of medicagenic acid, hederagenin, or possibly zahnic glycosides. These results indicate that it is possible to analyze impure saponin mixtures by varying the V_{cs} in ESI to decompose the impurities. This might reduce the burden of purifying crude saponin extracts and speed the characterization process.

Clover Seed Saponins (Samples IV-VII)

Cluster natriated molecular ions were observed, as from other saponin samples investigated in this study. For **IV**, up to seven molecules are clustered to form the natriated molecular ion. Although similar molecular weight information could be obtained from FAB/MS spectra, ESI/MS always uses less sample. This might be useful when the amount of sample is small. Sample **VI** was identified as soyasaponin **I**, having identical spectra to sample **I**; sample **V** was confirmed as soyasaponin **III**, and sample **VII** proved to be a new saponin.

CONCLUSION

Comparing the (+)FAB/MS, (–)FAB/MS, and ESI/MS spectra of the seven saponin samples investigated in this study yields the following conclusions.

1. ESI/MS yields simpler spectra. Distinct protonated and natriated molecular ions could readily provide molecular weight information, which could be further verified using clusters of molecular ions.

2. ESI/MS can analyze saponin mixtures by varying the V_{cs} to decompose impurities. It uses less sample than that required by FAB/MS. These characteristics are advantageous when sample purity is not high, sample amount is limited, or sample solubility is not good.

3. ESI/MS does not produce structural information. This shortcoming is generally overcome by using ESI in conjunction with the MS/MS techniques. The primary structure of saponins, which are mixtures of triterpenoid glycosides formed of certain aglycones and sugar moieties, might be deduced from the molecular weight information alone. The lack of structural information from ESI/MS may not be a serious drawback.

4. The FAB/MS spectra consist mainly of protonated and deprotonated molecular ions with limited structural information. (–)FAB/MS is more suitable for analyzing saponin samples.

ACKNOWLEDGMENTS

Financial support in part by the National Science Council of the Republic of China (NSC-84-2113-M-007-040 to Y.C. Ling) is gratefully acknowledged.

REFERENCES

1. H.J. Budzikiewicz, C. Djerassi, and D.H. Williams, *Structure Elucidation of Natural Products by Mass Spectrometry*, Holden-Day, San Francisco, Vol. 2, pp 204-240 (1964).

2. K. Hostettmann, J. Doumas, and M. Hardy, Desorption/chemical ionization mass spectrometry of naturally occurring glycosides, *Helv. Chim. Acta*, 64: 297 (1981).

3. H.-R. Schulten, T. Soldati, T. Kawasaki, T. Okuyama, and S. Shibata, Confirmation of new, high-mass saponins from *Gleditsia japonica* by field desorption mass spectrometry, *Planta Med.* 46: 67 (1982).

4. K.R. Price, J. Eagles, and G.R. Fenwick, Saponin composition of 13 varieties of legume seeds using fast atom bombardment mass spectrometry, *J. Sci. Food Agric.* 42: 183 (1988).

5. P.R. West, P.W. Geno, G.R. Waller, W. Oleszek, and M. Jurzysta, Liquid secondary ion mass spectrometry and linked scanning at constant B/E LSIMS/MS for structure confirmation of saponins in *Medicago sativa* (alfalfa), In*Saponins Used in Food and Agriculture,* G.R. Waller and K. Yamasaki, eds., Plenum Publishing Co., New York, NY. In press, 1996.

6. G.R. Waller, M. Jurzysta, and R.L.Z. Thorne, Root saponins from alfalfa (*Medicago sativa* L.) and their allelopathic activity on weeds and wheat, *Allelopathy J.* 2: 21 (1995).

7. R.M. Facino, M. Carini, P. Traldi, B. Belli, B. Gioia, and E. Arlandini, Confirmatory assay and quantitative determination of *Hedera helix* L. saponins (α-hederin, hederacosides B and C) in raw plant extracts and in cosmetic formulations by EI and CAD MIKES spectrometry, *Biomed. Environ. Mass Spectrom.* 14: 187 (1987).

8. M.P. Maillard and K. Hostettmann, Determination of saponins in crude plant extracts by liquid chromatography-thermospray mass spectrometry, *J. Chromatogr.* 647: 137 (1993).

9. C.E. Costello, Application of tandem mass spectrometry spectral approaches to sturctural determinations of saponins, In*Saponins Used in Food and Agriculture*, G.R. Waller and K. Yamasaki, eds., Plenum Publishing Co., New York, NY. In press, 1996.

10. K. Hostettman, Search for mulluscicidal and antifungal saponins from tropical plants, In*Saponins Used in Traditional and Modern Medicine,* G.R. Waller and K. Yamasaki, eds., Plenum Publishing Co., New York, NY. In press, 1996.

11. M. Jurzysta, K.R. Price, C. Ridout, and R. Fenwick, The structures of four triterpenoid saponins isolated from the seed of *Trifolium incarnatum* L., *Acta Soc. Bot. Pol.* 58: 575 (1989).

12. M.K. Lee, Y.C. Ling, P.J. Huang, and S.M. Wang, Ionization efficiency of dicopper complex in electrospray ionization mass spectrometry, *J. Chin. Chem. Soc.* 41: 711 (1994)

ANALYSIS, HEAT STABILITY AND PHYSIOLOGICAL EFFECTS OF SAPONINS FROM OATS

Gunilla Önning and Nils-Georg Asp

Department of Applied Nutrition and Food Chemistry
Lund University, PO Box 124, 221 00 Lund, Sweden

INTRODUCTION

In 1953 a bitter-tasting compound was isolated from oats (Mohr, 1953). The compound had saponin features such as hemolytic activity and foaming ability. Rohrlich and Train (1954) determined the sugar components to be rhamnose and glucose and found that the aglycon had a steroid structure. Tschesche and Schmidt (1966) proposed a name for the saponins in oats - avenacosides. Later the structure was elucidated for two oat saponins, avenacoside A and B (Tschesche et al., 1969; Tschesche and Lauven, 1971), both of which are bisdesmosidic (Fig. 1). The sugar residues containing glucose and rhamnose are linked to a steroid aglycon, nuatigenin. Avenacoside B contains one more glucose unit than avenacoside A.

Oat leaves also contain these avenacosides. A specific β-glucosidase, occurring naturally in oat leaves, is able to remove the C-26-bound glucose moiety (Fässler et al., 1984; Grünweller and Kesselmeier, 1985). The desglucoavenacosides formed are monodesmosidic with less bitterness but with a higher antibiotic and hemolytic activity than the avenacosides (Tschesche and Wiemann, 1977).

Saponins in oats have been analyzed by TLC (Tschesche et al., 1969; Fenwick and Oakenfull, 1983). In the first study the avenacoside A and B content in oat kernel was determined to be 0.040% and in the second study the total saponin content in rolled oats was 0.1%. Avenacosides isolated from oat leaves have been analyzed with HPLC (Kesselmeier and Strack, 1981) where they were separated on a silica gel C8-column with a water:acetonitrile solvent system and detected with a UV detector.

The researchers have so far concentrated on the saponins in oat leaves and oat roots. The interest in using oats in foods has increased in recent years, mainly due to the serum cholesterol lowering properties. It is therefore also important to get a better knowledge of the saponins in oat kernels. We have in this study developed a method for analysis of saponins in oat kernels. With this method the content in different oat varieties and in different fractions of the oat kernel was determined. Since almost all oats for human consumption are heat-treated the stability of the avenacosides at higher temperatures was studied. Finally the oat saponins effects on the lipid metabolism and intestinal permeability were investigated by using animal models.

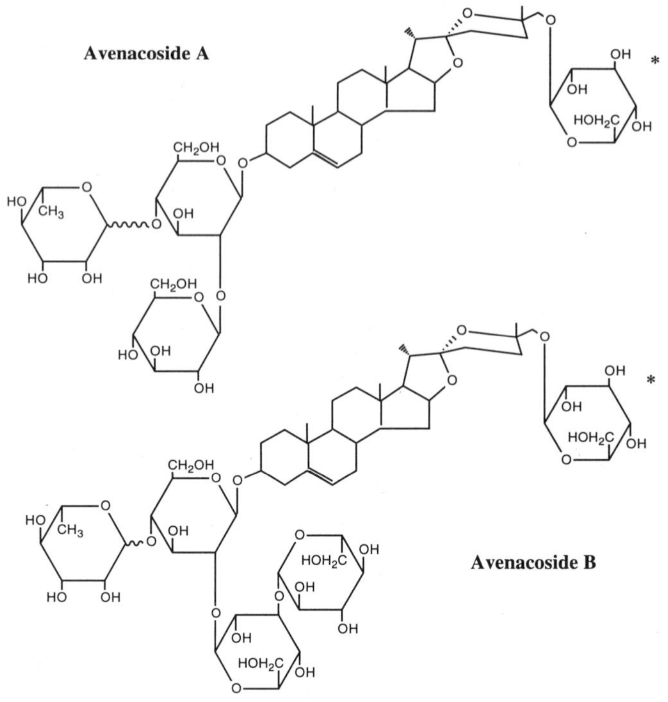

Figure 1. Structures of avenacoside A and B. The corresponding desglucoavenacoside is formed when the glucose marked * is removed.

ANALYSIS FOR AVENACOSIDES IN OAT KERNELS

For quantitative determination the oatmeal was first defatted with petroleum ether (bp 60−80 °C) for 16 h in a Soxhlet apparatus. Then the saponins were extracted with methanol for 24 h. The HPLC method of Kesselmeier and Strack (1981) was thereafter used to separate the avenacosides. The extract was injected onto a LiChrosorb 100 RP 8 column (125 × 4 mm, 5 μm, Merck) and separation was accomplished using gradient elution with 25 to 40% acetonitrile in water during 15 min. The UV detector was set at 200 nm.

Sixteen oat cultivars (Svalöf AB, Svalöv) were analyzed for saponin content (Table 1) (Önning et al., 1993). The UV detector was in this case calibrated with desgluco-avenacosides. There were significant differences in saponin content. The varieties (Svea, Vital 2, Sv 90677) with the highest level of avenacosides, 0.050%, contained over twice as much as the low-saponin variety NZ 89-703. The mean avenacoside content for the sixteen varieties was 0.040%. The avenacoside A content was 2−4 times the avenacoside B content. The two varieties with the lowest avenacoside levels were grown in New Zealand. The saponin content could be influenced by growing conditions (climate, soil), but genetic factors are probably also important.

The study confirms that, as in alfalfa (Livingston et al., 1984), legumes (Price et al., 1986), and quinoa (Ng et al., 1994), there are large differences in saponin content between cultivars.

As a way to investigate the location of the avenacosides in the kernel, two oat varieties (Magne, Selma) were milled and separated into fractions according to particle size. The avenacoside content in different fractions of Selma is presented in Figure 2. The saponin distribution for Magne was similar. For Selma, the meal fraction with the smallest particle sizes (<250 μm) contained the highest concentration of avenacosides, 0.068%, while the

Table 1. Content of avenacoside A and B in 16 oat varieties (%, dry matter basis)†

Oat variety	Avenacoside A*	Avenacoside B*
Svea	0.037^{ab}	0.013^{b}
Vital 2	0.034^{bcd}	0.016_{a}
Sv 90677	0.039^{a}	0.011^{bcd}
Sv 842097	0.036^{abc}	0.013^{b}
Adamo	0.036^{abc}	0.012^{bcd}
Magne	0.031^{de}	0.013^{b}
Sang	0.033^{cde}	0.011^{bcd}
Chihuahua	0.031^{de}	0.012^{bc}
Magne 8II	0.034^{cd}	0.009^{d}
Sv 90634	0.029^{ef}	0.009^{d}
Sv 843675	0.026^{fg}	0.011^{bcd}
Selma	0.026^{g}	0.010^{cd}
Vital 1	0.024^{g}	0.011^{bcd}
Å 83165	0.021^{h}	0.010^{d}
NZ89-701	0.015^{i}	0.009^{d}
NZ89-703	0.015^{i}	0.005^{e}
Mean	0.029	0.011
CV(%)	26	21

* means from at least duplicate determinations.
a-i means within columns with different superscript are significantly different ($P<0.05$).
† adopted from Önning et al. (1993) with permission from Elsevier Science Publishers Ltd.

content in the most coarse fraction (>1050 μm) was lowest, 0.013%.

The saponins are supposed to have an antibiotic effect, and in that context it is surprising that the avenacoside concentration is lower in the outer parts (corresponding to the coarser meal fractions) of the oat kernel. In quinoa the saponins are located mainly in the outer layers of the seeds. The content of two quinoa saponins was 0.9% in whole seeds and 2.3% in the bran (Ruales and Nair, 1993).

In legumes the saponins seem to be associated with proteins and are therefore concentrated in protein-rich fractions (Fenwick and Oakenfull, 1983; Curl et al., 1985). The

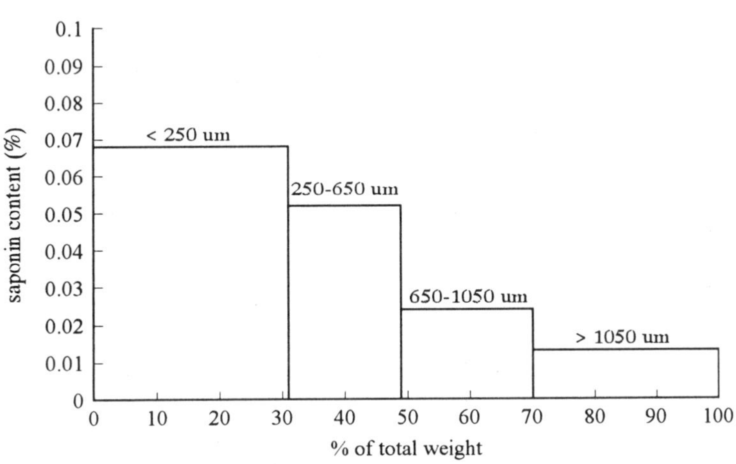

Figure 2. Avenacoside content in various particle-size fractions and the fraction sizes for the variety Selma. Reprinted from Önning et al. (1993) with permission from Elsevier Science Publishers Ltd.

highest concentration of protein in Selma was found in the fraction >1050 μm (Frølich and Nyman, 1988), which indicates that the saponins in oats are not associated in the same way with proteins.

EFFECTS OF HEAT PROCESSING

Avenacoside A and B were isolated from oatmeal by using silica gel SI-60, Sephadex LH-20, and silica gel RP18 chromatography (Önning et al., 1994). These isolates (> 98% pure) were then dissolved in buffers (pH 4−7, 0.3 mg saponins/ml) and heated in 100 μl glass vials up to 100 °C in a water bath. For heat treatment at 100 and 140 °C the saponin solutions were portioned (50 μl) in 5−6-mm glass bulbs. The bulbs were sealed with a flame and thereafter heated in an oil bath.

It was found that the avenacosides were stable at room temperature at pH 4−7. Heating for 3 h at 100 °C also did not give any degradation. The reduction in saponin content after ordinary cooking has also been modest in earlier studies. Cooking of chick pea and black gram reduced the content by 7−17% (Jood et al., 1986; Kataria et al., 1988). If the seeds were soaked before the heat treatment, the reduction was more extensive. Cooking of soaked faba bean seeds reduced the saponin content by 35% (Sharma and Sehgal, 1992).

Some reduction was seen when avenacoside A was heated at 140 °C. After 3 h heat treatment at pH 4, 5, 6, and 7, the saponin content was reduced by 50%, 19%, 12%, and 13%, respectively (Fig. 3). At pH 4 and 5 a degradation product could be detected by HPLC, eluting 0.7 min after avenacoside A. Mass spectrometric analysis of this product gave a molecular ion at *m/z* 916. This indicates that avenacoside A had lost one sugar unit, rhamnose. Very few previous studies have investigated the stability of saponins at higher temperatures. Autoclaving of soaked faba bean seeds for 15 min gave a reduction of 40% (Sharma and Sehgal, 1992). The reduction was 72% when quinoa meal was cooked and roller dried (Gee et al., 1993).

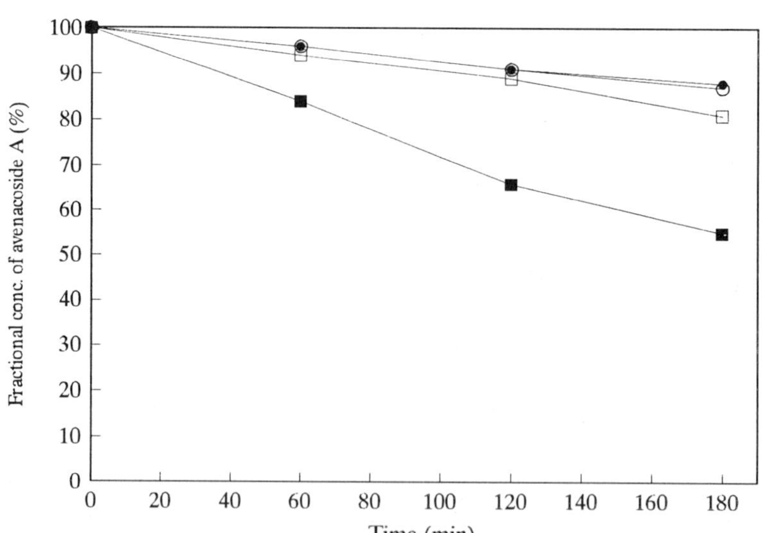

Figure 3. Effect of heat treatment at 140 °C on avenacoside A content, $n=2$.
(■) pH 4; (□) pH 5; (●) pH 6; (○) pH 7
Reprinted from Önning et al. (1994) with permission from American Chemical Society.

EFFECTS ON PLASMA AND LIVER LIPIDS IN GERBILS AND RATS

Oats have well documented blood-cholesterol-lowering effects (Ripsin et al., 1992). From dose-response studies these effects seem to be related primarily to the water-soluble fraction of dietary fibre, i.e. mainly β-glucans (Davidson et al., 1991), but other substances like the saponins could also be involved.

Saponins are able to bind cholesterol in vitro (Price et al., 1987) and in vivo studies have shown that some saponins have cholesterol-lowering effects in rats (Oakenfull et al., 1984, soya saponins; Southon et al., 1988, *Gypsophila* saponins; Stark and Madar, 1993, fenugreek saponins); gerbils (Potter et al., 1993, quillaja saponins); hamster (Harwood et al., 1993, β-tigogenin cellobioside); chicken (Jenkins and Atwal, 1994, *Gypsophila* and quillaja saponins); and man (Potter et al., 1980, soya saponins). The saponins act either directly, by inhibiting absorption of cholesterol from the small intestine, or indirectly, by inhibiting the reabsorption of bile acids (Oakenfull and Sidhu, 1990).

To investigate if the oat saponins could influence the lipid metabolism we gave gerbils and rats oat diets with three different saponin contents: negligible (ethanol-extracted oats), normal (unextracted oats), and twice normal (ethanol-extracted oats + a mixture of avenacoside A and B) (Önning and Asp, 1995).

Gerbils were choosen as an animal model since they, compared with rats, react more like humans to lipids in the diet. Saturated fatty acids increase the blood cholesterol level (Nicolosi et al., 1981) and dietary cholesterol, even at low concentrations, induce hypercholesterolemia (Temmerman et al., 1988). Thus, gerbils react more readily than e.g. rats to dietary lipid alterations, a reason why this species is preferred as a model for human cholesterol metabolism (Beynen, 1990). On the other hand, hypercholesterolemic rats have been used extensively to study dietary effects on plasma lipids. For these reasons, feeding experiments were carried out both in rats and gerbils.

Two animal groups were used as controls. One was fed cellulose, a type of dietary fibre known not to affect the lipid metabolism, and the other group was given guar gum, one of the most hypocholesterolemic types of fibre. The diets were balanced carefully and contained 65 g dietary fibre and 200 g fat/kg. To obtain increased blood lipid levels cholesterol was added: 1 g/kg diet for gerbils and 5 g/kg diet for rats. The gerbils were given the diets for 21 days and the rats for 19 days.

The plasma cholesterol values obtained after the feeding period are presented in Table 2. Non-HDL-cholesterol was determined as total cholesterol minus HDL-cholesterol. In the fasting state, this difference represents mainly LDL (low-density-lipoprotein) cholesterol. The total plasma cholesterol in the gerbils (4.3−5.1 mmol/l) was not significantly different between the groups. The rats had lower plasma cholesterol values, around 2.0 mmol/l for all groups, except for the group fed guar gum, that had a concentration of 2.5 mmol/l. There was one extra control group for the rats, that was given pellets with a negligible cholesterol content. This group had a plasma cholesterol level similar to the groups fed cholesterol-containing diets. Thus, 5 g cholesterol/kg did not increase the plasma cholesterol level in the rats. One way to obtain increased plasma cholesterol values in rats is to add cholic acid as well as cholesterol (Shinnick et al., 1990).

The HDL-cholesterol values for the different oat groups were similar regardless of the saponin content, about 2.5 mmol/l for the gerbils and about 1.2 mmol/l for the rats. On the other hand, the groups given cellulose had significantly lower HDL-cholesterol values, while the guar gum groups had similar levels compared with the oat groups. For the gerbils, non-HDL-cholesterol values were lowest for the guar gum group, intermediate in the oat groups, and highest in the cellulose group, confirming an LDL-cholesterol-lowering effect of both oats and guar gum. The rats had lower and more similar values.

Fasting plasma triglyceride levels were not significantly different between the dietary groups. Similar values were found in the gerbils and rats, about 1.0 mmol/l.

Table 2. Plasma lipid concentrations (mmol/l) for gerbils and rats after the feeding period (Mean values with their standard errors)†

Diet		Total cholesterol		HDL-cholesterol		Non-HDL cholest.	
	n	Mean	SE	Mean	SE	Mean	SE
Gerbils							
Cellulose	7	5.0[a]	1.5	1.9[a]	0.2	3.1[a]	0.6
Guar gum	13	4.3[a]	1.1	2.3[b]	0.2	1.9[b]	0.2
EtOH-extd. oats	14	4.8[a]	1.0	2.5[b]	0.1	2.3[b]	0.1
Oats	14	5.1[a]	1.0	2.6[b]	0.1	2.5[ab]	0.2
EtOH-extd. oats + saponins	14	4.7[a]	1.0	2.5[b]	0.1	2.2[b]	0.2
Rats							
Pellets	8	1.9[b]	0.1	1.1[bc]	0.1	0.9[bc]	0.06
Cellulose	8	2.0[b]	0.1	0.9[c]	0.06	1.1[ac]	0.08
Guar gum	8	2.5[a]	0.1	1.3[a]	0.07	1.2[a]	0.10
EtOH-extd. oats	8	2.1[b]	0.1	1.3[a]	0.1	0.8[b]	0.07
Oats	8	2.1[ab]	0.1	1.2[a]	0.09	0.9[bc]	0.04
EtOH-extd. oats + saponins	5	2.1[b]	0.04	1.1[ac]	0.04	1.0[abc]	0.06

[a-c]Means within columns with the same superscript are not significantly different.
† Adopted from Önning and Asp (1995) with permission from Cambridge University Press.

Mean liver weight and liver lipid content for the dietary groups are presented in Table 3. For gerbils no differences in liver weight and relative liver weight were found. In rats the relative liver weight was lowest for the group fed pellets. This group had also the lowest liver lipid content. The rat group that was given oats with the highest amount of avenacosides had a liver lipid level similar to that in the guar gum group, which was significantly lower than in the other oat groups and the cellulose group. For gerbils the liver lipid contents did not differ significantly between the groups. For both species the cellulose group had about twice the amount of liver cholesterol compared with the oat groups, that had similar levels regardless of the avenacoside content. Free liver cholesterol levels were highest for the cellulose group (about 2.0 mg/g wet tissue) in both gerbils and rats.

From these results it can be concluded that oat saponins do not play a significant role in the hypocholesterolemic effects of oats. Avenacoside A and B could nevertheless have some minor effects on the lipid metabolism, as indicated by the effect on the liver lipid contents. Lipid metabolism in the body is complex, and measuring plasma and liver lipids provides some information. To obtain a better understanding, other parameters should also be investigated, for example the bile acid excretion.

The saponin levels in the diets were chosen to be realistic compared with those found normally in oats. Earlier studies have used diets with a higher saponin content. For example, Oakenfull et al. (1979) fed rats a diet containing 10 g cholesterol/kg, and the addition of 10 g saponins from soapwort/kg diet lowered the cholesterol accumulation, especially in the liver. The same results were obtained with 10 g soya saponins/kg diet (Oakenfull et al., 1984). In another study rats were given soy flour high (22 g/kg) or low (4 g/kg) in saponins and with the same cholesterol content as in our study (5 g/kg diet) (Topping et al., 1980). Both diets lowered plasma cholesterol, but there was no difference related to the saponin content. In one study (Rotenberg and Eggum, 1986), addition of a low amount (2 g/kg diet) of saponins from *Gypsophila* roots to a diet also containing pectin did not reduce plasma or liver cholesterol in rats, compared with the saponin-free pectin diet.

Table 3. Liver weight and liver lipid values for gerbils and rats after the feeding period (Mean values with their standard errors)†

Diet		g liver/100 g body weight		Total lipid content, mg/g wet tissue		Total cholesterol, mg/g wet tissue	
	n	Mean	SE	Mean	SE	Mean	SE
Gerbils							
Cellulose	7	3.4^{a}	0.10	117^{a}	10.2	10.4^{a}	0.6
Guar gum	13	3.7^{a}	0.05	123^{a}	8.0	5.5^{b}	0.5
EtOH-extd. oats	14	3.5^{a}	0.07	116^{a}	6.9	6.9^{c}	0.2
Oats	14	3.4^{a}	0.08	115^{a}	6.4	6.5^{bc}	0.3
EtOH-extd. oats + saponins	14	3.5^{a}	0.08	111^{a}	4.8	6.5^{c}	0.3
Rats							
Pellets	8	3.7^{b}	0.1	50^{c}	2.8	3.8^{c}	0.3
Cellulose	8	4.3^{a}	0.1	163^{a}	11.1	25.8^{a}	1.5
Guar gum	8	4.0^{ab}	0.1	99^{b}	7.5	13.0^{b}	1.2
EtOH-extd. oats	8	4.0^{ab}	0.1	146^{a}	9.9	11.5^{b}	0.7
Oats	8	4.2^{a}	0.1	150^{a}	13.5	12.3^{b}	0.7
EtOH-extd. oats + saponins	5	3.9^{ab}	0.2	101^{b}	18.1	11.7^{b}	1.5

[a-c]Means within columns with the same superscript are not significantly different.

† Adopted from Önning and Asp (1995) with permission from Cambridge University Press.

EFFECTS ON INTESTINAL PERMEABILITY IN RATS

Saponins can combine irreversibly with plasma membranes and thereby increase the permeability of cells. In the intestine, this could lead to an increased uptake of passively transported compounds. Thus, Johnson et al. (1986) observed an increased uptake of L-glucose and polyethylene glycol 4000 in vitro in presence of *Gypsophila* saponins. The glycinin (a soya protein) uptake in vitro in intestine from rabbit was increased when saponins from soya was added (Alvarez and Torres-Pinedo, 1982). Increased sensitization to food allergens in rats, probably due to an increased uptake of antigens, has been demonstrated (Atkinson et al., 1994).

A decreased uptake of compounds in presence of saponins has also been reported. A reduced uptake of galactose has been observed in vitro in presence of *Gypsophila* saponins using rat small intestine (Johnson et al., 1986). In this case the saponins are probably acting on the systems that actively transport compounds from the intestine into the cells. Saponins can also induce morphological changes in the intestine such as increased villus and crypt lengths (Gee and Johnson, 1988).

We studied the effect of the avenacosides on the intestinal permeability both in vitro with a method of Pantzar et al. (1993), and in vivo using a method of Wang et al. (1994).

In vitro, segments of rat small intestine were mounted in modified Ussing chambers and the passage of ^{3}H-methyl-D-glucose, ^{51}Cr-EDTA, and ovalbumin was measured during 120 min. A mixture of avenacoside A and B was added to every second chamber (0.5 or 1 mg/ml).

The passage of ^{3}H-methyl-D-glucose, an actively transported monosaccharide, was the same with and without saponins. Other saponins have been demonstrated to reduce the passage of actively transported compounds in rats. In an in vivo study (Sidhu et al., 1987), the transport of glucose was decreased in the presence of soya saponins (2 mg/ml). A saponin extract from quinoa reduced the passage of glucose in vitro at a concentration of 2.5 mg/ml (Gee et al., 1993). These two studies used a higher saponin concentration than in our study.

The passage of ^{51}Cr-EDTA in the proximal intestine seemed to be increased in the late time interval studied when 1 mg saponins/ml were added but not with a concentration of 0.5 mg saponins/ml. No differences in passage of ^{51}Cr-EDTA in the distal small intestine were found.

There was a prominent effect on the permeability for ovalbumin when saponins were added. At a concentration of 1 mg/ml the saponins increased the passage of ovalbumin significantly in the proximal intestine at 100 min. The difference was more pronounced in the distal intestine, where the saponins increased the permeability for ovalbumin at all time points after 40 min (Fig. 4). With the lower saponin concentration (0.5 mg/ml) the difference was not significant any longer, indicating a dose-response relation. Saponins are probably increasing the permeability of cells by combining irreversibly with discrete sites within the plasma membrane (Price et al., 1987).

In vivo, rats were fed a solution (20 ml/kg body weight) containing ^{51}Cr-EDTA and for half of the animals 2 mg avenacoside A/ml 0.9% saline. Blood, urine, and feces were sampled at different time intervals. The mean concentration of ^{51}Cr-EDTA in the blood was lower, up to 8 h after the feeding, for the rats given saponins than for controls. Later, the blood concentration of ^{51}Cr-EDTA was low and similar for both groups. No significant differences at any time point were however found.

A significantly lower amount of ^{51}Cr-EDTA was excreted in the urine in the time interval 8-24 h for the animals given saponins. But the total amount of ^{51}Cr-EDTA excreted in 48 h, 2.2% of the fed dose for the saponin group and 2.7% for the control group, did not differ significantly.

A large part of the fed ^{51}Cr-EDTA (about 75% for both groups) was excreted in the feces, which was also analyzed for saponin content. No avenacosides could be detected (<0.8 μg/g) indicating that these compounds have been degraded in some part of the digestive system.

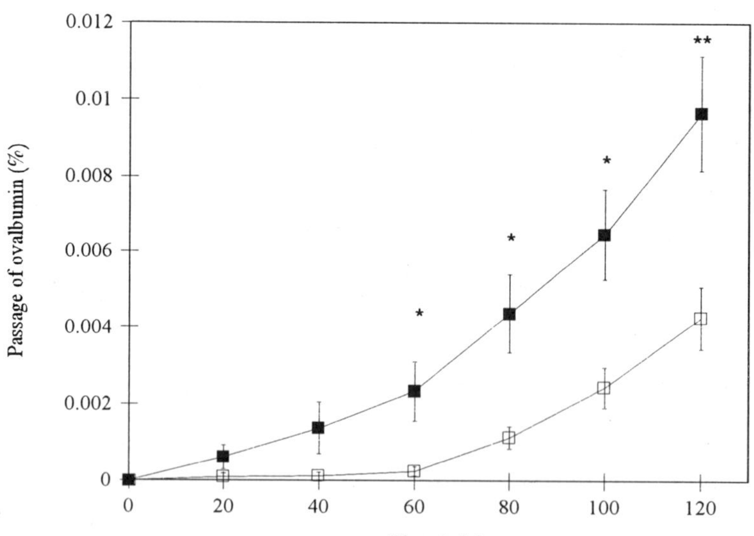

Figure 4. Passage (mean ± SE) of ovalbumin from the mucosal to the serosal side of distal (n = 8) rat intestine, without saponins (□) and with saponins (■) (1 mg/ml). t-test: * $P<0.05$, ** $P<0.01$

With the in vitro model there was a tendency to increased passage of ^{51}Cr-EDTA in the presence of saponins, but in the in vivo model the absorption seemed to be decreased. One explanation for this could be that cells interacting with saponins are replaced with new ones in vivo. Thus, saponins could have induced a more rapid exfoliation of mucosal cells into the lumen, and in this way a reduced absorption could occur. Another explanation for the result could be that compounds in the intestine can influence the membranolytic effect of the saponins. For example, bile salts, which were present in the in vivo study but not in the in vitro study, could reduce this effect (Gee and Johnson, 1988).

CONCLUDING REMARKS

Saponins are often regarded as antinutritional or toxic components. However, the present study does not give reasons for limiting the intake of oats due to saponins. The highest avenacoside level found in oats was 0.050% of the dry matter. This is a relatively low value compared with other foods that contain saponins. For example, soya contains ten times as much (Ireland et al., 1986). The oat saponins seem to have only minor effects at concentrations realistic in relation to the levels in oats. But the influence of saponins on intestinal permeability, especially for proteins, needs further investigation.

REFERENCES

Alvarez, J.R. and Torres-Pinedo, R., 1982, Interactions of soybean lectin, soya saponins and glycinin with rabbit jejunal mucosa in vitro, *Pediatr. Res.* 16:728.

Atkinson, H.A.C., Grigoriadou, F. and Miller, K., 1994, Enhancement of oral sensitisation to food allergens by the bioactive plant constituent *Gypsophila* saponin in the brown Norway rat, in: *Biochemical Biomarkers in Environmental Toxicology*, abstract booklet, University of Cambridge, Cambridge, U.K.

Beynen, A.C., 1990, The mongolian gerbil in cholesterol metabolism research, in: *Symposium über Hamster und Gerbil*, Detmold.

Curl, C.L., Price, K.R. and Fenwick, G.R., 1985, The quantitative estimation of saponin in pea (*Pisum sativum* L.) and soya (*Glycine max*), *Food Chem.* 18:241.

Davidson, M.H., Dugan, L.D., Burns, J.H., Bova, J., Story, K. and Drennan, K.B., 1991, The hypocholesterolemic effects of β-glucan in oatmeal and oat bran, *J. Am. Med. Assn.* 265:1833.

Fässler, L., Lichtenthaler, H.K., Gruiz, K. and Biacs, P., 1984, Accumulation of saponins and sterols in Avena seedlings under high-light and low-light growth conditions, *Developm. Plant Biol.* 9:225.

Fenwick, D.E. and Oakenfull, D., 1983, Saponin content of food plants and some prepared foods, *J. Sci. Food Agric.* 34:186.

Frølich, W. and Nyman, M., 1988, Minerals, phytate and dietary fibre in different fraction of oat-grain, *J. Cereal Sci.* 7:73.

Gee, J.M. and Johnson, I.T., 1988, Interactions between hemolytic saponins, bile salts and small intestinal mucosa in the rat, *J. Nutr.* 118:1391.

Gee, J.M., Price, K.R., Ridout, C.L., Wortley, G.M., Hurrell, R.F. and Johnson, I.T., 1993, Saponins of quinoa (*Chenopodium quinoa*): effects of processing on their abundance in quinoa products and their biological effects on intestinal mucosal tissue, *J. Food Sci. Agric.* 63:201.

Grünweller, S. and Kesselmeier, J., 1985, Characterization of a membrane-bound β-glucosidase responsible for the activation of oat leaf saponins, *Phytochemistry* 24:1941.

Harwood, H.J., Chandler, C.E., Pellarin, L.D., Bangerter, F.W., Wilkins, R.W., Long, C.A., Cosgrove, P.G., Malinow, M.R., Marzetta, C.A., Pettini, J.L., Savoy, Y.E. and Mayne, J.T., 1993, Pharmacologic consequences of cholesterol absorption inhibition: alteration in cholesterol metabolism and reduction in plasma cholesterol concentration induced by the synthetic saponin β-tigogenin cellobioside (CD-88818; tiqueside), *J. Lipid Res.* 34:377.

Ireland, P.A., Dziedzic, S.Z. and Kearsley, M.W., 1986, Saponin content of soya and some commercial soya products by means of high-performance liquid chromatography of the sapogenins, *J. Sci. Food Agric.* 37:694.

Jenkins, K.J. and Atwal, A.S., 1994, Effects of dietary saponins on fecal bile acids and neutral sterols, and availability of vitamins A and E in the chick, *J. Nutr. Biochem.* 5:134.

Johnson, I.T., Gee, J.M., Price, K., Curl, C. and Fenwick, G.R., 1986, Influence of saponins on gut permeability and active nutrient transport in vitro, *J. Nutr.* 116:2270.

Jood, S., Chauhan, B.M. and Kapoor, A.C., 1986, Saponin content of chickpea and black gram: varietal differences and effects of processing and cooking methods, *J. Sci. Food Agric.* 37:1121.

Kataria, A., Chauhan, B.M. and Gandhi, S., 1988, Effect of domestic processing and cooking on the antinutrients of black gram, *Food Chem.* 30:149.

Kesselmeier, J. and Strack, D., 1981, High performance liquid chromatographic analysis of steroidal saponins from *Avena sativa* L, *Z. Naturforsch.* 36c:1072.

Livingston, A.L., Knuckles, B.E., Teuber, L.R., Hesterman, O.B. and Tsai, L.S., 1984, Minimizing the saponin content of alfalfa sprouts and leaf protein concentrates, in: *Nutritional and Toxicological Aspects of Food Safety*, Plenum Press, New York.

Mohr, W., 1953, Über die Qualitätsbeeinflussung von Haferzeugnissen (III), *Dtsch. Lebensmittel-Rundsch.*. 49:127.

Ng, K. R., Price, K.R. and Fenwick, G.R., 1994, A TLC method for the analysis of quinoa (*Chenopodium quinoa*) saponins, *Food Chem.* 49:311.

Nicolosi, R.J., Marlett, J.A., Morello, A.M., Flanagan, S.A. and Hegsted, D.M., 1981, Influence of dietary unsaturated and saturated fat on the plasma lipoproteins of mongolian gerbils, *Atherosclerosis* 38:359.

Oakenfull, D.G., Fenwick, D.E., Hood, R.L., Topping, D.L., Illman, R.L. and Storer, G.B., 1979, Effects of saponins on bile acids and plasma lipids in the rat, *Br. J. Nutr.* 42:209.

Oakenfull, D., Topping, D.L., Illman, R.J. and Fenwick, D.E., 1984, Prevention of dietary hypercholesterolaemia in the rat by soya bean and quillaja saponins, *Nutr. Rep. Int.* 29:1039.

Oakenfull, D. and Sidhu, G.S., 1990, Could saponins be a useful treatment for hypercholesterolaemia?, *Eur. J. Clin. Nutr.* 44:79.

Önning, G., Asp, N-G. and Sivik, B., 1993, Saponin content in different oat varieties and in different fractions of oat grain, *Food Chem.* 48:251.

Önning, G., Juillerat, M.A., Fay, L. and Asp, N-G., 1994, Degradation of oat saponins during processing – effect of pH, stainless steel and iron at different temperatures, *J. Agric. Food Chem.* 42:2578.

Önning, G. and Asp, N-G., 1995, Effect of oat saponins on plasma and liver lipids in gerbils and rats, *Br. J. Nutr.* 73:275.

Pantzar, N., Weström, B.R., Luts, A. and Lundin, S., 1993, Regional small-intestinal permeability in vitro to different-sized dextrans and proteins in the rat, *Scand. J. Gastroenterol.* 28:205.

Potter, J.D., Illman, R.J., Calvert, G.D., Oakenfull, D.G. and Topping, D.L., 1980, Soya saponins, plasma lipids, lipoproteins and fecal bile acids: a double blind cross-over study, *Nutr. Rep. Int.* 22:521.

Potter, S.M., Jimenez-Flores, R., Pollack, J., Lone, T.A. and Berber-Jimenez, M.D., 1993, Protein-saponin interaction and its influence on blood lipids, *J. Agric. Food Chem.* 41:1287.

Price, K.R., Curl, C.L. and Fenwick, G.R., 1986, The saponin content and sapogenol composition of the seed of 13 varieties of legume, *J. Sci. Food Agric.* 37:1185.

Price, K.R., Johnson, I.T. and Fenwick, G.R., 1987, The chemistry and biological significance of saponins in foods and feedingstuffs, *CRC Crit. Rev. Food Sci. Nutr.* 26:27.

Ripsin, C.M., Keenan, J.M., Jacobs, D.R., Elmer, P.J., Welch, R.R., Van Horn, L., Liu, K., Turnbull, W.H., Thye, F.W., Kestin, M., Hegsted, M., Davidson, D.M., Davidson, M.H., Dugan, L.D., Demark-Wahnefried, W. and Beling, S., 1992, Oat products and lipid lowering, a meta-analysis, *J. Am. Med. Assn.* 267:3317.

Rohrlich, M. and Train, G., 1954, Isolierung eines bitterschmeckenden Glycosids aus Hafer, *Z. Lebensmittel - Unters. u. Forschung* 345.

Rotenberg, S. and Eggum, B.O., 1986, The effect of purified pectins with and without saponins in the diet on selected lipid parameters in liver and blood plasma of rats, *Acta Agric. Scand.*. 36:211.

Ruales, J. and Nair, B.M., 1993, Saponins, phytic acid, tannins and protease inhibitors in quinoa (*Chenopodium quinoa*, Willd) seeds, *Food Chem.* 48:137.

Sharma, A. and Sehgal, S., 1992, Effect of processing and cooking on the antinutritional factors of faba bean (*Vicia faba*), *Food Chem.* 43:383.

Shinnick, F.L., Ink, S.L. and Marlett, J.A., 1990, Dose response to a dietary oat bran fraction in cholesterol-fed rats, *J. Nutr.* 120:561.

Sidhu, G.S., Upson, B. and Malinow, M.R., 1987, Effects of soy saponins and tigogenin cellobioside on intestinal uptake of cholesterol, cholate and glucose, *Nutr. Rep. Int.* 35:615.

Southon, S., Johnson, I.T., Gee, J.M. and Price, K.R., 1988, The effect of *Gypsophila* saponins in the diet on mineral status and plasma cholesterol concentration in the rat, *Br. J. Nutr.* 59:49.

Stark, A. and Madar, Z., 1993, The effect of an ethanol extract derived from fenugreek (*Trigonella foenum-graecum*) on bile acid absorption and cholesterol levels in rats, *Br. J. Nutr.* 69:277.
Temmerman, A.M., Vonk, R.J., Niezen-Koning, K., Berger, R. and Fernandes, J., 1988, Long-term and short-term effects of dietary cholesterol and fats in the mongolian gerbil, *Ann. Nutr. Metab.* 32:177.
Topping, D.L., Trimble, R.P., Illman, R.J., Potter, J.D. and Oakenfull, D.G., 1980, Prevention of dietary hypercholesterolemia in the rat by soy flour high and low in saponins, *Nutr. Rep. Int.* 22:513.
Tschesche, R. and Schmidt, W., 1966, Zwei neue Saponine der oberirdischen Teile des Hafers (*Avena sativa*) mit Nuatigenin als Aglykon, *Z. Naturforsch.* 21b:896.
Tschesche, R., Tauscher, M., Fehlhaber, H-W. and Wulff, G., 1969, Avenacosid A, ein bisdesmosidisches Steroid-saponin aus *Avena sativa*, *Chem. Ber.* 102:2072.
Tschesche, R. and Lauven, P., 1971, Avenacosid B, ein zweites bisdesmosidisches Steroid-saponin aus *Avena sativa*, *Chem. Ber.* 104:3549.
Tschesche, R. and Wiemann, W., 1977, Desgluco-avenacosid-A und -B, biologisch aktive Nuatigeninglycoside, *Chem. Ber.* 110:2416.
Wang, Q., Pantzar, N., Jeppsson, B., Weström, B.R. and Karlsson, B.W., 1994, Increased intestinal marker absorption due to regional permeability changes and decreased intestinal transit during sepsis in the rat. *Scand. J. Gastroenterol.* 28:1001.

BIOLOGICAL EFFECTS OF FEED AND FORAGE SAPONINS AND THEIR IMPACTS ON ANIMAL PRODUCTION

P.R. Cheeke

Department of Animal Sciences
Oregon State University
Corvallis, OR 97331

Saponins occur in numerous forages and crop plants used to provide feed for domestic animals. The objective of this paper is to review the significance of saponins to animal production, with particular emphasis on new findings. In some cases, saponins are being revisited after a period of many years of inactivity. This applies, for example, to the roles of saponins in bloat and photosensitization.

Forage saponins of concern have traditionally been those of alfalfa and to a lesser extent clovers. Recently, saponins in tropical grasses have been found to have important implications in performance of grazing animals and a dermatitic condition called secondary photosensitization. Pasture and range weeds containing saponins have been implicated in photosensitization and reproductive disorders. Crop plants containing saponins include soybeans and minor grains such as quinoa and amaranth. The saponin-rich extracts of *Yucca schidigera* have been of extensive interest in animal and poultry production, because of their ammonia-binding properties.

The physiological effects of saponins have been reviewed previously (Applebaum and Birk, 1979; Birk and Peri, 1980; Cheeke, 1971, 1976, 1983; Fenwick *et al.*, 1991; Oakenfull and Sidhu, 1989; Price *et al.*, 1987). This paper will focus on aspects of saponins not emphasized in these previous reviews.

BIOLOGICAL EFFECTS OF SAPONINS

Some of the major biological effects of saponins include:

1. irritation of mucous membranes
2. erythrocyte hemolysis
3. effects on growth
4. ruminant bloat
5. enzyme inhibition
6. effects on feed intake and diet palatability
7. effects on nutrient absorption
8. effects on cholesterol and bile acid metabolism

9. effects on rumen microbiology and rumen metabolism
10. antifungal activity
11. insect feeding deterrence
12. secondary photosensitization
13. reproductive effects
14. ammonia-binding properties
15. effects on carcinogenesis

Saponins may have significant effects on all phases of animal metabolism, from the ingestion of feed to the excretion of wastes. In fact, their effects are significant even in the production of animal feed, as they have a role in alfalfa and other plants in protection against insects and diseases. Alfalfa saponins protect the plant against numerous pathogenic fungi, particularly in the roots (Cheeke, 1983). The effects of saponins will be discussed in a logical progression beginning with the ingestion of feed.

Effects on Feed Intake

Nonruminant animals such as swine and poultry display an aversion to dietary alfalfa. For example, Leamaster and Cheeke (1979) showed that at all dietary levels of alfalfa exceeding 0.5% of the diet, pigs preferred an alfalfa-free control diet. Various avian species (goose, turkey, quail, chicken) likewise showed an aversion to low dietary levels of alfalfa meal (Cheeke *et al.*, 1983). In these and other studies reviewed previously (Cheeke, 1983), the acceptability of low-saponin alfalfa was greater than for commercial alfalfa meal or alfalfa selected for high saponin content. In humans, high saponin alfalfa has an irritant effect on the tissues of the mouth and throat (Cheeke, 1983; Oleszek *et al.*, 1994). It is likely that irritant effects contribute substantially to the anti-palatability properties of alfalfa saponins. The presence of saponins significantly limits the extent to which alfalfa can be used in diets for swine and poultry.

Although alfalfa is primarily used as a feed for ruminants rather than nonruminants, the effects of alfalfa saponins on feed intake and diet preferences of ruminants have received little study. Sheep are less aversive than cattle to bitter substances (Cheeke, 1994), so there may be species differences in response. Dairy cattle often show aversion to particular batches of alfalfa hay. The role of saponins in palatability of alfalfa to dairy cattle warrants investigation, because of the importance of alfalfa in dairy cow feeding, and the critical factor of maximizing feed intake in high-producing dairy cattle.

Effects of Saponins in the Digestive Tract

Because dietary saponins are poorly absorbed, their major biological effects generally occur in the digestive tract. As indicated above, they influence feed intake. They continue to have an influence on the digestion and utilization of consumed feed as it enters the digestive tract.

Saponin Effects in the Rumen. Saponins have long been of interest as contributing factors to ruminant bloat. Because of their detergent properties, they contribute to the formation of a stable foam in the rumen. Their significance in bloat has been called into question by the observations of Majak *et al.* (1980), who found no difference in bloat incidence of cattle fed either high saponin or low saponin alfalfa. While their data provide no evidence that saponins contribute to ruminant bloat, their study was conducted with stall-fed animals. The impact of saponins on bloat under normal grazing conditions has not been assessed. A speculative role of saponins affecting rumen protozoa and thus bloat is provided below. Another possible mode of action of saponins in bloat is their pronounced inhibition of forestomach motility (Klita *et al.*, 1996).

Alfalfa saponins have an influence on rumen fermentation. Lu *et al.* (1987) showed that alfalfa saponins reduced total volatile fatty acid (VFA) and microbial protein synthesis in continuous flow cultures of rumen bacteria. Similar results were seen *in vivo* by Lu and Jorgensen (1987). Bacterial nitrogen flow to the duodenum was reduced 20-30% in sheep administered alfalfa saponins intraruminally. Of particular interest was a marked reduction in rumen protozoa numbers, particularly with high-concentrate diets. They speculated that defaunation (removal of protozoa) could be a result of the disruption of protozoal membranes by reaction of saponins (e.g. medicagenic acid) with membrane sterols. The susceptibility of rumen protozoa and lack of susceptibility of rumen bacteria to saponins is likely explained by the presence of cholesterol in eukaryotic membranes (including protozoa) but not in prokaryotic bacterial cells (Klita *et al.*, 1996). The defaunation effect of saponins could be of considerable significance in ruminant nutrition. Defaunation has been widely studied as a means of increasing microbial efficiency in the rumen (Yang and Varga, 1993), particularly with high-concentrate diets. Defaunation may be especially useful in ruminant production in the tropics, with low-protein, high-energy diets based on sugar cane by-products (Preston and Leng, 1987). Diaz *et al.* (1993) investigated the use of saponin-containing plant material as a means of defaunating tropical ruminants. They fed the saponin-rich seed pericarp of *Sapindus saponaria*, at 2.5% and 5% of the diet. The total number of rumen protozoa was significantly reduced with the feeding of *S. saponaria*. The numbers of rumen bacteria and fungi were increased, mainly as a result of reduced protozoal numbers. It may be possible to increase the efficiency of ruminant animal production in the tropics by using saponin-rich plants as feed additives, to reduce rumen protozoal numbers and thus improve the efficiency of ruminal fermentation.

Leng (1973) reviewed the possible role of protozoa in bloat. Under certain conditions, rumen protozoa may increase to very high numbers, and then suddenly burst and disintegrate, releasing starch and protein. The protozoal protein readily foams in the rumen, while the released starch would support a burst of bacterial fermentation and gas production. Variations in forage saponin concentrations could conceivably give rise to situations in which protozoa are suddenly killed, tipping the balance from a non-bloating to a bloating situation. A role of protozoa in bloat is indicated by the observation that defaunation of cows reduces the incidence and severity of bloat (Clarke *et al.*, 1969).

Saponins may influence rumen fermentation by way of effects on nitrogen metabolism. Saponin-containing yucca extract has ammonia-binding properties (Headon, 1991). Large diurnal variations in rumen ammonia concentrations may adversely affect microbial growth (Owens and Zinn, 1988), especially when diets containing highly fermentable sources of nitrogen are fed. For example, with dietary urea, there is a rapid release of ammonia in the rumen soon after feed is ingested. The resulting high concentration of ruminal ammonia leads to ammonia absorption, hepatic synthesis of urea, and urinary urea excretion. Subsequently, the concentration of ruminal ammonia may decline to levels insufficient to support optimal microbial growth. Hussain and Cheeke (1995) investigated the use of dietary yucca extract as a means of modulating diurnal fluctuations in ruminal ammonia, especially when non-protein nitrogen (NPN) as urea was fed. With urea-containing high-roughage diets, dietary yucca extract tended to reduce the peak concentration of ruminal ammonia nitrogen at 3 h post-feeding, and reduced plasma ammonia and urea concentrations. They concluded that dietary yucca extract could lead to improved efficiency of utilization of nitrogen in the rumen, particularly with high-roughage diets containing high levels of fermentable nitrogen. They proposed that yucca extract could bind ammonia when ruminal ammonia concentrations are high, and release it when ruminal ammonia is low, thus modulating ruminal ammonia concentrations to provide continuously adequate ammonia for microbial metabolism. The active ammonia-binding component of yucca extract has not been conclusively identified. In earlier studies, it has been assumed to be sarsaponin, but Headon (1991) has suggested that glycoproteins may be the active components. Wu *et al.* (1994) also investigated effects of yucca extract on ruminal nitrogen concentrations. They observed no significant effect, perhaps because of high rumen ammonia concentrations in their animals. They cited results indicating an ammonia-binding capacity of yucca extract of 34 μg of ammonia-N per g of yucca extract, which they suggested would be too low to be useful in modulating high rumen ammonia concentrations. However, this ammonia-binding capacity might be useful in serving as a slow-release source of ammonia when ruminal ammonia concentrations are low.

Wallace *et al.* (1994) examined the effect of yucca extract on rumen ammonia concentrations, ammonia-binding capacity, and rumen microbiology. They observed a decrease in rumen ammonia with addition of yucca extract, which they attributed to reduced proteolysis. They found little effect of yucca extract on ruminal bacteria, but protozoal activity was abolished. They concluded that yucca extract was unlikely to influence rumen ammonia concentrations directly, but by suppressing ciliate protozoa, may benefit ruminal fermentation and thus indirectly lead to lower ruminal ammonia concentrations.

To summarize effects of saponins on rumen fermentation, it appears that the major effect is defaunation. The detergent action of saponins kills rumen protozoa. Under some dietary conditions, especially high-energy, low-protein diets (e.g. sugar cane or molasses-based diets), defaunation improves protein utilization and animal performance.

Saponins are in some cases metabolized by rumen microbes. This is of particular significance in the secondary photosensitization syndrome, discussed below.

Intestinal Effects of Saponins. In non-ruminants, the primary digestive effects of saponins occur in the small intestine. Johnson *et al.* (1986) determined that some saponins increase the permeability of intestinal mucosal cells, thereby inhibiting active nutrient transport, and facilitate the uptake of substances to which the gut would normally be impermeable. They proposed that saponins react with cholesterol in membranes of the microvilli. There are several possible implications of such interactions. Increased gut permeability could lead to the uptake of antigens by the small intestine, causing allergenic reactions (Gee *et al.*, 1993). Structural lesions (membranolytic activity) increase the turn-over rate of intestinal mucosal cells, thus increasing endogenous energy and protein losses. These enhanced nutrient losses could partially account for the growth-decreasing effects of saponins. Story *et al.* (1984) noted damage to the intestinal villi of rats fed alfalfa, while the lesions were less severe in rats fed saponin-free alfalfa. Interestingly, the effects were greatest with non-cholesterol-binding saponins, suggesting that the effect of saponins might occur by a mechanism other than binding to membrane cholesterol.

Dietary saponins are poorly absorbed, because of their binding to the intestinal mucosa. This binding can be viewed as a protective mechanism to prevent saponin toxicity (Gee and Johnson, 1988). Absorbed saponins have marked hemolytic effects, so a means of preventing absorption of saponins and other surface-active compounds is virtually essential in herbivores.

Saponins inhibit active transport of nutrients (Johnson *et al.*, 1986). They may also reduce nutrient absorption by way of binding bile acids, thus interfering with the formation of lipid micelles. This is probably the mechanism by which saponins reduce the absorption of vitamin E and vitamin A (Jenkins and Atwal, 1994).

Saponins have been implicated in increased excretion of minerals such as sodium (Freeland *et al.*, 1985b). Chronic induction of negative mineral balance may be an important mode of action of dietary toxins (Foley *et al.*, 1995). Saponins may reduce the toxicity of some toxins, such as tannins, by binding them in the gut (Freeland *et al.*, 1985a). This interaction warrants further study and confirmation.

Saponins and Photosensitization

Photosensitization occurs when photodynamic compounds in the blood react with UV light, producing free radicals that react with tissue proteins and nucleic acids, causing severe skin lesions. In primary photosensitization, the photodynamic agents are consumed as such in the diet. An example is the compound hypericin, occurring in the common pasture weed *Hypericum perforatum* (Cheeke and Shull, 1985). Secondary photosensitization occurs when the excretion of bile is impaired, causing chlorophyll metabolites such as phylloerythrin to be present in the blood. Phylloerythrin is activated by UV light, producing a reactive free radical. Secondary photosensitization is a consequence of liver damage or damage to bile ducts, thus impeding the excretion of phylloerythrin, allowing it to occur in the blood. Saponins have recently been demonstrated to be involved in several livestock toxicoses involving secondary photosensitization. Some of the plants implicated include *Tribulus terrestris*, *Agave lecheguilla*, *Narthecium ossifragum*, and a number of tropical grasses including *Brachiaria* and *Panicum* spp.

Tribulus terrestris or puncture vine is a prostrate annual weed common on semi-arid rangelands in Australia, South Africa and the U.S. It has been responsible for major animal health problems in South Africa, causing a photosensitization condition known as geeldikkop (yellow thick head). A similar condition occurs in Australia (Glastonbury *et al.*, 1984). Signs of photosensitization are accompanied by hepatic lesions and crystalline deposits in the bile ducts (Kellerman *et al.*, 1988). *Tribulus* spp. contain a number of steroidal saponins. Kellerman *et al.* (1991) induced geeldikkop in sheep by administering a crude mixture of *Tribulus* saponins. Miles *et al.* (1994a,b) also induced the condition by oral administration to sheep of crude saponins from *Tribulus terrestris*, and demonstrated that the crystalline deposits in bile ducts were calcium salts of the glucuronides of the steroidal sapogenins epismilagenin and episarsasapogenin. Miles (1994b) studied ruminal and hepatic metabolism of *Tribulus* saponins in sheep. Ingested saponins were rapidly hydrolyzed in the rumen, releasing free sapogenins. Most of the sapogenins were rapidly converted by ruminal microflora to epismilagenin and episarsapogenin, which are absorbed. These are conjugated with glucuronic acid in the liver, and excreted in the bile as calcium salts of the sapogenin glucuronides. These salts are insoluble and precipitate in the bile ducts as crystals. The disruption of bile excretion results in retention of phylloerythrin and induction of photosensitization.

Several other plants cause photosensitization, with steroid saponins implicated in each case as the causative agents. These include the Norwegian pasture weed *Narthecium ossifragum*; calcium salts of steroid sapogenin glucuronides in bile crystals have been identified (Miles *et al.*, 1993). The succulent desert plant *Agave lecheguilla* also contains steroid saponins and causes bile crystals and photosensitization (Camp *et al.*, 1988).

Numerous tropical and warm-season grasses cause secondary photosensitization in livestock. These include various *Brachiaria* and *Panicum spp.* (Graydon *et al.*, 1991; Miles *et al.*, 1992). Liver damage characterized by deposition of crystals in and around the bile ducts is observed. Affected animals develop photophobia and severe dermatitis. The crystals consist of calcium salts of steroid saponin glucuronides (Miles *et al.*, 1991, 1992; Munday *et al.*, 1993). Experimental induction of photosensitization by administration of isolated saponins takes much larger quantities than would be obtained from consumption of the plants (Kellerman *et al.*, 1991). Miles *et al.* (1992) suggest that other hepatoxins such as mycotoxins (e.g. sporidesmin) may have synergistic effects with saponins. Kleingrass (*Panicum coloratum*) is widely grown in Texas, where it has caused outbreaks of photosensitization in sheep (Bridges *et al.*, 1987) and horses (Cornick *et al.*, 1988). Saponins were identified as causative agents, but the sporadic incidence of the condition suggests that other factors such as mycotoxins could contribute to the problem.

Photosensitization and liver damage occur in cattle grazing signalgrass (*Brachiaria decumbens*) in the tropics. Cattle on lush signalgrass pasture often exhibit poor growth, photosensitization, liver damage, and a marked reluctance to consume the grass (Low *et al.*, 1993). The lack of palatability, as well as the other signs, could be caused by saponins. Lajis *et al.* (1993) demonstrated that the rumen contents of sheep fed signalgrass contain the sapogenins episarsasapogenin and epismilagenin. Sheep are apparently more sensitive than cattle to signalgrass, suggesting differences in rumen metabolism of the saponins (Lajis *et al.*, 1993).

Thus steroid saponins in a wide variety of plants can induce photosensitization in ruminant animals. The saponins are metabolized in the rumen, and absorbed sapogenins are excreted in the bile as calcium salts of the sapogenin glucuronides. In some cases, these are insoluble, forming biliary crystals which prevent the excretion of the photodynamic compound phylloerythrin. Within a plant, there may be both lithogenic and non-lithogenic saponins. For example, *Tribulus terrestris* contains diosgenin and yamogenin which are metabolized to lithogenic glucuronides of epismilagenin and episarsasapogenin, while other saponins in the plant such as tigogenin, neotigogenin, gitogenin and neogitogenin are non-lithogenic (Miles *et al.*, 1994b). Variations in these saponins due to plant growth stage and environmental conditions could account for the sporadic incidence of photosensitization. As an historical note, Henrici (1952) in South Africa suggested that saponins in *Tribulus* were the causative agents of geeldikkop. This suggestion was not followed up, and the confirmation of the role of saponins did not occur for over forty years until the work of Miles *et al.* (1994a,b).

SAPONINS AS FEED ADDITIVES

Yucca extracts containing sarsaponins are extensively used as feed additives in animal production, to reduce the emission of ammonia from animal excreta. Yucca extract has ammonia-binding properties that result in ammonia in animal wastes being bound in a non-volatile form. In studies involving use of yucca extract, a growth-promotant effect is often observed (e.g. Johnston *et al.*, 1981; Al-Bar *et al.*, 1992). This effect has not been exploited commercially, primarily because the expense required to obtain approval for such use does not make this non-proprietary product of interest to private companies.

SAPONINS AND HUMAN HEALTH

Although not the focus of this paper, the possible roles of saponins in human health warrant brief mention. The hypocholesterolemic effects of saponins have been known for many years. They act in several ways. The most simple is the direct reaction of saponin with cholesterol in the intestine, yielding an insoluble complex which is excreted. In addition, some saponins interfere with micelle size and structure, thus altering bile acid absorption as well as inhibiting cholesterol absorption (Oakenfull and Sidhu, 1989). Harwood *et al.* (1993) demonstrated that a synthetic saponin, tiqueside, inhibits cholesterol absorption, which through a cascade of events in the liver results in reduction of plasma non-HDL cholesterol. They propose that this saponin or derivatives thereof have the potential to become drugs for control of hypercholesterolemia.

While interest in the hypocholesterolemic effects of saponins is of long standing, there is evidence of other beneficial effects of saponins on human health. Rao and Sung (1995) reviewed potential effects of saponins as anticarcinogens. Some saponins, such as those in ginseng, have a direct cytotoxic effect on tumor cells. Soybean and gypsophilla saponins inhibit growth of colon carcinoma cells in culture (Sung *et al.*, 1995). There is evidence that saponins are immune-stimulating agents. Rao and Sung (1995) suggest that saponins affect the cell membrane by binding to it, leading to changes in transmembrane signals. Because saponins bind bile acids (Oakenfull and Sidhu, 1989), they could reduce the availability of bile acids to intestinal microbes, reducing the formation of secondary bile acids which are implicated in colon cancer.

A final example of saponin effects on human health is the production of cholesterol-free milk. Dairy products can be passed through a column of polymer-supported saponins to remove cholesterol (Micich *et al.*, 1992).

CONCLUSIONS

Saponins are secondary compounds that function in plant defense against herbivory. Forage saponins have long been of interest in animal production, because of possible roles in pasture bloat and their growth-inhibiting properties. Areas of current emphasis involving saponins and animal production concern saponin effects on rumen protozoa, and saponin-induced photosensitization. Saponins inhibit rumen protozoa, probably by reaction with membrane sterols. In many situations, removal of rumen protozoa (defaunation) improves efficiency of rumen fermentation. Saponins may have potential application as defaunating agents. Recent research has demonstrated that steroidal saponins in several toxic plants and forage grasses are metabolized in the rumen, with the absorption of sapogenins. Absorbed sapogenins are conjugated with glucuronic acid, and excreted as calcium salts in the bile. With some sapogenins, the calcium salts are insoluble, causing biliary crystals that interfere with excretion of bile. Accumulation in the blood of phylloerythrin, a photodynamic metabolite of chlorophyll, causes secondary photosensitization. Finally, the saponins in *Yucca schidigera* are widely used in animal production as feed additives to control ammonia volatilization in animal excreta. Saponins are of additional interest in human health, because of hypocholesterolemic and anti-cancer activity.

REFERENCES

Al-Bar, A., A. Ismail, P.R. Cheeke and H. Nakaue, 1993, Effect of dietary *Yucca shidegera* extract (Deodorase) on environmental ammonia and growth performance of chickens and rabbits, *Proc. West. Sect. Am. Soc. Anim. Sci.* 44:106.

Applebaum, S.W., and Birk, Y., 1979, Saponins, pp. 539-566 in: *Herbivores. Their Interaction With Secondary Plant Metabolites,* G.A. Rosenthal and D.H. Janzen, Eds., Academic Press, New York.

Birk, Y., and Peri, I., 1980, Saponins, pp. 161-182 in: *Toxic Constituents of Plant Foodstuffs*, I.E. Liener, Ed., Academic Press, New York.

Bridges, C.H., Camp, B.J., Livingston, C.W., and Bailey, E.M., 1987, Kleingrass (*Panicum coloratum*) poisoning in sheep, *Vet. Pathol.* 24:525.

Camp, B.J., Bridges, C.H., Hill, D.W., Patamalai, B., and Wilson, S., 1988, Isolation of a steroidal sapogenin from a sheep fed *Agave lecheguilla*, *Vet. Human Tox.* 30:533.

Cheeke, P.R., 1971, Nutritional and physiological properties of saponins: a review, *Can. J. Anim. Sci.* 51:621.

Cheeke, P.R., 1976, Nutritional and physiological properties of saponins, *Nutr. Rep. Int.* 13:315.

Cheeke, P.R., 1983, Biological properties and nutritional significance of legume saponins, pp. 396-414 in: *Leaf Protein Concentrates*, L. Telek and H.D. Graham, Eds., AVI Publ. Co., Westport, CT.

Cheeke, P.R., 1994, A review of the functional and evolutionary roles of the liver in the detoxification of poisonous plants, with special reference to pyrrolizidine alkaloids, *Vet. Human Toxicol.* 36:240.

Cheeke, P.R., and Shull, L.R., 1985, Natural Toxicants In Feeds and Poisonous Plants, AVI Publishing Co., Inc., Westport, CT.

Cheeke, P.R., Powley, J.S., Nakaue, H.S., and Arscott, G.H., 1983, Feed preference responses of several avian species fed alfalfa meal, high and low saponin alfalfa, and quinine sulfate, *Can. J. Anim. Sci.* 63:707.

Clarke, R.T.J., Reid, C.S.W., and Young, P.W., 1969, Bloat in cattle. XXXII. Attempts to prevent legume bloat in dry and lactating cows by partial or complete elimination of the rumen holotrich protozoa with dimetridazole, *New Z. J. Agric. Res.* 12:446.

Cornick, J.L., Carter, G.K., and Bridges, C.H., 1988, Kleingrass-associated hepatotoxicosis in horses, *J. Am. Vet. Med. Assoc.* 193:932.

Diaz, A., Avendano, M., and Escobar, A., 1993, Evaluation of *Sapindus saponaria* as a defaunating agent and its effects on different ruminal digestion parameters, *Livestock Res. Rural Dev.* 2:1.

Fenwick, G.R., Price, K.R., Tsukamoto, C., and Okubo, K., 1991, Saponins, pp. 285-327 in: *Toxic Substances in Crop Plants*, J.P.F. D'Mello, C.M. Duffus and J.H. Duffus, Eds., The Royal Society of Chemistry, Cambridge, UK.

Foley, W.J., McLean, S., and Cork, S.J., 1995, Consequences of biotransformation of plant secondary metabolites on acid-base metabolism in mammals - A final common pathway?, *J. Chem. Ecol.* 21:721.

Freeland, W.J., Calcott, P.H., and Anderson, L.R., 1985a, Tannins and saponin: Interaction in herbivore diets, *Biochemical Systematics Ecol.* 13:189.

Freeland, W.J., Calcott, P.H., and Geiss, D.P., 1985b, Allelochemicals, minerals and herbivore population size, *Biochemical Systematics Ecol.*, 13:195.

Gee, J.M., and Johnson, I.T., 1988, Interactions between hemolytic saponins, bile salts and small intestinal mucosa in the rat, *J. Nutr.* 118:1391.

Gee, J.M., Price, K.R., Ridout, C.L., Wortley, G.M., Hurrell, R.F., and Johnson, I.T., 1993, Saponins of quinoa (*Chenopodium quinoa*): Effects of processing on their abundance in quinoa products and their biological effects on intestinal mucosal tissue, *J. Sci. Food Agric.* 63:201.

Gee, J.M., Wortley, G.M., and Johnson, I.T., 1994, Effects of some purified alfalfa (*Medicago sativa*) saponins on transmural potential difference in mammalian small intestine, *J. Sci. Food Agric.* 65:35.

Glastonbury, J.R.W., Doughty, F.R., Whitaker, S.J., and Sergeant, E., 1984, A syndrome of hepatogenous photosensitisation, resembling geeldikkop, in sheep grazing *Tribulus terrestris*, *Aust. Vet. J.* 61:314.

Graydon, R.J., Hamid, H., Zahari, P., and Gardiner, C., 1991, Photosensitisation and crystal-associated cholangiohepatopathy in sheep grazing *Brachiaria decumbens*, *Aust. Vet. J.* 68:234.

Harwood, H.J., Jr., Chandler, C.E., Pellarin, L.D., Bangerter, F.W., Wilkins, R.W., Long, C.A., Cosgrove, P.G., Maliow, M.R., Marzetta, C.A., Pettini, J.L., Savoy, Y.E., and Mayne, J.T., 1993, Pharmacologic consequences of cholesterol absorption inhibition: alteration in cholesterol metabolism and reduction in plasma cholesterol concentration induced by the synthetic saponin β-tigogenin cellobioside (CP-88818; tiqueside), *J. Lipid Res.* 34:377.

Headon, D.R., 1991, Glycofractions of the yucca plant and their role in ammonia control, pp. 95-108 in: *Biotechnology in the Feed Industry. Proc. 7th Alltech Symposium*, Alltech Technical Publications, Nicholasville, KY.

Henrici, M., 1952, Comparative study of the content of starch and sugars of *Tribulus terrestris*, lucerne, some *Gramineae* and *Pentzia incana* under different meteorological, edaphic and physiological conditions. 2. Carbohydrate nutrition. *Onderstepoort J. Vet. Res.* 25:45.

Hussain, I., and Cheeke, P.R., 1995, Effect of dietary *Yucca schidigera* extract on rumen and blood profiles of steers fed concentrate- or roughage-based diets, *Anim. Feed Sci. Tech.* 51:231.

Jenkins, K.J., and Atwal, A.S., 1994, Effects of dietary saponins on fecal bile acids and neutral sterols, and availability of vitamins A and E in the chick, *J. Nutr. Biochem.* 5:134.

Johnson, I.T., Gee, J.M., Price, K., Curl, C., and Fenwick, G.R., 1986, Influence of saponins on gut permeability and active nutrient transport *in vitro*, *J. Nutr.* 116:2270.

Johnston, N.L., Quarles, C.L., Fagerberg, D.J., and Caveny, D.D., 1981, Evaluation of yucca saponin on broiler performance and ammonia suppression, *Poul. Sci.* 60:2289.

Kellerman, T.S., Erasmus, G.L., Coetzer, J.A.W., Brown, J.M.M., and Maartens, B.P., 1991, Photosensitivity in South Africa. VI. The experimental induction of geeldikkop in sheep with crude steroidal saponins from *Tribulus terrestris*, *Onderstepoort J. Vet. Res.* 58:47.

Klita, P.T., Mathison, G.W., and Fenton, T.W., 1996, Effect of alfalfa root saponins on digestive function in sheep, *J. Anim. Sci.* 74:(in press).

Lajis, N.H., Abdullah, S.H., Salim, S.J.S., Bremner, J.B., and Khan, M.N., 1993, *Epi*-sarsasapogenin and *epi*-smilagenin: two sapogenins isolated from the rumen content of sheep intoxicated by *Brachiaria decumbens*, *Steroids* 58:387.

Leamaster, B.R., and Cheeke, P.R., 1979, Feed preferences of swine: Alfalfa meal, high and low saponin alfalfa, and quinine sulfate, *Can. J. Anim. Sci.* 59:467.

Leng, R.A., 1973, Ruminal fermentation and bloat: The possible role of protozoa in the development of bloat, pp. 57-62 in: *Proceedings of a Symposium Held at the University of New England,*

Armidale, NSW, Australia, R.A. Leng and J.R. McWilliam, Eds., University of New England Press.

Low, S.G., Bryden, W.L., Jephcott, S.B., and McL. Grant, I., 1993, Photosensitization of cattle grazing signal grass (*Brachiaria decumbens*) in Papua New Guinea, *New Z. Vet. J.* 41:220.

Lu, C.D., and Jorgensen, N.A., 1987, Alfalfa saponins affect site and extent of nutrient digestion in ruminants, *J. Nutr.* 117:919.

Lu, C.D., Tsai, L.S., Schaefer, D.M., and Jorgensen, N.A., 1987, Alteration of fermentation in continuous culture of mixed rumen bacteria by isolated alfalfa saponins, *J. Dairy Sci.* 70:799.

Majak, W., Howarth, R.E., Fesser, A.C., Goplen, B.P., and Pedersen, M.W., 1980. Relationships between ruminant bloat and the composition of alfalfa herbage, II. Saponins. *Can. J. Anim. Sci.* 60:699.

Micich, T.J., Foglia, T.A., and Holsinger, V.H., 1992, Polymer-supported saponins: An approach to cholesterol removal from butteroil, *J. Agric. Food Chem.*, 40:1321.

Miles, C.O., Munday, S.C., Holland, P.T., Smith, B.L, Embling, P.P., and Wilkins, A.L., 1991, Identification of a sapogenin glucuronide in the bile of sheep affected by *Panicum dichotomiflorum* toxicosis, *New Z. Vet. J.* 39:150.

Miles, C.O., Wilkins, A.L., Munday, S.C., Holland, P.T., Smith, B.L., Lancaster, M.J., and Embling, P.P., 1992, Identification of the calcium salt of epismilagenin β-d-glucuronide in the bile crystals of sheep affected by *Panicum dichotomiflorum* and *Panicum schinzii* toxicoses. *J. Agric. Food Chem.* 40:1606.

Miles, C.O., Wilkins, A.L., Munday, S.C., Flaoyen, A., Holland, P.T., and Smith, B.L., 1993, Identification of insoluble salts of the β-D-glucuronides of episarsasapogenin and epismilagenin in the bile of lambs with alveld, and examination of *Narthecium ossifragum*, *Tribulus terrestris*, and *Panicum miliaceum* for sapogenins. *J. Agric. Food Chem.* 41:914.

Miles, C.O., Wilkins, A.L., Erasmus, G.L., Kellerman, T.S., and Coetzer, J.A.W., 1994a, Photosensitivity in South Africa. VII. Chemical composition of biliary crystals from a sheep with experimentally induced geeldikkop, *Onderstepoort J. Vet. Res.* 61:215.

Miles, C.O., Wilkins, A.L., Erasmus, G.L., and Kellerman, T.S., 1994b, Photosensitivity in South Africa. VIII. Ovine metabolism of *Tribulus terrestris* saponins during experimentally induced geeldikkop, *Onderstepoort J. Vet. Res.* 61:351.

Munday, S.C., Wilkins, A.L., Miles, C.O., and Holland, P.T., 1993, Isolation and structure elucidation of dichotomin, a furostanol saponin implicated in hepatogenous photosensitization of sheep grazing *Panicum dichotomiflorum*, *J. Agric. Food Chem.* 41:267.

Oakenfull, D., and Sidhu, G.S., 1989, Saponins, pp. 97-141 in: *Toxicants of Plant Origin Vol. II*, P.R. Cheeke, Ed., CRC Press, Boca Raton, FL.

Oleszek, W., Nowacka, J., Gee, J.M., Wortley, G.M., and Johnson, I.T., 1994, Effects of some purified alfalfa (*Medicago sativa*) saponins on transmural potential difference in mammalian small intestine. *J. Sci. Food Agric.* 65:35-39.

Owens, F.N., and Zinn, R., 1988, Protein metabolism of ruminant animals, pp. 227-249 in: *The Ruminant Animal. Digestive Physiology and Nutrition*, D.C. Church, Ed., Prentice Hall, Englewood Cliffs, N.J.

Preston, T.R., and Leng, R.A., 1987, *Matching Ruminant Production Systems With Available Resources in the Tropics and Subtropics*, Penambul Books, Armidale, Australia.

Price, K.R., Johnson, I.T., and Fenwich, G.R., 1987, The chemistry and biological significance of saponins in foods and feedingstuffs, *CRC Crit. Rev. Food Sci. Nutr.* 26:27.

Rao, A.V., and Sung, M.-K., 1995, Saponins as anticarcinogens, *J. Nutr.* 125:717S.

Story, J.A., LePage, S.L., Petro, M.S., West, L.G., Cassidy, M.M., Lightfoot, F.G., and Vahouny, G.V., 1984, Interactions of alfalfa plant and sprout saponins with cholesterol in vitro and in cholesterol-fed rats, *Am. J. Clin. Nutr.* 39:917.

Sung, M.-K., Kendall, C.W.C., Koo, M.M., and Rao, A.V., 1995, Effect of soybean saponins and gypsophilla saponin on growth and viability of colon carcinoma cells in culture, *Nutr. Cancer* 23:259.

Wallace, R.J., Arthaud, L., and Newbold, C.J., 1994, Influence of *Yucca shidigera* extract on ruminal ammonia concentrations and ruminal microorganisms, *Appl. Environ. Microbiol.* 60:1762.

Wu, Z., Sadik, M., Sleiman, F.T., Simas, J.M., Pessarakli, M., and Huber, J.T., 1994, Influence of *Yucca* extract on ruminal metabolism in cows, *J. Anim. Sci.* 72:1038.

Yang, C.-M.J., and Varga, G.A., 1993, The effects of continuous ruminal dosing with dioctyl sodium sulphosuccinate on ruminal and metabolic characteristics of lactating Holstein cows. *British J. Nutr.* 69:397.

EFFECT OF QUILLAJA SAPONINS ON *IN VITRO* RUMEN FERMENTATION

Harinder P. S. Makkar and Klaus Becker

Institute for Animal Production in the Tropics and Subtropics
University of Hohenheim (480), D-70593 Stuttgart, Germany

INTRODUCTION

Saponins are glycosides of an aglycone (genin or sapogenin) linked to one or more sugar chains. These are generally divided into three classes: i) steroid glycosides, ii) steroid alkaloid glycosides, and iii) triterpene glycosides. Saponins have a wide distribution in the plant kingdom[1,2] and are generally grouped among antinutritional factors. We investigated effects of quillaja bark saponin on rumen fermentation characteristics using the Hohenheim *in vitro* gas method.

EXPERIMENTAL

Materials

Saponin (source: Quillaja bark) was obtained from Roth, Karlsruhe, Germany. The saponin was free of tannins.

Methods

Effect on digestion parameters. The *in vitro* gas production method was essentially according to Menke et al[3]. For determination of rate (c) and potential extent of fermentation ($a + b$) using the model ($y = a + b(1 - e^{-ct})$) where y is the gas produced at

Saponins Used in Food and Agriculture
Edited by Waller and Yamasaki, Plenum Press, New York, 1996

time *t*) of Ørskov and McDonald[4], 200 mg (190 mg on dry matter basis; 95 % dry matter) of hay was incubated in triplicate in syringes containing 30 ml of the *in vitro* medium containing rumen liquor[3]. The gas values were recorded at 4, 6, 8, 10, 12, 24, 30, 36, 48, 54, 60, 72 and 96 h. The saponin was added in different concentrations to the incubation medium. The sequence of addition was as follows. An aliquot (0.5 to 1 ml) of the saponin solution in distilled water was added to the syringe containing the hay sample, followed by 30 ml of the *in vitro* medium. The time during which the saponin solution came in contact with the hay before the injection of the *in vitro* medium was less than 1 min. For determination of true and apparent digestibilities, 400 mg (380 mg on dry matter basis; 95 % dry matter) of the hay was incubated in 40 ml of the double-strength *in vitro* medium containing rumen liquor[5] instead of 30 ml in each syringe. The amount of hay was increased in order to avoid errors due to small amounts of fermented material which was analyzed for neutral detergent fiber (NDF) content after incubation. Both true and apparent digestibilities were determined at 24 h of fermentation. Gas released in the syringes was also recorded before subjecting syringe contents to determination of true or apparent digestibilities. True digestibility was determined by treating the syringe contents with neutral detergent solution to obtain NDF residue. Truly digested substrate was the difference in weights between the hay taken for incubation and the NDF residue following 24 h fermentation. The apparent digestibility was the difference in weights between the hay taken for incubation and weight of the dried residue (after correcting for the weight of corresponding blank) obtained following centrifugation (20×10^3 *g* for 30 min) of syringe contents, washing the residue with normal saline and re-centrifugation as above. This residue was lyophilized and then kept at 50 °C for about 12 h. This residue was called the apparent residue.

For measurement of ^{15}N incorporation, the fermentation was carried out in the usual buffer containing ammonium carbonate consisting of highly enriched (96.5 %) ^{15}N. The ^{15}N was measured with a mass spectrophotometer (Finnigan MAT 251)[6] in the apparent residue obtained as above. The incorporation of ^{15}N was measured after correcting for ^{15}N in the corresponding blank.

Other analyses. The NDF was determined according to Goering and Van Soest [7]. The short-chain fatty acids were determined in the supernatants obtained on centrifuging (20×10^3 *g*) the syringe contents at 24 h of fermentation, with a Hewlett Packard gas chromatograph[8].

The significance of difference between means was compared by using Duncan Multiple Range Test after ANOVA for one-way classified data with the aid of SAS/STAT programme[9].

RESULTS

Effect on rate and potential extent of digestion, apparent and true digestibilities and gas production

Saponin from Quillaja bark at 0.2, 0.4 and 0.6 mg/ml in the medium decreased the rate by about 7, 11 and 15 % respectively. The potential extent of digestion was not affected up to a concentration of 0.6 mg/ml (Table 1). The apparent digestibility and the gas production at 24 h decreased significantly at 1.2 mg/ml (Table 2).

Table 1. Effect of saponin on rate of gas production *(c)* and potential extent of gas production $(a + b)$ of hay

	% Decrease in	
	c	$(a + b)$
Saponin (0.2 mg/ml)	7.0	0
Saponin (0.4 mg/ml)	10.7	0
Saponin (0.6 mg/ml)	15.1	0

The c, and $a + b$ of hay were 0.0702 $\pm$ 0.001/h and 47.39 $\pm$ 0.51 ml respectively (mean $\pm$ SD, $n = 3$).

Table 2. Effect of saponin on apparent (AD) and true (TD) digestibilities and gas production at 24 h of fermentation of hay

	Percent		ml Gas
	AD	TD	
Control	50 ^{a}	70.5 ^{a}	74.7 ^{a}
Saponin (0.4 mg/ml)	49.2 ^{ab} (1.6)	70.6 ^{a} (+0.1)	74.1 ^{a} (0.8)
Saponin (0.8 mg/ml)	48.9 ^{ab} (2.3)	68.0 ^{b} (3.5)	73.5 ^{a} (1.6)
Saponin (1.2 mg/ml)	46.3 ^{b} (7.4)	69.3 ^{ab} (1.7)	67.9 ^{b} (9.1)

Values in parentheses are the % decrease compared to the control. Values with different superscripts in a column differ at $P < 0.05$ ($n = 3$).

Effect on short-chain fatty acid (SCFA) production and their molar proportions

The SCFA levels at 24 h of fermentation and their molar proportions are presented in Table 3. The saponin at a level up to 1.2 mg/ml did not change the level of SCFA production or their molar proportions appreciably.

Table 3. Effect of saponin on the production of short-chain fatty acids (SCFA)

	SCFA (µmol per syringe)			
	C2	C3	C4	C5
Control	916.0 [a] (66.0)	323.9 [a] (23.3)	118.5 [a] (8.5)	30.1 [a] (2.2)
Saponin (0.4 mg/ml)	891.9 [a] (66.3)	279.8 [a] (23.8)	102.2 [a] (7.6)	30.8 [a] (2.3)
Saponin (0.8 mg/ml)	876.7 [a] (65.0)	335.1 [a] (24.8)	101.7 [a] (7.5)	35.8 [a] (2.7)
Saponin (1.2 mg/ml)	843.6 [a] (67.1)	315.4 [a] (25.1)	76.9 [b] (6.1)	21.4 [a] (0.8)

Values in parentheses are molar proportions. Values with different superscripts in a column differ at $P < 0.05$ ($n = 3$)

Efficiency of microbial protein synthesis

There was an increase in the incorporation of ^{15}N into the rumen microbes in the presence of saponin. The content of ^{15}N of the total nitrogen was relatively constant (23-24 %) (Table 4). The higher nitrogen content of the apparent residue (Table 4) appears to be due to decrease in the protein digestibility of the substrate and increase in the microbial protein synthesis. The efficiency of microbial protein synthesis, EMPS (mg^{15}N incorporation/mmol SCFA production) was higher for the saponin.

DISCUSSION

Saponin from quillaja bark decreased rate of digestion of feed. The lower intake of saponin-rich feeds is generally attributed to the bitter taste of saponins. Besides unpleasant taste, the decrease in rate of digestion of feeds by saponins could be another mechanism by

Table 4. Effect of saponin on total nitrogen and ^{15}N contents of the apparent residue in a syringe following 24 h fermentation, and efficiency of microbial protein synthesis (EMPS; mg ^{15}N incorporation/mmol SCFA production)

	mg Nitrogen	mg ^{15}N	EMPS
Control	5.89 ± 0.10	1.33 0± 0.032 (22.6)	0.958
Saponin (0.4 mg ml)	6.10 ± 0.41	1.414 ± 0.090 (23.2)	1.084
Saponin (0.8 mg/ml)	6.80 ± 0.09	1.649 ± 0.100 (24.3)	1.222
Saponin (1.2 mg/ml)	7.22 ± 0.26	1.833 ± 0.065 (23.0)	1.458

Values in parentheses are the % ^{15}N of the total nitrogen. Values are mean ± SD (n = 3)

which feed intake is decreased. The lower the rate of digestion (higher rumen fill), the lower the feed intake[10]. The saponin affected the rate to a greater extent than the potential extent of digestion and the digestibilities (Tables 1 & 2). Therefore, the threshold level at which the rate is affected will be lower as compared to the digestibility. Our similar studies on interaction of tannins and saponins have shown that the effects of both tannins and quillaja saponin are additive. These results do not support the hypothesis[11,12,13] that joint presence of different antinutrients alleviates the effects of one another. The results could be different with saponins or tannins from different sources, which should be examined. The biological effects of both tannins[14] and saponins[15,16] are dependent on their chemical structure. The results of such studies may extend the understanding of the nutritional role of complex mixtures of antinutrients in feed, as generally antinutrients do not exist in isolation. Furthermore, grazing animals interact with a great number of antinutrients daily.

The increase in the efficiency of microbial protein synthesis by the saponin is noteworthy. Lower SCFA production (Table 3), higher ^{15}N incorporation into the microbes (Table 4), and virtually no change in the true digestibilty (Table 2) suggest that nutrients are partitioned more towards microbial protein synthesis in the presence of the saponin. These results are in consonance with the concept that the efficiency of microbial growth in the rumen is highly variable depending on the feeding conditions[17]. Saponins have antibiotic activity. Avoparcin and monensin possessing antibiotic activity have been demonstrated to have growth-promoting activity but unlike saponins these uncouple fermentation and microbial growth[18]. About 15 % lower degradability of feed protein and 32 % decrease in protozoal number at a level of 1.2 mg saponin/ml in the medium (results not shown) are also interesting to note. Inhibition of rumen proteolytic activity and antiprotozoal activity of *Yucca* saponins have also been demonstrated[19]. Saponins have varied biological properties, some of which are deleterious, but many of which are beneficial[19,20] . The oral toxicity of saponins to warm-blooded organisms is very low, though they are highly toxic when given intravenously. The adverse effect of saponins ingested by ruminants is limited. The role of saponins in causing bloat is not clear. Some workers suggest that saponins may contribute to bloat whereas others do not support this contention. Saponins may reduce rumen motility and affect salivation (see review[21]). Saponins from quillaja are widely used as food additives and did not show any significant toxic effects in short-term feeding studies on rats and long-term toxicity studies on mice (see review[22]). These saponin have been classified as 'generally recognized as safe' (GRAS) for permitted food use in the US[23]. The nontoxic nature of the saponin studied in the present study, the increase in the synthesis of microbial protein and lack of effect on digestibilities suggest the potential of this saponin as feed additives, which should be tried *in vivo*. Saponins from a variety of sources have also been shown to assist the absorption of nutrients by increasing the permeability of small intestinal mucosa (see review[1]). It would be interesting to investigate the mechanisms involved in increasing the efficiency of microbial protein synthesis by these saponins. The primary biological effect of saponins is the interaction with cellular and membrane components. The increase in the efficiency of microbial protein synthesis by saponins or its degraded products in the rumen (saponins may be degraded by rumen bacteria[1]) may be mediated via increase in the cellular transport of substrates for microbial protein synthseis. Saponins from ginseng (*Panax ginseng*; Araliaceae and related species) are perhaps the most studied saponins whose actions both at cellular and molecular levels are well understood (see reviews:[1,2]). It is worth mentioning that saponin from Korean ginseng stimulated growth of *Escherichia coli* at lower concentrations[2]. There is some evidence that saponin supplements in feeds may be beneficial to fattening lambs[24] and steers[25]. In light of the above findings saponins may not be considered as antinutritional. These groups of chemical compounds could have a beneficial role by increasing the efficiency of feed utilization, as the saponin decreased rate of digestion (hence feed intake), decreased degradability of feed protein and increased the efficiency of microbial protein synthesis.

ACKNOWLEDGEMENTS

We would like to thank Mr. Herrmann Baumgärtner for excellent technical assistance.

REFERENCES

1. Price, K. R., Johnson, I. T., and Fenwick, G. R., 1987, The chemistry and biological significance of saponins in foods and feedstuffs. *CRC Crit Rev Food Sci Nutr* **26** : 27-127.
2. Mahato, S. B., Sarkar, S. K., and Poddar, G., 1988, Triterpenoid saponins. *Phytochemistry* **27** : 3037-3067.
3. Menke, K. H., Raab, L., Salewski, A., Steingass, H., Fritz, D., and Schneider, W., 1979, The estimation of the digestibility and metabolizable energy content of ruminant feedstuffs from the gas production when they are incubated with rumen liquor. *J Agric Food Sci* **93** : 217-222.
4. Ørskov, E. R., and McDonald, I., 1979, The estimation of protein degradability in the rumen from incubation measurements weighted according to rate of passage. *J Agric Sci, Camb* **92** : 499-503.
5. Makkar, H. P. S., Blümmel, M., and Becker, K., 1995, Formation of complexes between polyvinyl pyrrolidones and polyethylene glycols with tannins and their implications in gas production and true digestibility in *in vitro* techniques. *Brit J Nutr* **73** : 897-913.
6. Abel, H-J. V., Coenen, G., and Immig, I., 1990, Untersuchungen zum Einfluß von Fett- und Stärkezulagen auf den mikrobiellen Stoffwechsel im Pansensimulationssystem RUSITEC. *J Anim Physiol & Anim Nutr* **64** : 62-73.
7. Goering, H.G., and Van Soest, P. J. 1970, *Forage fiber analysis*. ARS. USDA Washington, DC. Agriculture Handbook, no. 379, 20pp.
8. Cafantaris, B., 1981, Über die Wirkung von Antibiotikazusätzen auf die mikrobielle Gärung im Pansensaft in vitro. PhD Thesis, University of Hohenheim, Stuttgart, Germany.
9. SAS, 1988, *SAS/STAT Programme*, SAS Institute Inc, Cary, NC, USA
10. Van Soest , P. J., 1982, Intake. In: *Nutritional Ecology of the Ruminant,* O & B Books, Corvallis, Oregon, USA, pp 276-293.
11. Freeland, W. J., Calcott, P. H., and Anderson, L. R., 1985, Tannins and saponins: interaction in herbivore diets. *Biochemi Systematics Ecolo* **13** : 89-193.
12. Goldstein, W. S., and Spencer, K. C., 1985, Inhibition of cyanogenesis by tannins. *J Chem Ecol* **11** : 847-857.
13. Fish, B. C., and Thompson, L. U., 1991, Lectin-tannin interactions and their influence on pancreatic amylase activity and starch digestibility. *J Agric Food Chem* **39** : 727-731.
14. Makkar, H. P. S., 1993, Antinutritional factors in foods for livestock. In: *Animal production in developing countries,* Occasional publication no. 16, Gill, M., Owen, E., Pollott, G. E., and Lawrence, T. L., eds., British Society of Animal Production, pp 69-85.
15. Øleszek, W. 1993, Allelopathic potentials of alfalfa (*Medicago sativa*) saponins: their relation to antifungal and hemolytic activities. *J Chem Ecol* **19** : 1063-1074.
16. Potter, S. M., Jimenez-Fores, R., Pollack, J., Lone, T. A., and Berber-Jimenez, D. M., 1993, Protein-saponin interactions and its influence on blood lipids. *J Agric Food Chem* **41** : 1287-1291.
17. Leng, R. A., 1993, Quantitative ruminant nutrition - A green science. *Aust J Agric Res* **44** : 363-380.
18. Jouany, J. P., and Thivend, P., 1986, In vitro effects of avoparcin on protein degradability and rumen fermentation. *Anim Feed Sci Technol* **15** : 215-229.

19. Wallace, R.J., Arthaud, L., and Newbold, C.J., 1994, Influence of *Yucca schidigera* extract on ruminal ammonia concentrations and ruminal microorganisms. *Appl Environ Microbiol* **60** : 1762-1767.
20. Hussain, I., and Cheeke; P.R., 1995; Effect of dietary *Yucca schidigera* extract on rumen and blood profiles of steers fed concentrate- or roughage-based diets. *Anim Feed Sci Technol* **51** : 231-242.
21. Bondi, A., and Alumot, E., 1987, Anti-nutritional factors in animal feedstuffs and their effects on livestock. *Prog Food Nutr Sci* **11** : 115-151.
22. Mahato, S. B., and Nandy, A. K., 1991, Triterpenoid saponins discovered between 1987 and 1989. *Phytochemistry* **30** : 1357-1390.
23. Fenwick, G. R., Price, K R., Tsukamoto, C., and Okubo, K., 1992, Saponins. In : *Toxic Substances in Crop Plants,* D'Mello J P F, Duffus C M, Duffus J H, eds., The Royal Society of Chemistry, London, U.K, pp 285-327.
24. Hale, W. H., Sherman, W. C., Reynold, W. M., and Luther, H. G., 1961, The value of certain steroidal sapogenins in rations of fattening lambs and cattle. *Proc Soc Exp Biol Med* **106** : 486.
25. Goodall, S. R., Braddy, P., Horton, D., and Beckner, B., 1982, Steam flaked vs. high moisture content rations with and without sarsasaponin for finishing steers. *Proc West Sec Am Soc Anim Sci* **33** : 45-48.

DO STEROIDAL SAPONINS HAVE A ROLE IN HEPATOGENOUS PHOTOSENSITIZATION DISEASES OF SHEEP?

Arne Flåøyen

Department of Reproduction and Forensic Medicine
Norwegian College of Veterinary Medicine
P.O. Box 8146 Dep.
N-0033 Oslo, Norway

INTRODUCTION

Photosensitization of sheep is a problem of great economic importance in various parts of the world. For instance, more than half a million sheep have been reported to be photosensitized in a severe outbreak of geeldikkop, a photosensitization associated with ingestion of the plant *Tribulus terrestris* in South Africa (Steyn, 1949). In some years in some flocks in Norway, up to 30 to 50% of the lambs will suffer from alveld, a photosensitization occurring after ingestion of *Narthecium ossifragum*.

PHOTOSENSITIZATION

The clinical signs of all photosensitization diseases are similar regardless of the course of the disease (Fig. 1). Even though the clinical signs are almost identical, it is important to separate the disease into three major groups according to the pathogenesis of the diseases.

1. Hepatogenous photosensitization: a toxin normally produced by a plant, fungus, or alga, causes liver damage or liver dysfunction, resulting in retention of the photosensitizing agent phylloerythrin. Phylloerythrin is produced through metabolism of chlorophyll by rumen microorganisms (Rimington and Quin, 1933). All photosensitization diseases associated with ingestion of steroidal saponins belong to this group.
2. Primary photosensitization: diseases caused by a photosensitizing agent absorbed directly from the digestive tract and transported with the blood to the skin where it causes the characteristic lesions when exposed to light.
3. Photosensitization due to aberrant pigment synthesis: the photosensitizing agent is produced endogenously by an aberrant metabolic process.

The two latter forms of photosensitization do not involve ingestion of steroidal saponins and will not be further discussed in this paper.

Due to chemical reactions of the photosensitizers after exposure to sunlight, an acute inflammatory response in the skin is induced. In general, only unpigmented animals or animals with unpigmented areas become photosensitized. In sheep, the most salient clinical signs are increasing restlessness, head shaking, scratching of the face and ears with the hind feet, and rubbing of the irritated skin against the ground. The skin changes develop rapidly, with swelling and reddening. The eyelids, muzzle, and lips become swollen and turgid. However, the most obvious signs are the thickened edematous, heavily drooping ears, followed by seepage of sticky, honey-colored serum from the thickened skin which then forms extensive scabs that mat the covering hair after one or two days. In addition, signs of jaundice are often seen in animals suffering from hepatogenous photosensitization.

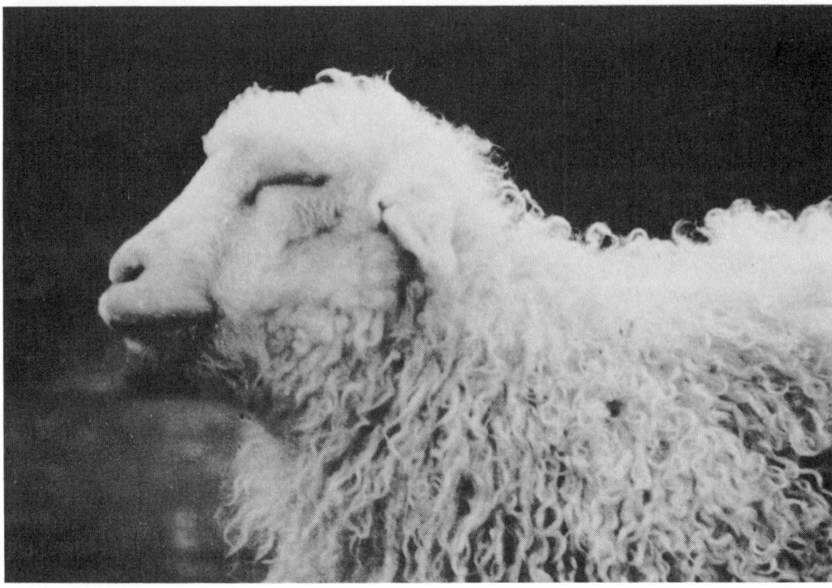

Figure 1. Ten-weeks-old lamb photosensitized after ingesting *Narthecium ossifragum.*

SAPONIN-CONTAINING PLANTS IMPLICATED IN PHOTOSENSITIZATION DISEASES

In the mid 1930s, Mathews reported the occurrence of photosensitization of sheep grazing *Agave lecheguilla* in Texas, United States, and he suggested that steroidal saponins in the plant caused the disease (Mathews, 1937). Since then, more than 8 other diseases have been suggested to have a similar etiology. Typical for all these diseases are their sporadic occurrence, the difficulty of reproducing symptoms during dosing experiments, and the fact that the saponin-containing plants associated with the diseases seem not to be always toxic to the grazing sheep. Another common finding for this group of diseases is the accumulation of birefringent crystals in hepatocytes and in and about bile ductuli and bile ducts. For the diseases that have been well studied, the crystals have been reported to be insoluble salts of sapogenin glucuronides originating from the saponins present in the corresponding plants.

Agave lecheguilla

A. lecheguilla is a conspicuous, long-lived perennial with a thick fibrous toothed crown bearing a cluster of thick, fleshy basal leaves and a tall flower stalk (Mathews, 1937). The plant grows thickly over a wide area of southwestern Texas, where it is reported to cause photosensitization of sheep (Mathews, 1937; Wall et al., 1962). The leaves contain, on average, 1% smilagenin (dry basis) as the sole sapogenin constituent (Wall et al., 1962). Crystals from the bile of sheep fed *A. lecheguilla* have been analyzed and suggested to contain either smilagenin or sarsasapogenin (Camp et al., 1988).

Tribulus terrestris

T. terrestris is a prostrate creeping plant with a semi-perennial underground stem and root system. Each year a mat of aerial branches emerges bearing compound leaves with five to eight pairs of hairy leaflets. The small flowers are yellow (Kellerman and Coetzer, 1984). The plant grows in different parts of the world and is reported to cause photosensitization of sheep in South Africa, United States, Argentina, Australia, and Iran (Amjadi et al., 1977; Camp et al., 1988; Coetzer et al., 1983; Glastonbury et al., 1984; Tapia et al., 1994). As well as causing photosensitization of sheep, *T. terrestris* has been reported to cause photosensitization in goats (Glastonbury and Boal, 1985). *T. terrestris* contain several steroidal sapogenins. These include diosgenin, yamogenin, epismilagenin, tigogenin, neotigogenin, gitogenin, and neogitogenin, ruscogenin, and neoruscogenin (Miles et al., 1994b). Biliary crystals from sheep photosensitized after ingesting *T. terrestris* have been found to consist of the insoluble calcium salts of epismilagenin ß-D-glucuronide and episarsasapogenin ß-D-glucuronide, in the ratio 6:1 (Miles et al., 1994b).

Narthecium ossifragum

N. ossifragum is a loosely to densely clonal, perennial herb, up to 40 cm tall, with a creeping rhizome. The plant occurs on oligotrophic, mesotrophic, and eutrophic peat deposits in Scandinavia to 69° 42' N, in the British Isles, the Netherlands, Belgium, northwestern Germany, western and central France, northern Spain, and eastern Portugal (Summerfield, 1974). Photosensitization of sheep grazing this plant has been reported from Norway, the Faroe Island, and the British Isles (Ender, 1955; Flåøyen, 1993; Flåøyen et al., 1995b; Ford, 1964). Alveld, literally "elf-fire", is the Norwegian name of the photosensitization. In Britain the disease is called saut, yellowsees, or plocteach. Photosensitization occurs normally only in lambs 2-6 months of age, and is rarely seen in adult sheep (Flåøyen, 1993). Breed differences in susceptibility to the disease have been demonstrated (Flåøyen, 1991). In addition to photosensitization of sheep *N. ossifragum* is also causing nephrotoxicity to cattle (Flåøyen et al., 1995a; Malone et al., 1992). Unpublished results from ongoing dosing experiments indicate that the hepatotoxicity and the nephrotoxicity are caused by different toxins (Flåøyen and others). *N. ossifragum* has been found to contain four different sapogenins (Ceh and Hauge, 1981; Miles et al., 1993). These are sarsasapogenin, smilagenin, yamogenin, and another spirostanol, tentatively identified as neotigogenin, in a ratio 82:9:5:4 (Miles et al., 1993). Bile from sheep photosensitized after grazing *N. ossifragum* has been found to contain insoluble salts of a 4:1 mixture of episarsasapogenin ß-D-glucuronide and epismilagenin ß-D-glucuronide (Miles et al., 1993) (Fig. 2).

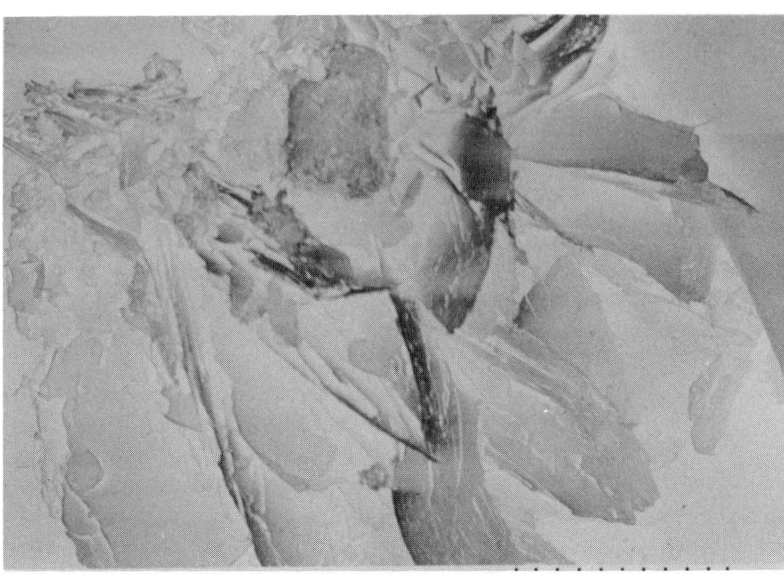

Figure 2. Characteristic birefringent crystals from the bile of a lamb photosensitized after ingesting *Narthecium ossifragum*.

Plants belonging to the *Panicum* spp.

The *Panicoid* grasses are annual or perennial grasses, either tufted or rhizomatous. The leaves are broad or narrow, and the ligule may be a membrane, a ciliate membrane, a rim of hairs, or sometimes absent. The silica bodies are cross- or dumbbell-shaped and microhairs are always present. *Panicum* plants are widely distributed from the tropics to the warm temperate regions. Many of the *Panicum* species are important pasture grasses or cereals and some are aggressive weeds (Dahlgren et al., 1985).

Panicum dichotomiflorum. *P. dichotomiflorum* is thought to have been introduced accidentally to New Zealand via aircraft during the Second World War, and it is considered to be a weed of importance (Miles et al., 1991). Photosensitization of sheep due to grazing of *P. dichotomiflorum* is recorded from New Zealand (Clare, 1952; Holland et al., 1991; Miles et al., 1991). As well as causing photosensitization in sheep, it is well known to cause this disorder in cattle and goats (Holland et al., 1991). The only sapogenin reported to be found in *P. dichotomiflorum* is diosgenin (Miles et al., 1991; Miles et al., 1992b), and the biliary crystals from photosensitized sheep are composed of the calcium salts of epismilagenin ß-D-glucuronide.

Panicum schinzii. Photosensitization of sheep grazing *P. schinzii* has been reported to occur in Australia (Button et al., 1987; Lancaster et al., 1991; Miles et al., 1992a; Miles et al., 1992b). The only sapogenin that has been found in *P. schinzii* is diosgenin, and, as in *P. dichotomiflorum*-intoxicated sheep, the calcium salt of epismilagenin ß-D-glucuronide was the only component in the biliary crystals of sheep affected by *P. schinzii* toxicosis (Lancaster et al., 1991; Miles et al., 1992a; Miles et al., 1992b).

Panicum miliaceum. *P. miliaceum* is a grass that has its primary centre in northern China, where it has been used for producing cereals (Zeven and Zhukovsky, 1975). *P. miliaceum*-associated photosensitization of sheep is known to occur in New Zealand (Clare, 1952; Holland et al., 1991; Miles et al., 1991). The plant contains a mixture of two sapogenins, diosgenin and yamogenin, in a 4:1 ratio (Miles et al., 1993). Biliary crystals from sheep intoxicated after ingestion of *P. miliaceum* have not been studied for the presence of sapogenin salts.

Panicum coloratum. This grass originates from southern Africa. The grass was introduced to the United States in the 1950s because it is considered to produce good quality forage (Patamalai et al., 1990). *P. coloratum* is known to cause photosensitization of sheep in the United States and in South Africa (Bridges et al., 1987; Kellerman and Coetzer, 1984). Two steroidal sapogenins have been found in *P. coloratum*. These are diosgenin and yamogenin (Patamalai et al., 1990). The content of biliary crystals of sheep photosensitized after ingestion of *P. coloratum* has not been reported.

Panicum virgatum. *P. virgatum* has been reported to be implicated in photosensitization of sheep in the United States (Puoli et al., 1992). The author has found no reports on the content of sapogenins in *P. virgatum* or in bile crystals from sheep grazing the plant.

Brachiaria decumbens

B. decumbens belongs to the grass family and is the major species used to improve pasture in many places in the tropics due to its aggressive growth habit, efficient nitrogen utilization, ability to withstand heavy grazing, drought resistance and relative freedom from pests and diseases. *B. decumbens* intoxication of sheep has been reported to occur in Australia, Malaysia, Indonesia, Nigeria, and Brazil (Abas Mazni et al., 1983; Graydon et al., 1991; Opasina, 1985). As well as causing photosensitization of sheep, the plant has been reported to cause photosensitization in goats and cattle (Abas Mazni et al., 1985; Low et al., 1993; Opasina, 1985) and neurological disorders in sheep (Abdullah et al., 1989). The saponins reported to be present in *B. decumbens* are yamogenin and diosgenin (Smith and Miles, 1993). In the ruminal contents of sheep suffering from *B. decumbens* intoxication, episarsasapogenin and epismilagenin have been isolated (Lajis et al., 1993).

ABSORPTION, METABOLISM AND EXCRETION OF SAPONINS AND SAPOGENINS

The saponins exist as glycosides in the plant (Fig. 3). Upon ingestion, the sugars from the saponins are hydrolyzed by the ruminal microorganisms into free sapogenins (Miles et al., 1994a) (Fig. 3). Some of the free sapogenins will thereafter be transported unchanged via the omasum and abomasum to the duodenum and jejunum. In work on *T. terrestris* intoxication, it was found that most of the liberated diosgenin and yamogenin in the rumen was converted to epismilagenin and episarsasapogenin respectively before being transported further down the digestive tract. A similar finding has been made in work on *N. ossifragum* metabolism (Flåøyen, Miles and Wilkins, unpublished) where some of the free sarsasapogenin was converted to free episarsasapogenin in the rumen before being transported further. This conversion of diosgenin, yamogenin, and sarsasapogenin into epismilagenin and episarsasapogenin is due, at last in part, to metabolism by rumen microorganisms.

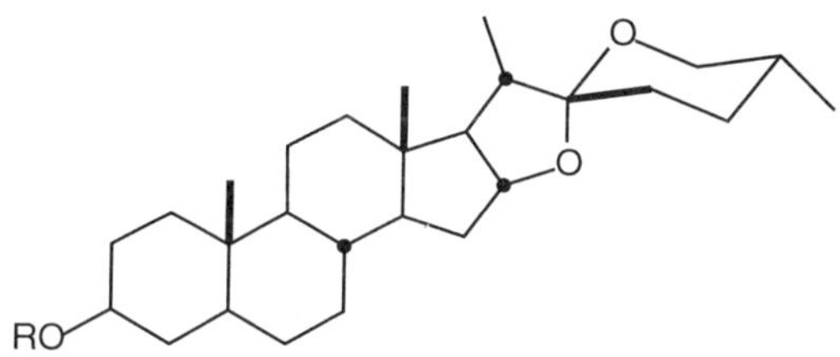

Figure 3. Typical structure of a steroidal saponin (R = sugars) and a steroidal sapogenin (R = H).

What happens to the sapogenins after they depart from the rumen is not well studied, but unpublished results from work on *N. ossifragum* (Flåøyen, Miles and Wilkins) indicate that the hydrolysis and conversion into epi-forms that occur in the rumen continues in the omasum, and ceases in the abomasum.

The major site for absorption of free sapogenins is probably the jejunum, whence they are transported via the portal vein to the liver where further metabolism takes place (Flåøyen, Miles and Wilkins, unpublished). A proportion of the free sapogenins will not be absorbed, and will remain unchanged in the intestines and be excreted in the feces.

All biliary crystals from sheep photosensitized after ingestion of plants containing steroidal saponins so far studied have been shown to be insoluble salts of ß-D-glucuronides of either episarsasapogenin or epismilagenin. Only those sapogenins capable of being metabolized to epismilagenin or episarsasapogenin are capable of forming biliary crystals (Fig. 4). Indeed conjugation of other sapogenins (e.g. tigogenin) are often abundant in the bile of affected animals, but are not found in the biliary crystals isolated from these animals.

Only trace amounts of sapogenin conjugates seem to be excreted into the urine, indicating that the metabolism and excretion of sapogenin conjugates in the liver is the most important metabolic pathway for these compounds (Flåøyen et al., 1993; Miles et al., 1994a).

The glucuronides of episarsasapogenin and epismilagenin emptied into the duodenum (from the ductus choledochus) will probably not be reabsorbed from the jejunum and they will be transported unchanged to the colon (Flåøyen, Miles and Wilkins, unpublished results). In the colon and cecum, most of the episarsasapogenin and epismilagenin glucuronides are deconjugated into free episarsasapogenin and epismilagenin before being excreted in the feces (Miles et al., 1994a; Flåøyen, Miles and Wilkins, unpublished results). The deconjugation that is taking place in the colon and cecum is probably due to microbial metabolism, but it cannot be excluded that intestinal cells play an important role in this metabolism.

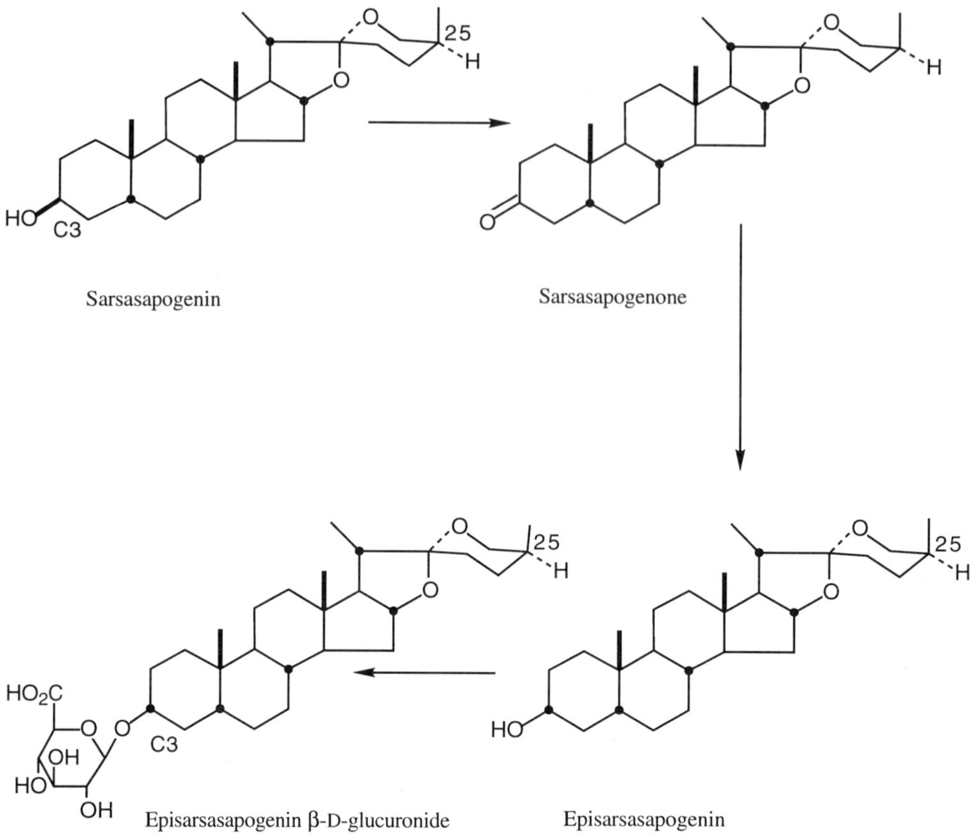

Figure 4. Possible metabolic transformation involved in the conversion of sarsasapogenin into episarsasapogenin β-D-glucuronide (Miles et al., 1993). The epimerisation takes place at C-3.

ARE THE SAPOGENINS LIVER TOXIC ?

Hepatogenous photosensitization has been reproduced in sheep by dosing them with crude saponins from *N. ossifragum* (Abdelkader et al., 1984; Ender, 1955) and *T. terrestris* (Kellerman et al., 1991), but the doses necessary for causing the typical symptoms have been unrealistically large. In all these experiments the sheep got doses that were far beyond those which they possibly could have ingested by grazing. Results from other experiments have shown that the saponins alone hardly could cause the liver damage resulting in photosensitization. It has been shown that dosing with relatively large, but not unrealistically large, doses of *N. ossifragum* daily for 21 days did not cause liver damage in lambs (Flåøyen et al., 1991b). An attempt to reproduce the typical lesions in the liver by dosing with pure sarsasapogenin and diosgenin has also failed (Flåøyen et al., 1993).

The difficulties of reproducing geeldikkop and alveld in dosing experiments, and the fact that the pastures containing *T. terrestris* or *N. ossifragum* seem not to be always toxic to sheep, have prompted some workers to suggest that mycotoxins might be involved in the etiology of the diseases (di Menna et al., 1992; Kellerman et al., 1980; Aas and Ulvund, 1989). However, such a connection has not been established. Another possible explanation for the difficulty of reproducing the diseases in dosing experiments and differences in toxicity observed on different pastures is the possibility for variations in the quantitative and qualitative saponin content of the plants, possibly in combination with variation in the activity of the ruminal flora (Miles et al., 1994a).

The strange features of these diseases make researchers in different parts of the world still discuss whether the saponins alone are the sole cause of these diseases or not. Even though it is still uncertain that the saponins are the sole cause of the diseases, there seems to be a consensus that saponins are in some way involved in these diseases, at least as co-factors enhancing liver lesions primarily caused by other toxic compounds. Further investigations have to be carried out before a final conclusion can be made on this matter.

The typical liver lesions in these diseases are varying degrees of necrosis of hepatocytes, portal fibroplasia, and proliferation of the bile duct epithelium, as well as accumulation of the sapogenin crystals in the bile ducts and the bile duct epithelium (Bridges et al., 1987; Button et al., 1987; Flåøyen et al., 1991a; Kellerman and Coetzer, 1984). Results from work on *T. terrestris* intoxication indicates that the retention of phylloerythrin causing photosensitization is due to occlusion and/or obstruction of the biliary system (Kellerman and Coetzer, 1984), which is contradictory to results from work on *N. ossifragum* intoxication indicating that the damage to the hepatocytes is more severe than the damage to the biliary system (Flåøyen et al., 1991a). Assuming that both observations were correct, it can be suggested that the sapogenins can cause or enhance liver damage, resulting in retention of phylloerythrin, in two different ways.

1. The sapogenins can cause damage to the hepatocytes resulting in reduced or stopped conjugation and/or excretion of phylloerythrin into the biliary system.
2. The sapogenin crystals (the calcium salts of the episarsasapogenin and epismilagenin [Fig. 5]) accumulate in the biliary system, causing fibroplasia, proliferation of the biliary epithelium, and/or obstruction of the bile canaliculi and bile ducts.

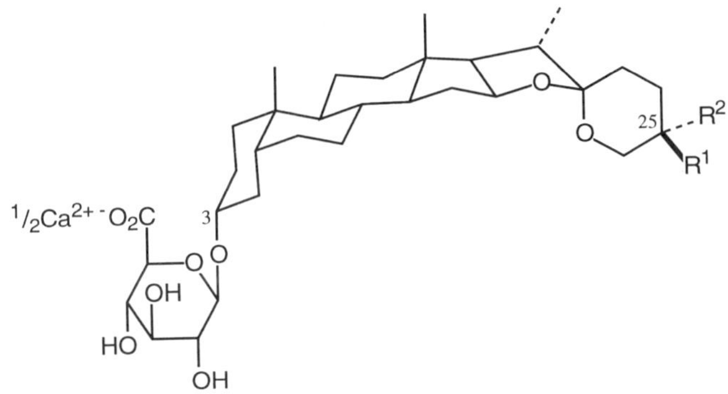

Figure 5. Chemical structure of the components of the biliary sediment from sheep photosensitized after ingesting plants containing steroidal saponins.
$R^1 = CH_3$, $R^2 = H$: Eposmilagenin ß-D-glucuronide calcium salt.
$R^1 = H$, $R^2 = CH_3$: Episarsasapogenin ß-D-glucuronide calcium salt.

CONCLUSIONS

- Plants containing steroidal saponins are associated with hepatogenous photosensitization diseases of sheep in many countries of the world.
- The saponins from the plants are hydrolyzed by the rumen microorganisms into free sapogenins.
- Free sapogenins are probably absorbed from the jejunum and transported via the portal vein to the liver.
- Only certain sapogenins are converted in the rumen or liver into epi forms of episarsasapogenin or epismilagenin.
- The episarsasapogenin and epismilagenin are conjugated with glucuronic acid before being excreted into the bile canaliculi.
- Only trace amounts of sapogenin metabolites are excreted in the urine.
- Varying amounts of calcium salts of episarsasapogenin and epismilagenin glucuronides are found in and about bile ducts in sheep suffering from hepatogenous photosensitization after ingestion of plants containing steroidal saponins.
- The accumulation of sapogenin crystals in and about bile ducts may either be the sole cause of the retention of phylloerythrin in photosensitized sheep, or may exacerbate the liver lesions caused by another toxin.

REFERENCES

Aas, O., and Ulvund, M.J., 1989, Do microfungi help to induce the phototoxic disease alveld in Norway? *Vet. Rec.* **124**: 563.

Abas Mazni, O., Mohd. Khusahry, Y., and Sheikh Omar, A.R., 1983, Jaundice and photosensitization in indigenous sheep of Malaysia grazing on *Brachiaria decumbens*, *Malays. Vet. J.* **7**: 254.

Abas Mazni, O., Sharif, H., Mohd. Khusahry, M.Y., and Vance, H.N., 1985, Photosensitization in goats grazed on *Brachiaria decumbens*, *Mardi Res. Bull.* **13**: 203.

Abdelkader, S.V., Ceh, L., Dishington, I.W., and Hauge, J.G., 1984, Alveld-producing saponins. II. Toxicological studies, *Acta Vet. Scand.* **25**: 76.

Abdullah, A.S., Noordin, M.M., and Rajion, M.A., 1989, Neurological disorders in sheep during signal grass (*Brachiaria decumbens*) toxicity, *Vet. Hum. Toxicol.* **31**: 128.

Amjadi, A.R., Ahourai, P., and Baharsefat, M., 1977, First report of geeldikkop in sheep in Iran, *Arch. Inst. Razi.* **29**: 71.

Bridges, C.H., Camp, B.J., Livingston, C.W., and Bailey, E.M., 1987, Kleingrass (*Panicum coloratum* L.) poisoning in sheep, *Vet. Pathol.* **24**: 525.

Button, C., Paynter, D.I., Shiel, M.J., Colson, A.R., Paterson, P.J., and Lyford, R.L., 1987, Crystal-associated cholangiohepatopathy and photosensitisation in lambs, *Aust. Vet. J.* **64**: 176.

Camp, B.J., Bridges, C.H., Hill, D.W., Patamalai, B., and Wilson, S., 1988, Isolation of a steroidal sapogenin from the bile of a sheep fed *Agave lecheguilla*, *Vet. Hum. Toxicol.* **30**: 533.

Ceh, L., and Hauge, J.G., 1981, Alveld-producing saponins. I. Chemical studies, *Acta Vet. Scand.* 22: 391.

Clare, N.T., 1952, *Photosensitization in Diseases of Domestic Animals*, CAB, Farnham Royal.

Coetzer, J.A.W., Kellerman, T.S., Sadler, W., and Bath, G.F., 1983, Photosensitivity in South Africa. V. A comparative study of the pathology of the ovine hepatogenous photosensitivity diseases, facial eczema and geeldikkop (*Tribulosis ovis*), with special reference to their pathogenesis, *Onderstepoort J. Vet. Res.* **50**: 59.

Dahlgren, R.M.T., Clifford, H.T., and Yeo, P.F., 1985, *The Families of the Monocotyledons. Structure, Evolution, and Taxonomy*, Springer-Verlag, Berlin.

di Menna, M.E., Flåøyen, A., and Ulvund, M.J., 1992, Fungi on *Narthecium ossifragum* leaves and their possible involvement in alveld disease of Norwegian lambs, *Vet. Res. Commun.* **16**: 117.

Ender, F., 1955, Undersøkelser over alveldsykens etiologi. [Eng.: Etiological studies on "alveld"- a disease involving photosensitization and icterus in lambs], *Nord. Vet. Med..* **7**: 329.

Flåøyen, A., 1991, A difference in susceptibility of two breeds of sheep to the alveld toxin, *Vet. Res. Commun.* **15**: 455.

Flåøyen, A., 1993, *Studies on the Aetiology and Pathology of Alveld with some Comparisons to Sporidesmin Intoxication*, Dr.med.vet.-thesis, Norwegian College of Veterinary Medicine, Oslo.

Flåøyen, A., Binde, M., Bratberg, B., Djønne, B., Fjølstad, M., Grønstøl, H., Hassan, H., Mantle, P.G., Landsverk, T., Schönheit, J., and Tønnesen, M.H., 1995a, Nephrotoxicity of *Narthecium ossifragum* in cattle in Norway, *Vet. Rec.* **137**: 259.

Flåøyen, A., Borrebæk, B., and Nordstoga, K., 1991a, Glycogen accumulation and histological changes in the livers of lambs with alveld and experimental sporidesmin intoxication, *Vet. Res. Commun.* **15**: 443.

Flåøyen, A., Hjorth Tønnesen, H., Grønstøl, H., and Karlsen, J., 1991b, Failure to induce toxicity in lambs by administering saponins from *Narthecium ossifragum*, *Vet. Res. Commun.* **15**: 483.

Flåøyen, A., Jóhansen, J., and Olsen, J., 1995b, *Narthecium ossifragum* associated photosensitization in sheep in the Faroe Islands, *Acta Vet. Scand.* **36**: 277.

Flåøyen, A., Smith, B.L., and Miles, C.O., 1993, An attempt to reproduce crystal-associated cholangitis in lambs by the experimental dosing of sarsasapogenin or diosgenin alone and in combination with sporidesmin, *N. Z. Vet. J.* **41**: 171.

Ford, E.J.H., 1964, A preliminary investigation of photosensitization in Scottish sheep, *J. Comp. Pathol.* **74**: 37.

Glastonbury, J.R.W., and Boal, G.K., 1985, Geeldikkop in goats, *Aust. Vet. J.* **62**: 62.

Glastonbury, J.R.W., Doughty, F.R., Whitaker, S.J., and Sergeant, E., 1984, A syndrome of hepatogenous photosensitisation, resembling geeldikkop, in sheep grazing *Tribulus terrestris*, *Aust. Vet. J.* **61**: 314.

Graydon, R.J., Hamid, H., Zahari, P., and Gardiner, C., 1991, Photosensitisation and crystal-associated cholangiohepatopathy in sheep grazing *Brachiaria decumbens*, *Aust. Vet. J.* **68**: 234.

Holland, P.T., Miles, C.O., Mortimer, P.H., Wilkins, A.L., Hawkes, A.D., and Smith, B.L., 1991, Isolation of the steroidal sapogenin epismilagenin from the bile of sheep affected by *Panicum dichotomiflorum* toxicosis, *J. Agric. Food Chem.* **39**: 1963.

Kellerman, T.S., and Coetzer, J.A.W., 1984, *Hepatogenous Photosensitivity Diseases in South Africa*, Technical communication No. 193, Veterinary Research Institute, Onderstepoort.

Kellerman, T.S., Erasmus, G.L., Coetzer, J.A.W., Brown, J.M.M., and Maartens, B.P., 1991, Photosensitivity in South Africa. VI. The experimental induction of geeldikkop in sheep with crude steroidal saponins from *Tribulus terrestris*, *Onderstepoort J .Vet. Res.* **58**: 47.

Kellerman, T.S., van der Westhuizen, G.C.A., Coetzer, J.A.W., Roux, C., Marasas, W.F.O., Minne, J.A., Bath, G.F., and Basson, P.A., 1980, Photosensitivity in South Africa. II. The experimental production of the ovine hepatogenous photosensitivity disease geeldikkop (*Tribulosis ovis*) by the simultaneous ingestion of *Tribulus terrestris* plants and cultures of *Pithomyces chartarum* containing the mycotoxin sporidesmin, *Onderstepoort J. Vet. Res.* **47**: 231.

Lajis, N.H., Abdullah, S.H., Salim, S.J.S., Bremner, J.B., and Khan, M.N., 1993, *Epi*-Sarsasapogenin and *epi*-smilagenin: two sapogenins isolated from the rumen of sheep intoxicated by *Brachiaria decumbens*, *Steroids* **58**: 387.

Lancaster, M.J., Vit, I., and Lyford, R.L., 1991, Analysis of bile crystals from sheep grazing *Panicum schinzii* (sweet grass), *Aust. Vet. J.* **68**: 281.

Low, S.G., Bryden, W.L., Jephcott, S.B., and Grant, I.McL., 1993. Photosensitization of cattle grazing Signal grass (*Brachiaria decumbens*) in Papua New Guinea, *N. Z. Vet. J.* **41**: 220.

Malone, F.E., Kennedy, S., Reilly, G.A.C., and Woods, F.M., 1992, Bog asphodel (*Narthecium ossifragum*) poisoning in cattle, *Vet. Rec.* **131**: 100.

Mathews, F.P., 1937, *Lechuguilla (Agave lecheguilla) Poisoning in Sheep, Goats, and Laboratory Animals*, Bulletin 554, Texas Agricultural Experiment Station, College Station.

Miles, C.O., Munday, S.C., Holland, P.T., Lancaster, M.J., and Wilkins, A.L., 1992a, Further analysis of bile crystals from sheep grazing *Panicum schinzii* (sweet grass), *Aust. Vet. J.* **69**: 34.

Miles, C.O., Munday, S.C., Holland, P.T., Smith, B.L., Embling, P.P., and Wilkins, A.L., 1991, Identification of a sapogenin glucuronide in the bile of sheep affected by *Panicum dichotomiflorum* toxicosis, *N. Z. Vet. J.* **39**: 150.

Miles, C.O., Wilkins, A.L., Erasmus, G.L., and Kellerman, T.S., 1994a, Photosensitivity in South Africa. VII. Ovine metabolism of *Tribulus terrestris* saponins during experimentally induced geeldikkop, *Onderstepoort J. Vet. Med.* **61**: 351.

Miles, C.O., Wilkins, A.L., Erasmus, G.L., Kellerman, T.S., and Coetzer, J.A.W., 1994b, Photosensitivity in South Africa. VII. Chemical composition of biliary crystals from a sheep with experimental induced geeldikkop, *Onderstepoort J Vet Med.* **61**: 215-222.

Miles, C.O., Wilkins, A.L., Munday, S.C., Flåøyen, A., Holland, P.T., and Smith, B.L., 1993, Identification of insoluble salts of the ß-D-glucuronides of episarsasapogenin and epismilagenin in the bile of lambs with alveld and examination of *Narthecium ossifragum*, *Tribulus terrestris*, and *Panicum miliaceum* for sapogenins, *J. Agric. Food Chem.* **41**: 914.

Miles, C.O., Wilkins, A.L., Munday, S.C., Holland, P.T., Smith, B.L., Lancaster, M.J., and Embling, P.P., 1992b, Identification of the calcium salts of epismilagenin ß-D-glucuronides in the bile crystals of sheep affected by *Panicum dichotomiflorum* and *Panicum schinzii* toxicoses, *J. Agric. Food Chem.* **40**: 1606.

Opasina, B.A., 1985, Photosensitization jaundice syndrome in West African Dwarf Sheep and goats grazed on *Brachiaria decumbens*, *Trop. Grass.* **19**: 120.

Patamalai, B., Hejtmancik, E., Bridges, C.H., Hill, D.W., and Camp, B.J., 1990, The isolation and identification of steroidal sapogenins in Kleingrass, *Vet. Hum. Toxicol*, **32**: 314.

Puoli, J.R., Reid, R.L., and Belesky, D.P., 1992, Photosensitization in lambs grazing switchgrass, *Agron. J.* **84**: 1077.

Rimington, C., and Quin, J.I., 1933, Photosensitising agent in 'geel-dikkop' phylloerythrin, *Nature* **132**: 178.

Smith, B.L., and Miles, C.O., 1993, A role for *Brachiaria decumbens* in hepatogenous photosensitization of ruminants? *Vet. Hum. Toxicol.* **35**: 256.

Steyn, D.G., 1949,*Vergiftiging van Mens en Der*, Van Schaik, Pretoria.

Summerfield, R.J., 1974, *Narthecium ossifragum* (L.) Huds, *J. Ecol.* **62**: 325-339.

Tapia, M.O., Giordano, M.A., and Gueper, H.G., 1994, An outbreak of hepatogenous photosensitization in sheep grazing *Tribulus terrestris* in Argentina, *Vet. Hum. Toxicol.* **36**: 311.

Wall, M.E., Warnock, B.H., and Willaman, J.J., 1962, Steroidal sapogenins. LXVIII. Their occurrence in *Agave lecheguilla*, *Econ. Bot.* **16**: 266.

Zeven, A.C. and Zhukovsky, P.M., 1975. *Dictionary of Cultivated Plants and their Centres of Diversity. Excluding Onamentals, Forest Trees and Lower Plants*, Centre for Agricultural Publishing and Documentation, Wageningen.

STEROIDAL GLYCOALKALOIDS IN ANDEAN POTATOES

Isao Kubo and Katsuya Fukuhara

Department of Environmental Science, Policy and Management
University of California
201 Wellman Hall
Berkeley, California 94720-3112

INTRODUCTION

The Andes is the native soil of the "potato." There are more than one hundred species of wild starchy tubers in the Andes, although only about ten are commonly sold in the local marketplace. They appear different in many respects, and some of them are taxonomically very different. Besides the potato, there are four other starchy tubers that are of local importance. They are "oca", "arracacha", "ullucu", and "anu", each of which belongs to a different family (Popenoe *et al.*, 1989).

Throughout history, the potato – known as the "papa" – has been and is still one of the most important food crops for the inhabitants in the Andes. There are mainly two kinds of potatoes sold in the marketplace, namely "papa negra" and "papa blanca." The main physical difference between these two is that the papa blanca is the same (or similar) in appearance to our cultivated potato, *Solanum tuberosum*, while the papa negra has darker skin. Although both potatoes belong to the genus *Solanum*, the taxonomy of this genus is complicated since there are many hybrids of *Solanum*. Thus, more than one hundred wild species of tuber-bearing *Solanum* have been identified from South America (Hawkes, 1978). Moreover, all species of cultivated potatoes have been derived in some way through hybridization. *S. tuberosum* subsp. *andigena* is the most widespread cultigen in the Andes; subspecies *tuberosum* is the potato of global importance (Horton, 1987). Nevertheless, in the Andes' marketplace papa negra and papa blanca are the most common. Figure 1 shows the potatoes (a mixture of papa negra and papa blanca) sold in La Paz, Bolivia.

WAYS TO EAT POTATO CONTAINING STEROIDAL GLYCOALKALOIDS

The potato is also known to be poisonous (Kingsbury, 1964; Vernard and Trump, 1969). In the Andes the papa has also long been known to contain toxins. The presence of

Figure 1. Potato, sold in La Paz, Bolivia as papa (papa negra and papa blanca).

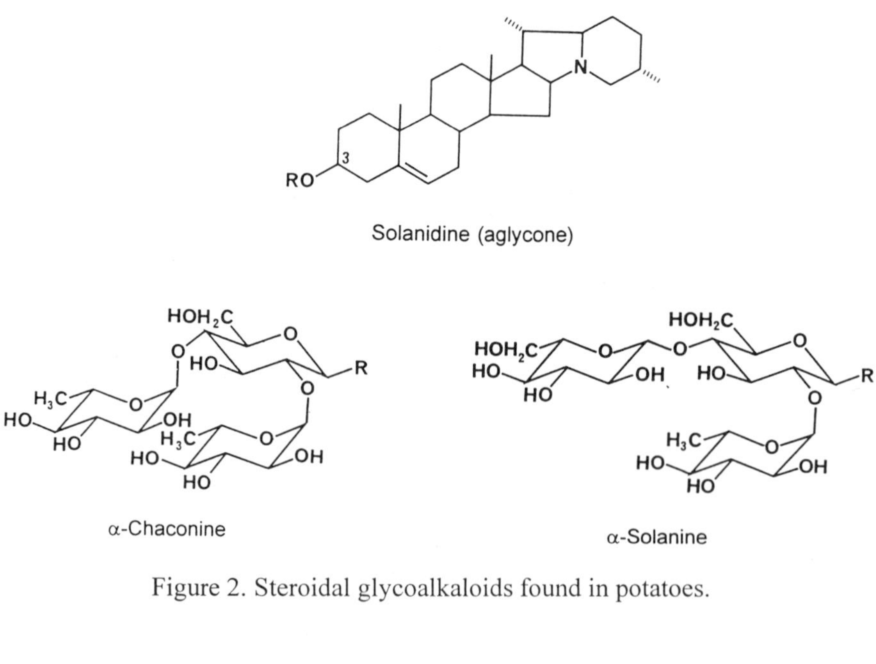

Figure 2. Steroidal glycoalkaloids found in potatoes.

toxic steroidal glycoalkaloids in various *Solanum* plants is well documented (Gregory *et al.*, 1981; Hsu and Tien, 1974; Sanford and Sinden, 1972). In our preliminary TLC analysis we found rather large amounts of alkaloids consisting, primarily, of two compounds which were positive to Dragendorff's reagent in the methanol extracts, especially in papa negra. They were identified as α-chaconine (**1**) and α-solanine (**2**) (Figure 2) by direct comparison with the authentic samples. However, the Andean people have discovered detoxification methods, probably through unexpected experiences. As a result, they have made several unique dishes and have also developed distinct preserving methods. Thus, there are several ways to eat the papa that contains large amounts of toxic glycoalkaloids. One way is to eat specific types of clay. This is known as "geophagy." Geophagy occurs before or after eating many potato cultigens, but especially when those eaten are from frost-resistant strains (e.g. *Solanum* X *juzepczukii* Buk. and *S.* X *curtilobum* Juz. et Buk.). This is done to absorb the polar glycoalkaloids and then excrete them from the body (Johns, 1985, 1986). Figure 3 shows the clay sold in La Paz, Bolivia.

Figure 3. Clay sold for geophagy.

Another process, is to make dried potatoes, which are more or less detoxified during the drying process. The dried potatoes are sometimes given different names that designate the process by which they were prepared. The chuño is usually made of papa negra, which is somehow detoxified in the process. First, the potato is freeze-dried naturally during the night. Subsequently, this frozen potato is dried in the sun. Second, the dark fluid which contains toxic steroidal glycoalkaloids is squeezed out. These processes are repeated depending on the size of the potato. Finally, the papa is dried until it becomes as hard as a stone, and yet very light, like a pumice rock. This dried potato is called chuño. Sometimes the potato, before its final drying process, is washed in a running body of water, usually a stream, for about two weeks. This step can remove any residual toxic steroidal glycoalkaloids more efficiently. The white, dried potato is called tunta. The steroidal glycoalkaloids can be almost completely removed from tunta. However, some amounts were still detected in chuño. Dried potatoes are usually soaked in water for a while (usually overnight) before cooking. Most of the remaining toxic glycoalkaloids in chuño can be removed by this process. It should be interesting to see if

Figure 4. Dried potato, sold as chuño in the Andes.

this is accurate based on quantitative analysis of the toxic glycoalkaloids. Figure 4 shows chuño purchased in Cuzco, Peru.

The climate of the Andes, especially the area around Lake Titicaca, allows the potato to thrive. The high altitude lowers night temperatures which can freeze the harvested potato. The proximity to the equator causes the day temperatures to be high and the air to be dry. It is natural that the dried potato be discovered there. It should be noted that many other dried vegetables and fruit are also produced in the Andes. For example, "oca seca" is a common dried tuber of "oca", *Oxalis tuberosa* Molina. Fresh oca tastes bitter owing to a large amount of calcium oxalate, which can be removed before consumption by a process resembling that used to prepare chuño. Oca seca is also called "caya" (Hodge, 1951).

Incidentally, the dark color of papa negra is caused by violet pigments which are often concentrated in the skin. Chuño is even darker (almost black in color) than papa negra because the pigments are unstable anthocyanins (Nerdal and Andersen, 1991) and easily decompose to a very black color as soon as the chuño is cut. Solanaceous alkaloids probably do not contribute to this color change since they are colorless and quite stable. However, it may be generalized that darker potatoes contain more toxic glycoalkaloids. Thus, the pigments may be used indirectly as an indicator of toxic glycoalkaloids in the Andean potatoes.

SEMIPREPARATIVE ISOLATION OF STEROIDAL GLYCOALKALOIDS BY RLCC

The steroidal glycoalkaloids have been isolated by various chromatographic methods (Fitzpatrick and Osman, 1974; Arnaldo and Georgina, 1982; Shibata *et al.*, 1982; Mahoto *et al.*, 1982). These methods have always required solid packing materials, which often cause the irreversible adsorption of large amounts of these polar saponins. Therefore, an efficient isolation method for the glycoalkaloids, possibly without using any solid packing material, has been awaited. In order to resolve this problem, countercurrent chromatography based on two-phase liquid-liquid partition chromatography was applied.

We have recently developed an efficient method of isolating steroidal glycoalkaloids by a combination of two countercurrent chromatographies – rotation locular countercurrent chromatography (RLCC) and droplet countercurrent chromatography (DCCC) (Conway, 1989; Hanke and Kubo, 1987; Kubo, 1991; Kubo *et al.*, 1988; Mandava and Ito, 1988). Thus, two glycoalkaloids, solasonine and solamargine, were efficiently isolated from the ripe fruit of an East African medicinal plant *S. incanum* without using any solid packing material (Fukuhara and Kubo, 1991). This combination of methods has proved to be very efficient when isolating polar steroidal glycoalkaloids. However, there are some limitations to the use of DCCC. First, the choice of solvent system is restricted because the system must be favorable to the formation of droplets, which is critical for efficient separation. Second, the low flow rate limits the speed of separation. For example, the above mentioned isolation of several hundred mg of solasonine and solamargine by the combination of RLCC and DCCC required from 6 to 8 days. In contrast to DCCC, RLCC does not require droplet formation. As a result, there is a larger variety of solvent systems to choose from and also gradients for elution of mobile phase. However, it has the limitation of a lower chromatographic resolution.

Two glycoalkaloids, α-chaconine (**1**) and α-solanine (**2**), were isolated from papa negra using only RLCC from the aqueous fraction obtained from the methanol extract. This was followed by removal of the solvent, suspension of the residue in water, and partitioning with organic solvents. Without further purification, the aqueous layer was injected into the sample loop of the RLCC apparatus in the ascending mode. The column was filled with water as the stationary phase. After elution with water-saturated ethyl ether and ethyl acetate as the mobile phase to remove nonpolar compounds, the solvent system was changed to the upper layer of ethyl acetate-butanol-water (4:1:1, v/v/v). This ratio was successively changed to 3:1:1 and 2:1:1 in order to increase the polarity of the mobile phase. As a result, 59.5 mg of α-chaconine and 57.9 mg of α-solanine were efficiently isolated without using any solid packing material. The total quantity of these two glycoalkaloids in one of the papa negra purchased from a marketplace in La Paz, Bolivia was 0.012% of the dry weight. Figure 5 illustrates the RLCC separation.

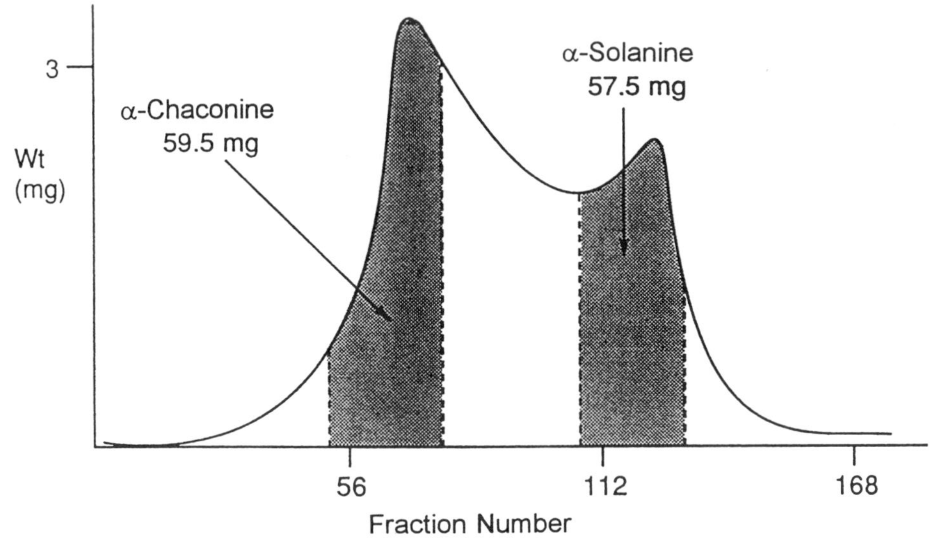

Figure 5. Semi-preparative purification of two steriodal glycoalkaloids by RLCC.

QUALITATIVE ANALYSIS OF STEROIDAL GLYCOALKALOIDS BY MS/MS

Prior to our attempts at quantitative analysis for α-chaconine and α-solanine in various potato samples by HPLC, a simple qualitative identification method was needed. This was established by utilizing tandem mass spectrometry (MS/MS). The collision-activated dissociation (CAD) of fragment ion by Ar gas and using positive FAB-MS with Xe gas provided the possibilities for the identification of the two glycoalkaloids, α-chaconine and α-solanine in the crude potato sample at microgram levels. In the FAB-MS spectrum of the crude methanol extract of the tuber's skin part of "papa negra", α-chaconine and α-solanine were displayed by the ions, *m/z* 868 ($[M+H]^+$) and 852 ($[M+H]^+$), respectively as shown in Figure 6. The ion of $[M+H]^+$ 868 *m/z* (by the positive FAB using the glycerol matrix) as *m/z* underwent to produce CAD several fragments, at 722 ($[M+H–146]^+$), 706 ($[M+H–162]^+$), 560 ($[M+H–146–162]^+$), and 398 ($[M+H–146–162–162]^+$) corresponding to losses of a rhamnose, a glucose, a rhamnose, and a glucose, and a rhamnose and two glucoses, shown in Figure 7. In the CAD spectrum at 853 of α-chaconine obtained by using the positive FAB-MS, two major peaks wer Chemical ecology of the Aymara of Western Bolivia e observed at *m/z* 720 and 398 corresponding to $[M+H–146]^+$ and $[M+H–146–146–162]^+$, respectively as shown in Figure 8. The latter fragment ion at *m/z* 398 arose from solanidine. The fragmentation analyses by MS/MS were found to be very useful for the detection and identification of these glycoalkaloids.

QUANTITATIVE ANALYSIS FOR STEROIDAL GLYCOALKALOIDS BY HPLC

During our ecological study of potatoes in the Andes, quantitative analyses for these toxic glycoalkaloids were needed. Although we have developed an efficient isolation of the two steroidal glycoalkaloids by RLCC without using any solid packing material, the RLCC method is not, despite its simplicity, an appropriate technique for quantitative analysis, especially since it requires rather large amounts of samples (potatoes). There is no doubt that HPLC is a most appropriate method for this purpose. Although a number of reports regarding HPLC analysis for glycoalkaloids have been published (Bushway and Storch, 1982; Bushway and Ponnampalam, 1981; Bushway *et al.*, 1979, 1983; Carman *et al.*, 1986; Eldridge and Hockridge, 1983; Sinden and Webb, 1972), we have also attempted to develop our own quantitative method. To begin with, it was necessary to establish the optimal conditions for column, solvent system, and sample preparation. In our preliminary search for appropriate conditions for HPLC analysis, such as column and solvent system, reversed-phase columns (ODS $C_{8,\ 18}$) were first examined. However, none of them were satisfactory for separation of the two polar glycoalkaloids, α-chaconine and α-solanine, because of their poor resolution. For example, the chromatographic peaks of these two glycoalkaloids appeared to have the same retention time on ODS-C_{18} columns eluted with methanol-water (85:15, v/v). The other ODS-type gel-filtration columns also failed to separate these two steroidal glycoalkaloids. In contrast, a SiO_2 column gave superior resolution. Therefore, the quantitative analyses for α-chaconine and α-solanine in potato extracts were carried out on a SiO_2 column.

For the quantitative analysis for these two steroidal glycoalkaloids in aqueous extracts with the SiO_2 column, pre-purification was necessary. Two ODS C_{18} cartridges packed with 0.1 g of ODS C_{18} (Bond Elut) and 1.0 g of ODS C_{18} (Sep-Pak) were used to remove other water-soluble compounds such as sugars and amino acids, before an extract was injected into the HPLC column. Otherwise, the SiO_2 column would have been damaged by irreversible absorption of highly polar components. The undesirable result of

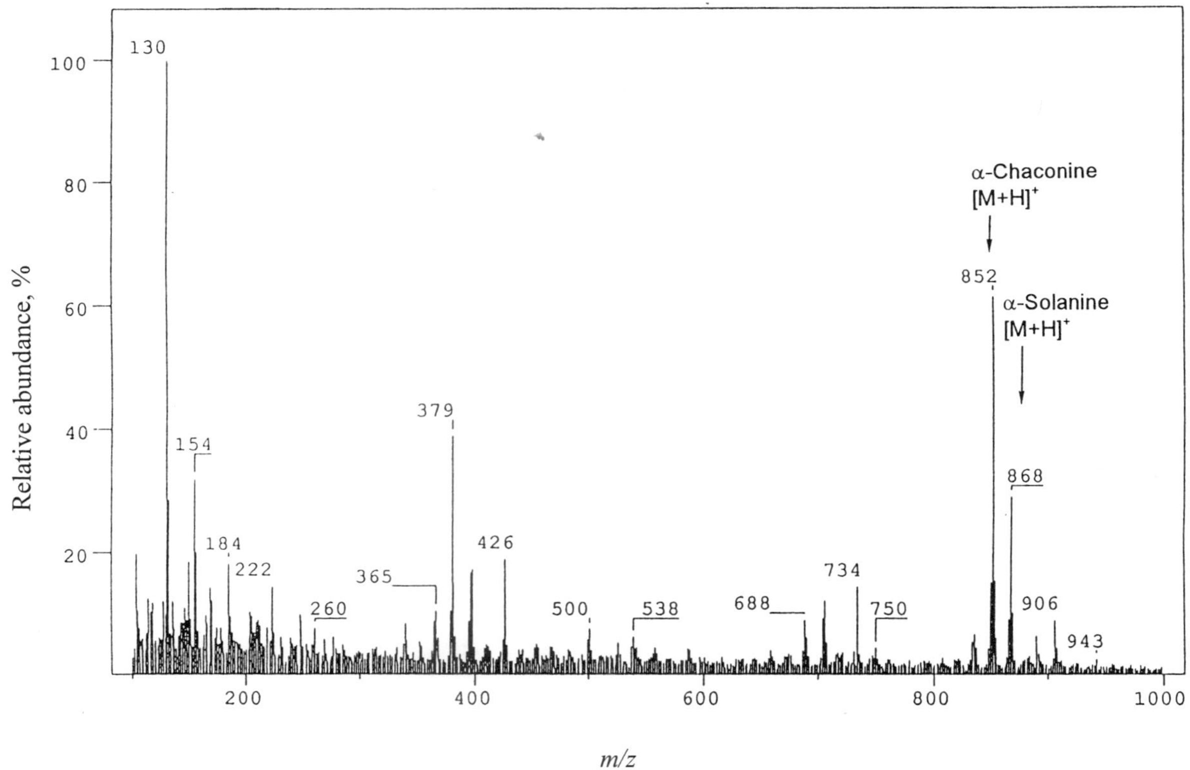

Figure 6. FAB-MS spectrum of the methanol extract of papa negra skin.

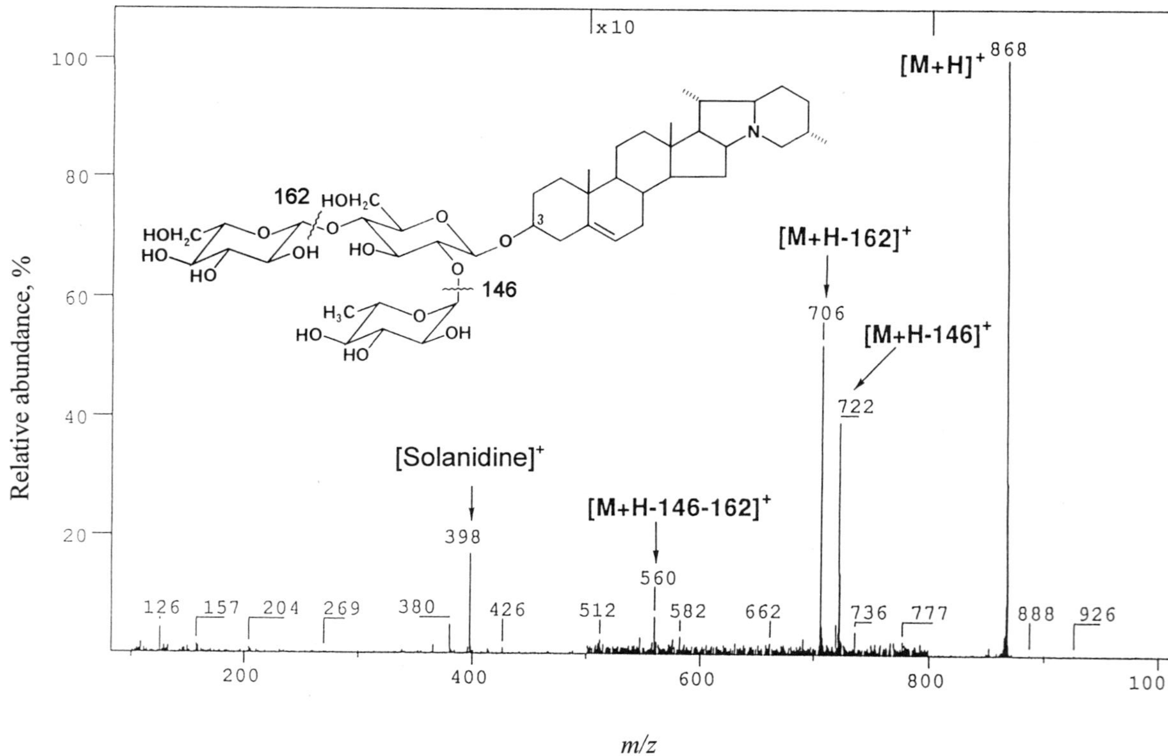

Figure 7. CAD spectrum of α-solanine.

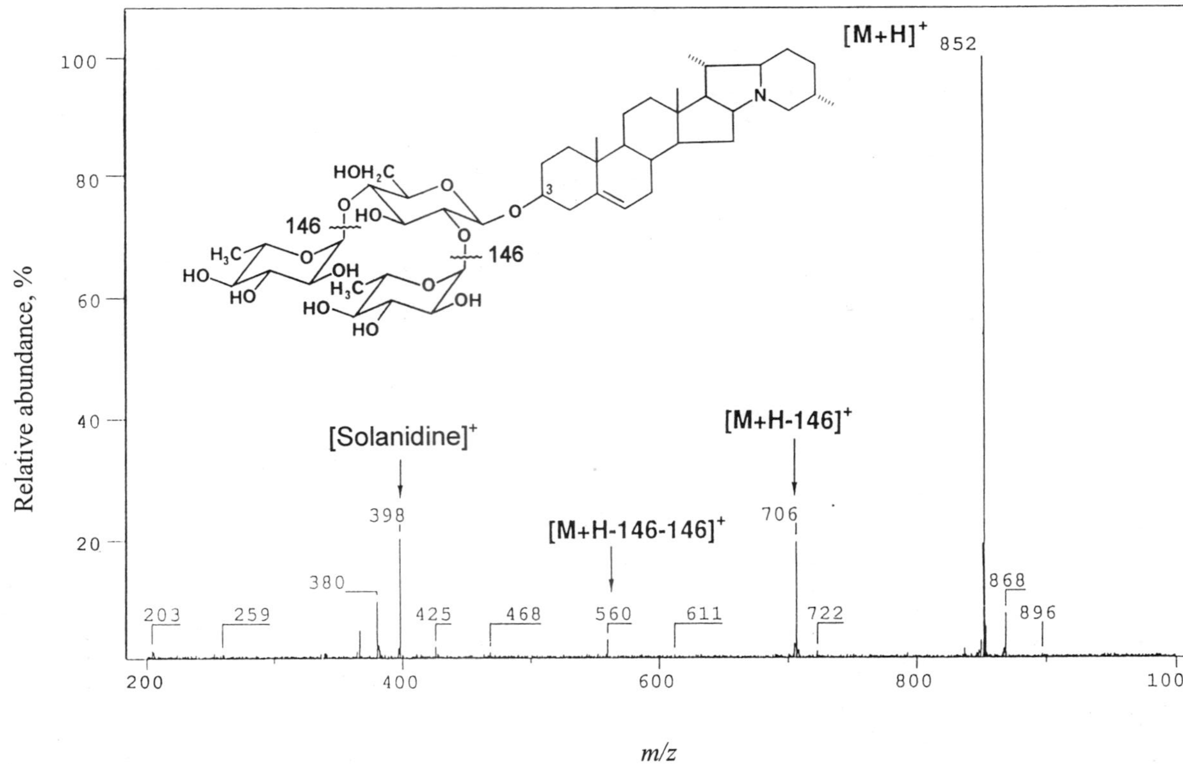

Figure 8. CAD spectrum of α-chaconine.

this would be peak broadening. The Bond Elut cartridge was used as a disposable pre-cartridge. This way, the Sep-Pak cartridge was reusable several times if washed with solvents such as methanol or ethanol. Figure 9 illustrates the preliminary purification process.

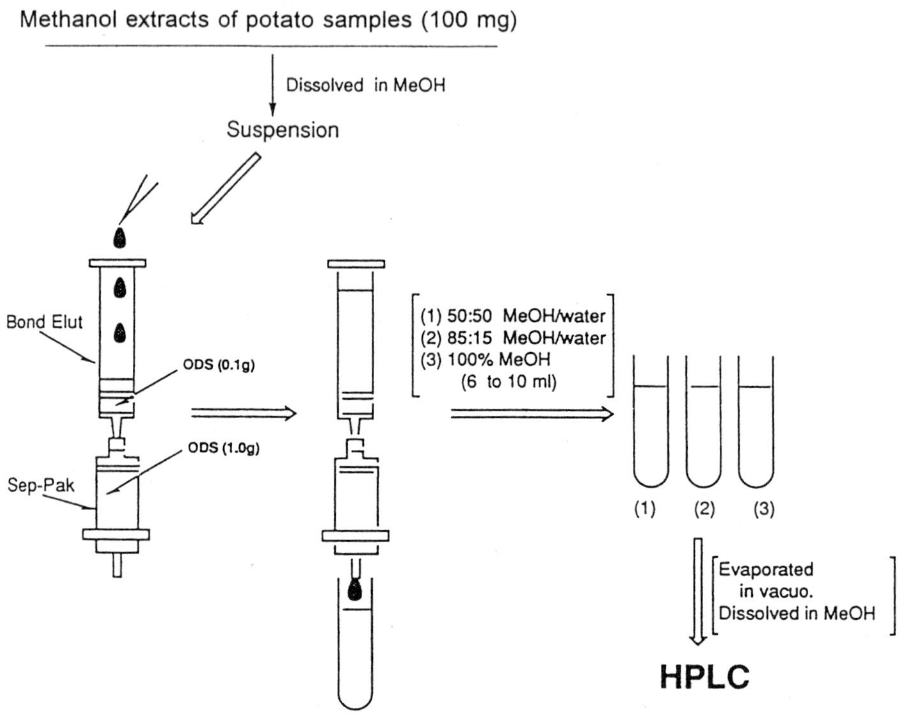

Figure 9. Preliminary purification prior to HPLC analysis.

Figure 10 shows the chromatogram of a pre-purified "papa negra" sample on the SiO_2 column eluted with chloroform-methanol-ammonium hydroxide (7:13:2, v/v/v). Although this solvent system on a SiO_2 column gave notable resolution between the two steroidal glycoalkaloids and the other peaks, in order to obtain superior optimal resolution, chloroform-methanol-ammonium hydroxide (10:10:1, v/v/v) for α-chaconine and chloroform-methanol-ammonium hydroxide (7:13:2, v/v/v) for α-solanine were chosen to avoid interference from peaks of less polar compounds. This was well illustrated in the case of α-chaconine. The retention times were 8.4 min for α-chaconine at the flow rate of 1.0 ml/min and 13.8 min for α-solanine at the flow rate of 1.2 ml/min. During the HPLC analyses, there were no interfering components and no change of the retention time under these conditions.

This method was applied to various Andean potato extracts. As a result, one of the skin extracts of papa negra was found to contain large amounts of both α-chaconine and α-solanine, 820 μg/g of α-chaconine and 476 μg/g of α-solanine in the skin, while the tuber contained 76 μg/g of α-chaconine and 42 μg/g of α-solanine (dry weight). The recovery rate of these two steroidal glycoalkaloids was generally around 85%. Therefore, their actual amounts in the potato should be higher.

The total glycoalkaloid content in the peel of papa negra was 30.2 mg per 100 g, while that of the tuber was 6 mg per 100 g (wet weight). This is in agreement with the

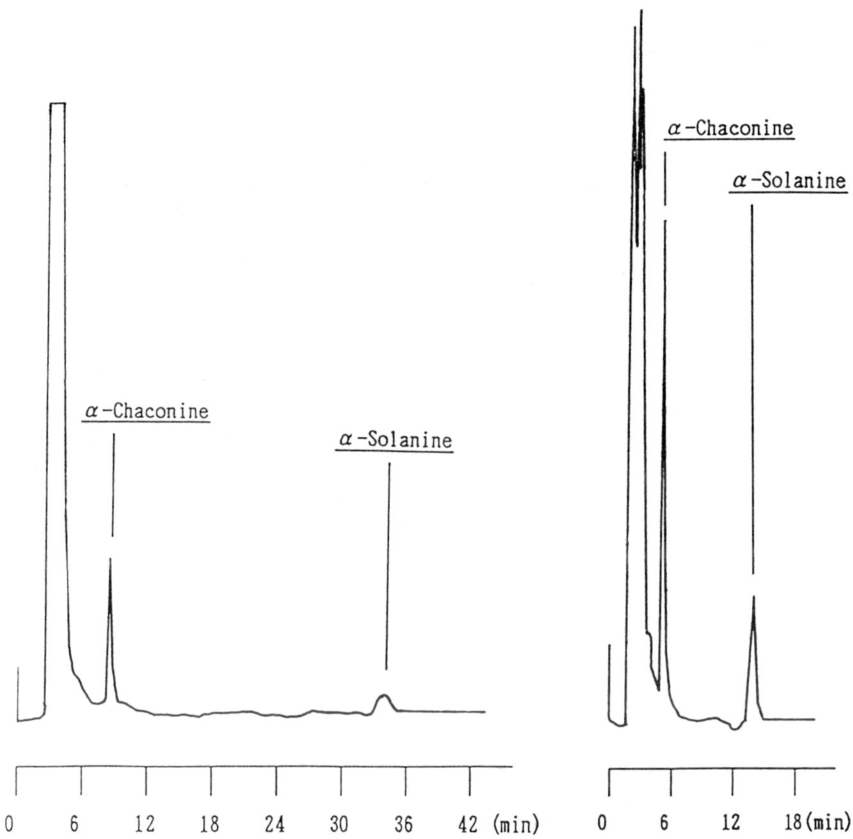

Column : YMC-Pack SiO_2 (6 mm i.d. x 20 cm)
+ Guard column (1 mm i.d. x 3 cm)
Solvent : $CHCl_3$:CH_3OH:NH_4OH=10:13:1 (v/v/v) for α-chaconine and 7:13:2 (v/v/v) for α-solanine
Flow rate : 1.0 ml/min for α-chaconine and 1.2 ml/min for α-solanine
Detection : UV 230 nm

Figure 10. Chromatography of two steroidal glycoalkaloids.

previous report (Schmiediche et al., 1980) that the total glycoalkaloid content of *S.* X *juzepczukii* ranged between 18 and 49 mg per 100 g (wet weight). In another common variety of Andean potato "papa blanca", the total alkaloid content was about one-seventh of that of papa negra, regardless of the location of sample collection. The result obtained indicates that genetic differences may be more important in total alkaloid content than environmental factors such as altitude. A similar result has also been reported in American potato species (Sinden and Webb, 1972).

CONCLUSION

In conclusion, semi-preparative isolation using only RLCC and a quantitative analytical method using HPLC have been established for the two primary steroidal glycoalkaloids in potato, α-chaconine and α-solanine. The fragmentation analysis by MS/MS has been very useful for the detection and identification of these glycoalkaloids. For HPLC analysis, it was necessary to pre-purify the crude methanol extract through the ODS C_{18} cartridges before injection into the HPLC column to protect the SiO_2 column and obtain reliable quantitative data. This quantitative analysis for polar glycoalkaloids by HPLC could be generally applicable to *Solanum* plants and many others with the limitation of small sample amounts. As a result of the HPLC analysis, we have confirmed the previous report (Sinden and Webb, 1972) that more frost-resistant varieties contain more steroidal glycoalkaloids. In addition, α-chaconine was generally found in higher concentrations than was α-solanine, and most importantly both are concentrated in the skin. Therefore, most of the steroidal glycoalkaloids can be removed prior to consumption by peeling off a rather thick layer of skin.

ACKNOWLEDGMENTS

The authors are grateful to Ms. I.O. Hinojosa for assistance in obtaining the potato samples, and Mr. C.S. Lunde for critical reading of the manuscript.

REFERENCES

Arnaldo, A., and Georgina, H., 1982, Molluscicidal steroid glycoalkaloids possessing stereoisomeric spirosolane structure, *Toxicol. Lett.* 12:151.

Bushway, R. J., Barden, E. S., Bushway, A. W., and Bushway, A. A., 1979, High-performance liquid chromatographic separation of potato glycoalkaloids, *J. Chromatogr.* 178:533.

Bushway, R. J., and Ponnampalam, R., 1981, α-Chaconine and α-solanine content of potato products and their stability during several modes of cooking, *J. Agric. Food Chem.* 29:814.

Bushway, R. J., Bureau, J. L., and Mcgann, D. F., 1983, α-Chaconine and α-solanine content of potato peels and potato peel products, *J. Food Sci.* 48:84.

Bushway, R. J., and Storch, R. H., 1982, Semi-preparative high-performance liquid chromatographic separation of potato glycoalkaloids, *J. Liq. Chromatogr.* 5:731.

Carman, A. S. Jr., Kaun, S. S., Ware, G. M., Francis, O. J. Jr., and Kirshenheuter, G. P., 1986, Rapid high-performance liquid chromatographic determination of the potato glycoalkaloids α-chaconine and α-solanine, *J. Agric. Food Chem.* 34:279.

Conway, W. D., 1989, *Countercurrent Chromatography*, VCH, New York.

Eldridge, A. C., and Hockridge, M. E., 1983, High-performance liquid chromatographic separation of Eastern black nightshade (*Solanum ptycanthum*) glycoalkaloids, *J. Agric. Food Chem*, 31:1218.

Fitzpatrick, T. J., and Osman, S. F., 1974, A comprehensive method for the determination of total potato glucoalkaloids, *Am. Potato J.* 51:318.

Fukuhara, K., and Kubo, I., 1991, Isolation of steroidal glycoalkaloids from *Solanum incanum* by two countercurrent chromatographic methods, *Phytochemistry* 30:685.

Gregory, P., Sinden, S. L., Osman, S. F., Tingey, W. M., and Chessin, D. A. Glycoalkaloids of wild, tuber-bearing *Solanum* species. *J. Agric. Food Chem.* 29:1212.

Hanke, J. F., and Kubo, I., 1987, Recent examples of the use of droplet countercurrent chromatography in the chemistry of natural products, *LC&GC* 5:248.

Hawkes, J. G., 1978, In *The Potato Crop*, Harris, P. M., Ed., Chapman and Hall, London, pp. 1-14.

Hodge, W. H., 1951, Three native tuber foods of the Andes. *Econ. Bot.* 5:185.

Horton, D., 1987, *Potatoes; Production, Marketing, and Programs for Developing Countries*, Westview Press, Boulder, CO.

Hsu, P. M., and Tien, H. J., 1974, Studies on the components of Formosan *Solanum* species, *J. Taiwan Pharm. Assoc.* 26:28.

Johns, T. 1985, Chemical ecology of the Aymara of Western Bolivia: Selection for glycoalkaloids in the *Solunum* X *ajanhuiri* domestication complex. *Ph.D. Dissertation*, University of Michigan, Ann Arbor.

Johns, T., 1986, Detoxification function of geophagy and domestication of the potato, *J. Chem. Ecol.* 12:635.

Kingsbury, J. M., 1964, *Poisonous Plants of the United States and Canada*; Prentice-Hall, Englewood Cliffs, pp. 293-294.

Kubo, I., 1991, Recent applications of counter-current chromatography to the isolation of bioactive natural products, *J. Chromatogr.* 538:187.

Kubo, I., Marshall, G. T., and Hanke, J. F., 1988, Rotation locular countercurrent chromatography for natural products isolation, In: *Countercurrent Chromatography*, Mandava, N. B.; Ito, Y., Eds., Dekker, New York, pp. 493-507.

Mahato, S. B., Ganguly, A. N., and Sahu, N., 1982, Steroid saponins, *Phytochemistry* 21:959.

Mandava, N., and Ito, Y., 1988, *Countercurrent Chromatography*, Dekker, New York.

Nerdal, W. and Andersen, O. M., 1991, Evidence for self-association of the anthocyanin petanin in acidified, methanolic solution using two-dimensional nuclear Overhauser enhancement NMR experiments and distance geometry calculation, *Phytochem. Anal.* 2:263.

Popenoe, H., King, S. R., León, J., and Kalinowski, L. S., 1989, *Lost Crops of the Incas*, National Academy Press, Washington, D. C.

Sanford, L. L., and Sinden, S. L., 1972, Inheritance of potato glucoalkaloids, *Am. Potato J.* 49:209.

Schmiediche, P. E., Hawkes, J. G., and Ochoa, C. M., 1980, Breeding of the cultivated potato species *Solanum* X *juzepczukii* Buk. and *Solanum* X *curtilobum* Juz. et Buk. I. A study of the natural variation of *S.* X *juzepczukii*, *S.* X *curtilobum* and their wild progenitor, *S. acaule* Bitt, *Euphytica* 29:685.

Shibata, S., Itokawa, H., Sankawa, U., Shoji, J., and Takido, M., 1982, *Natural Products for Medicinal Use*; Nanzando, Tokyo, pp. 436-440.

Sinden, S. L., and Webb, R. E., 1972, Effect of variety and location on the glucoalkaloids content of potatoes., *Am. Potato J.* 49:334.

Vernard, B. and Trump, E. C., 1969, *Common Poisonous Plants of East Africa*, Collins, London, pp. 169-172.

APPENDIX I

PLANT SAPONIN DISEASES NETWORK

The Plant Saponin Diseases Network is an informal network for researchers who have a special interest in work on diseases of animals associated with grazing of plants containing steroidal saponins. Research on these diseases has been going on for many years at several places in the world, but most of the work has been done by a relatively small number of enthusiasts with limited resources.

The network was started in 1993 and has today about 30 members from all over the world.

It is hoped that the members of the network will share information and ideas, and perhaps start collaborating projects.

The work of the network is coordinated by one of its members and he or she is responsible for distribution of addresses and information of general interest. The coordinator will also help facilitate contact between different groups of researchers.

For more information, please contact the present coordinator:
Dr. Arne Flåøyen
Department of Reproduction and Forensic Medicine
Norwegian College of Veterinary Medicine
P.O.Box 8146 Dep.
N-0033 Oslo
Norway
Tele: 47 22 964 500
Fax: 47 22 964 962
Email: Arne.Flaaoeyen@veths.no

APPENDIX II

SAPONIN NETWORK FOR MASS SPECTROMETRY AND NUCLEAR MAGNETIC RESONANCE SPECTROMETRY

The saponin network for MS and NMR spectra is a part of the *Databases of Carbohydrate Spectra* which was conceived August 18-20, 1995 in Seattle, Washington for the *Complex Carbohydrate Center*, The University of Georgia, Athens, GA, by Prof. Peter Albersheim, and by the Department of Bio-Organic Chemistry, The Utrecht University, Utrecht, The Netherlands by Prof. Johannes F.G. Vliegenthart, and sponsored by the U.S. Department of Energy. The curator of the mass spectra and NMR spectra for saponins is:

Dr. Tadahiro Takeda
Professor of Kyoritsu College of Pharmacy
1-5-30, Shiba-Kouen, Minato-ku
Tokyo 105 JAPAN
Tel: +81-3-5400-2696
Fax: +81-3-5400-2666

At this time Dr. Takeda has just accepted the responsibility for collecting and disseminating the spectra of saponins, and he does not have his organizational scheme set-up as yet. The mass spectra would include fast atom bombardment (FAB), electrospray ionization (ESI), tandem MS/MS with collision-induced decomposition, matrix-assisted laser-desoription (MALDI), electron impact and/or chemical ionization spectrometry, and liquid chromatography/mass spectrometry (LC/MS). To be of value to all who use the mass spectra database the details of the instrument settings and experimental conditions used to acquire each spectrum must be submitted. These guidelines will be available from Dr. Takeda or published in the letters to the editors of the leading journals in the field.

Although there has not been time enough to prepare a proposal on mass spectrometry, Dr. J. Albert van Kuik and Professor Johannes F.G. Vliegenthart have put together a proposal on a NMR spectra database gathering and dissemination for complex carbohydrates and it is presented below:

Proposal on standardized information-exchange formats for spectral databases of complex carbohydrates.

J. Albert van Kuik and Johannes F.G. Vliegenthart
Department of Bio-Organic Chemistry, University of Utrecht, P.O. Box 80.075, 3508 T Butrecht, The Netherlands, Tel: +31-30-253-5184, Fax: +31-30-254-0980

Currently, no general database format exists for the storage of spectral data for these classes of bio-molecules. Some efforts have been made to collect NMR data for specific types of bio-molecules, such as proteins and peptides by BioMagResBank[1], or complex carbohydrates by SUGABASE[2]. These collections typically contain chemical-shift assignments, a description of the molecule, and experimental conditions, but no original spectral data. One major obstacle associated with collecting original spectral data is the variety of formats in which spectra are stored. Even with one NMR-instrument manufacturer, different machines and different generations of software result in many incompatible formats. Collections of chemical-shift assignments bring their own compatibility problems. To unequivocally assign a chemical shift in an arbitrary bio-molecule, this molecule needs a unique identifier, and, additionally, each atom in this molecule requires a unique identifier too. These identifiers must be universal and not class- or database-dependent. The lack of standards in this field leads to a specialized nomenclature for sub-classes of bio-molecules, e.g. proteins, carbohydrates, or saponins thereby excluding other classes, and making the inclusion of combined classes, e.g. glycoproteins or saponins, virtually impossible.

Fortunately some solutions for these problems are currently emerging[3]. To store spectral data of carbohydrate-containing bio-molecules in a format that is compatible with databases of other classes of biomolecules, we propose to extend current examples in the field, and to keep to the recommendations of IUPAC whenever possible.

Original spectral data should be stored in JCAMP-DX[4,5], a computer-readable exchange format, originally designed for infrared spectral data exchange. Data are represented as printable ASCII characters. This format has specific extensions for NMR data, and is recommended by IUPAC. It is, or will soon be, supported by major NMR manufacturers Bruker, Varian and Jeol. JCAMP-DX can also be used to store experimental conditions, e.g. type of instrument, solvent, concentration, original standard, conversion shift, temperature, and quality of the data.

Structures are to be submitted in the Standard Molecular Data (SMD) format[6], an organic structural information exchange format, which is currently under development by the IUPAC Committee of Chemical Databases (CCDB) and others in collaboration with the United States of America standards organization ASTM. This format will be supported by major organizations, e.g. *Chemical Abstracts* and Beilstein Institute.

For the electronic handling of data, in particular electronic submission of spectral data and storage of these data in a computer database, not a single clear-cut standard is available. For electronic handling, we propose to adopt the format used by crystallographers and by BioMagResBank. This Self-defining Text Archival and Retrieval (STAR) format utilizes tag-value pairs, and is suitable for electronic data submission and subsequent processing by the computer. It is also the format of choice to represent lists of chemical shift assignments. For each database, a specific data dictionary is required to clearly define each data tag and associated attribute. Crystallographers have developed CIF, the Crystallographic Information File Data Dictionary. This is now the format in which structure papers are submitted to *Acta Crystallographica.* BioMagResBank has defined a Data Disposition Form for submission of NMR spectra for their database. This form is specific for data stemming from proteins and peptides, but could in principle be extended to handle other bio-molecules as well. At the moment it does not handle complex branched structures such as some carbohydrates, and it doesn't mention JCAMP-DX or SMD formats.

1. B.R. Seavey, E.A. Farr, W.M. Westler, and J.L. Markley, A Relational Database for Sequence-Specific Protein NMR Data, *J.Biomolecular NMR* **1**, 217-236 (1991).

2. J.A. van Kuik and J.F.G. Vliegenthart, A NMR Spectroscopic Database of Complex Carbohydrate Structures, *Carbohydr. Europe* **10**, 31-32 (1994).

3. C.L. Wilkins, Guidelines on Nuclear Magnetic Resonance Computerized Databases, *Pure & Appl. Chem.* **67**, 593-596 (1995).

4. J.G. Grasselli, JCAMP-DX, A Standard Format for Exchange of Infrared Spectra in Computer Readable Form, *Pure & Appli Chem.,* **63**, 1781-1792 (1991).

5. A.N. Davies and P. Lampen, JCAMP-DX for NMR, *Appl. Spectrosc.* **47**, 1093 (1993).

6. F.H. Allen, J.M. Barnard, A.P.F. Cook and S.R. Hall, The Molecular Information File (MIF): Core Specifications of a New Standard Format for Chemical Data, *J. Chem. Inf. Comput. Sci.* **35**, 412-427 (1995).

CONTRIBUTORS

N. K. Abubakirov
Institute of the Chemistry of Plant Substances
Uzbek Academy of Sciences
Kh. Abdullaev Ave. 77
Tashkent 700170 UZBEKISTAN

P.K. Agrawal
Central Inst. of Medicinal & Aromatic Plants
Post Bag No. 1
R.S.M. Nagar
Lucknow 266016 INDIA

Yu.A. Akimov
Institute of Genetics
Academy of Sciences
Padurii str. 20
Kishinev 277002 MOLDOVA

Nils-Georg Asp
Chemical Center
University of Lund
Applied Nutrition & Food Chemistry
P.O. Box 124
Lund S-221 00 SWEDEN

Y. Baissac
Faculty of Medicine
Laboratoire de Pharmacologie
Universitie Montpellies II
9921 du CNRS
Montpellier 34060 FRANCE

Klaus Becker
Hohenheim University
Centre for Agriculture in the Tropics & Subtropics
Department of Animal Nutrition & Aquaculture
Universitat Hohenheim (480),
Stuttgart D-70593 GERMANY

Valentin A. Bobeyko
Institute of Genetics
Academy of Sciences
Padurii str. 20
Kishinev 277002 MOLDOVA

Z. Charrouf
Departement de Chimie
Faculté des Sciences
University of Mohammend V
B.P. 1014
Rabat MOROCCO

P.R. Cheeke
Department of Animal Science
Oregon State University
Corvallis, OR 97331

L.F. Chen
Department of Soil Science
National Chung Hsing University
Taichung TAIWAN

V.Ya. Chirva
Department of Organic Chemistry
Simferopol State University
Taltinskaya Str. 4
Simferopol, Crimea 333036 UKRAINE

C.H. Chou
Institute of Botany
Academia Sinica
Taipei TAIWAN

G.A. Cordell
PCPRS College of Pharmacy
University of Illinois at Chicago
833 S. Wood St.
Chicago, IL 60612-7231

Catherine E. Costello
Department of Biophysics
Boston University School of Medicine
Boston, MA 02118-2394

Marie-Geneviéve Dijoux
Centre National De La Recherche Scientifique
Faculte de Pharmacognosie
URA CNRS 492
51, rue Cognacq-Jay
51096 Reims-cedex, FRANCE

Arne Fláoyen
Norwegian College of Veterinary Medicine
Department of Reproduction and Forensic Medicine
P.O. Box 8146 Dep.
N-0033 Oslo, NORWAY

Katsuya Fukuhara
Department of Environmental Science,
Policy and Management
University of California
Berkeley, CA 94720-3112

T.V. Ganenko
Institute of Organic Chemistry
Siberian Division
Russian Academy of Science
Irbutsk 664033 RUSSIA

P.W. Geno
Department of Environmental Chemistry
Southwest Research Institute
San Antonio, TX 78228

Roberto R. Gil
PCPRS College of Pharmacy
University of Illinois at Chicago
833 S. Wood St.
Chicago, IL 60612-7231

Grace Gosmann
Universidade Federal do Rio Grande do Sul
Av. Ipiranga 2752
90.0610.00 Porto Alegre, RS BRASIL

V.I. Grishkovets
Institute of Genetics
Academy of Sciences
Padurii str. 20
Kishinev 277002 MOLDOVA

A.S. Gromova
Institute of Organic Chemistry
Siberian Division
Russian Academy of Science
Irbutsk 664033 RUSSIA

D. Guillaume
Laboratoire de Chimie Thérapeutique
URA 1310 du CNRS
Faculté des Sciences Pharmaceutique et Biologique
75270 Paris Cedex 06 FRANCE

Oleg N. Gutsu
Institute of Genetics
Academy of Sciences
Padurii str. 20
Kishinev 277002 MOLDOVA

Robert E. Hoagland
Southern Weed Science Laboratory
USDA/ARS
P.O. Box 250
Stoneville, MS 38776

K. Igarashi
Faculty of Agriculture
Yamagata University
1-23, Wakabamachi
Tsuruoka-shi 997 JAPAN

M. I. Isaev
Institute of the Chemistry of Plant Substances
Uzbek Academy of Sciences
Kh. Abdullaev Ave. 77
Tashkent 700170 UZBEKISTAN

Marian L. Jurzysta
Department of Biochemistry
Institute of Soil Science & Plant Cultivation
Pulawy 24-1000 POLAND

Edward J. Kennelly
Department of Medicinal Chemistry and
Pharmacognosy
College of Pharmacy
University of Illinois at Chicago
Chicago, IL 60612

E.A. Khamidullina
Institute of Organic Chemistry
Siberian Division
Russian Academy of Science
Irbutsk 664033 RUSSIA

G.V. Khodakov
Institute of Genetics
Academy of Sciences
Padurii str. 20
Kishinev 277002 MOLDOVA

D. Kim
Asian Vegetable Research and Development Center
Tainan TAIWAN

Douglas A. Kinghorn
PCRPS, Coll. Pharmacy (M/C 877)
University of Illinois at Chicago
833 S. Wood Street
Chicago, IL 60612-7231

Dr. P. Kintia
Institute of Genetics
Academy of Sciences
Padurii str. 20
Kishinev 277002 MOLDOVA

O. Kol
Laboratoire de Chimie Biologique et UMR n° 111.
Université des Sciences et Technologie de Lille
59655 Villeneuve d'Ascq Cedex FRANCE

A.E. Kondratenko
Department of Organic Chemistry
Simferopol State University
Taltinskaya Str. 4
Simferopol, Crimea 333036 UKRAINE

Isao Kubo
Department of Environmental Science,
Policy and Management
University of California
Berkeley, CA 94720-3112

Catherine Lavaud
Centre National De La Recherche Scientifique
Faculte de Pharmacognosie
URA CNRS 492
51, rue Cognacq-Jay
Reims-Cedex 51096 FRANCE

O. Leconte
Faculty of Medicine
Laboratoire de Pharmacologie
Universitie Montpellies II
9921 du CNRS
Montpellier 34060 FRANCE

J.S. Lee
National Chung-Hsing University
Department of Chemistry
Taichung 40227 TAIWAN

M.K. Lee
Department of Chemistry
National Tsing Hua University
Hsinchu 30043 TAIWAN

Maw-Rong Lee
National Chung-Hsing University
Department of Chemistry
Taichung 40227 Taiwan

Y. Leroy
Laboratoire de Chimie Biologique et UMR n° 111
Université des Sciences et Technologie de Lille
59655 Villeneuve d'Ascq Cedex FRANCE

Long-Ze Lin
PCPRS College of Pharmacy
University of Illinois at Chicago
833 S. Wood St.
Chicago, IL 60612-7231

Y.C. Ling
Department of Chemistry
National Tsing Hua University
Hsinchu 30043 TAIWAN

R.M. Liou
Department of Soil Science
National Chung Hsing University
Taichung TAIWAN

Galina A. Lupashku
Institute of Genetics
Academy of Sciences
Padurii str. 20
Kishinev 277002 MOLDOVA

V.I. Lutsky
Institute of Organic Chemistry
Siberian Division
Russian Academy of Science
Irbutsk 664033 RUSSIA

A. Nait M'Bark
Departement de Chimie
Faculté des Sciences
University of Mohammend V
B.P. 1014
Rabat MOROCCO

K. M. Madyastha
Department of Organic Chemistry
Indian Institute of Science
Bangalore 560012 INDIA

Harinder P.S. Makkar
Hohenheim University
Institute of Animal Production in the Tropics &
Subtropics
Universität Hohenheim (480),
Stuttgart D-70593 GERMANY

George S. Massiot
Centre National De La Recherche Scientifique
Faculte de Pharmacognosie
URA CNRS 492
51, rue Cognacq-Jay
Reims-Cedex 51096 FRANCE

Hitoshi Masuda
Maruzen Pharmaceuticals Co. Ltd.
Mukaihigashi-Cho, Onomichi 722 JAPAN

Kenji Mizutani
Maruzen Pharmaceuticals Co. Ltd.
Mukaihigashi-Cho, Onomichi 722 JAPAN

Jarbas A. Montanha
Universidade Federal do Rio Grande do Sul
Av. Ipiranga 2752
90.0610.00 Porto Alegre, RS BRASIL

Mugio Nishizawa
Faculty of Pharmaceutical Sciences
Tokushima Bunri University
Yamashiro-cho, Tokushima 770 JAPAN

Miriam Odoardi
Instituto Spermentale per le Colture Foraggere
29 viale Piacenza
Lodi 20075 ITALY

Kazuyoshi Okubo
Food Biotechnology, Faculty of Agriculture
Tohoku University
1-1 Tsutsumitori, Amamiyamachi, Aobaku
Sendai 981 JAPAN

Wieslaw Oleszek
Department of Biochemistry
Institute of Soil Science & Plant Cultivation
Pulawy 24-100 POLAND

Gunilla Önning
Chemical Center
University of Lund
Applied Nutrition & Food Chemistry
Lund S-221 00 SWEDEN

P. Petit
Laboratoire de Pharmacologie
Faculté de Médecine et UMR 9921 CNRS
Université Montpellier I
Montpellier 34095 FRANCE

Krishna N. Reddy
Southern Weed Science Laboratory
USDA/ARS
P.O. Box 250
Stoneville, MS 38776

J.A.A. Renwick
Boyce Thompson Institute
Tower Road
Ithaca, NY 14853

G. Ribes
Laboratoire de Pharmacologie
Faculté de Médecine et UMR 9921 CNRS
Université Montpellier I
Montpellier 34095 FRANCE

Yves Saivaire
Faculty of Medicine
Laboratoire de Pharmacologie
Universitie Montpellies II
9921 du CNRS
Montpellier 34060 FRANCE

E. P. Schenkel
Universidade Federal do Rio Grande do Sul
Av. Ipiranga 2752
90.0610.00 Porto Alegre RS BRASIL

A.A. Semenov
Institute of Organic Chemistry
Siberian Division
Russian Academy of Science
Irbutsk 664033 RUSSIA

A.S. Shashkov
Institute of Genetics
Academy of Sciences
Padurii str. 20
Kishinev 277002 MOLDOVA

Stephan A. Shvets
Institute of Genetics
Academy of Sciences
Padurii str. 20
Kishinev 277002 MOLDOVA

C.H. Su
Department of Soil Science
National Chung Hsing University
Taichung TAIWAN

Rutt Suttisri
Department of Pharmaceutical Botany
Faculty of Pharmaceutical Sciences
Chulalongkorn University
Bangkok 10330 THAILAND

Yukiyoshi Tamura
Maruzen Pharmaceuticals Co. Ltd.
Mukaihigashi-cho
Onomichi 722 JAPAN

Osamu Tanaka
Suzugamine Womens' College
Inokuchi, Nishi-ky
Hiroshima 733 JAPAN

Aldo Tava
Instituto Spermentale per le Colture Foraggere
29 viale Piacenza
Lodi 20075 ITALY

Almira E. Timbekova
Institute of the Chemistry of Plant Substances
Uzbek Academy of Sciences
Kh. Abdullaev Ave. 77
Tashkent 700170 UZBEKISTAN

Yumiko Toshiki
Food Biotechnology, Faculty of Agriculture
Tohoku University
1-1 Tsutsumitori, Amamiyamachi, Aobaku
Sendai 981 JAPAN

N.N. Trofimova
Institute of Organic Chemistry
Siberian Division
Russian Academy of Science
Irbutsk 664033 RUSSIA

Stephen C. Wagner
Southern Weed Science Laboratory
USDA/ARS
P.O. Box 250
Stoneville, MS 38776

George R. Waller
Oklahoma State University
Department of Biochemistry & Molecular Biology
246B Noble Research Center
Stillwater, OK 74078-0454

J.C Wang
National Chung-Hsing University
Department of Chemistry
250 Kuo-Kuang Road, Taichung
Taiwan 40227 CHINA

P.R. West
Structural Chemistry
Department 418, AP31
Abbott Laboratories
Abbott Park, IL 60064

S.C. Wu
Faculty of Pharmaceutical Sciences
Tokushima Bunri University
Yamashiro-cho, Tokushima 770 JAPAN

R.N. Yadava
Department of Chemistry
Dr. Harisingh Gour University
Sagar 470 003 INDIA

Hidetoshi Yamada
Faculty of Pharmaceutical Sciences
Tokushima Bunri University
Yamashiro-cho, Tokushima 770 JAPAN

C.F. Yang
Institute of Botany
Academia Sinica
Taipei TAIWAN

Y. Yoshiki
Food Biotechnology, Faculty of Agriculture
Tohoku University
1-1 Tsutsumitori, Amamiyamachi, Aobaku
Sendai 981 JAPAN

C.C. Young
Department of Soil Science
National Chung Hsing University
Taichung TAIWAN

R.M. Zablotowicz
USDA - ARS
Southern Weed Science Laboratory
Stoneville, MS 38776

LATIN NAME INDEX

Abrus precatorius 16, 17, 20-22, 25, 35
Abutilon theophrasti 61
Adonis aleppica 296
Adoxophyes orana 104-106, 164
Aeschynomene indica 147
Agapenthus inapertus 314
Agava americana 273, 279, 315
Agava lecheguilla 381, 383, 396, 402, 403
Agrobacterium tumefaciens 83, 85-88, 90, 92, 93, 177, 180
Ailanthus excelsa 283, 296, 297
Ailanthus integrifolia 283, 296
Albizzia lebbek 61
Alcaligenes faecalis 4, 8
Allium ampeloprasum 313
Allium giganteum 314
Allium macrostemon 314
Allium stipitatum 313
Allium suvorovii 313
Allium vineale 256, 294, 315
Alnus japonica 20
Alysicarpus vaginalis 147
Amaranthus palmeri 58, 71
Amaranthus retroflexus 58, 165
Amaranthus viridis 58, 72
Ambrosia artemisiifolia 71
Amorpha fruticosa 147
Amphicarpaea edgeworthii 147
Anthrobacater spp. 87
Aphis nerii 113, 120
Apios americana 145, 147
Arachis hypogaea 147
Arthrobacter sp. 86
Asclepias albicans 283, 295
Asclepias curassavica 113, 120
Asclepias syriaca 120
Ascochyta imperfecta 43
Aspergillus awamori 4, 8
Aspergillus flavus 163
Aspergillus fumigatus 163
Aspergillus niger 4, 8, 83, 162
Aapergillus ochraceus 270
Aspergillus oryzae 4, 8
Aspergillus sydowii 4, 8
Aspergus officinalis 314
Aspilia kotschyi 192
Aster batangensis 283, 296, 297
Aster yunnanensis 283, 297
Atylosia scarabaeoides 147
Avena sativa 58, 71, 72, 374, 375
Azadirachta indica 283, 286, 295

Bacillus cereus 87-89
Bacillus circulans 4, 8
Bacillus licheniformis 4, 8
Bacillus mesentericus 177, 180
Bacillus subtilis 4, 8, 87, 88, 163
Bacillus thuringiensis 85, 87-89, 92, 93
Beaumontia brevituba 120
Beta vulgaris 191, 192
Brachiaria decumbens 382, 384, 398, 402, 403
Bradyrhizobium japonicum 64, 84, 86, 87
Bromus secalinus 58, 164, 165
Brucea javanica 283, 296
Bryonia dioica 15, 23
Bufo vulgaris formosa 210
Bupleurum longicaule 283, 291
Bupleurum radix 6
Bupleurum rotundifolium 209, 210
Bupleurum wenchuanense 283, 291, 293, 296

Caenocoris nerii 114, 120, 121
Caesalpinia crista 147
Caesalpinia pulcherima 147
Cajanus cajan 147
Caltha palustris 50, 56
Camellia sinensis 233
Campsis chinensis 182
Canavalia gladiata 147
Canavalia maritima 147
Candida albicans 3-6, 8, 10, 39, 43, 48, 163, 177, 180
Candida famata 4, 5, 8
Candida guillermondii 43
Candida pseudotropicalis 43
Candida tropicalis 177, 180
Candida utilis 3, 177, 180
Capsicum annuum 76, 81, 82
Caraipa densiflora 283, 295
Caralluma umbellata 283, 295
Cassia alata 147
Cassia nomame 147

Cassia obtusifolia 61
Cassia occidentalis 147
Cassia tora 147
Castanospermum australe 167
Caulophyllum robustum 182
Cellulomonas cartae 88, 90
Cellulomonas sp. 87, 91
Centrosema pubescens 147
Cephalosporium gramineum 163
Ceratonia siliqua 147
Cercis siliquastrum 147, 221
Cestrum nocturnum 315
Chenopodium quinoa 373, 374, 384
Christia obcordata 147
Chrysolina coerulans 114
Chrysolina fuliginosa 114, 120
Chrysolina herbacea 114
Cicer arietinum 61, 73, 147
Citrobacter diversus 87-89
Cladosporium cucumerinum 58
Cladosporium herbarum 270
Cordyline rubra 313
Corynebacterium insidiosum 177, 180
Corynebacterium michiganense 83, 177, 180
Creatonotos transiens 120
Cryptococcus laurentii 4, 8
Cryptococcus magnus 10
Cryptococcus spp. 8
Cucumis sativus 61
Cunninghamella elegans 314
Curtobacterium flaccumfaciens 85, 86, 88, 89, 93
Curtobacterium spp. 87
Cyamopsis tetragonoloba 147
Cyclocarya paliurus 23
Cycnia tenera 113

Danaus plexippus 113, 120
Datura stramonium 66
Davidsonia pruriens 283, 296
Debaryomyces hansenii 4, 5, 8, 10
Dechelostemma multiflorum 314
Desmodium heterophyllum 147
Desmodium triflorum 147
Desmodium uncinatum 147
Dianthus barbatus 295
Digitalis purpurea 76, 82
Dioscorea deltoidea 273, 279
Dioscorea tokoro 312
Dolichos kilimandscharicus 72, 167, 170
Dolichos lablab 147, 239
Dunbaria villosa 147

Echinochloa crus-galli 57, 165
Enterobacter asubriae 87, 90, 91
Enterobacter cloacae 86-90
Enterococus faecalis 4, 8
Epidermophyton floccosum 4, 8, 10
Erwinia caratovara 87, 90, 91, 177
Erwinia herbicola 88
Erysimum asperum 115
Erysimum cheiranthoides 115, 116, 119, 121, 121
Escherichia coli 4, 8, 48, 163, 392
Euphorbia cyparissias 284, 296
Euphorbia nicaeensis 284, 296

Flemingia prostrata 147
Fructus momordicae 23, 35
Funkia ovata 76, 82
Fusarium avensceum 58, 72, 73
Fusarium graminearum 76-80, 82
Fusarium oxysporum 76, 83, 162
Fusarium solani 71, 83
Fusarium spp 127

Gaeumannomyces graminis 72, 163, 169
Galactia tashiroi 147
Gelsemium sempervirens 284, 286, 296
Gleditsia japonica 339, 364
Glycine max 57, 147, 148, 331, 337
Glycine soja 147, 148
Glycine tabacina 147
Glycine tomentella 147
Glycyrrhiza glabra 13, 22, 25
Glycyrrhiza inflata 14, 23
Glyptopetalum sclerocarpum 284, 294
Gossypium arboreum 165
Gossypium hirsutum 61
Gramineae incana 384
Guettarda angelica 56
Gymnema sylvestre 15, 23
Gymnema yunnanense 290, 294
Gypsophila spp. 369-373
Gypsophila arrostii 339, 352
Gypsophila paniculata 69, 71
Gyrinops walla 284, 296

Haloniopsis orientalis 314
Hansenula anomala 3-5, 8, 10
Hansenula spp. 4, 8
Hedera helix 337, 352, 364
Hedera nepalensis 182
Hedera taurica 209, 210, 222
Helix pomatia 196, 200, 202, 248
Helleboru purpurascens 314
Helminthosporum oryzi 58
Hemsleya panicis-scandens 22
Herniaria fontanesii 241, 242, 244, 245
Herniaria glabra 158, 169, 245, 314
Herniaria hirsuta 158, 169, 245
Heterodera schachtii 106

Hibiscus esculentus 68
Homeria spp. 113
Hordeum vulgare 58, 71, 73
Hovenia dulcis 15, 23, 24
Hypericum perforatum 381
Hystrix africae-australis 113

Ilex argentina 54, 56
Ilex dumosa 54, 56
Ilex integerrima 54, 56
Ilex paraguariensis 47, 49, 52, 54, 56
Ilex pseudobuxus 54, 56
Ilex taubertiana 54, 56
Ilex theezans 54, 56
Indigofera trifoliata 147

Jacaranda caucana 284, 296
Jucca aloifolia 182

Klebsiella pneumoniae 4, 8, 87, 88, 90
Kloeckera apiculata 4, 5, 8

Lactobacillus plantarum 4, 8
Lactobacillus rhamnosus 4, 8
Lactuca sativa 58, 125
Larrea tridentata 284, 295
Lcucaena leucocephala 147
Lens culinaris 147
Leptinotarsa decemlineata 164, 170
Lilium dauricum 314
Lilium henryi 313
Lilium pardarinum 313
Lilium regal 313
Lobesia botrana 104, 105, 164
Lotononis bainesii 147
Lupinus hirstus 147
Lupinus luteus 147

Maackia amurensis 147
Macaca fascicularis 163, 182
Macroptilium atropurpeum 147
Madhuca longifolia 192
Majorana hortensis 223, 224
Malassezia furfur 10
Marsdenia globifera 290, 296
Marsdenia koi 284, 297
Marsdenia tenacissima 284, 290, 296
Medicago arborea 98, 99, 102, 167
Medicago cretacea 167
Medicago falcata 98
Medicago granadensis 168
Medicago hybrida 167, 168
Medicago lupulina 58, 71, 147, 167
Medicago media 58, 98, 109, 168
Medicago monspeliaca 167
Medicago radiata 167
Medicago sativa L. 58, 72, 73, 95-99, 102, 103, 108, 109, 137, 139, 147, 155, 165, 167-171, 181, 182, 186, 192, 337, 340, 352, 353, 356, 364, 384, 385, 393
Medicago scutellata 98
Medicago sufruticosa 168
Medicago truncatula 98
Medicago varia 98
Melilotus albus 211, 215, 221
Melilotus neapolitanus 211
Melilotus officinalis 211, 221
Melilotus tauricus 211, 215, 218, 220, 221
Mimosa pudica 147
Momordica grosvenorii 25
Moraea spp. 113
Moraxella spp 260, 261, 263
Mucor griseocyanus 270
Mucor piriformis 259, 264-268
Mucor pusillus 4, 8, 259, 264-268, 270
Mycobacterium intracellulare 163

Narthecium ossifragum 381, 385, 395-397, 399-403
Nerium oleander 113
Nicandra physaloides 284, 295
Nicotiana tabacum L. 138, 247, 256
Nierembergia aristata 284, 288, 293
Nierembergia stricta 295
Nolina microcarpa 222
Nolina recurvata 314

Oncopeltus fasciatus 113, 120
Onobrychis viciaefolia 221
Ophiobolus graminis 76
Ophiopogon japonicus 314
Oreganum vulgare L. 239
Ornithoglossum spp. 113
Ostrinia nubilalis 104, 106
Oxalis tuberosa 408

Panax ginseng 352, 392
Panax quinquefolium 69
Panicum coloratum 381, 383, 385, 398, 402
Panicum dichotomiflorum 385, 398, 402, 403
Panicum miliaceum 398
Panicum schinzii 385, 398, 403
Panicum virgatum 398
Patrinia scabiosaefolia 51, 56, 209, 210
Pedetes capensis 113
Penicillium expansum 4, 8
Pentzia incana 384
Periandra dulcis 15, 16, 18-23
Peromyscus melanotis 114
Petersianthus macrocarpus 188, 192
Phaseolus coccineus 145, 147

Phaseolus lunatus 147
Phaseolus vulgaris 147
Phlomis betonicoides 25
Phoma sp. 83
Phoron humuli 164
Phycomyces blakesleeanus 267, 268, 270
Phytophthora cinnamoni 83, 95
Pichia carsonii 4, 5, 8
Pichia nakazawae 4, 5, 8, 10
Pieris napi oleracea 116, 117
Pieris rapae 115-117, 119, 121
Pisum sativum L. 58, 71, 73, 147, 192, 373
Plutella xylostella 119, 121
Poekilocerus bufonius 113
Polygonati officinalis 314
Polygonatum kingianum 313
Polypodium glycyrrhiza 20, 21, 23, 313
Polypodium occidentale 22
Polypodium vulgare 15, 21-23, 25-27, 30, 32, 33, 35, 36
Proteus vulgaris 4, 8
Pseudomonas aeruginosa 4, 8, 163
Pseudomonas fluorescens 4, 8, 85, 87-93
Pseudomonas lachrymans 83, 177, 180
Pseudomonas putida 87, 88, 90
Pseudomonas vesicularis 87, 90
Pterocarya paliurus 19, 21, 23
Pueraria lobata 147
Pycnospora lutescens 147
Pythium spp 127

Quillaja saponaria 317-330

Rapanea melanophloeos 58, 72
Raphanus sativus 61
Rhizobium meliloti 64, 86, 87
Rhizoctonia solani 162
Rhizoctonis spp 127
Rhizopus arrhizus 259
Rhizopus formosaensis 4, 8
Rhizopus mucco 83
Rhizopus nigricans 4, 8, 270
Rhodococcus corallinus 84, 87-90
Rhodococcus meliloti 84
Rhynchosia minima 147
Rhynchosia volubilis 147
Robinia pseudacacia 147, 221
Rosellinia necatrix 39, 42, 43,
Rudua aurea 147
Ruscus aculeatus 314

Sabouraudites canis 4, 8, 10
Saccharomyces cerevisiae 2-5, 8, 163, 180
Sapindus delavayi 10, 11
Sapindus mukurossi 1, 2, 5-7, 10, 11
Sapindus rarak 10
Sapindus saponaria 379, 384
Sapindus trifolius 10
Saponaria officialis 69, 71
Schefflera octophylla 209, 210
Sclerotinia rolfsii 39, 42, 43, 83, 162
Scopolia japonica 313
Secale cereale 58
Serratia plymutica 87-90
Sesbania exaltata 57, 165
Sesbania grandiflora 147
Sida spinosa 61
Simaba multiflora 284, 294
Siraitia grosvenorii 14, 16, 23
Siraitia sinamensis 15, 22
Smilax menispermoidea 313
Smilax sieboldii 313
Solanum dulcamara 314
Solanum incanum 409, 417
Solanum ptycanthum 416
Solanum tuberosum 405
Solanum X ajanhuiri 407
Solanum X curtilobum 407
Solanum X juzepczukii 407
Sorghum halepense 61
Sorghum vulgare 61
Soulamea soulameoides 284, 295
Spilostethus pandurus 120
Spodoptera littoralis 106
Staphylococcus aureus 4, 8, 48, 163
Staphylococcus epidermidis 4, 8
Stevia rebaudiana 25
Streptococcus mutans 4, 8
Strophanthus divaricatus 295
Stylosanthes guyanensis 147
Stylosanthes humilis 147
Stypholobim fomentosa 147
Stypholobim japonica 147
Syntomeida epilais 120
Syntomis mogadorensis 120

Tacca leontopetaloides 314
Taraxacum officinale 58
Taraxacum vulgare L. 165
Tenebrio molitor 98, 106, 109
Terminalia chebula 284, 296
Tetranychus urticae 164
Thalictrum appendiculatum 193
Thalictrum foetidum 193, 197, 199, 200, 206, 207
Thalictrum herba 207
Thalictrum minus 193-197, 200, 205-207
Thalictrum petaloideum 193
Thalictrum squarrosum 193, 200-203, 206, 207
Thlandiantha hooderi 6, 11
Tinospora cordifolia 313

Tribulus terrestris 76, 381, 382, 384, 385, 395-397, 399-403
Trichoderma harzianum 39, 42, 43
Trichoderma viride 39, 42, 43, 73, 83, 86, 95, 98, 109, 155, 156, 161-163, 165-167, 169
Trichophyton mentagrophytes 4, 8, 10, 163
Trichophyton rubrum 4, 6, 8,-10
Tridesmostemon claessenssi 184, 192
Trifolium incarnatum L. 353, 355, 356, 364
Trigonella foenum-graecum 37, 41, 46, 69, 375
Trigonella monantha 167
Trigonella monspeliaca 167, 170
Trigonella polyceratia 167
Trillium tschonoskii 314
Triteleva lacta 315
Triticum aestivum 57, 165
Tropaeolum majus 119

Uraria lagopodioides 147
Urginea spp. 113

Verticillium albo-atrum 72, 83
Vicia angustifolia 147
Vicia faba 147, 374
Vicia japonica 147
Vicia sepium 147
Vigna angularis 147, 156
Vigna radiata L. 61, 123, 129, 137, 147, 331, 335-337, 352, 353, 356
Vigna sinensis 147
Vigna vexillata 147

Wisteria floribunda 147

Xanthomonas campestris 83, 86-88, 90, 91, 177
Xylosma velutina 284, 294

Yucca gloriosa 2, 11
Yucca schidigera 2, 10, 377, 383, 385, 394

Zanha golugensis 158
Zea mays 61
Ziziphus jujuba 15, 23, 24
Zygiella x-notata 115
Zygosaccharomyces rouxii 4, 5, 8

SUBJECT INDEX

Abrusosides, 17-18, 25-26
Acetaldehyde, a promoter of chemiluminesence, 149-154, 231-238
3β-Acetoxy-19-hydroxycholest-5-ene
  19-HCA and other metabolites
    conversion to estrone, 260-261
    structure, 260-261
Active oxygen species, 231-237
Agavoside C, 273, 275
Alianthinone, 283
Alfalfa, 171, 186
  breeding for lower saponin content, 97
  saponins
    aglycones, 102
    hydrolysis, 102, 172
    isolation, 99, 101, 171
    mass spectra, 354, 357, 360-361, 363
    root saponins, 357
    structures, 99, 100, 340-341, 355
    toxicity, 98, 171
Allelopathic properties of saponins, 58, 70, 124, 155, 171
Alveld, 395, 397, 400
Androstane derivatives, 260, 262-263, 267
Anthocyanin, as superoxide and hydroxyl radical scavengers, 149, 152, 231, 237, 239
Androstenediones, 267
Antibacterial action, 4, 6, 8, 83-84, 177-184
Anticancer agents, 282-283, 290
Antidermatophytic activity, 10
Antifungal activity, 2, 4-6, 8-9, 39, 58, 83, 162-163, 177-179
Antimicrobial activity, 9
Antinutritional factors, effect on
  rumen fermentation, 387-394
  rate of gas production, 388-389
  digestibility of hay, 388-387
  short-chain fatty acids, 390
  microbial synthesis, 390
Antioxidants, 149-150, 233, 237
Antitumor activity, 203-204
Antiviral effects
  mechanisms of action, 253-254
  nicotianosides, 252-253
Antiyeast activity, 1-10, 39, 83, 180
Apioglycyrrhizin, araboglycyrrhizin, 14
Arjungenin, 284
Arjunglucoside, 284
Artifacts, formed in saponins research, 183-192, 203
Asterbatonosides, 285
Asteryunnanosides, 283
Avenacosides
  effects on intestinal permeability, 371-372
  effects of heat treatment, 368
  effects on lipids in animals, 369-371
  isolation and stability, 368
  levels in oats, 367
  structures, 365-366

Baiyunoside, 25-26
Barbatosides, 283
Bassic acids, as artifacts, 184
BE LSIMS/MS, 339-352
BE CAD LSIMS/MS 348-350
Betulin
  effect on bacterial esterases, 85
  phytotoxicity, 57, 61-62, 67
  structure, 59, 84
Bile, bile ducts, 377-383, 395-399, 401
Bitter taste, oats, 365
α, α'-Bipyridyl, 261-262
Bufadienolides, 111-112

Cabbage, 178-179, 181
Cabbage butterflies, 115-117
Carbohydrate portions of saponins of
  alfalfa, 346-351
  biological activity, 293
  Bupleurun species, 292
  Heriaria fontanensis, 242
  Majorana hortensis, 223-225, 229
  Milelotus species, 212-221
  Nicotiana tabacum, 248-250
  Quillaja saponaria, 323-328
  Thalictrum species, 194-203
  various types, 308-312
Cardenolides, 111-119
  biological activity, 290

plants producing, 112
presence in crucifers, 115-116
structure, 111-112, 117-118, 288-290
toxicity, 112-116
Carumbellosides, 283
Catechins as superoxides and hydroxyl radical scavengers, 149-150, 235
α-Chaconine, CAD spectrum, 413
α-Chaconine, α-solanine
detection and identification by mass spectrometry, 410, 411
determination by HPLC, 410, 415
separation by column chromatography, 410, 415
Chaparrinone, 283-284
Chemiluminesmence, 146-152, 231-237
Chemotaxonomic significance of alfalfa saponins, 167-168
Cholesterol, 39, 104-105, 141, 259-260, 369-371
Cholesterol-free milk, 383
Chromatography
countercurrent, RLCC and DLCC, 408-409
column, 194, 202, 215, 219, 224, 254, 279, 414
gas, 102, 172, 254
HPLC, 99-101, 127, 146, 414, 416
thin layer, 48-49, 52-53, 99, 125-128, 171, 203, 218-219, 247, 257, 279
CID Spectra, 323-326
CID MS/MS spectra, 325
Circular dichroism spectrum, 200
Clover seed saponins, 353, 356-357, 360-361
Commercial uses, 1-11
beverages, 1
cosmetics, 1
flavor enhancer, 1
foam agents for five oxtinguishers, 5
foods, 1
liquid detergents, 1
photosensitized film, 1
shampoos, 1
sweeteners, 1
tooth paste, 1
Contraception, 204
Convallatoxin, 118
Convallamaraogenin, 2
Correlations of ^{1}H[^{13}C] in NMR spectra, 304, 311
Cosmetics, 1-11
Cotton, 177-178
^{51}Cr-EDTA, 372-373
Crop species, effect of saponins, 57, 61-70
Crucifers, cardenolides in, 115
Cucurbitane glycosides, 15
Cycloartane saponins, 17, 21, 194-199, 201, 205
Cyclocariosides, 19
Cyclofoetigenin B, 199
Cyclofoetosides, 197-200, 203-204
Cymarin, 117

Dammnarane saponins, 15, 19
Dehydration of saponin components, 184-190
Dehydrogenases, 262-264
2,3-Dehydro-2,5-dihydroxy-6-methyl-4H-pyran-4-one (DDMP) saponins, 231
as antioxidants, 151
distribution in leguminous seeds, 147
isolation from plants, 145
structures, 143-144, 232
Deltosides, deltinine, 273, 275
Deoxy mimbolide, 287
Dermatophytic species, 1, 5-6, 8, 10
Dietary effects on farm animals, 377-388
feed intake, 378
digestive track, 378-380
intestinal effects, 380
as feed additives, 381
effects on human health, 382-383
Digestibility, 387-392
Digitoxigenin, 118
Digitoxin, 118
Digoxigenin, 118
Digoxin, 118
2α,3β-Dihydroxyolean-12-en-28-oic acid, 223, 225-226, 228
derivatives, 226, 228-229
Diosgenin, 38, 80215-216, 218, 273, 275, 305, 398-399

Echinocystic acid, 210
3-sulfate derivative, 210
Electron spin resonance, for measuring radical scaveging activity, 151
Electrolyte leakage, 57, 60, 66-67, 76
Electrospray ionization, 353-363
17-Epi-lla-hydroxy-6,7-dehydrostrophanthidin-3-O-β-biovinopyranoside,
HOHAHA spectrum, 289
ROSEY spectrum, 289
Episarsasapogenin, epismilagenin, 392, 398, 400-401
crystals from lamb, 397
purified chemical structure, 401
β-Escin and/or chloramphenicol, effects on proteins and enzyme levels on *P. fluorescens* cultures, 91-92
β-Escin
actions on bacterial growth, 84-94
in culture, protein leakage and structure, 59
hydrolytic activity, 93
phytotoxicity, 57, 61-70

in soil, microbial population and hydrolytic activity, 94
Erychroside, 117
Erycordin, 117
Erysimoside, 117
ESI mass spectra, 318, 353-376
ESI/MS mass spectra, 361-365
Esterase activity, 57-58, 68-69, 76
Estrone, formation from 19-HCA, 260-261
European corn borer, toxicity of alfalfa saponins for, 104-106
European grape moth, toxicity of alfalfa saponins for, 104-105

FAB-MS, 319-328, 331-336
spectrum of methanol extract of papa negra skin, 411
Fenugreek
pharmacologic affects, 37
steroid sapogenin of, 37, 38
steroid saponins
effects of fungi, 39, 42-43
effects of rats, 44-45
hypocholesterolemic effects, 39, 43
isolation, 39, 41
structures 38-40
Flavanoids, as superoxide and hydroxyl radical scavengers, 231-238
Flavor enhancers, 1-11
Foetoside, 197, 203-204
Foods, saponins added to, 1-11
Furostanol glycosides, 271, 273, 277, 279
Fusarium diseases, 75-81

3-O-[β-D-Galactopyranosyl-(1→2)-β-D-glucuronopyranosyl] sophradiol, 337
Gallic acid and derivatives, 233, 235
Gas production, 388-389
Geeldikkop, 395, 400
Gekogenin, 273, 275
Genes controlling soybean saponin sugar chains, 148
Geophagy, 407
Germination tests, 60-61, 78, 177-179, 181, 251-252
Gitoxigenin, 118
Gitoxin, 118
Glaucarubinone, 283
Glycosidation, 111, 311
Glycosides, steroidal, 75-78
β-Glycyrrhetinic acid
effect on bacterial esterases, 85
phytotoxicity, 57, 61-63, 66-69
structure, 84
Glycyrrhizin, 1, 13-14, 17-18, 20-21, 25-26, 237
structure, 59
Gymnemarsogenin
INEPT spectra, 291
Gymnernasaponins, 15
Gymnemic acid, 15
Greening experiments 57, 60, 67-68

Hecogenin
analysis for, 103
effect on bacterial esterases, 85
NMR spectrum, 305
phytotoxicity, 57, 61, 63, 66-68
structure, 59, 84
Hederagenin, 6-7, 10, 102-104, 355
Hederagenin glycosides, 7, 157, 175-177, 181, 355
Helveticoside, 117
Hemolytic activity of alfalfa saponins, 167
Herniaria saponin B, 241, 244-245
hydrolysis and methanolysis, 242
isolation, 242, 244
lithasis effect, 241-242
NMR spectra, 242-245
structure, 243
Hispidol, 284
History of NMR studies of steroids and triterpenes, 282-284
14α-Hydroxylase, 268
Huzhangoside, 9
Hydroxyl groups, determination of number and types of by NMR, 303-305
Hydroxylation, 261-262, 264-265
Hypercholesterolemic effect, 155, 382

INEPT spectra, 285-291
Insects *see also* Oviposition
production of cardenolides, 113
toxicity of cardenolides, 113
toxicity of saponins to, 104

Jacoumaric acid, 284
Jacarandic acid, 284
Javancin, 283

Leukemic cells, human, inhibition by alfalfa saponins, 107
Leguminous seeds, classification by DDMP saponins, 146-148
Lettuce growth, effect of mungbean saponins, 130, 132-134, 136
Licorice, root saponins, 1, 17-18
24-O-Linoleoyl soyasasapogenol B, 190
Lipids, in animals, effects of avenacosides, 370-371
Liver, involvement in photosensitization, 395-401
LSIMS, 320-332
Lucernic acid
as artifact, 186
structure, 186, 187

Maltol, 189, 222, 236-238

Markogenin, 3
Marsdekoside, 284
Marstenacigenins, 284
MALDI, 318
MALDI-TOF, 320-323
Mass spectra, 331
  of alfalfa saponins, 360-361, 363
  of clover saponins, 363
  of hernaria saponins 13, 242
  liquid secondary ionization, 318-328
  of matesaponins, 50-54
  tandem, 317-328
    acyl fragments, 324-327
    electrospray ionization, 318-319, 328
    matrix-assisted laser desorption ionization, 318-319, 328
  positive and negative ion, 320-328, 335-336, 341, 343, 354-364
Mass spectrometry
  electrospray ionization vs fast-atom bombardment, 354-364
  linked scanning in, 340-351
  tandem, 317-328, 331-332, 340-351
Maté
  antidemic effect, 47
  history and general information, 47
  saponins of, 47-48
    oleanolic and ursolic esters as, 48
    separation, 48, 49, 53
    spectroscopy, 50, 56
    structure, 48, 50- 52
Medicagenic acid,
  analysis for, 102-103, 159-160, 165
  concentration of, 166
  fragmentation in mass spectrometry, 334, 342-345
  glycosides, 172-175, 177-181, 340-341, 355
    allelopathic activity, 161, 164-165, 171, 179-181
    animal effects, 162-166, 171
    antimicrobial, 171, 179-181
    fungi effects, 161-162
    hemolytic activity, 161
    insect effects, 164, 177
  LSIM and EI spectrum, 342-343, 347-348
  structure, 157, 173, 341, 355
Medicosides, 155-156, 171
Melilotosides
  carbohydrate components, 212-215
  NMR spectra, 212-214
  structure, 212-215
Melilotus species
  distribution and components, 211
  triterpenoid saponins of, 211-221
Metabolism of sarsasapogenin, 399-400
Microbes, transformation of steroids by, 259-268
Microbial protein synthesis, 390-392
Mogrosides, 14-15, 21, 25-26
Mogroside, 14, 26
Moldstim®, 76-77, 79, 81
Molecular structure determination of steroidal sapogenins moiety
  parent skeleton, ^{13}C NMR shift, 302-303
  substitution pattern, 303-306
  stereochemistry at A/B ring junction, 306-308
Molecular structure of oligosaccharide moiety
  number and types of sugars, 308-309
  interglycosidic linkage and sequence, 309-310
  location of sapogenin-sugar linkage, 310-311
Monarch butterfly, 113-114
Mukuroizioside, 7
Mungbeans
  effect of mungbean saponins, 131-136
  growth stages, 124, 127
Mungbean saponins
  allelopathic effects, 121
  bioassay, 126, 131-132
  biosynthesis, 127-128, 131
  general information, 123-124, 335
  isolation and purification, 124, 128
  mass spectra, 357-359
  mechanism of action, 135-136
  structures, 129

Murwaoside, 223-226
  reactions, 224-225, 229-230

Natriated ions, 342, 346-351, 354, 357-359, 361-363
Neotigogenin, 273, 275, 397
Neogitogenin, 38
Nicotianosides, 242-249
  antiviral action, 252-254
  effect on germination of tobacco seeds, 251-252
  hydrolysis, 248, 255
  isolation, 254-255
  leaf protoplasts, 254
  methylation and oleanene, 248, 255
  periodate oxidation, 255
  structures, 251, 256
Nimbolides, 286-287
Nicoderdrone, 284
Nimbolide, 283
  spectra, 287
NMR
  determination of oligosaccharide moiety characteristics, 308
  determination of parent skeleton, 302-303, 311
  determination of substitution patterns, 303

^1H, assignment of resonances, 304
one-dimensional vs two-dimensional, 285, 301
techniques applied to steroidal saponins 285, 301, 311
NMR spectra
of alfalfa saponins, 188-189
of herniarina saponin B, 242
of hederagenin glycosides, 175-177
for locating OH groups in triterpenes, 184
of maté saponins, 50-55
of medicagenic acid and derivatives, 172-175, 181
of metelosides, 212-214
of nicotianosides, 247-249
of saponins and derivatives of *Majorana hortensis*, 223-237
of steroids and triterpenes, 281-293
of thalicosides and hydrolysis products, 194-203
NOE connectivity, 305, 307, 310
Nomenclature of soyasaponins, 142

Oleanene glycosides, 15, 18, 20-21, 205
Oleandrin, 59, 118
Oleanolic acid
analysis for, 108
formation from saponins, 102, 196, 205
phytotoxicity, 61-62, 66
structure, 59
Oleandrin
phytotoxicity, 57, 61-62, 66
structure, 59
Olefin carbon and carbonyl carbon resonances in NMR, 305
Osladin, 15, 20-21, 25-27
structure, 26, 31-32
sweetness, 25-26,31
synthesis, 26-33
Ouabain, 118
Oviposition, factors influencing, 115, 119

Pavstim®, 76-77, 79-80
Periandrims, 18-19, 21
Peroxides, 149-152, 232-237
Photosensitization diseases, 395-401
Photsensitization, 381, 395-404
lamb following ingestion of *N. ossifragum*, 396
Phylloerythrin, 395, 401
Physiological effect, 66-70
Phytotoxicity of saponins, 57-70
Plant growth responses, 61-66, 77-81, 129-130, 156-169, 172-176
Plasmalema permeability, 77, 79-80
Polypodogenin, 20-21
Polypodosides, 20-21
Potato
avoidance of toxicity, geophagy, 407-408
native varieties and relatives, 405-406
steroidal glycoalkaloids in, 405-406
Potassium sorbate, 5
Predators, effects of cardenolides on, 114
Pregnenolones, 267-268
Progesterone derivatives, 264-266, 268
Prosapogenins, 216-220
Protobassic acids, dehydration, 184
Proton and Carbon-13 assignment of structure, unambiguous by NMR, 285-293
Pseudoaldosteronism, 14
Pterocaryosides, 19, 21

Quanoa saponin, 8-9
Quassinoids, biological activity in, 293
Quillaja saponin, 1, 3, 318-328, 387-392

Radical scavengers, in chemiluminescence, 149-152
Retro-Diels-Alder reaction, 344-354
Rhizosphere bacteria, effect of β-escin
in growth, 87
on hydrolytic activity, 88-91
on esterase activity, 89
on GPNA-α-glutamyl transpeptidase activity, 90
on extracellular protein levels, 90-92
Root rots, *see Fusarium* diseases
Rumen, 378-380, 387-392

Saikosaponins, 9, 291-293
biological activity, 293
Sapindosides, 7
Sapogenin
gas chromatography determination, 102-103
levels in alfalfa species, 102
thermal behavior, 273, 276
Sapogenins causing photosensitization disease, 401
Saponin disease network, 419
Saponin network for
mass spectrometry, 421
NMR, 411
Saponins of *Thalictrum* genus
isolation and purification, 194-195
structures, 195, 198-199, 201
Saponins of *Thalictrum squarrosum*, 200-203
Saponins and sapogenols of *Thalictrum minus*
reactions, 194-197
structure, 194-197
Saponin containing tiglic and nilic acid,

structure, 188
Saponins,
biological effects in feeds and forage, 377-383, 387-392
in human health, 382-383
nutrients, 380
on feed intake, 378
on intestine, 380
on rumen, 378-380, 383, 387-392
photosensitization, 381-383
of soybeans
groupings, 143-144, 232
interconversion, 144
by NaOH, 144
by $FeCl_3$, 145
sugar chains genes controlling, 148
steroid, of *Melilotus tauricus*, 215-218, 220-221
of *Thalictrum foetidum*, 197-200
triterpenoid, of *Melilotus albus*, 211-215
hydrolysis, 219
structure, 215-218, 221
Saponins or sapogenins, effects on cellular hydrolytic activity and protein leakage in rhizosphere bacteria, 82-83
Sarsasapogenin, 2-3, 38, 400
Secodammaranes, 19-21
Senecoiyloxychaparrinone, 284
Senecoiyloxychaparrine, 284
Sequestration and tolerance of cardenodides, by insects, 113, 116
Sheep, photosensitization disease, 395-401
Short-chain fatty acids, 388, 390
Siamenoside I, 14
Smilagenin, 3, 38, 305, 396-397
Simalactone, 284
Skeletal types of steroidal sapogenins, 308
α-Solanine, CAD spectrum, 407, 412
determination by HPLC, 410
separation by column, 415
chromatography, 410, 415
Solanidine, 406
Solamarzine, 409
analysis by DCCC and HLCC, 409
Solasodine, 33-34
Soyasapogenols, 103, 305
analysis for, 102-103
presence in melilotosides, 212-213
structure and reactions, 135-136, 143
Soyasapogenol B, 158, 332-333, 340, 342-346
Soyasaponin I, H^+ form, 357-359
Soyasaponin I, Na^+ form, 357-359
Soyasaponin I and derivative, structure and reaction, 185-189, 332-336, 356
Soyasapogenol III, 128
Soyasaponin VI, structure and degradation, 189-190
Soyasaponins
participation in chemiluminescence, 235-237
structure, 142, 143, 232
Soybean seeds
distribution of DDMP saponins in, 146, 231
dependence on variety and organ, 148
Spin density distribution theory, 237-238
Spin-spin interaction constants, 248-249
Spirrstane and furostane series of nicotianosides, 247-248
Spirostanoid saponins, 2, 273, 278, 305-307
Squarrofuric acid, 201
Squarogenesis, squarrosides, 201-203
Steroidal glycoalkaloid isolation, 408-409
Steroidal glycosides
biological activity, 293
thermal behavior, 271-279
Steroidal saponins, 399
classification by structure, 215-219, 299-303
determination of stereochemistry, 306-307
involvement in photosensitization, 401
^{13}C NMR chemical shift ranges for C-22, 302
structure determination by NMR, 299-312
Steroids,
microbial transformations of, 259-268
NMR studies, 281-293
Stevioside, 25-30
Strophanthidin, 117
Strophanthidol, 117
K-Strophanthosides, 117
Structure-antifungal relationship, 9
Sulfur trioxide-dimethyl sulfoxide complex, 209-210
Summer fruit tortrix moth, toxicity of alfalfa saponins for, 104-106
Sweet clover, *see Melilotus*
Sweeteners, 13-18
technique for study, 16
Sweetness inhibitors, 15-16

Tandem mass spectrometry for studying saponin structures, 317-328, 331-336, 340-351, 410
Tenacigenin derivatives, 290
Testosterones, 266-267
Thalictoside, A antitumor activity, 203-204
hydrolysis, to produce thalicogenin, 194
structure, 195-196
Thalicosides, 196-198
Thalictrum genus, Siberian chemorace,

saponins of, 193-205
physiological activity of, 283-305
Thlandioside, 9
Tiglic acid, 2-methyl-3-hydroxybutyric acid, 188
Tigogenin, 38
Tingenone, biological activity, 293
Thermal behavior of steroid glycosides, 271-274
Tomatin, tomatidin, tomatoside, 272-273
Toxicity of saponins of sheep, 400-401
Transformations of steroids, 264-268
Trigofoenosides, 38
Trilline, 215
Triterpenes, NMR structure, 281-293
Triterpenoid-3-sulfates, preparation, 209-210
Triticale, 75-77, 79-80

Uzarigenin, 283

Velutnic acid, 283

Wallenone, 284
Weed control with saponins, 57-59, 61-70
Wheat, 75-79, 177-178

Yamogenin, 38, 398-399
Yuccagenin, 38
Yucca saponin, 2-5, 8, 377-379, 383

Zanhic acid
analysis for, 103-104
as genin of herniaria saponins B, 242
structure, 158, 186-187
Ziziphin, 15